UPDATED FOURTH EDITION

PEARSON EDEXCEL A LEVEL GEOGRAPHY 2

CAMERON DUNN, KIM ADAMS, MICHAEL CHILES, DAVID HOLMES, SIMON OAKES, SUE WARN, MICHAEL WITHERICK

Endorsement Statement

In order to ensure that this resource offers high-quality support for the associated Pearson qualification, it has been through a review process by the awarding body. This process confirms that this resource fully covers the teaching and learning content of the specification or part of a specification at which it is aimed. It also confirms that it demonstrates an appropriate balance between the development of subject skills, knowledge and understanding, in addition to preparation for assessment.

Endorsement does not cover any guidance on assessment activities or processes (e.g. practice questions or advice on how to answer assessment questions), included in the resource nor does it prescribe any particular approach to the teaching or delivery of a related course.

While the publishers have made every attempt to ensure that advice on the qualification and its assessment is accurate, the official specification and associated assessment guidance materials are the only authoritative source of information and should always be referred to for definitive guidance.

Pearson examiners have not contributed to any sections in this resource relevant to examination papers for which they have responsibility.

Examiners will not use endorsed resources as a source of material for any assessment set by Pearson. Endorsement of a resource does not mean that the resource is required to achieve this Pearson qualification, nor does it mean that it is the only suitable material available to support the qualification, and any resource lists produced by the awarding body shall include this and other appropriate resources.

Although every effort has been made to ensure that website addresses are correct at time of going to press, Hachette Learning cannot be held responsible for the content of any website mentioned in this book. It is sometimes possible to find a relocated web page by typing in the address of the home page for a website in the URL window of your browser.

Hachette UK's policy is to use papers that are natural, renewable and recyclable products and made from wood grown in well-managed forests and other controlled sources. The logging and manufacturing processes are expected to conform to the environmental regulations of the country of origin.

To order, please visit www.hachettelearning.com or contact Customer Service at education@hachette.co.uk / +44 (0)1235 827827.

ISBN: 978 1 0360 1146 8

© Cameron Dunn, Kim Adams, Michael Chiles, David Holmes, Simon Oakes, Sue Warn, Michael Witherick 2025

Fourth edition published in 2021. This updated fourth edition published in 2025 by

Hachette Learning (a trading division of Hodder & Stoughton Limited),

An Hachette UK Company

Carmelite House

50 Victoria Embankment

London EC4Y 0DZ

www.hachettelearning.com

The authorised representative in the EEA is Hachette Ireland, 8 Castlecourt Centre, Dublin 15, D15 XTP3, Ireland (email: info@hbgi.ie)

Impression number   10 9 8 7 6 5 4 3 2 1

Year   2029 2028 2027 2026 2025

All rights reserved. Apart from any use permitted under UK copyright law, no part of this publication may be reproduced or transmitted in any form or by any means, electronic or mechanical, including photocopying and recording, or held within any information storage and retrieval system, without permission in writing from the publisher or under licence from the Copyright Licensing Agency Limited. Further details of such licences (for reprographic reproduction) may be obtained from the Copyright Licensing Agency Limited, www.cla.co.uk

Cover photo © anekoho – stock.adobe.com

Illustrations by Aptara, Inc and Barking Dog

Typeset in India by Aptara, Inc

Printed and bound in Great Britain by Bell & Bain Ltd, Glasgow

A catalogue record for this title is available from the British Library.

CONTENTS

Introduction　　　　　　　　　　　　　　　　　　　　　　　　　　　　　　　　　　iv

Unit 3　Physical Systems and Sustainability

Topic 5　The Water Cycle and Water Insecurity
Chapter 1: The operation and importance of the hydrological cycle　　　　　2
Chapter 2: Short- and long-term variations in the hydrological cycle　　　　24
Chapter 3: Water security – is there a crisis?　　　　　　　　　　　　　　　48

Topic 6　The Carbon Cycle and Energy Security
Chapter 4: The carbon cycle and planetary health　　　　　　　　　　　　　86
Chapter 5: Consequences of the increasing demand for energy　　　　　　106
Chapter 6: Human threats to the global climate system　　　　　　　　　　126

Unit 4　Human Systems and Geopolitics

Topic 7　Superpowers
Chapter 7: What are superpowers?　　　　　　　　　　　　　　　　　　　148
Chapter 8: Superpower impacts　　　　　　　　　　　　　　　　　　　　　163
Chapter 9: Superpower spheres of influence　　　　　　　　　　　　　　　178

Topic 8　Global Development and Connections

Option 8A: Health, Human Rights and Intervention
Chapter 10: Human development　　　　　　　　　　　　　　　　　　　　194
Chapter 11: Human rights　　　　　　　　　　　　　　　　　　　　　　　213
Chapter 12: Interventions and human rights　　　　　　　　　　　　　　　228
Chapter 13: The outcomes of geopolitical interventions　　　　　　　　　　246

Option 8B: Migration, Identity and Sovereignty
Chapter 14: The impacts of globalisation on international migration　　　　　261
Chapter 15: Nation states in a globalised world　　　　　　　　　　　　　　277
Chapter 16: Global organisations and their impacts　　　　　　　　　　　　295
Chapter 17: Threats to national sovereignty　　　　　　　　　　　　　　　310

Chapter 18: Synoptic themes　　　　　　　　　　　　　　　　　　　　　　324
Chapter 19: Independent investigation　　　　　　　　　　　　　　　　　339

Glossary　　　　　　　　　　　　　　　　　　　　　　　　　　　　　　　　　368
Index　　　　　　　　　　　　　　　　　　　　　　　　　　　　　　　　　　　377

INTRODUCTION

This book has been written specifically for the Edexcel specification introduced for first teaching in September 2016. The writers are all experienced authors, teachers and subject specialists with experience of examining. This book has been designed to cover the specification in a comprehensive, interesting and informative way.

Edexcel A level Geography Book 2 covers the content which is only tested at A level. For most of you, this will mean what you cover in class in Year 13. For your A level exams you also need to cover the content in Book 1. All of the compulsory topics, as well as all of the options, are covered in this book. There is also a section on the Independent Investigation coursework (Chapter 19) to help you plan and complete that piece of work. Chapter 18 covers Synoptic Themes. These include the three synoptic themes in the specification:

- Actions and attitudes
- Players
- Futures and uncertainties

Chapter 18 also explains some specialist geographical concepts which cut across different topics.

At the end of your two-year A level course you will sit three examinations and submit your Individual Investigation coursework. The tables below summarise these exams and show how they link to Books 1 and 2:

A level Paper 1

- 2 hours, 15 minutes
- 105 marks

Section A Topic 1: Tectonic Processes and Hazards	16 marks	Book 1
Section B Topic 2: Landscape Systems, Processes and Change (either Topic 2A Glaciated Landscapes and Change or Topic 2B Coastal Landscapes and Change)	40 marks	Book 1
Section C Topic 5: The Water Cycle and Water Insecurity	49 marks	Book 2
Topic 6: The Carbon Cycle and Energy Security		Book 2

Summary of the specification and its coverage in *Edexcel A level Geography Book 1* and *Book 2*

Book 1	
Year 12 Compulsory content	Topic 1: Tectonic Processes and Hazards Topic 3: Globalisation
Year 12 Optional content	Topic 2: Landscape Systems, Processes and Change Study one of these topics: • Topic 2A Glaciated Landscapes and Change • Topic 2B Coastal Landscapes and Change Topic 4: Shaping Places Study one of these topics: • Topic 4A Regenerating Places • Topic 4B Diverse Places
Book 2	
Year 13 Compulsory content	Topic 5: The Water Cycle and Water Insecurity Topic 6: The Carbon Cycle and Energy Security Topic 7: Superpowers Individual Investigation
Year 13 Optional content	Topic 8: Global Development and Connections Study one of these topics: • Topic 8A Health, Human Rights and Intervention • Topic 8B Migration, Identity and Sovereignty

A level Paper 2
- 2 hours, 15 minutes
- 105 marks

| Section A
Topic 3: Globalisation
Topic 7: Superpowers | 32 marks | Book 1
Book 2 |
|---|---|---|
| Section B
Topic 4: Shaping Places
(either Topic 4A Regenerating Places or Topic 4B Diverse Places) | 35 marks | Book 1 |
| Section C
Topic 8: Global Development and Connections
(either Topic 8A Health, Human Rights and Intervention or Topic 8B Migration, Identity and Sovereignty) | 38 marks | Book 2 |

A level Paper 3
- 2 hours, 15 minutes
- 70 marks

This exam paper is an Issues Analysis using an unseen resource booklet. It tests your understanding of different, linked parts of the whole two-year A level course	70 marks	Book 2, Chapter 18 and the synoptic links in Books 1 and 2

A level Paper 4
- Individual Investigation coursework
- 70 marks

The coursework component is your own Individual Investigation. Your will choose your own topic, meaning it could link to either your Year 12 or Year 13 studies, or both	70 marks	Book 2, Chapter 19 and relevant topics in Books 1 and/or 2

The three exams and your Individual Investigation contribute to your A level in this way:

Paper 1 (Physical geography)	30% of the total marks
Paper 2 (Human geography)	30% of the total marks
Paper 3 (Synoptic issues analysis)	20% of the total marks
Paper 4 (Coursework)	20% of the total marks

Each chapter in this book covers one enquiry question in a topic. Within each chapter, each of the key ideas from the specification is covered in detail using a combination of text, figures and easy-to-access tables. There is also a range of features in each chapter designed to boost your skills, understanding and confidence, presented in an interesting and accessible way. These will also help you revise and prepare for the exams:

- **An introduction** to each chapter gives you an overview of what is covered in that chapter.
- **Key terms** are defined throughout to help increase your geographical vocabulary.
- **Key concepts** explain important ideas and theories and how to apply them.
- **Skills focus** features cover the compulsory skills in the specification, linked to a particular sub-topic. These skills can be tested in the exams so are important to understand and practice.
- **Synoptic themes** in all topics indicate when content and examples need to be related to the synoptic themes of Actions and attitudes, Players, and Futures and uncertainties. This means thinking across topics so that you see links to other areas of study.
- **Place contexts** are indicated with a globe symbol and show how an idea or theme can be applied to a particular place example. These link to the place contexts in the specification.
- **Fieldwork opportunities** suggest ideas for carrying out fieldwork which could form part of your Individual Investigation.
- A range of **photographs**, **maps** and **graphs** help you develop your data response skills.
- **Further research** ideas are provided at the end of each chapter in the form of websites you could use to take some ideas further, perhaps as part of your Individual Investigation or to deepen your understanding.
- **Review questions** at the end of each chapter are designed to enhance your understanding of key ideas and allow you to test that understanding.
- **Exam-style questions** have been designed to provide practice with exam questions in the format in which they will be presented in your final exams.

The Publishers would like to thank the following for permission to reproduce copyright material.

Photo credits
p. 1 © Greenshoots Communications/Alamy Stock Photo; **fig. 2.8** © Kerry Whitworth/Alamy Stock Photo; **fig. 3.10** © Greenshoots Communications/Alamy Stock Photo; **fig. 3.14** © NASA Earth Observatory /MODIS/Jesse Allen; **p. 85** © Dunrobin Studios/Alamy Stock Photo; **fig. 4.12** © Aquapix/stock.adobe.com; **fig. 5.6** *a* © Jason Redmond/Reuters/Alamy Stock Photo, *b* © Dunrobin Studios/Alamy Stock Photo, *c* © Clive Gee/PA Images/Alamy Stock Photo; **fig. 5.16** *a* © Dan_prat/iStock/Getty Images, *b* © Goss Images/Alamy Stock Photo, *c* © Leo Francini/Alamy Stock Photo; **fig. 6.1** © Romeo Gacad/AFP/Getty Images; **fig. 6.12** © Rachel Platt; **p. 147** © Ezra Acayan/Getty Images; **fig. 9.2** *t* © Chris Ratcliffe/Bloomberg/Getty Images, *b* © REUTERS/Alamy Stock Photo; **fig. 9.5** © Ezra Acayan/Getty Images; **fig. 9.8** © NurPhoto/Getty Images; **p. 193** © James Brunker News/Alamy Stock Photo; **fig. 10.4** © ERIC LAFFORGUE/Alamy Stock Photo; **fig. 10.5** © Ministerio de La Presidencia/LatinContent WO/Getty Images; **fig. 10.15** © ChameleonsEye/Shutterstock.com; **fig. 10.17** © UN Sustainable Development Goals; **fig. 11.1** © Amnesty International UK, "Do the Human Right Thing" 2015; **fig. 11.5** © ASSOCIATED PRESS; **fig. 11.6** © Patrick Robert/Corbis/Sygma/Getty Images; **fig. 11.7** © Jeff Widener/AP/Shutterstock.com; **fig. 11.8** © Press Information Bureau/Planet Pix Via Zuma Wire/Shutterstock; **fig. 11.10** © Ye Aung Thu/AFP/Getty Images; **fig. 11.11** © Alexander Joe/AFP/Getty Images; **fig. 11.14** © Ricardo Mazalan/AP/Shutterstock.com; **fig. 11.15** © Zuma Press/Alamy Stock Photo; **fig. 11.16** © Parwiz Sabawoon/Anadolu Agency/Getty Images; **fig. 11.17** © James Brunker News/Alamy Stock Photo; **fig. 11.18** © EPA/epa european pressphoto agency b.v./Alamy Stock Photo; **fig. 12.5** © Arindam banerjee/123RF.com; **fig. 12.6** © THONY BELIZAIRE/AFP via Getty Images; **fig. 12.8** © Kjell Nilsson-Maki/Cartoonstock; **fig. 12.10** © Polyp.org.uk; **fig. 12.11** *l* © EPA/epa european pressphoto agency b.v./Alamy Stock Photo, *r* © American Photo Archive/Alamy Stock Photo; **fig. 12.12** © EPA/epa european pressphoto agency b.v./Alamy Stock Photo; **fig. 12.14** © Oli Scarff/Getty Images; **fig. 12.16** © Thaier Al-Sudani/REUTERS/Alamy Stock Photo; **fig. 12.17** © Pm1 Shane T. Mccoy/Planet Pix via ZUMA Wire/ZUMA Press, Inc. / Alamy Stock Photo; **fig. 13.2** © Gang – Fotolia; **fig. 13.4** © EPA/epa european pressphoto agency b.v./Alamy Stock Photo; **fig. 13.5** © Per-Anders Pettersson/Getty Images; **fig. 13.7** © Craig F. Walker/The Boston Globe/Getty Images; **fig. 13.8** © Beha el Halebi/Anadolu Agency/Getty Images; **fig. 13.11** © Olivier Asselin/Alamy Stock Photo; **p. 259** © Richard Levine/Alamy Stock Photo; **fig 14.2** © Alex Tihonov/stock.adobe.com; **fig. 14.5** © Cameron Dunn; **fig. 14.8** © Alex Segre/Alamy Stock Photo; **fig. 14.13** © Deborah Vernon/Alamy Stock Photo; **fig. 15.1** © Richard Levine/Alamy Stock Photo; **fig. 15.2** © Tuul & Bruno Morandi/The Image Bank/Getty Images; **fig. 15.7** © Sean pavone/123RF.com; **fig. 15.8** © vladimir/stock.adobe.com; **fig. 15.10** © Bettmann/Getty Images; **fig. 15.12** © U.S. Air Force photo; **fig. 15.14** © Droneandy/Shutterstock.com; **fig. 15.16** © PSL Images/Alamy Stock Photo; **fig. 15.17** © Jeffrey Blackler/Alamy Stock Photo; **fig. 15.18** © StringerAL/Shutterstock.com; **fig. 16.1** © Purestock/Thinkstock/Getty Images; **fig. 16.4** © Epa european pressphoto agency b.v./Alamy Stock Photo; **fig. 16.5** © Ed Darack/RGB Ventures/SuperStock/Alamy Stock Photo; **fig. 16.7** © Jim Pruitt/123RF.com; **fig. 16.10** © Jake Lyell/Water Aid/Alamy Stock Photo; **fig. 16.13** © Wusuowei/Fotolia; **fig. 16.15** © Chain45154/Moment/Getty Images; **fig. 16.17** © kkaplin/iStock/Thinkstock; **fig. 17.1** © Tramp57/Shutterstock.com; **fig. 17.2** © Ben Stansall/AFP/Getty Images; **fig. 17.3** © Archive Images/Alamy Stock Photo; **fig. 17.5** © Filmfoto/123RF.com; **fig. 17.6** © Yoshikazu Tsuno/AFP/Getty Images; **fig. 17.7** © Peter Forsberg/Alamy Stock Photo; **fig. 17.10** © Marek Stepan/Alamy Stock Photo; **fig. 17.12** © K_Z/Shutterstock.com; **fig. 19.13** © David Brabiner/Alamy Stock Photo.

Acknowledgements
Research to update Topic 5 was conducted with the assistance of Charlotte Howells BA.

Figure 2.11 Smith, K. (2013). Environmental hazards: assessing risk and reducing disaster. Routledge. Copyright by Taylor & Francis; **Figure 2.22** World Water Assessment Programme. 2009. The United Nations World Water Development Report 3: Water in a Changing World. Paris: UNESCO, and London: Earthscan; **Figure 3.15** Reproduced from The Atlas of Water (second edition), 2nd Edition by Maggie, Black, published by Routledge. Reproduced by arrangement with Taylor & Francis Group; **Figure 5.14** WCA, 2009: The Coal Resource: A Comprehensive Overview of Coal. World Coal Association, London, UK; **Figure 6.10** Climate change: processes, characteristics and threats. Used by permission of GRID-Arendal; **Figure 6.14** Cornell, H. 2011. Driver: Climate Change in the Salish Sea Ecosystem in Puget Sound Science Review. Puget Sound Partnership. Tacoma, Washington, U.S.A. Accessed from www.eopugetsound.org/science-review/section-2-driver-climate-change-salish-sea-ecosystem; **Figure 9.7** United Nations, 2021. Retrieved from http://www.sais-cari.org/data-china-africa-trade; **Figure 14.3** Andrew Taylor 2008, Workers of the world on the move FT.com 24th June. Used under license from the Financial Times. All Rights Reserved; **Figure 14.10** The great migration south: 80% of new private sector jobs are in London, Copyright Guardian News & Media Ltd, 2016; **Figure 14.16** Source: 2015, Paris attacks renew questions about Europe's border-free travel, Financial Times, 16, November. Used under license from the Financial Times. All rights reserved; **Figure 15.5** Source: 2014, As winter nears, Russia strains to feed and heat Crimea, Financial Times, 11, November. Used under license from the Financial Times. All rights reserved; **Table 6.1** Source: Food and Agriculture Organization of the United Nations,2020 [http://www.fao.org/3/ca8642en/ca8642en.pdf]; **Table 8.4** © Interbrand Best Global Brands, 2019; **Table 10.1** Happy Planet Index, Explore the data, 2016 (devised by Nic Marks); text **p.307** Claim no easy victories. Paris was a failure, but a climate justice movement is rising by Martin Lukacs. Copyright Guardian News & Media Ltd, 2021.

Every effort has been made to trace all copyright holders, but if any have been inadvertently overlooked, the Publishers will be pleased to make the necessary arrangements at the first opportunity.

Topic 5
The Water Cycle and Water Insecurity

Chapter 1: The operation and importance of the hydrological cycle

Chapter 2: Short- and long-term variations in the hydrological cycle

Chapter 3: Water security – is there a crisis?

1 The operation and importance of the hydrological cycle

What are the processes operating within the hydrological cycle from global to local scale?
By the end of this chapter you will:
- understand the importance of the hydrological cycle in supporting life on Earth and how it operates at a range of spatial and temporal scales
- know that the global hydrological cycle is a closed system and that the drainage basin, a subsystem within it, is an open system
- understand how processes, stores and flows operate within systems
- understand how processes, stores and flows contribute to contrasting water budgets, river regimes and storm hydrographs at a more local scale.

1.1 The operation of the hydrological cycle at a global scale

In order to understand the operation of the hydrological cycle (also known as the natural water cycle) a **systems approach** is useful. Three concepts are key to understanding how water cycling operates:

1. **Stores** (stocks), which are reservoirs where water is held, such as the oceans.
2. **Fluxes**, which measure the rate of flow between the stores.
3. **Processes**, which are the physical mechanisms which drive the fluxes of water between the stores.

The global hydrological cycle

The global hydrological cycle is an example of a closed system driven by solar energy and gravitational potential energy. In a closed system there is a fixed amount of water in the Earth–atmosphere system (estimated at 1385 million km³). A closed system does not have any external inputs or outputs, so this total volume of water is constant and finite. However, the water can exist in different states within the closed system (liquid water, water vapour gas and solid ice) and the proportions held in each state can vary for both physical and human reasons.

For example, in the last Ice Age more water was held within the **cryosphere** in a solid form as snow and ice; as less was held in the oceans, sea levels dropped considerably – over 140 m lower than they are today. Recent climate warming is beginning to reverse this with major losses of ice in Greenland and, more recently, Antarctica, and significant rises in sea level (see page 37 for the impacts of climate warming). At a small scale, humans have built numerous water storage reservoirs to complement natural lakes in order to increase the security of their water supplies.

Figure 1.1 shows how the global hydrological cycle works. Essentially there are four major stores of water, of which the oceans are by far the largest: they contain an estimated 96.5 to 97 per cent of the world's total water. The next largest stores occur in the cryosphere (1.9 per cent), and then shallow groundwater. The atmosphere is by far the smallest of the significant stores.

Table 1.1 shows a recent estimate of the size of these stores.

> **Key terms**
>
> **Systems approach:** Systems approaches study hydrological phenomena by looking at the balance of inputs and outputs, and how water is moved between stores by flows.
>
> **Stores:** Reservoirs where water is held, such as the oceans.
>
> **Fluxes:** The rate of flow between the stores.
>
> **Processes:** The physical mechanisms that drive the flux of material between stores.
>
> **Cryosphere:** Areas of the Earth where water is frozen into snow or ice.

The Water Cycle and Water Insecurity

① In the oceans the vast majority of water is stored in liquid form, with only a minute fraction as icebergs

② In the cryosphere water is largely found in a solid state, with some in liquid form as melt water and lakes.

③ On land the water is stored in rivers, streams, lakes and groundwater in liquid form. It is often known as **blue water**, the visible part of the hydrological cycle. Water can also be stored in vegetation after interception or beneath the surface in the soil. Water stored in the soil and vegetation is often known as **green water**, the invisible part of the hydrological cycle.

④ Water largely exists as vapour in the atmosphere, with the carrying capacity directly linked to temperature. Clouds can contain minute droplets of liquid water or, at a high altitude, ice crystals, both of which are a precursor to rain.

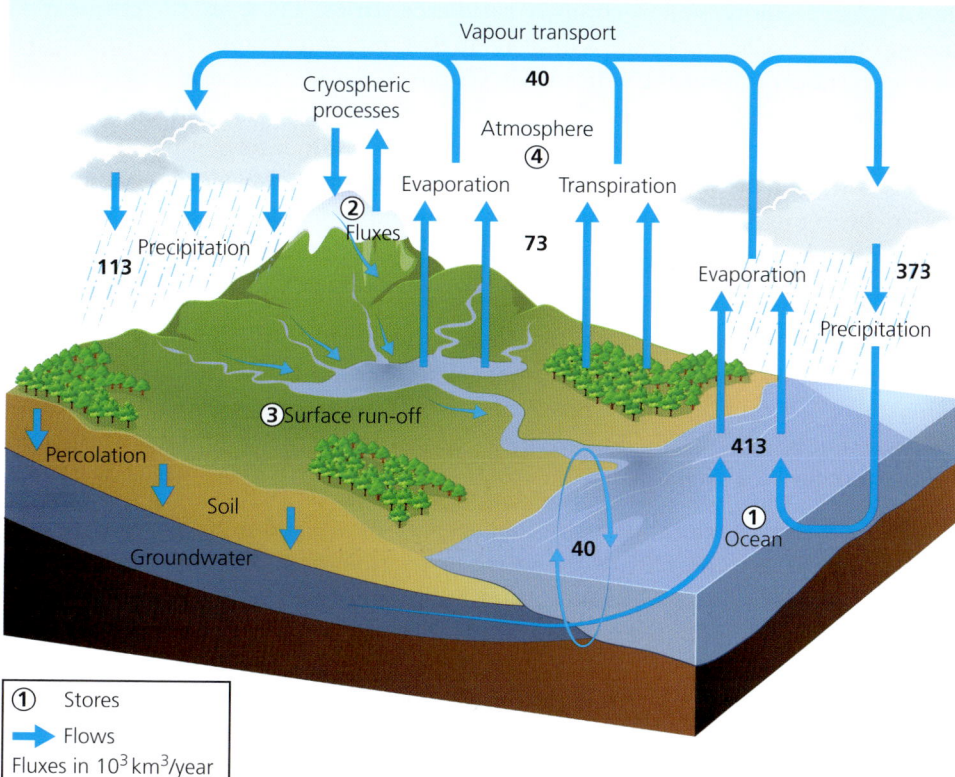

Figure 1.1 The global hydrological cycle (water cycle)

Key terms

Blue water: Water is stored in rivers, streams, lakes and groundwater in liquid form (the visible part of the hydrological cycle).

Green water: Water stored in the soil and vegetation (the invisible part of the hydrological cycle).

Table 1.1 Details of the main global water stores; note that numbers are rounded so the totals may not add up to 100

Store	Volume (10^3 km³)	Percentage of total water	Percentage of freshwater	Residence time
Oceans	1,335,040	96.9	0	3,600 years
Icecaps	26,350	1.9	68.7	15,000 years depending on size
Groundwater	15,300	1.1	30.1	Up to 10,000 years for deep groundwater; 100–200 years for shallow groundwater
Rivers and lakes	178	0.01	1.2	2 weeks to 10 years; 50 years for very large scale
Soil moisture	122	0.01	0.05	2–50 weeks
Atmospheric moisture	13	0.001	0.04	10 days

1 The operation and importance of the hydrological cycle

> **Key terms**
>
> **Precipitation:** The movement of water in any form from the atmosphere to the ground.
>
> **Evaporation:** The change in state of water from a liquid to a gas.
>
> **Residence time:** The average time a water molecule will spend in a reservoir or store.
>
> **Fossil water:** Ancient, deep groundwater from former pluvial (wetter) periods.
>
> **Transpiration:** The diffusion of water from vegetation into the atmosphere, involving a change from a liquid to a gas.
>
> **Groundwater flow:** The slow transfer of percolated water underground through pervious or porous rocks.

In Figure 1.1 (page 3) the major fluxes are shown, driven by key processes such as **precipitation**, **evaporation**, cryospheric exchange, and run-off generation (both surface and groundwater). These fluxes have been quantified, with the most important being evaporation from the oceans and precipitation on to land and the oceans.

Table 1.1 (page 3) allows you to compare **residence times**. These are the estimates of the average times a water molecule will spend in that reservoir or store. Residence times impact on turnover within the water cycle system. Groundwater, if it is deep seated, can spend over 10,000 years beneath the Earth's surface. Some ancient groundwater, such as that found deep below the Sahara Desert – the result of former pluvial (wetter) periods – is termed **fossil water** and is not renewable or reachable for human use. Major ice sheets too (such as Antarctica and Greenland) store water as ice for very long periods, so the figures in the table represent an average. Ice core dating has suggested that the residence time of some water in Antarctic ice is over 800,000 years.

Conversely, some very accessible stores, such as soil moisture, and small lakes and rivers, have much shorter residence times. Water stored in the soil, for example, remains there very briefly as it is spread very thinly across the Earth. Because of its accessibility it is easily lost to other stores by evaporation, **transpiration**, **groundwater flow** or recharge.

Atmospheric water has the shortest residence time of all, about ten days, as it soon evaporates, condenses and falls to the Earth as precipitation. There is a strong link between residence times and levels of water pollution: stores with a slower turnover tend to be more easily polluted as the water is *in situ* for a longer length of time.

Accessible water for human life support

Figure 1.2a–c summarises where the Earth's total global water is stored, with an overwhelming 96 to 97 per cent stored in the oceans – only around 2.5 per cent occurs as freshwater.

Figure 1.2b looks at the Earth's freshwater supply. Around 69 per cent is locked up in snowflakes, ice sheets, ice caps and glaciers found in high latitudes and high-altitude locations. This water supply is largely inaccessible for human use, although some streams in mountain areas are 'fed' from ice and snow as melt water. Another

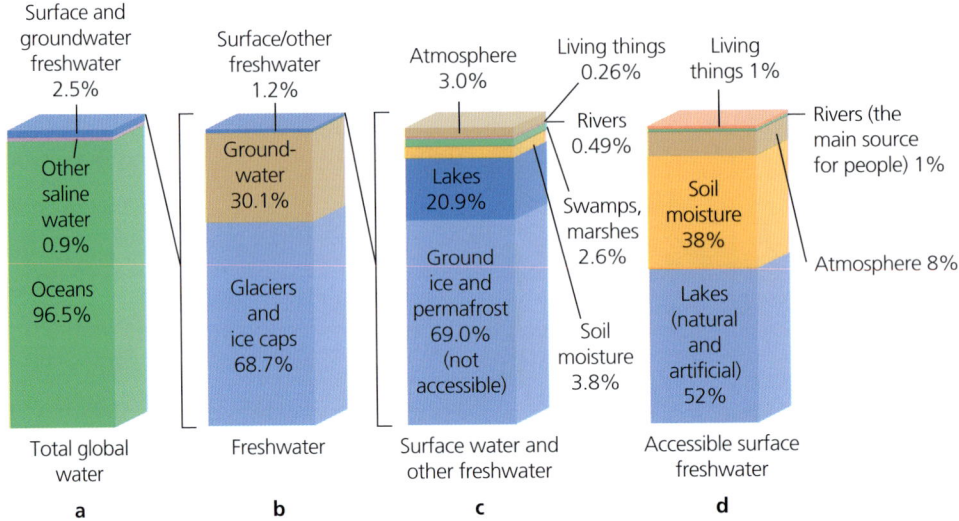

Figure 1.2 Where is the Earth's water? (Source: Adapted from Igor Shiklomanov's 'World freshwater resources' in Peter Gleick (editor), 1993, *Water in Crisis: A Guide to the World's Freshwater Resources*)

30 per cent occurs as groundwater, some of which is very deep seated as fossil water and, therefore, also inaccessible. This leaves only around 1 per cent of freshwater which is easily accessible for human use.

Figure 1.2c includes all sources of surface water, including ground ice and permafrost, which are very difficult to access.

Figure 1.2d shows only freshwater that is accessible to humans with current levels of technology – note the importance of lakes and soil moisture. Rivers, which are currently the main source of surface water for humans, constitute only 0.007 per cent of *total* water. It is not surprising that there are so many concerns and disputes about the usage of this tiny, precious fraction. As with any global overview, the differences between places are masked and, in terms of availability of water, it is a very unequal world. It is also notable that technology is being used widely to extend the availability of freshwater supplies, for example, by desalination of ocean water.

1.2 The operation of drainage basins as open systems

Drainage basins

On a smaller scale (variable from regional to local, depending on the size of the drainage basin) a drainage basin is a subsystem within the global hydrological cycle. It is an open system as it has external inputs and outputs that cause the amount of water in the basin to vary over time. These variations can occur at different temporal scales, from short-term hourly through to daily, seasonal and annual (Figure 1.3).

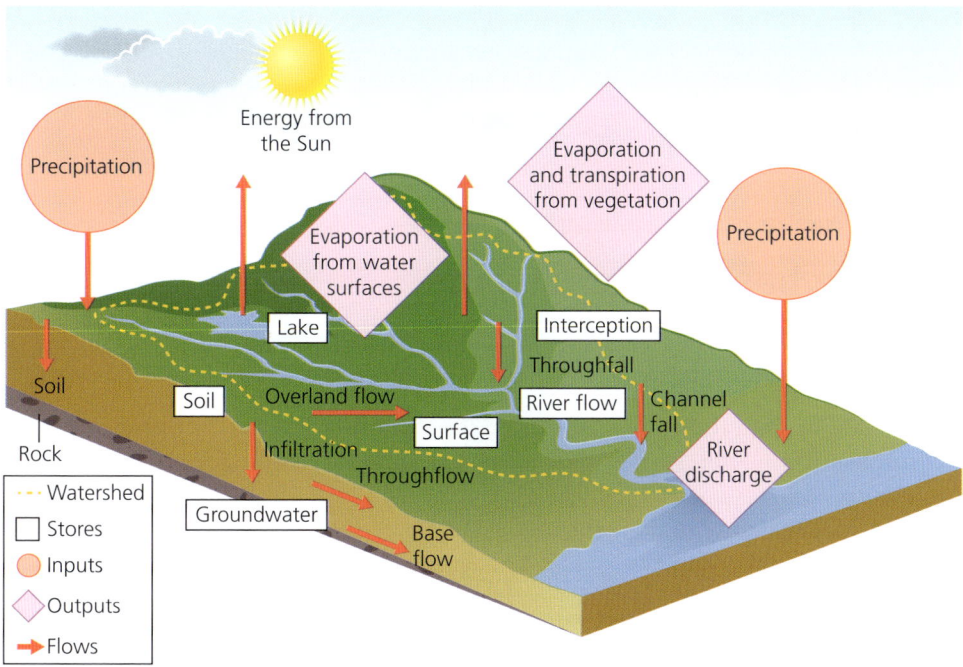

Figure 1.3 The drainage basin system

> **Key terms**
>
> **Catchment:** The area of land drained by a river and its tributaries.
>
> **Watershed:** The high land which divides and separates waters flowing to different rivers.
>
> **Condensation:** The change from a gas to a liquid, such as when water vapour changes into water droplets.
>
> **Dew point:** The temperature at which dew forms; it is a measure of atmospheric moisture.

A drainage basin can be defined as the area of land drained by a river and its tributaries, and is frequently referred to as a river **catchment**. The boundary of a drainage basin is defined by the **watershed**, which is usually a ridge of high land which divides and separates waters flowing to different rivers.

Drainage basins can be of any size, from that of a small stream possibly without tributaries up to a major international river flowing across borders of several countries.

Figure 1.4 shows how a drainage basin works. It has the advantage of showing the inter-linkages between components of the system.

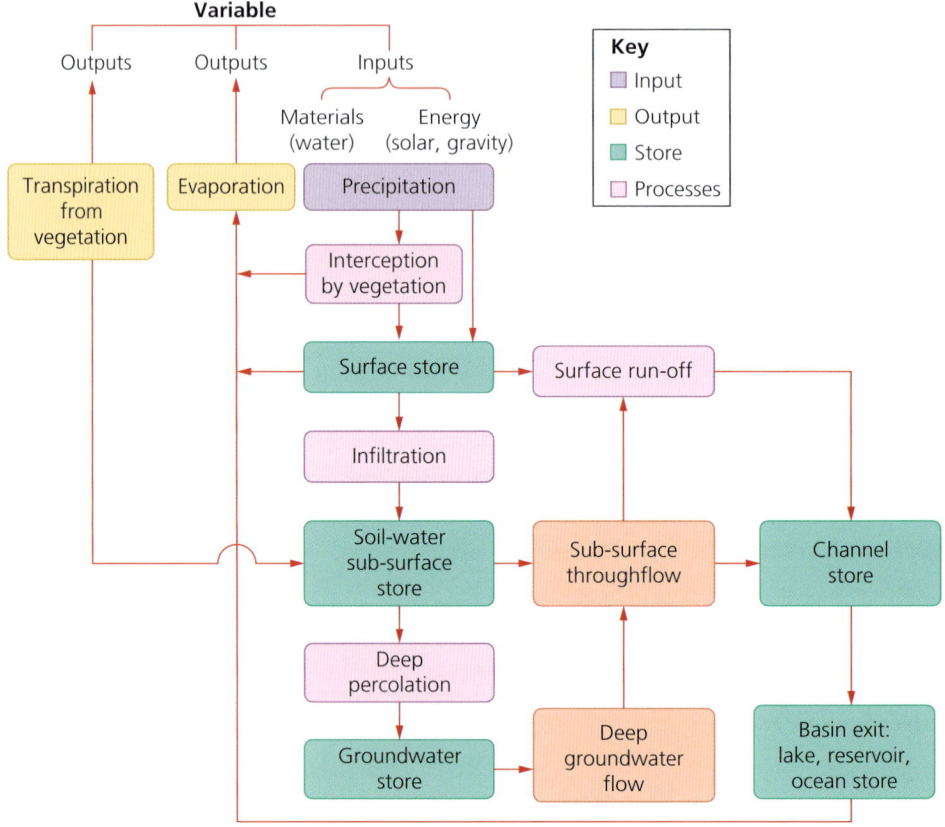

Figure 1.4 A drainage basin as a hydrological system

Drainage basin system inputs

Precipitation patterns

For precipitation (rain, snow, hail) to form, certain conditions are needed:

- air cooled to saturation point with a relative humidity of 100 per cent
- **condensation** nuclei, such as dust particles, to facilitate the growth of droplets in clouds
- a temperature below **dew point**.

There are three main triggers for the development of rainfall, all of which involve uplift and cooling and condensation.

As far as the impacts on the drainage basin hydrological system are concerned, there are six key influencing factors:

1. The amount of precipitation, which can have a direct impact on drainage discharge: as a general rule, the higher the amount the less variability in its pattern.

2. The type of precipitation (rain, snow or hail): the formation of snow, for example, can act as a temporary store and large fluxes (flows) of water can be released into the system after a period of rapid melting resulting from a thaw.
3. Seasonality. In some climates, such as monsoon, Mediterranean or continental climates, strong seasonal patterns of rainfall or snowfall will have a major impact on the physical processes operating in the drainage basin system.
4. Intensity of precipitation is also a key factor as it has a major impact on flows on or below the surface. It is difficult for rainfall to infiltrate if it is very intense, as the soil capacity is exceeded.
5. Variability can be seen in three ways:
    - Secular variability happens long term, for example, as a result of climate change trends.
    - Periodic variability happens in an annual, seasonal, monthly or diurnal context.
    - Stochastic variability results from random factors, for example, in the localisation of a thunderstorm within a basin.
6. The distribution of precipitation within a basin. The impact is particularly noticeable in very large basins such as the Rhone or the Nile, where tributaries start in different climatic zones. At a local scale and shorter time scale the location of a thunderstorm within a small river basin can have a major impact temporarily as inputs will vary, with contrasting storm hydrographs for different stream tributaries.

Precipitation data

It is important to recognise that data on precipitation may not always be reliable. In the UK, 200 automated weather stations spaced about 40 km apart continuously collect precipitation data. In the semi-arid Sahel countries of Mali, Chad and Burkina Faso roughly 35 weather stations collect data across an area of 2.8 million km^2 (more than 10 times the area of the UK). Major storms can easily fall between these weather stations because rainfall is geographically patchy. Understanding rainfall patterns and trends is critical in semi-arid areas (see page 27) but data reliability in these regions is often low.

Fluxes (flows) in the drainage basin

Interception

Interception is the process by which water is stored in the vegetation. It has three main components: **interception loss**, **throughfall** and **stem flow**.

Interception loss from the vegetation is usually greatest at the start of a storm, especially when it follows a dry period. The interception capacity of the vegetation cover varies considerably with the type of tree, with the dense needles of coniferous forests allowing greater accumulation of water. There are also contrasts between deciduous forests in summer and in winter – interception losses are around 40 per cent in summer for certain Chiltern beech forests, but under 20 per cent in winter. Coniferous forest intercepts 25–35 per cent of annual rainfall, whereas deciduous forest only 15–25 per cent and arable crops 10–15 per cent.

Meteorological conditions also have a major impact. Interception varies by vegetation cover. Wind speeds can decrease interception loss as intercepted rain is dislodged, and they can also increase evaporation rates. The intensity and duration of rainfall is a key factor too. As the amount of rainfall increases, the relative importance of interception losses will decrease: as the tree canopies become saturated, so more excess water will reach the ground. There are also variations for agricultural crops, with interception rates increasing with crop density.

Key terms

Interception loss: This is water that is retained by plant surfaces and later evaporated or absorbed by the vegetation and transpired. When the rain is light, for example, drizzle, or of short duration, much of the water will never reach the ground and will be recycled by this process (it's the reason you can stand under trees when it is raining and not get wet).

Throughfall: This is when the rainfall persists or is relatively intense, and the water drops from the leaves, twigs, needles, etc.

Stem flow: This is when water trickles along twigs and branches and then down the trunk.

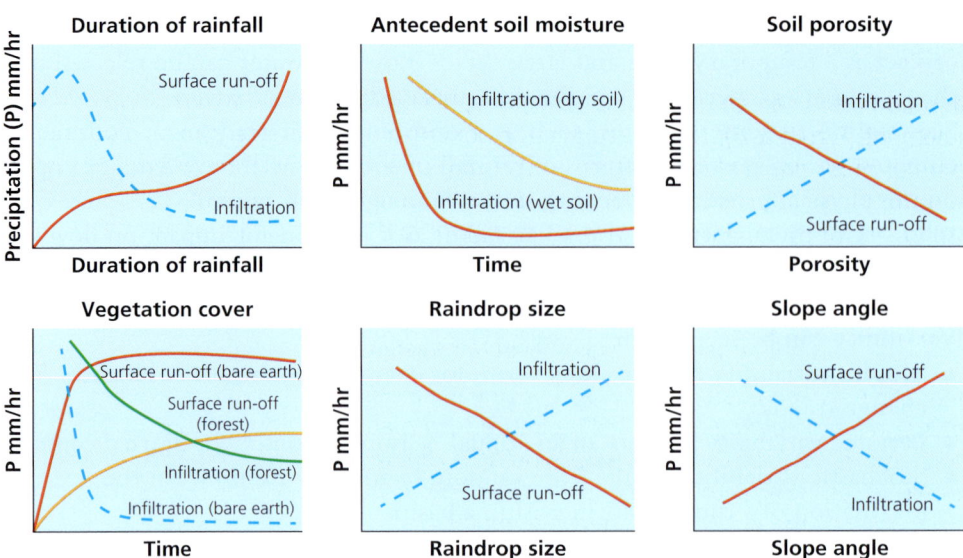

Figure 1.5 Some of the factors influencing the rate of infiltration

Infiltration

Infiltration is the process by which water soaks into (or is absorbed by) the soil. The **infiltration capacity** is the maximum rate at which rain can be absorbed by a soil in a 'given condition' and is expressed in mm/hr. The rate of infiltration depends on a number of factors, as shown in Figure 1.5.

- Infiltration capacity decreases with time through a period of rainfall until a more or less constant low value is reached.
- The rate of infiltration also depends on the amount of water already in the soil (antecedent soil moisture) as surface or overland flow will take place when the soil is saturated.
- Soil texture – whether sand, silt, loam or clay – also influences soil porosity, with sandy soils having an infiltration capacity of 3–12 mm/hr and less permeable clays 0–4 mm/hr.
- The type, amount and seasonal changes in vegetation cover are a key factor, with infiltration far more significant in land covered by forests (50 mm/hour) or moorland (42 mm/hour), hence the recent drive to vegetate upland catchments that flow into areas liable to flooding. Permanent pasture has infiltration rates of 13–23 mm/hour depending on grazing density and soil type.
- The nature of the soil surface and structure is also important. Compacted surfaces inhibit infiltration (around 10 mm/hour), especially when rain splash impact occurs.
- Slope angle can also be significant: very steep slopes tend to encourage overland run-off, with shallower slopes promoting infiltration.

As Figure 1.6 shows, infiltration is inversely related to **surface run-off** (overland flow), which is also influenced by similar factors.

> ### Key terms
>
> **Infiltration:** The movement of water from the ground surface into the soil.
>
> **Infiltration capacity:** The maximum rate at which rain can be absorbed by a soil.
>
> **Surface run-off:** The movement of water that is unconfined by a channel across the surface of the ground. Also known as overland flow.

Fieldwork opportunity

Infiltration rates can be measured fairly cheaply and easily using home-made equipment (Figure 1.6). Sink a bottomless container made from a plastic pipe (a diameter of 20 cm is ideal) 10 cm into the ground. Fill the container with water until the water measures 15 cm above the ground. Record the time it takes for the water level to drop by 5 cm. Keep topping up the water and record the times until they are constant for three successive periods. Calculate the results in mm/second:

$$\text{Infiltration rate} = \frac{\text{difference in levels at 5 cm}}{\text{time taken for water level to be reached}}$$

The factors you can usually test for by sampling a comparatively small area include: type of surface cover, soil moisture (using a probe), soil texture (mechanical analysis), angle of slope (using a clinometer), soil compaction and rainfall pattern over a given time (rain gauge, etc.). You can also use a geology map to look at the impact of underlying geology.

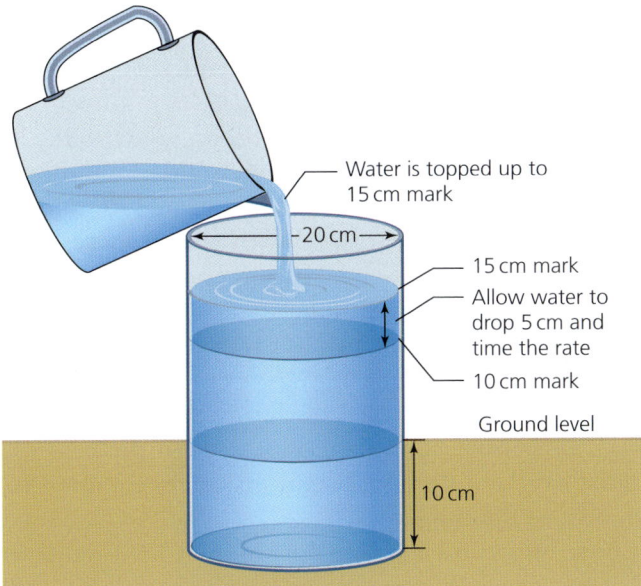

Figure 1.6 Measuring infiltration rates

Flows and transfers (see Figure 1.7)

1 Overland flow (variously known as surface run-off or direct overland flow on account of its rapidity in reaching the river channel) is a concept developed by Horton. He saw this flow as the main way that rainwater was transferred to the river channel. For this type of flow to occur, precipitation intensity must exceed the infiltration rate. Circumstances include an intense torrential storm, persistently high levels of precipitation over a longer period, or the release of very large quantities of melt water from the rapid melting of snow. Alternatively, bare, 'baked' unvegetated surfaces, which commonly occur in arid or semi-arid regions, also lend themselves to overland flow as this type of ground has very limited infiltration capacity.

This type of flow is the primary agent of soil erosion as sediment is removed by a range of erosive processes: rain splash, sheet, rill and gully erosion. Direct overland flow occurs once depression storage capacity in puddles has been exceeded. Overland flow is also a feature of many urban areas (see page 18), especially when the capacity of storm drains and sewers has been exceeded.

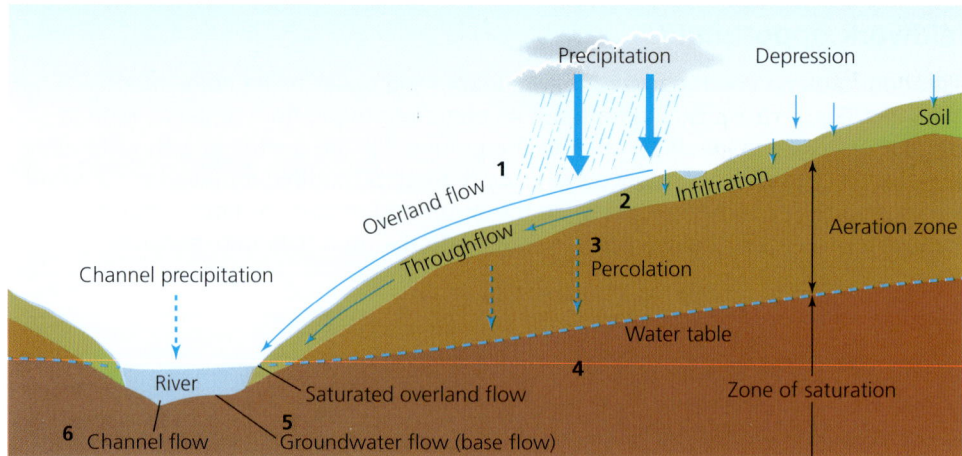

Figure 1.7 How the various flows operate within the drainage basin system

2 **Throughflow** refers to the lateral transfer of water down slope through the soil via natural pipes and **percolines** (lines of concentrated water flow between soil horizons to the river channel). While slower than direct overland flow, this shallow transfer can occur quite rapidly in porous, sandy soils.

3 **Percolation** can be regarded as a continuation of the infiltration process; it is the deep transfer of water into permeable rocks – those with joints (pervious rocks such as carboniferous limestone) or those with pores (porous rocks such as chalk and sandstone). The throughflow percolation route is much more likely to be associated with humid climates with vegetated slopes.

4 **Saturated overland flow** is a much slower transfer process as it results from the upward movement of the water table into the evaporation zone. After a succession of winter storms (for example, in the UK during the winters of 2015 or 2019) the water table rises to the surface in depressions and at the base of hill sides. This leads to saturated overland flow making a major contribution to **channel flow** and is a component of flooding.

5 Groundwater flow (also known as base flow or interflow) is the very slow transfer of percolated water through pervious or porous rocks. It is a vital regulatory component in maintaining a steady level of channel flow in droughts and other varying weather conditions.

6 Channel flow takes place in the river once water from the three transfer processes – overland flow, throughflow or groundwater flow – reaches it. Direct channel precipitation is added to **channel storage**.

> ### Key terms
>
> **Throughflow:** The lateral transfer of water down slope through the soil via natural pipes and percolines.
>
> **Percolines:** Lines of concentrated water flow between soil horizons to the river channel.
>
> **Percolation:** The transfer of water from the surface or from the soil into the bedrock beneath.
>
> **Saturated overland flow:** The upward movement of the water table into the evaporation zone.
>
> **Channel flow:** The flow of water in streams or rivers.
>
> **Channel storage:** The storage of water in streams or rivers.

Drainage basin system outputs

Evaporation

Evaporation is the physical process by which moisture is lost directly into the atmosphere from water surfaces (the largest transfer) and soil. Evaporation results from the effects of the Sun's heating and air movement, so rates increase in warm, windy and dry conditions. Climatic factors influencing evaporation rates include temperature, hours of sunshine, humidity and wind speed, although temperature is the most important factor. Other factors include the size of the water body, depth of water, water quality, type of vegetation cover and the colour of the surface (which determines the **albedo** or reflectivity of the surface).

Transpiration

Transpiration is a biological process by which water is lost from plants through minute pores (stomata) and transferred to the atmosphere. Transpiration rates depend on the time of year, the type and amount of vegetation cover, the degree of availability of moisture in the atmosphere and the length of growing season.

Evapotranspiration (EVT) is the combined effect of evaporation and transpiration. EVT represents the most important aspect of water loss to the atmosphere, accounting for the removal of nearly 100 per cent of the annual precipitation in arid and semi-arid areas, and around 75 per cent in humid areas. Obviously over ice/snow fields, bare rock slopes and soils, desert areas and the majority of water surfaces, the losses are purely evaporative.

Potential evapotranspiration (PEVT) is the water loss that would occur if there was an unlimited supply of water in the soil for use by vegetation. Therefore, the difference between PEVT and EVT is much greater in arid areas than in humid areas.

> **Key terms**
>
> **Albedo:** A measure of the proportion of the incoming solar radiation that is reflected by the surface back into the atmosphere and space.
>
> **Evapotranspiration (ET or EVT):** The combined effect of evaporation and transpiration.
>
> **Potential evapotranspiration (PET or PEVT):** The water loss that would occur if there was an unlimited supply of water in the soil for use by vegetation.

Physical factors that influence the drainage basin cycle

As can be seen when studying the inputs, flows and outputs within the drainage basin system, their relative importance is determined by a number of physical factors. Table 1.2 summarises some of these key influences.

Table 1.2 Physical factors within the drainage basin system and effect on inputs, flows and outputs

Climate	Climate has a role in influencing the type and amount of precipitation overall and the amount of evaporation, i.e. the major inputs and outputs. Climate also has an indirect impact on the vegetation type.
Soils	Soils determine the amount of infiltration and throughflow and, indirectly, the type of vegetation.
Geology	Geology can impact on subsurface processes such as percolation and groundwater flow (and, therefore, on aquifers). Indirectly, geology alters soil formation.
Relief	Altitude can impact on precipitation totals. Slopes can affect the amount of run-off.
Vegetation	The presence or absence of vegetation has a major impact on the amount of interception, infiltration and occurrence of overland flow, as well as on transpiration rates.

Figure 1.8a and Figure 1.8b show contrasting hydrological cycles in two different areas with completely different physical factors. This leads to contrasting inputs, stores, flows and outputs.

Skills focus: Analysing contrasting hydrological cycles

Explain how physical factors have led to contrasts in the hydrological cycles shown in Figure 1.8.

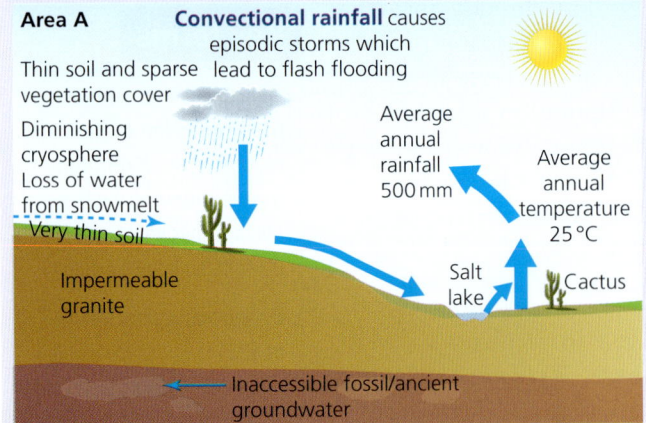

Figure 1.8a The impact of physical factors on two contrasting hydrological cycles (Area A)

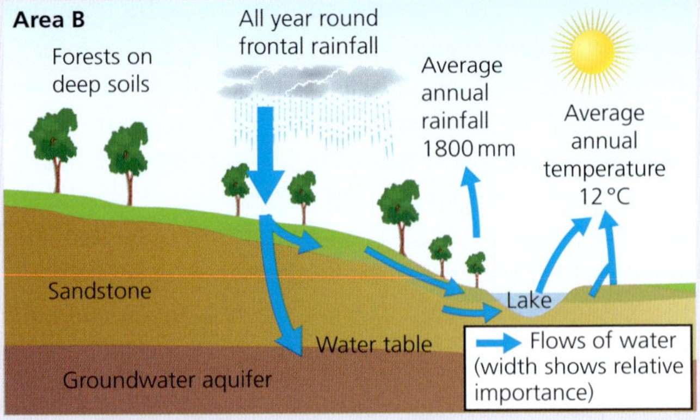

Figure 1.8b The impact of physical factors on two contrasting hydrological cycles (Area B)

- Area A is a semi-arid area, for example, on the fringe of the Atacama Desert in northern Chile. It has a low level of water security as there is very little storage potential, and outputs exceed inputs. Other sources are not accessible (fossil water, melt water from the cryosphere).
- Area B is a temperate rainforest area in southern Chile with a high level of water security; inputs of precipitation exceed outputs and there is also abundant groundwater storage.

Human factors that influence the drainage basin system

Human impact on precipitation

Human activity can affect precipitation by cloud seeding: the introduction of silver iodide pellets, or ammonium nitrate, to act as condensation nuclei to attract water droplets. The aim is to increase rainfall in drought-stricken areas. It has variable results. Pollution also provides condensation nuclei.

Human impact on evaporation and evapotranspiration

Changes in global land use, for example, deforestation, are a key influence. Also important is the increased evaporation potential resulting from the enormous artificial reservoirs behind mega dams, for example, the Aswan Dam and Lake Nasser in southern Egypt. Conversely, the channelisation of rivers in urban areas into conduits cuts down surface storage and, therefore, evaporation.

Human impact on interception

As interception is largely determined by vegetation type and density, **deforestation** and **afforestation** both have significant impacts.

Deforestation leads to a reduction in evapotranspiration and an increase in surface run-off. This increases flooding potential, leads to a decline of surface storage and a decrease in the lag time between peak rainfall and peak discharge. In other words, it speeds up the cycle.

Research on deforestation in Nepal shows a range of negative impacts that have been linked to deforestation, including increases in the sediment load downstream

> **Key terms**
>
> **Convectional rainfall:** Often associated with intense thunderstorms, which occur widely in areas with ground heating such as the Tropics and continental interiors.
>
> **Deforestation:** The cutting down and removal of all or most of the trees in a forested area.
>
> **Afforestation:** The planting of trees in an area that has not been forested in recent times.

The Water Cycle and Water Insecurity

in northern Nepal. Figure 1.9 summarises possible impacts of deforestation in the Himalayas in Nepal.

In theory, afforestation should have the reverse impact by trapping silt and slowing up the hydrological cycle by lengthening lag times. However, as a recent research project in the Plynlimon area of the catchment of the River Severn in Mid Wales showed, there is a period of time just after the planting of young trees where there is an increase in run-off and sediment loss as a result of compaction of soil by tractors and planting equipment, which only stops after 30 years when the trees are more fully grown.

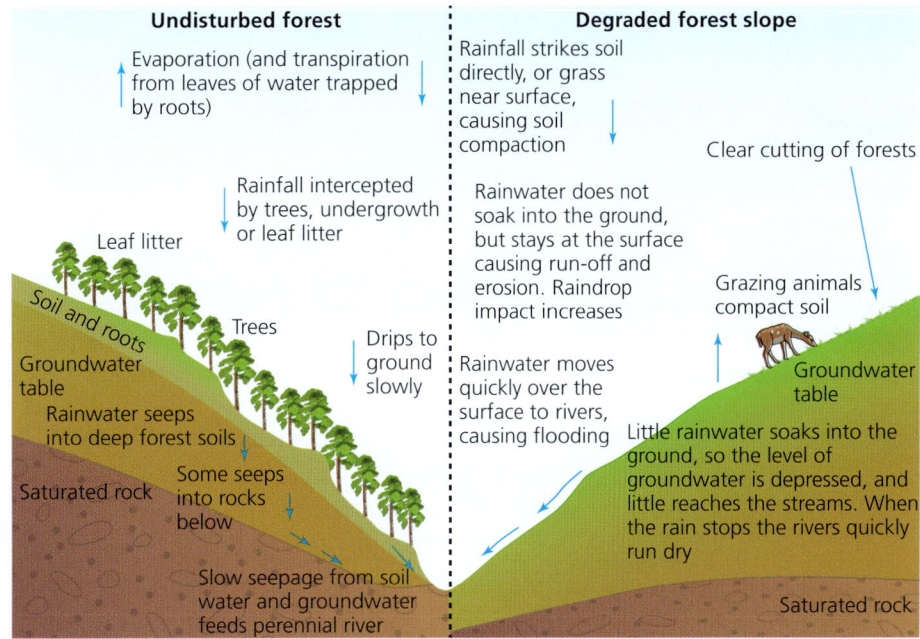

Figure 1.9 The possible impacts of deforestation in the Himalayan foothills of Nepal

Human impact on infiltration and soil water

Human impacts on infiltration largely result from a change in land use. Infiltration is up to five times greater under forests when compared with grassland. With conversion to farmland there is reduced interception, increased soil compaction and more overland flow. This impact is summarised in Figure 1.10. Land-use practices are also important: while grazing cows leads to soil compaction by the trampling of animals, ploughing increases infiltration by loosening and aerating the soil. Waterlogging and salinisation are common if there is poor drainage, so installing drainage mitigates these problems.

Deforestation issues in Amazonia

The environmental impacts are likely to be severe because of the sheer scale of the deforestation in Amazonia. Over 20 per cent of the forest has been destroyed, at an accelerating rate in the last 50 years, by a combination of cattle ranching, large-scale commercial agriculture for biofuels and soya beans, general development of towns and roads, as well as legal and illegal logging. Whilst former president Jair Bolsorano encouraged exploitation of the Amazon, records show that deforestation rates have fallen under the presidency of Lula da Silva since 2022.

As the Amazon forests contain 60 per cent of the world's rainforests, the environmental impact on global life support systems is bound to be highly significant. The trees act as 'green lungs' by removing CO_2 as they photosynthesise and act as carbon sinks. Destruction of forests reduces this capacity, so adding to the global greenhouse gas emissions, especially in times of drought.

There is also an enormous impact on water cycling. In a forest environment 75 per cent of intercepted water is returned by EVT to the atmosphere, which reduces to around 25 per cent when the forest is cleared. Ultimately, the drier climate can lead to desiccation and further rainforest degradation. The El Niño–Southern Oscillation (ENSO) (see page 26) can lead to significant occurrence of droughts in Amazonia, which can exacerbate forest fires and further destruction.

The sheer scale of Amazonian destruction can have a very significant impact on the water cycle. As more water runs off into the Amazon drainage system, not only does this exacerbate the possibility of severe flooding and mudslides, it also leads to aquifer depletion, as less water infiltrates to recharge them. Overland flow also increases the amount of soil erosion and degradation as nutrients are 'washed away'.

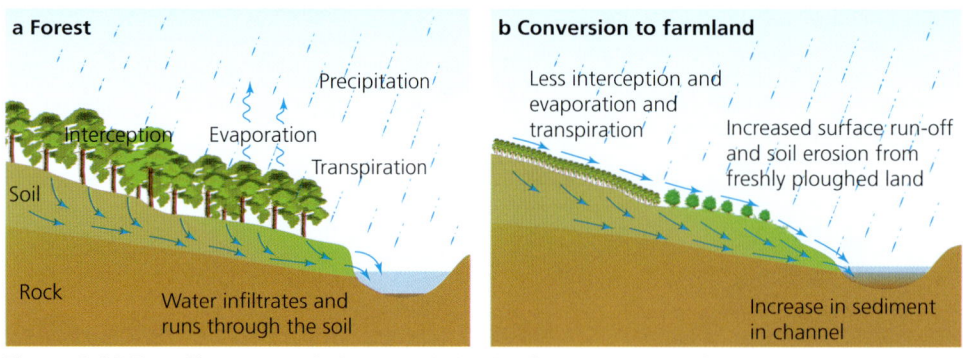

Figure 1.10 The effect on the drainage cycle basin of converting land from forest to farmland

Human impact on groundwater

Human use of irrigation for extensive cereal farming has led to declining water table levels in areas such as the Texan aquifers. The Aral Sea, between Kazakhstan and Uzbekistan, is an example of the damaging effects of the overextraction of water. The Aral Sea began shrinking in the 1960s when Soviet irrigation schemes for the growth of cotton took water from the Syr Darya and Amu Darya rivers, which greatly reduced the amount of water reaching the Aral Sea. By 1994, levels had fallen by 16 m, the surface area had declined by 50 per cent, the volume by 75 per cent, and salinity levels had increased by 300 per cent, with major ecological consequences.

In many British cities, including London, recent reductions in water-using manufacturing activity have led to less groundwater being abstracted. As a result groundwater levels have begun to rise, leading to a different set of problems, such as surface water flooding, flooding of cellars and basements in houses, and increased leakage into tunnels such as those used by the London Underground. The water supplies are also more likely to become polluted.

1.3 The operation of the hydrological cycle at contrasting scales

Water budgets

Water budgets, the balance between precipitation, evaporation and run-off, can be useful at global, regional and local scales.

Figure 1.11 depicts the global water balance in 10-degree latitudinal steps. The graph indicates the importance of the distribution of the atmosphere circulation and, to an extent, land and sea bodies. Only two zones – A and B, temperate and tropical equatorial – show a positive balance of run-off; they mark zones of convergence and uplift and subsequent precipitation. Rivers flowing from these zones are vital in supplying zones of deficit (for example, the Nile supplies Egypt's deserts with vital water). Both climate change and human activities such as deforestation have the capacity to modify the situation in the long and short term.

As you can see in Table 1.3 (page 14), the water balance varies considerably between continents, with South America the most well-endowed continent and Africa the least. Run-off is divided into surface flow and base flow. This is an important distinction because in some places there are severe seasonal differences in surface flow (for example, monsoonal areas):

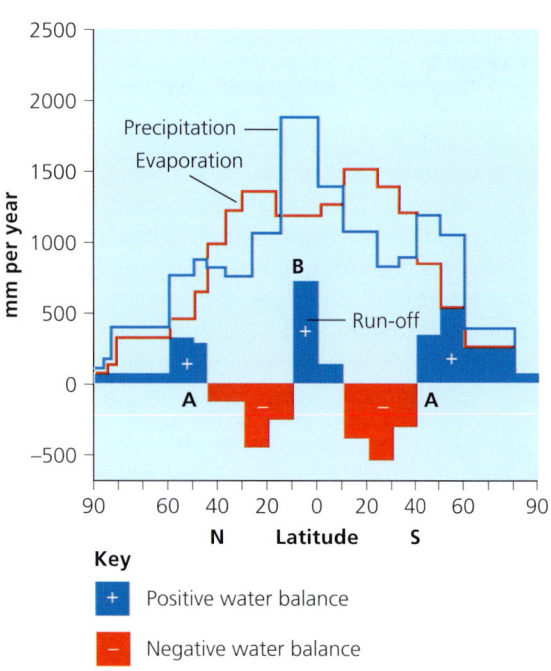

Figure 1.11 The variation of global water budgets with latitude

Table 1.3 Continental scale variation in water balance

Continent	Precipitation (mm/year)	Evapotranspiration (mm/year)	Difference (mm/year)	Run-off (mm/year)	
				Surface	Base flow
Europe	657	375	283	185	97
Asia	696	420	276	205	71
Africa	696	582	114	74	40
Australia (and Oceania)	803	534	269	205	64
North America	645	403	242	171	71
South America	1,564	946	618	395	223

at certain times of the year there may be a shortage of water but at other times a surplus. The base flow represents the usually available water. Although very generalised, Table 1.3 does point to water supply problems on some continents; for example, in Africa precipitation and evapotranspiration are very similar, leaving little water to enter rivers as surface run-off. In contrast, South America has a large precipitation/evapotranspiration difference leading to high surface run-off. Water budgets at a country or regional scale provide a more useful indication of available water supplies (see page 56, water poverty index).

At a more local scale, water budgets show the annual balance between inputs (precipitation) and outputs (EVT), and how this can impact on soil water availability.

The soil moisture budget is a subsystem of the catchment water balance and is of vital importance to agriculturalists. Drainage basin water budgets are usually called water balances and are usually expressed using the following formula:

$P = Q + E \pm S$

Where:

P = precipitation
Q = discharge (stream flow)
E = evapotranspiration
S = changes in storage

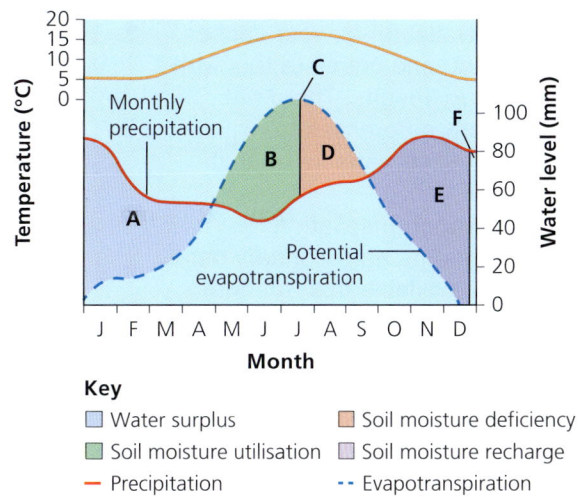

Key
- Water surplus
- Soil moisture deficiency
- Soil moisture utilisation
- Soil moisture recharge
- Precipitation
- -- Evapotranspiration

A Precipitation > potential evapotranspiration. Soil water store is full and there is a soil moisture surplus for plant use. Run-off and groundwater recharge.

B Potential evapotranspiration > precipitation. Water store is being used up by plants or lost by evaporation (soil moisture utilisation).

C Soil moisture store is now used up. Any precipitation is likely to be absorbed by the soil rather than produce run-off. River levels fall or rivers dry up completely.

D There is a deficiency of soil water as the store is used up and potential evapotranspiration > precipitation. Plants must adapt to survive, crops must be irrigated.

E Precipitation > potential evapotranspiration. Soil water store starts to fill again (soil moisture recharge).

F Soil water store is full, field capacity has been reached. Additional rainfall will percolate down to the water table and groundwater stores will be recharged.

Figure 1.12 A water budget graph for southern England showing soil moisture status

Figure 1.12 shows a water budget for southern England. In the UK, the annual precipitation exceeds evaporation in most years and in most places. Therefore, precipitation inputs exceed evaporation losses, so there will be a positive water balance. However, in some years of drought (for example, 1975–6 and 1995–6), and in some summer months, England has a temporary negative water balance.

Fieldwork opportunity

Use the following method to measure water balances within a catchment:

1. Measure stream discharge at a number of points (cross-sectional area × velocity) or make use of a discharge flume within a defined drainage basin to calculate run-off.
2. Use multiple rain gauges located next to discharge sites to measure the spatial variation of rainfall over a given time.
3. Use secondary rainfall and run-off data (if available) for your chosen catchment, and correlate this with the primary results. There are often many differences to explain in a climate as variable as that of the UK.
4. Obtain evapotranspiration measurements as secondary data (possibly available from the Met Office).

Primary measurement of evaporation can be measured using open saucers or pans located across your chosen catchment area. If it is possible to co-operate with a university department or field study research centre, you may be able to borrow a lysometer, which measures EVT.

Key terms

River regime: The annual variation in discharge or flow of a river at a particular point or gauging station, usually measured in cumecs.

Rising limb: The part of a storm hydrograph in which the discharge starts to rise.

Peak discharge: The time when the river reaches its highest flow.

River regimes

A **river regime** can be defined as the annual variation in discharge or flow of a river at a particular point or gauging station, usually measured in cumecs. Much of this river flow is not from immediate precipitation or run-off, but is supplied from groundwater between periods of rain, which feeds steadily into the river system from base water flow. This masks the fluctuations in stream flow caused by immediate precipitation. British rivers flowing over chalk, for example, the River Kennet, show this feature as well, as they maintain their flow even in very dry conditions, which is a result of base flow from the chalk aquifers.

The character of a regime of the resulting stream or river is influenced by several variable factors:

- The size of the river and where measurements are taken in the basin: many large rivers have very complex regimes resulting from varied catchments.
- The amount, pattern and intensity of the precipitation: regimes often reflect rainfall seasonal maxima or when the snow fields or glaciers melt (for snow the peak period is in spring, for glaciers it is early summer).
- The temperatures experienced: evaporation will be marked in summer as the temperatures are warmer.
- The geology and overlying soils, especially their permeability and porosity: water is stored as groundwater in permeable rocks and is gradually released into the river as base flow, which tends to regulate the flow during dry periods.
- The amount and type of vegetation cover: wetlands can hold the water and release it very slowly into the system.
- Human activities, such as dam building, which can regulate the flow.

Overall the most important factor determining stream flow is climate. Figure 1.13 (page 16) shows how these factors lead to a variety of regimes.

Storm hydrographs

Storm hydrographs show the variation of discharge within a short period of time, normally an individual storm or a group of storms not more than a few days in length. Before the storm starts the main supply of water to the river or stream is through groundwater or base flow but, as the storm develops, water comes to the stream by a number of routes. Some water infiltrates into the soil and becomes throughflow, while some flows over the surface as overland flow. This water reaches the river in a comparatively short time so is known as quick flow. The storm hydrograph records the changing discharge of a river or stream in response to a specific input of precipitation.

Figure 1.14 (page 16) shows the main features of a storm hydrograph.

- Once the rainfall input begins the discharge starts to rise; this is shown on the **rising limb**.
- **Peak discharge** is eventually reached some time after the peak rainfall because the water takes time to move through the system to the gauging station of the basin.
- The time interval between peak rainfall and peak discharge is known as **lag time**.
- Once the storm input has ceased the amount of water in the river starts to decrease; this is shown by the **falling or recessional limb**.
- Eventually the discharge returns to its normal level or **base flow**.

The shape of a storm hydrograph may vary from event to event on the same river (temporal variation), usually closely linked to the pattern of the storm event, or from one river to another (spatial variation – often related to basin characteristics) (Table 1.4, page 17). Some hydrographs have very steep limbs, especially rising limbs with a high peak discharge and a very short lag time – usually called 'flashy' hydrographs. At the other end of the spectrum, some storm hydrographs have a very gentle rising limb, a lower peak discharge and a long lag time – usually called delayed or attenuated hydrographs.

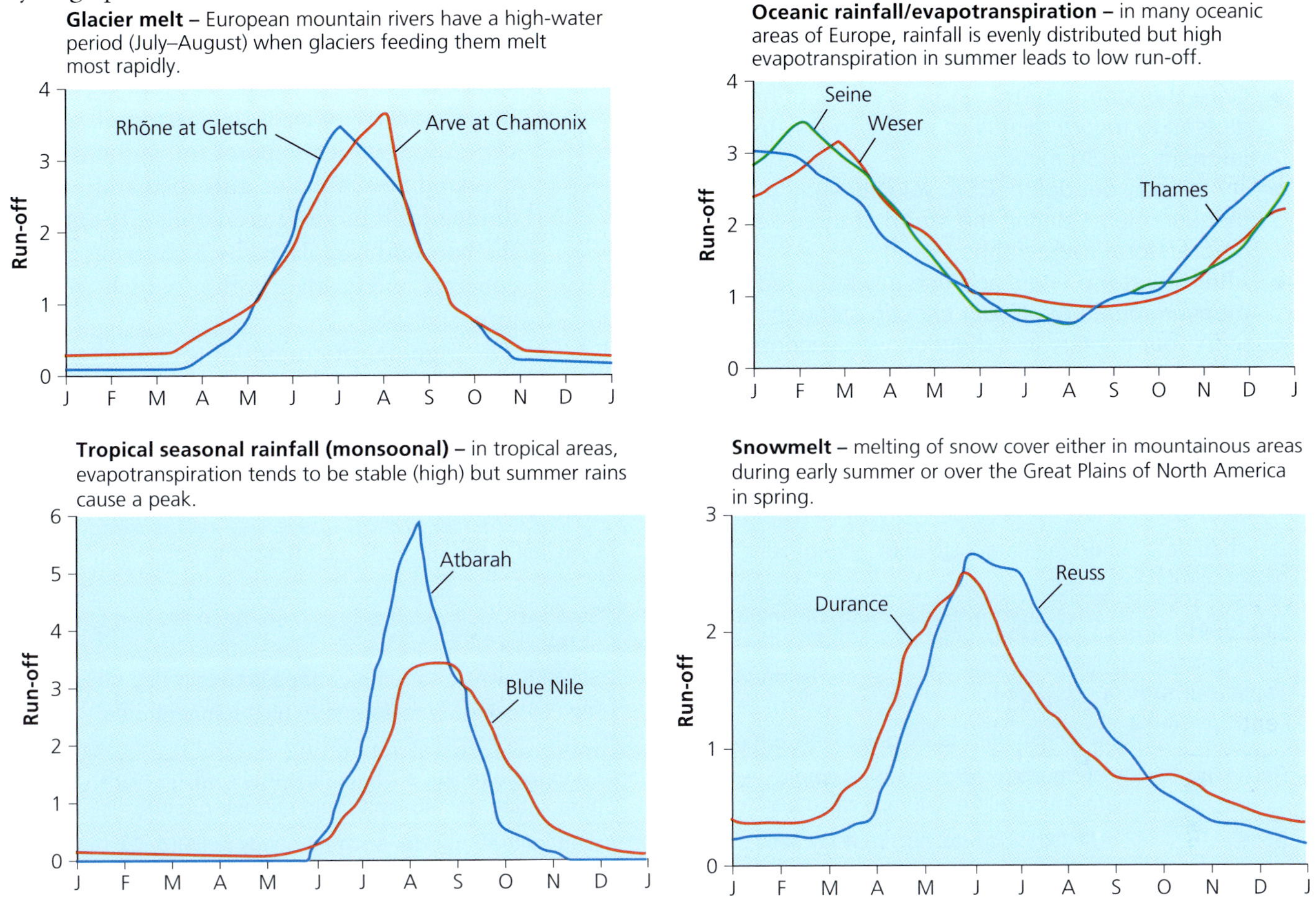

Glacier melt – European mountain rivers have a high-water period (July–August) when glaciers feeding them melt most rapidly.

Oceanic rainfall/evapotranspiration – in many oceanic areas of Europe, rainfall is evenly distributed but high evapotranspiration in summer leads to low run-off.

Tropical seasonal rainfall (monsoonal) – in tropical areas, evapotranspiration tends to be stable (high) but summer rains cause a peak.

Snowmelt – melting of snow cover either in mountainous areas during early summer or over the Great Plains of North America in spring.

Figure 1.13 River regimes in four climate zones

Key terms

Lag time: The time interval between peak rainfall and peak discharge.

Falling or recessional limb: The part of a storm hydrograph in which the discharge starts to decrease.

Base flow: The normal, day-to-day discharge of the river.

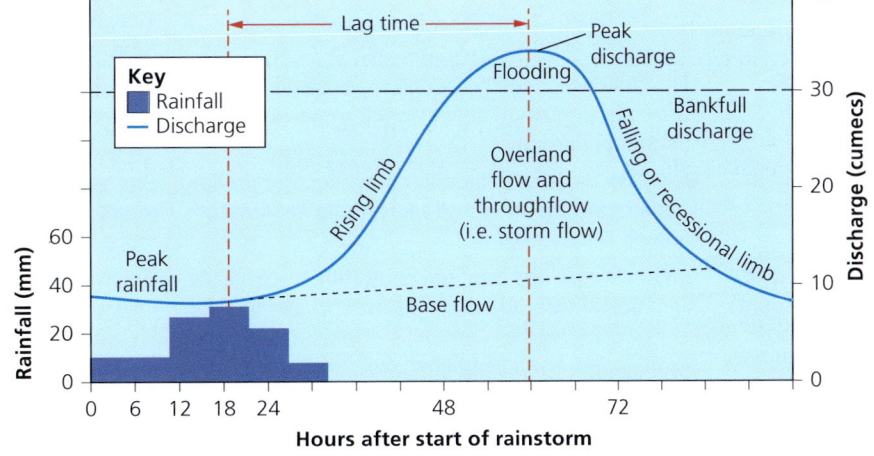

Figure 1.14 Features of a storm hydrograph

1 The operation and importance of the hydrological cycle

> **Fieldwork opportunity**
>
> Investigate the impact of a storm on a small stream:
> - Use the weather forecast to identify the likely occurrence of a storm in your local area, having researched an accessible stream of 2–3 km in length.
> - Develop a sampling strategy for ten to fifteen measuring points from source to mouth.
> - Before the storm arrives, develop a set of baseline results for width, depth, velocity, cross-sectional area (CSA), discharge, stone shape and size.
> - Once the storm starts, set up a series of homemade rain gauges to record rainfall at key intervals. Try to calculate intensity per 30 minutes.
> - Try to carry out depth, CSA, velocity and discharge measurements during the storm at three key stations to create storm hydrographs.
> - After the storm ends, and one day later, redo all the measurements at all your chosen stations on the stream.
>
> Support your primary investigation with secondary data from the Met Office on the storm, and also any measurements of discharge in neighbouring rivers and streams.

There is little control over the physical factors; however, effective planning and management can help to mitigate catchment flooding.

The impact of urbanisation on hydrological processes

Urbanisation is probably the most significant human factor that leads to increased flood risk. Figure 1.15 summarises the effect of urbanisation on hydrological processes.

- Building activity leads to clearing of vegetation, which exposes soil and increases overland flow. Piles of disturbed and dumped soil increase erodability. Eventually the bare soil is replaced by a covering of concrete and tarmac, both of which are impermeable.
- The high density of buildings means that rain falls on to roofs and is then swiftly dispatched into drains by gutters and pipes.

Table 1.4 The range of factors that interact to determine the shape of a storm hydrograph

Factor	'Flashy' river	'Flat' river
Description of hydrograph	Short lag time, high peak, steep rising limb	Long lag time, low peak, gently sloping rising limb
Weather/climate	Intense storm which exceeds the infiltration capacity of the soil Rapid snowmelt as temperatures suddenly rise above zero Low evaporation rates due to low temperatures	Steady rainfall which is less than the infiltration capacity of the soil Slow snowmelt as temperatures gradually rise above zero High evaporation rates due to high temperatures
Rock type	Impermeable rocks, such as granite, which restrict percolation and encourage rapid surface run-off	Permeable rocks, such as limestone, which allow percolation and so limit rapid surface run-off
Soils	Low infiltration rate, such as clay soils (0–4 mm/h)	High infiltration rate, such as sandy soils (3–12 mm/h)
Relief	High, steep slopes that promote surface run-off	Low, gentle slopes that allow infiltration and percolation
Basin size	Small basins tend to have more flashy hydrographs	Larger basins have more delayed hydrographs; it takes time for water to reach gauging stations
Shape	Circular basins have shorter lag times	Elongated basins tend to have delayed or attenuated hydrographs
Drainage density	High drainage density means more streams and rivers per unit area, so water will move quickly to the measuring point	Low drainage density means few streams and rivers per unit area, so water is more likely to enter the ground and move slowly through the basin
Vegetation	Bare/low density, deciduous in winter, means low levels of interception and more rapid movement through the system	Dense, deciduous in summer, means high levels of interception and a slower passage through the system; more water lost to evaporation from vegetation surfaces
Pre-existing (antecedent) conditions	Basin already wet from previous rain, water table high, soil saturated so low infiltration/percolation	Basin dry, low water table, unsaturated soils, so high infiltration/percolation
Human activity	Urbanisation producing impermeable concrete and tarmac surfaces Deforestation reduces interception Arable land, downslope ploughing	Low population density, few artificial impermeable surfaces Reforestation increases interception Pastoral, moorland and forested land

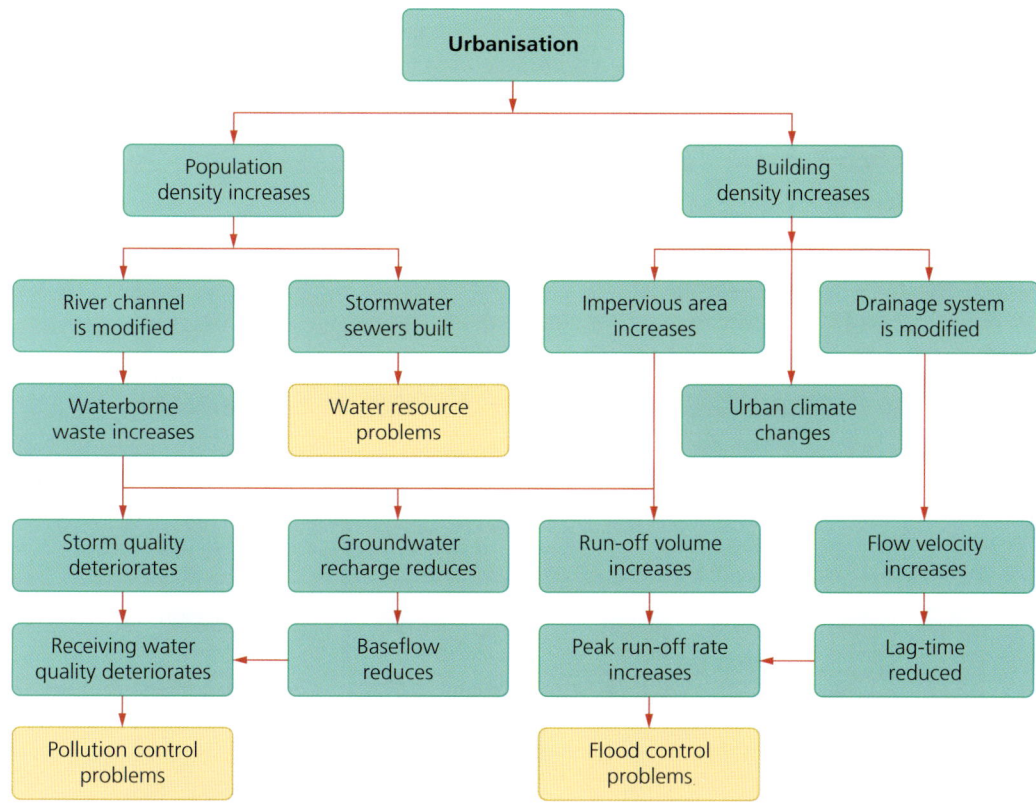

Figure 1.15 The impact of urbanisation on hydrological processes

- Drains and sewers are built, which reduce the distance that storm water must travel before reaching a channel. The increase in the velocity occurs because sewers generate less friction than natural pathways: sewers are designed to drain water quickly.
- Urban rivers tend to be channelised with embankments to guard against flooding. When floods occur they can be more devastating as the river overtops defences in a very confined space.
- Bridges can restrain the free discharge of floodwaters and act as local dams for upstream floods.
- In extreme weather events, urban areas such as Manchester, Leeds and York are highly vulnerable. They have to manage flood control problems with a higher, quicker peak discharge, as well as pollution problems from the storm water which washes off the roads, containing toxic substances.

Figure 1.16 (page 19) shows that as urbanisation intensifies, it has a major impact on the working of the hydrological cycle. Decision makers and planners therefore have a number of options that involve managing the catchment as a whole, for example, developing appropriate land use, such as forestry and moorlands, in the upper areas, and managing development in the lower part of catchments by land use zoning, and by limiting building on the flood plains so 'making space for water to flood'. At the same time they have to defend high-value properties and installations against agreed flood recurrence levels. There are also a number of grants for comparatively low-cost strategies that can be used to lower flood risks, such as semi-permeable surfaces for car parks and high level wiring systems in houses, as well as the government developing affordable insurance (Flood Re). Additionally, building regulations can be tightened to ensure flood-proof property designs.

Synoptic themes:

Players

Environmental managers and planners increasingly look at catchment management both upstream (e.g. afforestation) and in the lower course (e.g. flood defences) in order to manage the impacts of urbanisation and changes in land use which have exacerbated flood risk.

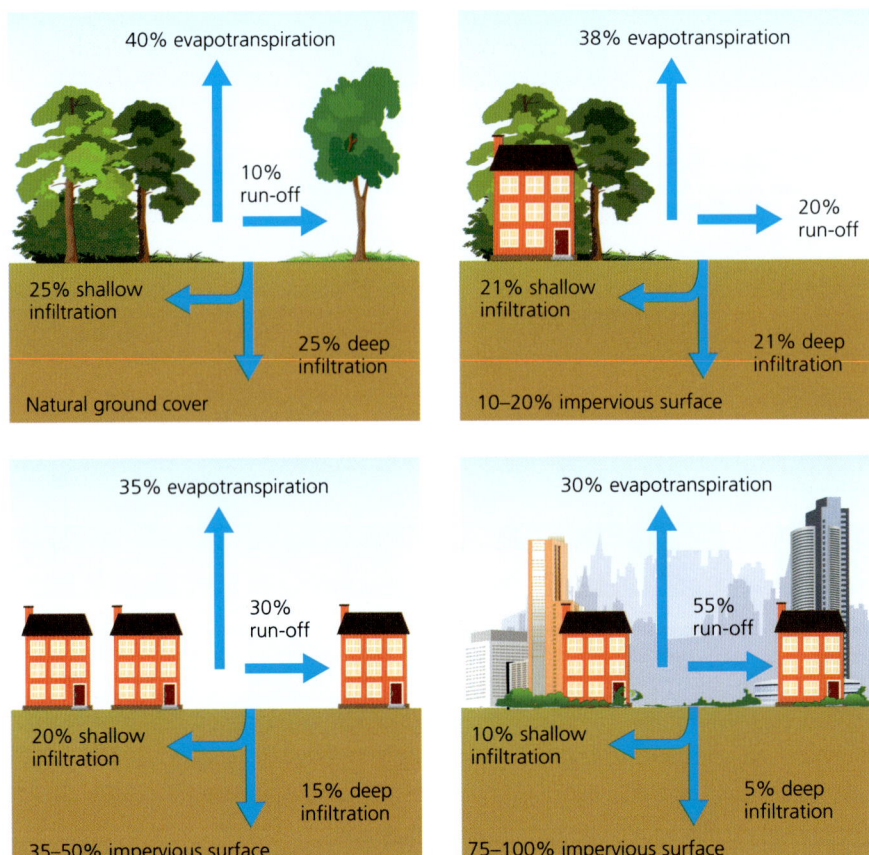

Figure 1.16 The impact of the intensity of urbanisation on hydrological processes (Source: US Department of Agriculture)

Skills focus: Regimes and storm hydrographs

1. Using data from Table 1.5, draw two graphs to compare rainfall and run-off in the two neighbouring catchments of Austwick Beck and Clapham Beck in North Yorkshire.

Table 1.5 Yearly figures for two drainage basins

	Austwick Beck		Clapham Beck	
	Rainfall (mm)	Channel run-off or stream flow (mm)	Rainfall (mm)	Channel run-off or stream flow (mm)
Oct	74	6	66	22
Nov	88	6	96	22
Dec	170	18	136	26
Jan	148	102	176	38
Feb	12	30	16	26
Mar	122	42	122	36
Apr	90	34	90	28
May	136	24	100	26
Jun	168	16	130	20
Jul	208	44	182	22
Aug	92	24	114	20
Sep	204	26	210	22
TOTAL	1,512	372	1,408	306

The Water Cycle and Water Insecurity

2. Study the map in Figure 1.17. Using your own knowledge, describe and suggest reasons for any differences shown on your graphs from Question 1.
3. Describe the differences in the storm hydrographs shown in Figure 1.17. They are both for the same storm.
4. Suggest reasons for the differences you have described. You should refer back to page 16 (river regimes and storm hydrographs).

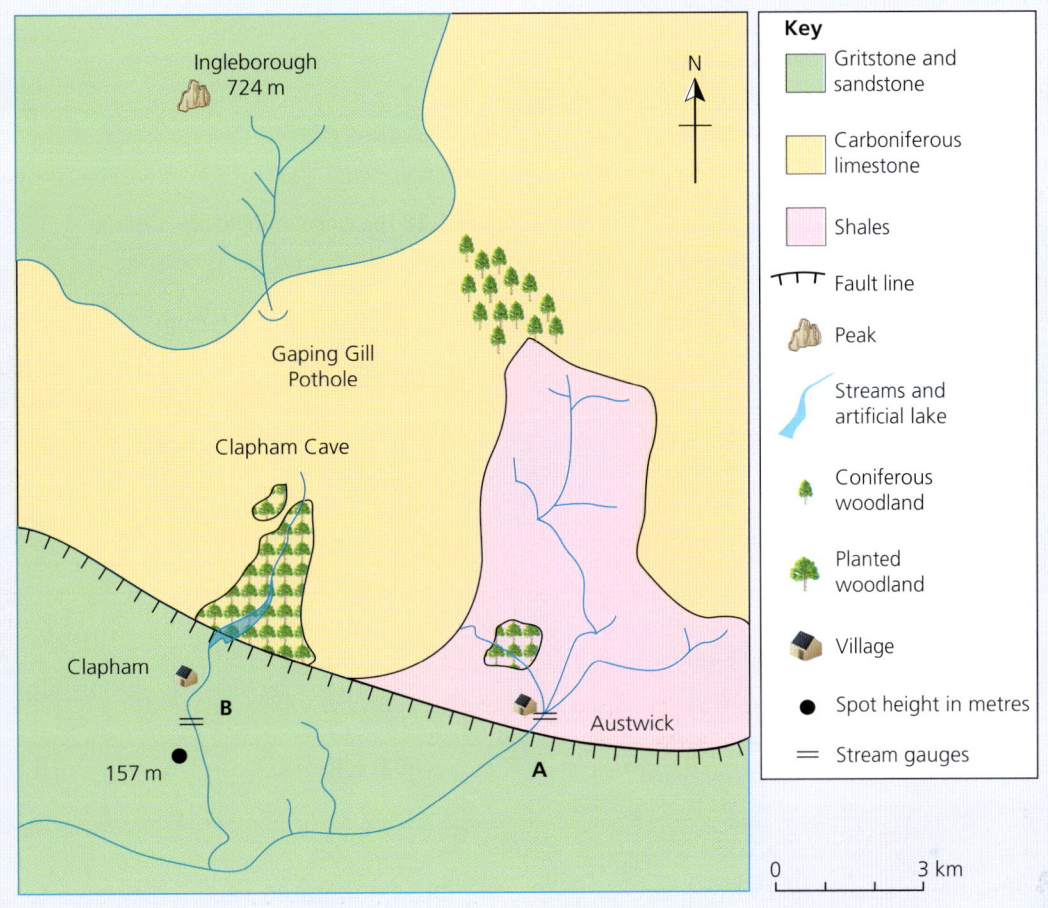

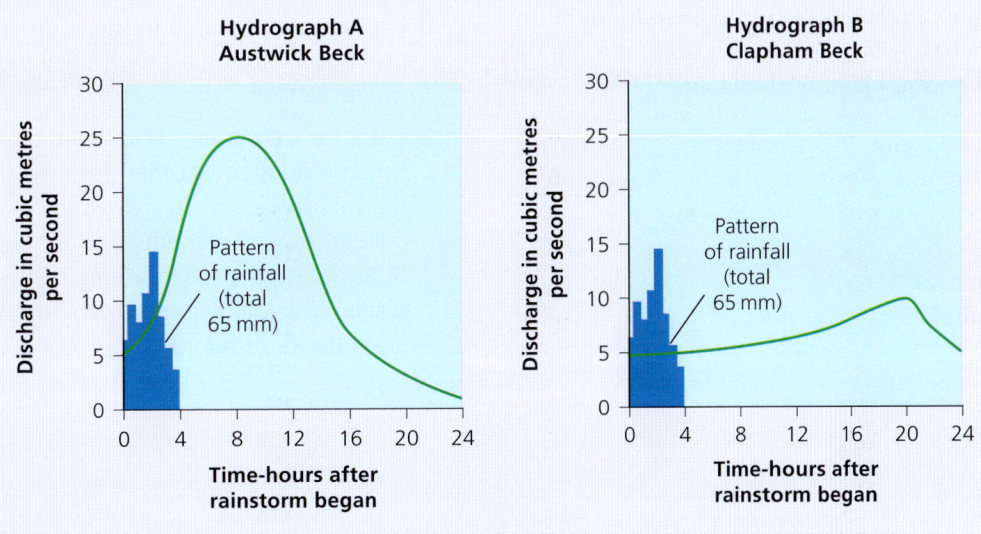

Figure 1.17 Contrasting catchments in North Yorkshire

1 The operation and importance of the hydrological cycle

Review questions

1 Using Figure 1.18 as a framework, use Figure 1.1 on page 3 to annotate the transfers between stores with the correct flux measurements. Provide a brief commentary on your annotations.

2 Study Table 1.6 below, showing changes in run-off and soil erosion after deforestation. Comment on the impact of different land uses on run-off and erosion rates. Are similar changes apparent in all five locations, or do slope and rainfall totals also play a part?

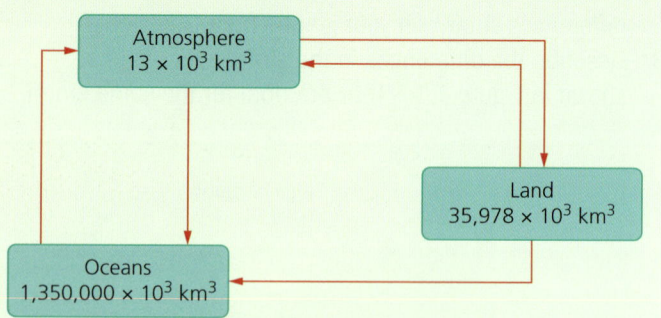

Figure 1.18 The global hydrological system

Table 1.6 Changes in run-off and erosion after deforestation; A = forest or ungrazed thicket, B = crops, C = barren soil

Location	Average annual rainfall (mm)	Slope (%)	Annual run-off (%)			Erosion (tonnes per hectare per year)		
			A	B	C	A	B	C
Ouagadougou, Burkina Faso	850	0.5	2.5	2–32	40–60	0.1	0.6–0.8	10–20
Sefa, Senegal	1300	1.2	1.0	21.2	39.5	0.2	7.3	21.3
Bouaké, Ivory Coast	1200	4.0	0.3	0.1–26	15–30	0.1	1–26	18–30
Abidjan, Ivory Coast	2100	7.0	0.4	0.5–20	38	0.03	0.1–90	108–170
Mbapwa, Tanzania	c570	6.0	0.4	26.0	50.4	0	78	146

3 Carry out further research and write a short scientific article to assess the impact of deforestation in Nepal and the rivers Indus and Brahmaputra. You should weigh up the range of evidence on the scale of the issue.

4 Figure 1.19 (page 22) shows the classification of river regimes across the world. Research the regimes of the following rivers:
- the Yukon
- the Amazon
- the Indus.

Draw annotated sketch diagrams to explain their patterns.

5 Explain the changes in the hydrological processes during a storm shown in Figure 1.20 (page 22).

6 Select a Catchment Management Plan or River Basin Management Plan for any UK river from the following website: www.gov.uk/government/collections/catchment-flood-management-plans. Summarise the key features of the plan and explain how decision makers are working to manage its problems.

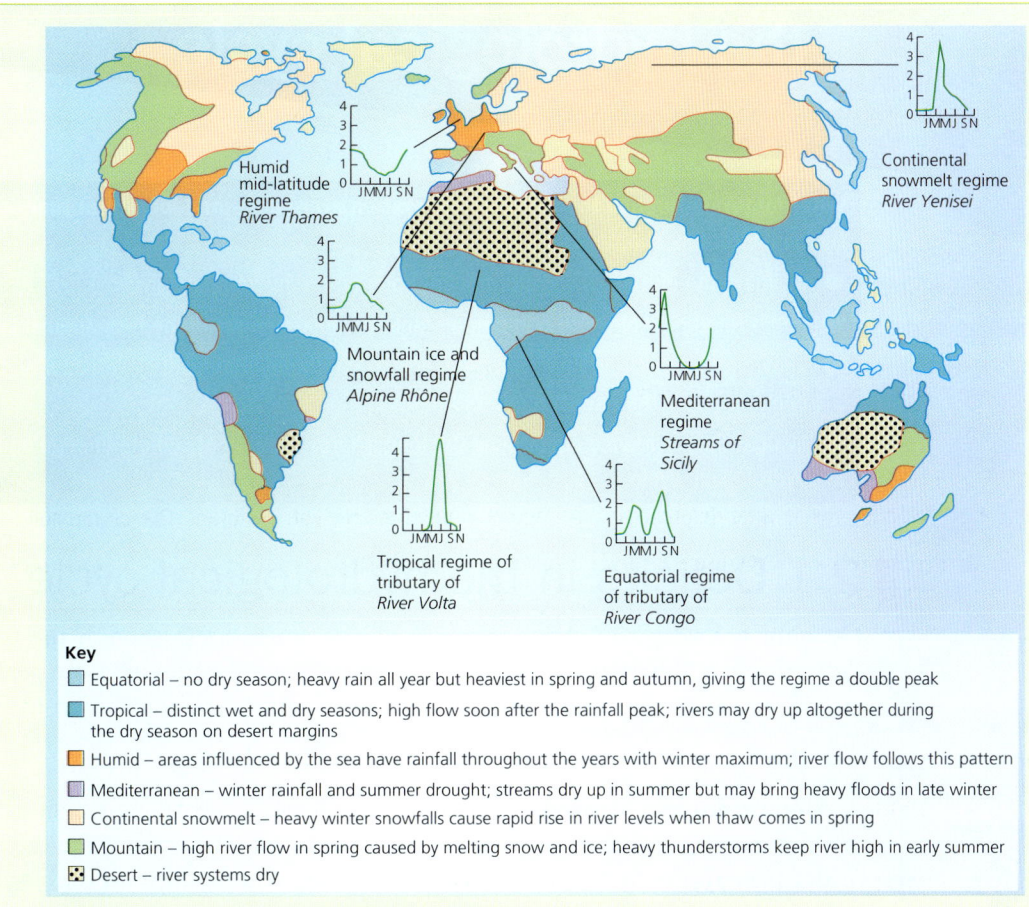

Figure 1.19 Classification of river regimes

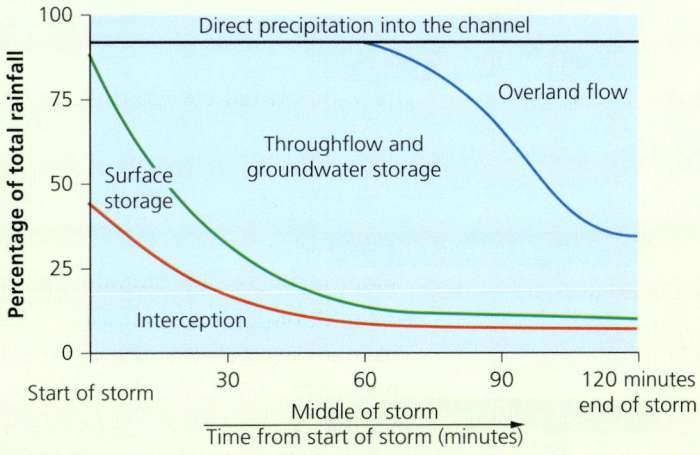

Figure 1.20 The changes in hydrological processes during a storm

Further research

Find out more about the hydrological system:
http://www.water-research.net
http://water.usgs.gov/edu/watercycle.html

2 Short- and long-term variations in the hydrological cycle

> **What factors influence the hydrological system over short- and long-term timescales?**
> By the end of this chapter you will be able to:
> - evaluate the impact of short-term variations on the hydrological cycle
> - understand how physical and human factors cause deficits in the hydrological cycle
> - know and understand how surpluses of water in the hydrological cycle occur and how they can impact on the environment and people
> - evaluate the evidence concerning the longer-term impacts of climate change on the hydrological cycle.

2.1 Deficits in the hydrological cycle

Drought is a very complex geographical phenomenon. Although the fundamental definition is a 'shortfall' or deficiency of water over an extended time period, usually at least a season, there are a wide variety of drought types, as shown in Figure 2.1.

The three major types of drought are sequentially related:

1. **Meteorological drought** is defined by shortfalls in precipitation as a result of short-term variability, or longer-term trends, which increase the duration of the dry period. Precipitation deficiency is usually combined with high temperatures,

> **Key term**
>
> **Meteorological drought:** Defined by shortfalls in precipitation as a result of short-term variability within the longer-term average overall, as shown in many semi-arid and arid regions such as the Sahel. Drought has become almost a perennial problem in recent years as longer-term trends have shown a downward movement in both rainfall totals and the duration and predictability of the rainy season (see Figure 2.4 on page 27), or the occurrence of megadroughts in California in 2017–19.

Types of drought and their characteristics		
	Major features	**Major impacts**
Meteorological drought — Rainfall deficit	Low precipitation High temperatures Strong winds Increased solar radiation Reduced snow cover	Loss of soil moisture Supply of irrigation water declines
Hydrological drought — Stream flow deficit	Reduced infiltration Low soil moisture Little percolation and groundwater recharge	Reduced storage in lakes and reservoirs Less water for urban supply and power generation – restrictions Poorer water quality Threats to wetlands and wildlife habitats
Agricultural drought — Soil moisture deficit	Low evapotranspiration Plant water stress Reduced biomass Fall in groundwater levels	Poor yields from rain-fed crops Irrigation systems start to fail Pasture and livestock productivity declines Rural industries affected Some government aid required
Famine drought — Food deficit	Loss of natural vegetation Increased risk of wild fires Wind-blown soil erosion Desertification	Widespread failure of agricultural systems Food shortages on seasonal scale Rural economy collapses Rural–urban migration Increased malnutrition and related mortality Humanitarian crisis International aid required

(Drought duration and severity increases downward)

Figure 2.1 A model of drought development and its impacts

high winds, strong sunshine and low relative humidity, all of which increase evaporation. The causes of rainfall deficiency can be natural variations in atmospheric conditions or desiccation caused by deforestation, or longer term such as occurrence of El Niño events and climate change.

Figure 2.1 (page 24) shows how these conditions impact on the hydrological cycle, with decreases in infiltration, percolation and groundwater recharge, and increases in evaporation and transpiration.

2 Over time **agricultural drought** ensues. Some farming practices such as overgrazing can accelerate the onset of agricultural drought. The rainfall deficiency leads to deficiency of soil moisture and soil water availability which has a knock-on effect on plant growth and reduces biomass. Soil moisture budgets (see page 15) can show if the deficit stage is protracted and more severe than normal. The outcomes, with falling groundwater levels, are poor yields from rain-fed crops, failure of irrigation systems, decline in pasture quality and livestock well-being, and a knock-on effect on the economy of rural areas with many subsistence farmers requiring government aid.

3 **Hydrological drought** is associated with reduced stream flow and groundwater levels, which decrease because of reduced inputs of precipitation and continued high rates of evaporation. It also results in reduced storage in any lakes or reservoirs, often with marked salinisation and poorer water quality. There are also major threats to wetlands and other wildlife habitats. Hydrological droughts are also linked to decreasing water supplies for urban areas, often in developed countries, which inevitably results in water-use restriction to control abstraction rates, as in 1976 in the UK when a Minister for Drought was actually appointed to manage the crisis!

It can be a particular problem in areas such as rural northeastern Brazil, where there are no permanent rivers and water supplies depend on seasonal rainfall stored in shallow reservoirs and ponds. As well as leaving many rural dwellers with less access to water, the quality of the water declines, leading to ill health and a reliance on high-cost water distributed by road tankers.

Resulting from these droughts, a fourth type of drought can occur, often called **famine drought**. With widespread failure of agricultural systems, food shortages develop into famines that have severe social, economic and environmental impacts. Humanitarian crises such as those associated with the Horn of Africa in 2012–14 or in eastern Africa in 2019 require international solutions.

As populations grow and become wealthier, their demand for water also increases. At the same time, natural variability in climate can cause a temporary decline in supply, and stores are not replenished. To this can be added more long-term susceptibility to drought brought about by **ENSO** and climate change associated with global warming.

Higher temperatures lead to increased evaporation. Areas that are severely affected by drought have doubled to include more than 30 per cent of the world's land area in the last 30 years – especially in southern Europe, many parts of USA such as California, parts of the Asian landmass and eastern Australia, as well as the Sahel in Africa.

Droughts (known as 'creeping hazards') typically have a long period of onset, sometimes several years, which makes it difficult to determine whether a drought has begun or whether it is 'just a dry period'.

> **Key terms**
>
> **Agricultural drought:** The rainfall deficiency from meteorological drought leads to deficiency of soil moisture and soil water availability, which has a knock-on effect on plant growth and reduces biomass.
>
> **Hydrological drought:** Associated with reduced stream flow and groundwater levels, which decrease because of reduced inputs of precipitation and continued high rates of evaporation. It results in reduced storage in lakes and reservoirs, often with marked salinisation and poorer water quality.
>
> **Famine drought:** A humanitarian crisis in which the widespread failure of agricultural systems leads to food shortages and famines with severe social, economic and environmental impacts.

Table 2.1 Measurement of drought

Palmer Drought Severity Index (PDSI)	This applies to long-term drought and uses current data as well as that of the preceding months, as drought is dependent on previous conditions. It focuses on monitoring the duration and intensity of large-scale, long-term, drought-inducing atmospheric circulation.
Crop Moisture Index (CMI)	This is a measure of short-term drought on a weekly scale and is useful for farmers to monitor water availability during the growing season.
Palmer Hydrological Drought Index (PHDI)	The hydrological system responds slowly to drought, both in reacting to drought and recovering from it, so different models need to be developed for rivers, lakes, etc.

Key concept: The El Niño–Southern Oscillation (ENSO)

Figure 2.2 shows normal conditions in the Pacific basin, and then conditions during an El Niño event when the cool water normally found along the coast of Peru is replaced by warmer water. At the same time the area of warmer water further west, near Australia and Indonesia, is replaced by cooler water. La Niña is preceded by a build-up of cooler-than-normal subsurface water in the tropical Pacific – an extreme case of the normal situation. El Niño events usually occur every three to seven years and usually last for 18 months. A La Niña episode *may*, but does not always, follow an El Niño event.

In terms of drought occurrence, El Niño can trigger very dry conditions throughout the world, usually in its second year, especially in Southeast Asia, India, eastern Australia, southeastern USA, Central America and northeastern Brazil, as well as further afield in parts of Africa (Kenya and Ethiopia). In India, El Niño years always lead to relatively weak monsoon rains, exacerbating drought by monsoon failure. La Niña can also lead to severe drought conditions, but these are usually localised on the western coasts of South America. Cooler than normal ocean temperatures can generate anticyclonic weather and, therefore, very dry conditions associated with descending air.

A normal year
Warm, moist air rises, cools and condenses, forming rain clouds

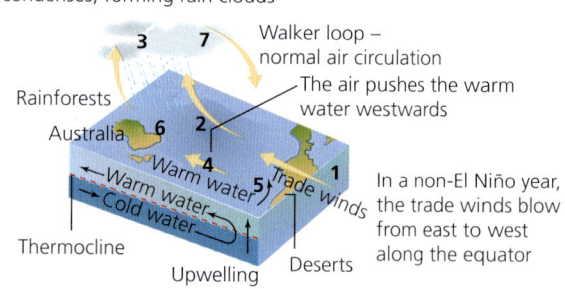

(1) The trade winds blow equator-wise and westwards across the tropical Pacific
(2) The winds blow towards the warm water of the western Pacific
(3) Convectional uplift occurs as the water heats the atmosphere
(4) The trade winds push the warm air westwards. Along the east coast of Peru, the shallow position of the thermocline allows winds to pull up water from below
(5) This causes upwelling of nutrient-rich cold water, leading to optimum fishing conditions
(6) The pressure of the trade winds results in sea levels in Australasia being 50 cm higher than Peru and sea temperatures being 8 °C higher
(7) The Walker loop returns air to the eastern side of the Pacific

An El Niño year

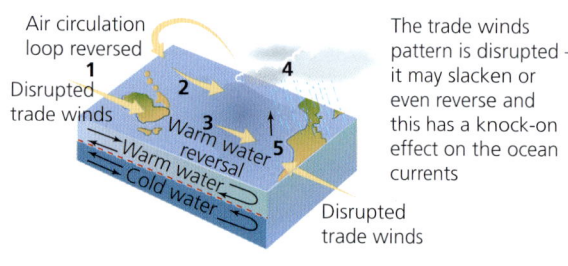

(1) The trade winds in the western Pacific weaken and die
(2) There may even be a reverse direction of flow
(3) The piled-up water in the west moves back east, leading to a 30 cm rise in sea level in Peru
(4) The region of rising air moves east with the associated convectional uplift. Upper air disturbances distort the path of jet steams, which can lead to teleconnections all around the world
(5) The eastern Pacific Ocean becomes 6–8 °C warmer. The El Niño effect overrides the cold northbound Humboldt Current, thus breaking the food chain. Lack of phytoplankton results in a reduction in fish numbers, which in turn affects fish-eating birds on the Galapagos Islands
(6) Conditions are calmer across the whole Pacific

A La Niña year

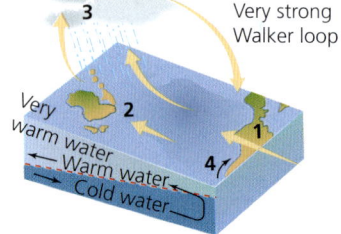

This an exaggerated version of a normal year, with a strong Walker loop.
(1) Extremely strong trade winds
(2) The trade winds push warm water westwards, giving a sea level up to 1 m higher in Indonesia and the Philippines
(3) Low pressure develops with very strong convectional uplift as very warm water heats the atmosphere. This leads to heavy rain in Southeast Asia
(4) Increase in the equatorial undercurrent and very strong upwelling of cold water off Peru results in strong high pressure and extreme drought. This can be a major problem in the already semi-arid areas of northern Chile and Peru

The most recent El Niño major event was 2016, although 2019 was a possible as the hottest year

Figure 2.2 The workings of the ENSO

Human influences on drought

A study of the Sahel region of Africa shows how, although there are physical factors associated with drought development, human activity plays a major role in making droughts even more severe.

Figure 2.3 shows rainfall trends in the Sahel, the name given to the vast semi-arid region on the southern edge of the Sahara, which stretches right across the African continent from Mauritania to Eritrea. It contains several of the poorest developing countries in the world (LDCs).

As Figure 2.4 shows, the Sahel has high variability of rainfall at all climate scales.

- Seasonally – the African Sahel is drought sensitive as it occupies a transitional climate zone. Under so-called normal conditions, the mean annual rainfall (around 85 per cent) is nearly all concentrated in the summer. It varies from 100 mm (very arid) on the edge of the Sahara to 800 mm along its southern margins.
- Annually – from year to year there is huge variability, especially on the Saharan fringe. Unusually warm sea surface temperatures (SSTs) in tropical seas favour strong convectional uplift over the ocean that, in turn, weakens the West African monsoon and contributes to drought in the Sahel.
- Decadal anomalies are very clear in Figure 2.3.

Human factors do not cause drought but they act like a positive feedback loop in enhancing its impacts. In the 1999–2000 Ethiopian–Eritrean drought/famine crisis, about 10 million people needed food assistance. The drought impacts were increased by socio-economic conditions faced by nomadic communities. Nomadic communities surrounding the Abyssinian plateaus had experienced the encroachment of farmland on grazing areas and obstructions to pastoral migration. The development of cotton growing on state farms in the Awash valley had reduced grazing areas. Any agriculture was rain-fed, making it vulnerable to droughts. In addition, poverty and civil war in Ethiopia and Eritrea drove people out of the Sahel and onto marginal land. As a result, rural population densities increased, putting further pressure on limited food supply. Communication blockages in the Tigray region in late 2020 continues to impact food security.

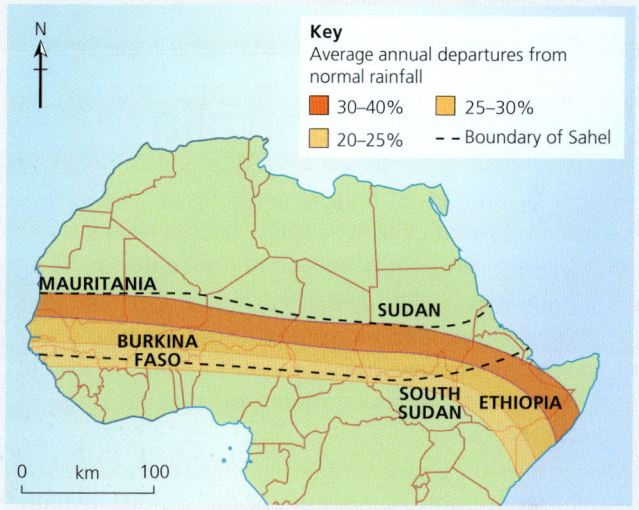

Figure 2.4 The Sahel region

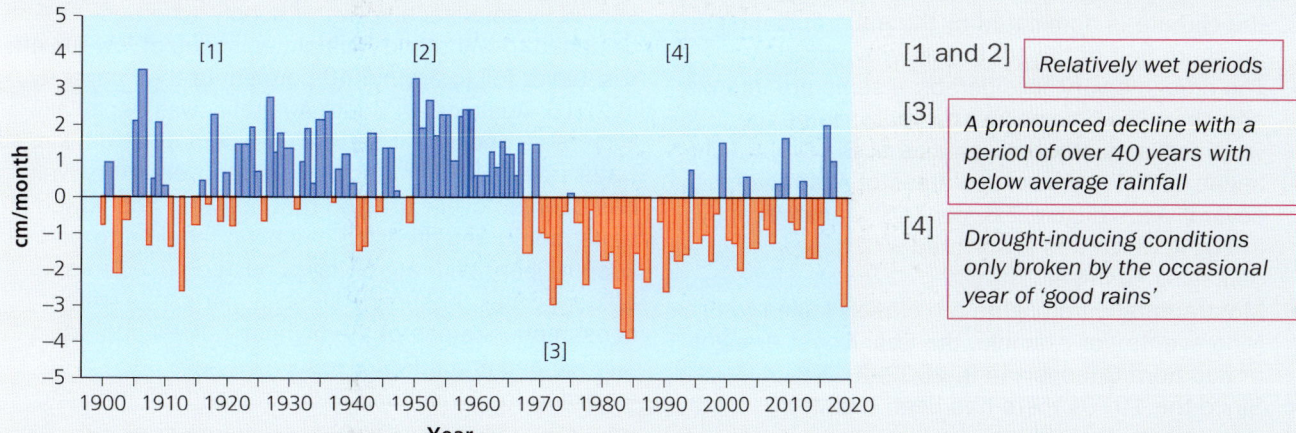

Figure 2.3 Rainfall trends in the Sahel region. Note the figures are for June to October each year and show trends above or below the average. Average annual precipitation is calculated from 30 years' worth of monthly precipitation data, but any single year's precipitation could vary considerably from this average

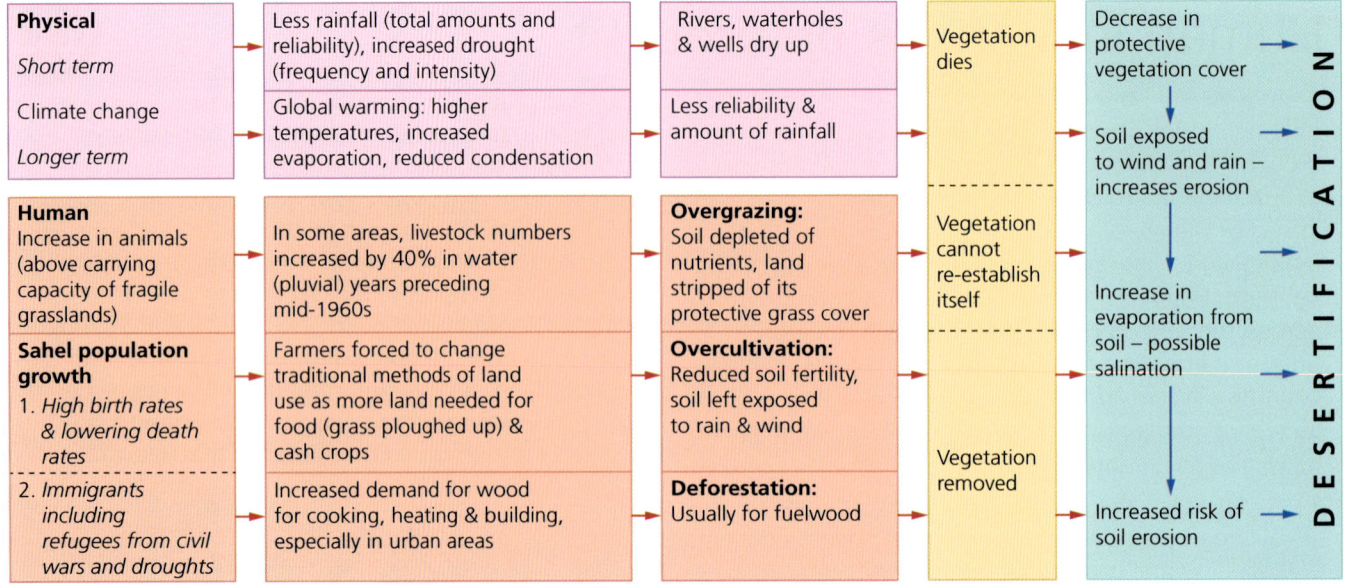

Figure 2.5 The complex causes of desertification in the Sahel

Drought in Australia

Drought is a recurrent feature in Australia. The Australian Bureau of Meteorology recognises two main types of drought based on rainfall criteria:

- Serious deficiency – rainfall totals within 10 per cent of values recorded for at least three months.
- Severe deficiency – rainfall totals within the lowest 5 per cent of values on record for at least three months.

There are major physical reasons why Australia is so drought-prone, with 30 per cent of the country usually affected in any one year:

- Low, highly variable rainfall totals occur because the climate is dominated by the sub-tropical high-pressure belt of the southern hemisphere.
- The droughts vary considerably – some are intense and short lived; some last for years; some are very localised; others, such as the 'Big Dry' which began in 2006, cover huge areas of Australia for several years (Figure 2.6, page 29), most recently in 2019, which was associated with hugely damaging bushfires.
- Most Australian droughts are closely linked to El Niño events, for example, the East Coast drought of southern Queensland in 2002–3.
- Since the 1970s there has been a shift in rainfall patterns with the eastern area, where most people live, becoming drier compared to northwestern areas.
- The 'Big Dry' in 2019 is thought to have been associated with longer-term climate change, leading to a trend of a warmer, drier climate for south-eastern Australia, so the Big Dry may be a recurrent event.

The 'Big Dry' was assessed as a 1-in-1000-year event as it spread nationwide. It has affected more than half the farmlands, especially in the Murray–Darling Basin (the agricultural heartland), which provides 50 per cent of the nation's agricultural outputs. This has had disastrous impacts on Australia's food supplies and wool, wheat and meat exports. Farmers also rely on water for their irrigated farming of rice, cotton and fruits (newer crops).

Despite the fact that most of Australia's cities are served by sophisticated water supply schemes designed to withstand multiple episodes of low run-off, reservoirs fell to around 40 per cent of their capacity in 2016. Adelaide, in South Australia, was especially vulnerable because it drew 40 per cent of its drinking water from the River Murray. In recent years the river has been so over-extracted that no water has flowed at its mouth and dredging has been required to keep it open. With a growing population used to an affluent water-consuming lifestyle, per capita water consumption is one of the highest in the world. Many surface and groundwater resources have been over-extracted for agricultural, industrial and urban usage.

With future demands likely to exceed supply (threatened by more El Niño events and climate change), new schemes for urban areas must be developed to include desalination plants, large-scale recycling of grey water and sewage, and more strategies for water conservation.

The Water Cycle and Water Insecurity

Figure 2.6 The Murray–Darling Basin, southeastern Australia

With limited supplies of water, inevitably there is competition between farmers and urban dwellers. The farmers claim they need the water for vital irrigation, but they need to look towards smart irrigation. The over-abstraction of water in the past, during normal periods, was the root cause of the severity of the drought's impact.

Key concept: Day Zero

Describes the situation when any very large urban area, usually in the developing world, is approaching a water supply crisis with only a limited number of days of water supply left. It was first used when describing the situation in Cape Town 2014, but is now used widely for cities such as Chennai (India) 2019, São Paulo 2014/15. Drought induced by Climate Warming + El Niño + population growth and economic development = a potential Day Zero water crisis. Beijing, Bangalore, Jakarta and Mexico City are all possibilities in the future.

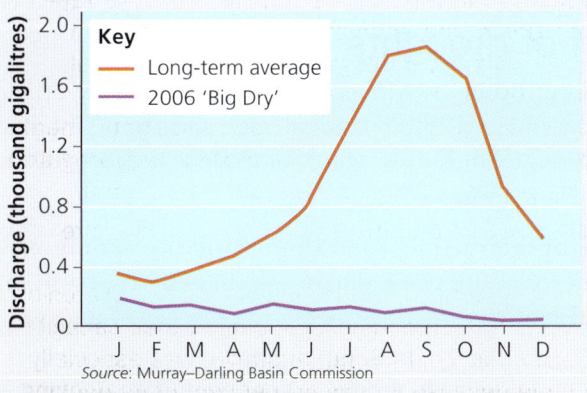

Figure 2.7 River discharge in the Murray–Darling Basin: long-term average and during the 'Big Dry' 2006 (Source: Murray–Darling Basin Commission)

Figure 2.8 Drought-ridden area

South Africa's drought

The average rainfall across South Africa is only 500 mm (compared to the global average of 860 mm); thus, like Australia, South Africa is vulnerable to any deficits in precipitation. While drought is therefore a regular and recurrent feature of the South African climate, droughts are becoming more frequent, and the five-year drought 2013–17 was by far the worst.

While climate change is the underlying factor, El Niño events, as in 2016 and 2019, can exacerbate the water supply situation. While South Africa does have adequate supplies of groundwater, drought-induced shortages of surface water have led to falling levels of this supply.

Droughts as in Australia can be linked to demographic and economic factors (a rapidly rising population in West Cape Province, especially in Cape Town, and rising demands from a growing agricultural and industrial economy) – South Africa is a member of the BRICS group. These pressures were combined with a lack of forward planning in building increased reservoir storage capacity. The effect of the drought can be summarised as hydrological, ecological, economic and social.

South Africa has responded with both short- and long-term solutions. Short term this involved restricting water use (especially in Cape Town) which in 2017 faced a **Day Zero** (page 29) situation, for example shorter showers, fewer flushes and the use of grey water.

Longer term, a threefold response of predict, prepare and plan is being developed at national, regional and local scales.

> **Key terms**
>
> **Teleconnection:** In atmospheric science, refers to climate anomalies which relate to each other at large distances.
>
> **Desertification:** Land degradation in arid, semi-arid and dry sub-humid regions resulting from various factors, including climatic variations and human activities.
>
> **Wetland:** An area of marsh, fen, peatland or water, whether natural or artificial, permanent or temporary, with water that is static or flowing, fresh, brackish or salt.

Table 2.1 (page 26) shows the various indices used to define drought. Most use a water balance approach: inputs of precipitation and losses due to evapotranspiration, and run-off (where applicable).

The physical causes of droughts are only partially understood. Climate dynamics is the study of the interlocked systems of the atmosphere, oceans, cryosphere, biomass and land surface, all of which interact to produce the global climate. Individual research on various drought occurrences has suggested that sea surface temperature anomalies are a very important factor. **Teleconnections** mean that development of the ENSO within the Pacific Ocean has an impact on climates around the world.

Figure 2.5 (page 28) shows how the combined processes of drought-induced environmental fragility can combine with poverty-induced human vulnerability to contribute to a very high risk from **desertification**.

The ecological impact of droughts

Wetland is a term that covers a multitude of different landscape types. Wetlands currently cover about 10 per cent of the Earth's land surface and, until 50 years ago, were considered as worthless wastelands only good for draining, dredging and infilling, and as the habitat of malarial mosquitoes.

However, increasing knowledge of their value, their importance as a component of the global biosphere, and concerns about their alarming rate of destruction, has made their conservation a global priority.

Wetlands perform a number of key functions:

- They act as temporary water stores within the hydrological cycle, thus mitigating river floods downstream, protecting land from destructive erosion by acting as washlands, and recharging aquifers.
- Chemically, wetlands act like giant water filters by trapping and recycling nutrients, as well as pollutants, which helps to maintain water quality.
- They have very high biological productivity and support a very diverse food web, providing nursery areas for fish and refuges for migrating birds.
- All these functions contribute towards their value for human society, as providers of resources (fish, fuelwood, etc.), of services in terms of hydrology within the water cycle, and as carbon stores (peat) within the carbon cycle. Figure 2.9 summarises their value.

Drainage and destruction

Drought can have a major impact on wetlands – with limited precipitation, there will be less interception as vegetation will deteriorate, and less infiltration and percolation to the groundwater stores, causing water table levels to fall. The processes of evaporation will continue and might increase from the less-protected surface, while transpiration rates will decrease, making wetlands less functional. Desiccation can also accelerate destruction by wild fires.

In addition to the physical causes of wetland loss, perhaps 2.5 million square kilometres, mostly in the developed world, has been destroyed in Europe and the USA largely for agriculture and urban development (for example, in Florida).

There are many other schemes that have led to wetland drainage, including water transfer schemes such as the Jonglei Canal Project, which diverted the White Nile discharge away from the Sudd Swamp to the dry land areas of South Sudan, or the degradation of the Okavango Delta in Botswana for cattle rearing. Exploiting fuel resources, such as peat, is another reason for wetland habitat loss, with a major impact on the carbon cycle.

The marshland of southern Iraq has been almost completely destroyed by dams on the Tigris and Euphrates, substantially reducing their flow, and also by Saddam Hussein's drainage schemes, largely designed to destroy the lifestyle of the 250,000 Marsh Arabs who lived there. However, the government's reasoning was that drainage was needed to reclaim land for agriculture and wipe out a breeding ground for mosquitoes. Intervention here was not new, as this was also the reason why former British Mandate administrators were the first to attempt to drain the marshes in this area. For a time the marshes became largely desert with great ecological losses. Since the overthrow of Saddam Hussein in 2003, water has been allowed to flow again and some level of vegetation has spread over half of the original area, but so far it is of a lower ecological quality.

As a result of concerns about wetland habitat destruction, schemes have been developed at a local, national and international scale to protect them. The 1991 Ramsar Convention on Wetlands has listed over 1800 wetlands of international importance, covering 1.7 million square kilometres in 160 states, to promote their conservation.

As ecosystems play such a vital role within the hydrological cycle, it is clearly important to keep them in a pristine state where possible, and to ensure their sustainable use.

Ecosystem services of wetlands	
Supporting	**Regulating**
Primary production at a very high level Nutrient cycling Food chain support Carbon state within Life support systems of carbon cycles	Flood control Groundwater recharge/discharge Shorelines as change and to protect Water purification
Provisioning	**Cultural**
Fuelwood, peat Fisheries Mammals and birds as tourism	Aesthetic value Recreational use Cultural heritage

Figure 2.9 The value of wetlands

2.2 Surpluses in the hydrological cycle

A number of physical factors lead to very high flows of water in a drainage basin. If the discharge is of sufficient quantity to cause a body of water to overflow its channel and submerge the surrounding land, flooding is deemed to have occurred.

There are a number of environments which are more at risk:

- Low-lying parts of flood plains and river estuaries. These are not only subject to river flooding, but also to **groundwater flooding** after the ground becomes saturated from prolonged heavy rainfall.
- Where low-lying areas are partially urbanised with impermeable surfaces, there is a greater danger of temporary **surface water flooding** as intense rainfall has insufficient time to infiltrate the soil, so flows overland.

> **Key terms**
>
> **Groundwater flooding:** Flooding that occurs after the ground has become saturated from prolonged heavy rainfall.
>
> **Surface water flooding:** Flooding that occurs when intense rainfall has insufficient time to infiltrate the soil, so flows overland.

Key term

Flash flooding: A flood with an exceptionally short lag time – often minutes or hours.

- Small basins, especially in semi-arid and/or arid areas, are subject to **flash flooding**. These are floods with an exceptionally short lag time – often minutes or hours – which are therefore extremely dangerous. They are usually associated with very intense convectional storms, so again infiltration, especially on semi-impermeable surfaces, and steep, unvegetated slopes, is very limited, allowing surface overland flow to develop very rapidly.

Causes of flooding

Physical factors

The primary causes of floods are either meteorological (short-term weather events) or longer-term climatic causes such as changing rainfall patterns.

In areas such as the UK, the usual cause of flooding is the prolonged and heavy rain associated with the passage of low-pressure systems or depressions. The traditional

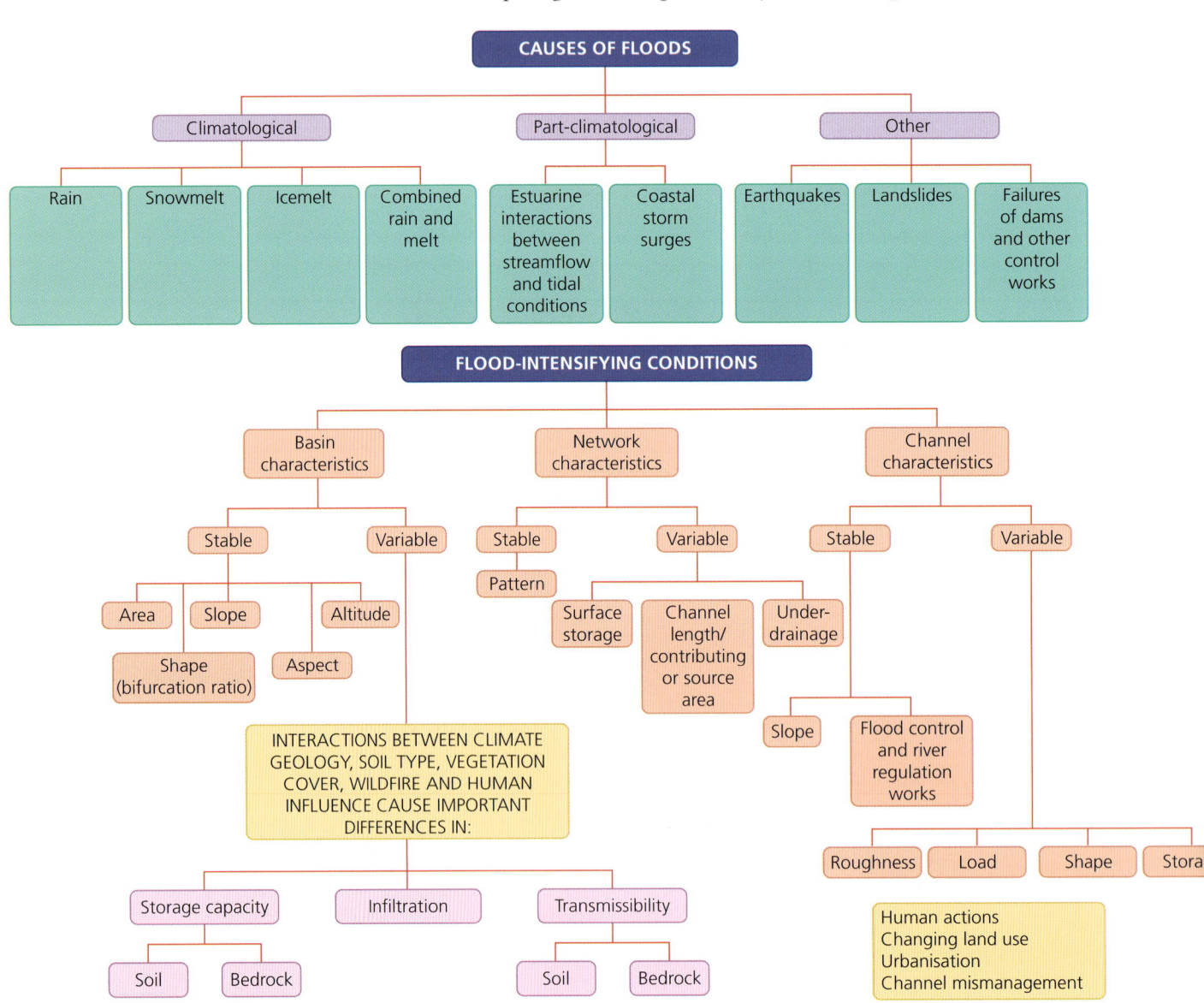

Figure 2.10 The causes of floods

time of year for this sequence, known as a progressive cycle, is autumn or early winter but, as a result of unusual positions in the jet stream, this sequence can occur at other times of the year (for example, the summer floods of 2007 and 2008). The degree of flooding also depends on the precise depression sequence – sometimes a succession of very intense storms, as occurred in the UK from October to December 2015, or in 2019, has a cumulative effect on the drainage system. This resulted from a very sinuous jet stream in a fairly constant track, which means that all high-pressure systems (anticyclones) were 'blocked'. Data for northern England from the Centre for Hydrology and Ecology reported many rivers with flows up to 50 times higher than normal, some experiencing their highest ever recorded flows.

In other areas, in particular southern and eastern Asia, intense seasonal monsoonal rainfall can result in widespread, damaging flooding. Around 70 per cent of the average annual rainfall occurs during 100 days – usually from July to September. The low-lying plains of the larger rivers in India, Pakistan, Bangladesh and Viet Nam, as well as China, are most at risk. Around 80 per cent of Bangladeshi people are exposed to flood risk. As Figure 2.11 shows, there are a variety of flood types in Bangladesh, with some areas affected by more than one type of flood. The highest flood risks are along the river courses and at the edge of the delta. This is hardly surprising as half of the country is less than 12.5 m above sea level.

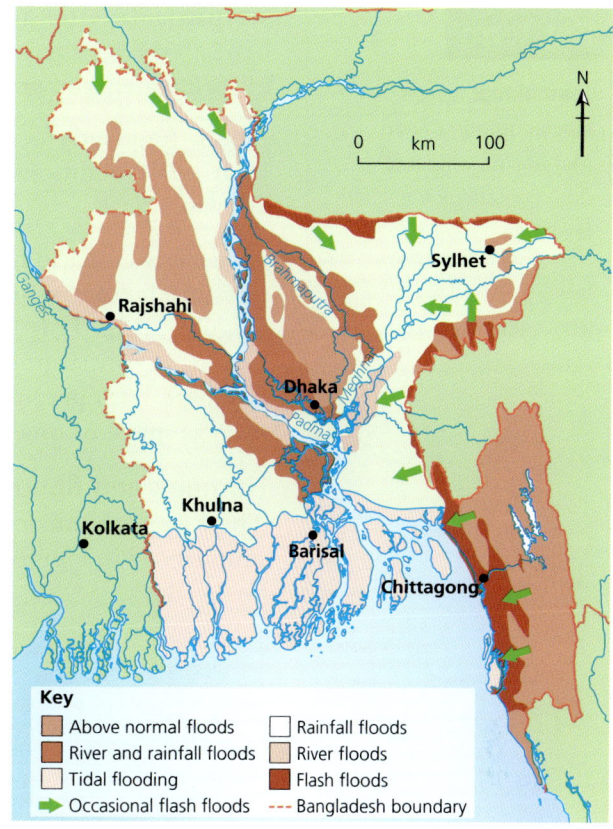

Figure 2.11 Types of flooding in Bangladesh

Pakistan suffered disastrous floods in August 2022. Rainfall totals were double those that normally fall between July and August and in the provinces of Baluchistan and Sindh, rainfall was nearly five times more. The flooding was exacerbated by seasonal run-off from melting glaciers, which was increased by the effects of climate change. As a result a third of the country was under water.

Excess rainfall across large river basins is also associated with tropical cyclones (mainly late summer in the sub-tropics). In southern Africa regular rainfall is sometimes supplemented by tropical cyclones, for example, in Mozambique in 2000, or 2019 when there was significant flooding from two major cyclones, Idai and Kenneth.

Snow and ice are responsible for many flood events, usually in higher latitudes or mountainous areas. Melting snow in late spring regularly causes extensive flooding in the continental interiors of Asia and America. The great north-flowing Siberian rivers, such as the Ob and Yenisei, cause vast annual flooding in the plains of Siberia. The quick transition from winter to spring upstream causes rapid snow melting, while their lower reaches remain frozen, with very limited infiltration. Flood water is often held up by temporary ice dams. Sometimes rain falls on melting snow when a rapid thaw occurs and this combination can cause heavy flooding. In the UK, spring floods in York are frequently intensified by rapid snowmelt in the higher parts of the River Ouse catchment.

In the Himalayas, glacial outburst floods (GOFs) occur as ice dams melt, leading to catastrophic draining of glacial lakes. Sometimes the flooding is exacerbated by landslides or earthquake-induced dam failure. In Iceland, glacial outburst floods

Key term

Jökulhlaup: A type of glacial outburst flood that occurs when the dam containing a glacial lake fails.

are particularly frequent because of volcanic activity, which generates melt water beneath the ice sheets and acts as a trigger for ice instability and the sudden release of melt water, known as a **jökulhlaup**.

Floods frequently occur in estuarine areas as very high river flows interact with high tidal conditions or coastal storm surges, so the causes here are only partly climatological.

As well as climatic causes of flooding (primary factors) there are also secondary physical factors which affect flooding levels, which tend to be basin specific (Figure 2.12). This makes certain river basins more flood prone than others. Geology, soil, topography and vegetation all play an important role as they combine with precipitation characteristics to determine key features of a flood, such as speed of onset, peak flow and flood duration.

Humans and flooding

There are a number of human actions that can exacerbate flood risk. These are summarised in Figure 2.13.

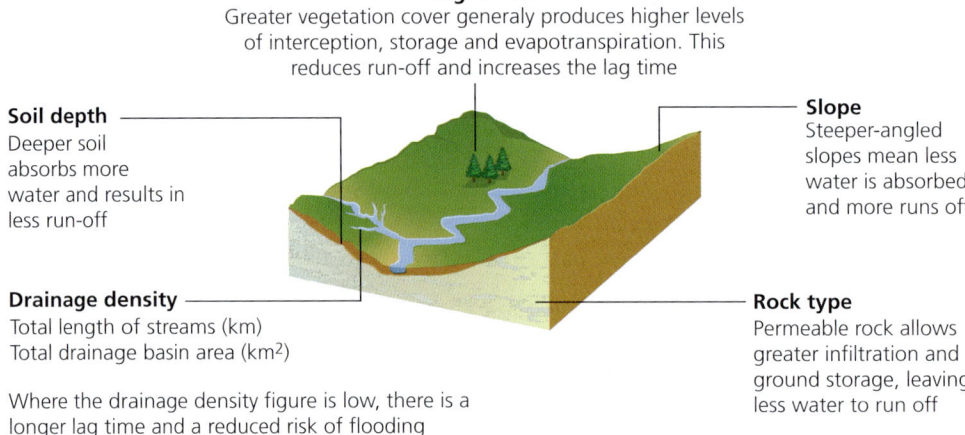

Figure 2.12 Physical factors affecting flooding levels

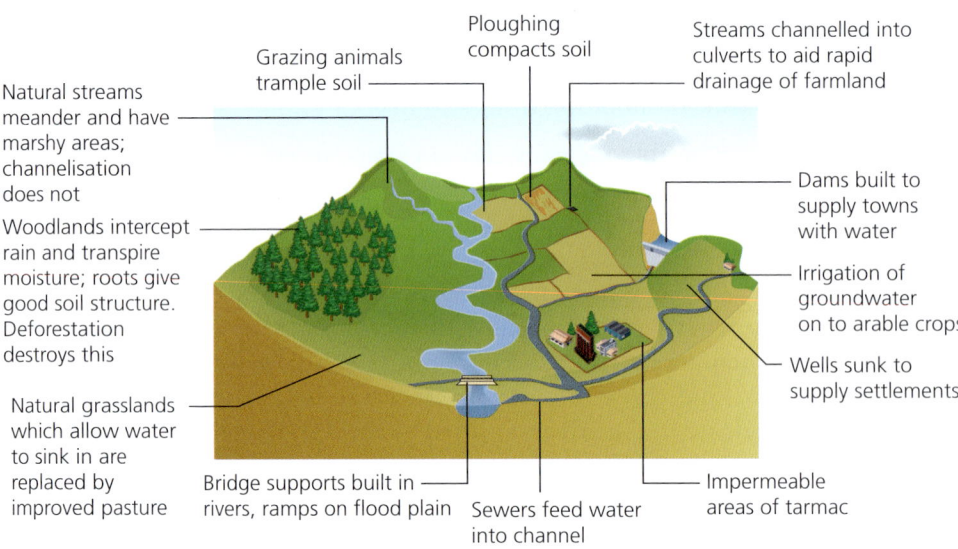

Figure 2.13 Human causes of flooding

The Water Cycle and Water Insecurity

A combination of economic growth and population movements throughout the twentieth century have caused many flood plains to be built on and many natural landscapes to be modified for agricultural, industrial and urban purposes.

Of these changes, many researchers have suggested that **urbanisation** is the key factor, for the following reasons:

- Creation of impermeable surfaces – roofs, pavements, roads and car parking space. It has been estimated that in London, the land taken up by car parks is around 25 square kilometres as around 60 per cent of residents have paved over their front gardens for car parking.
- Speeding up the drainage of water in built-up areas via artificial conduits, e.g. drains and sewers.
- Impeding channel flow by building alongside the river, e.g. building bridge supports and carrying out structural engineering by building levees.
- Straightening channels (realignment) to increase the flow, which results in flooding downstream – this could be regarded as mismanagement in some ways. Resectioning by dredging involves widening and deepening the channel to increase efficiency by increasing capacity and moving water away at a faster rate – but at considerable environmental cost.
- Changing land use associated with agricultural development. Deforestation, overgrazing, ploughing or drowning wetlands usually occurs upstream from urbanised flood plains, which has a knock-on effect downstream with increased run-off and increased levels of sediment (which are washed into rivers and block river channels).

The fact that urbanisation is concentrated on lower-lying land within drainage basins (especially on flood plains) means that natural and human factors coincide, which enhances both the frequency and the magnitude of **flood risk**. In 2015 the causes of recurrent flooding in areas in which flood defences were built only a decade ago (Carlisle, Cockermouth), and how to make them safe from future flooding, became a major topic for discussion in the UK and EU. It has also been noted that some of these flooding events (Cumbria) seem to have been of a higher magnitude than ever experienced previously.

In the British media both public and social blame for the floods of 2015 has been allocated to:

- extreme weather induced by climate warming
- budget cuts in the amount of money being spent on flood defences
- the green priorities of the EU Water Framework Directive, which puts environmental concerns before regular maintenance (i.e. dredging), although it does advocate making space for water to flood lowland areas
- poor land management, for example, blocking ditches or improving pasture then overgrazing it.

> **Key term**
>
> **Urbanisation:** The increase in the number of people living in towns and cities compared to the number of people living in the countryside.

> **Skills focus: Calculating flood frequency**
>
> The size of the largest flood event for each year for a particular location is placed in rank order, with Rank 1 being the largest for all available records for any given location.
>
> The following calculation is applied to calculate the time interval between floods of similar size:
>
> $$T = \frac{n + 1}{m}$$
>
> Where
> - T = recurrence interval
> - n = number of years of observation
> - m = rank order
>
> The calculated recurrence level indicates the number of years within which a flood of this size might be expected.
>
> However, it is a probability based on existing historic evidence and does not mean that similar floods will now occur more or less frequently, as with climate warming increased frequency is the more likely option. The recurrent interval is expressed as a 1 in 25, 50, 100, 500, 1000... event. The floods of highest magnitude will have much longer **return periods**: while they have the highest impact, they may be less likely to occur.
>
> Flood return periods are average recurrence intervals. It is possible to get two '100 year' floods very close together. This happened on many English rivers in 2007 and 2012. However, over centuries such floods should occur on average once every 100 years.

2 Short- and long-term variations in the hydrological cycle

> ### Key concept: Understanding flood risk and return periods
>
> Like all hazards, flooding has a frequency and a magnitude, both of which are important in assessing the risks involved. The flood return period, also known as the flood recurrence interval, is an estimate of the likelihood of a flood of a certain size recurring. A flood likely to happen once in ten years has a 10 per cent chance of happening in any one year. However, this is not a forecast and such a flood may happen more than once in the same interval or may not occur at all.
>
> A river may flood on average every two or three years, with significant flooding only every 50 to 100 years. One way to illustrate this is shown in Figure 2.14. Floods may only reach the furthest edge of the flood plain once in 500 years.
>
>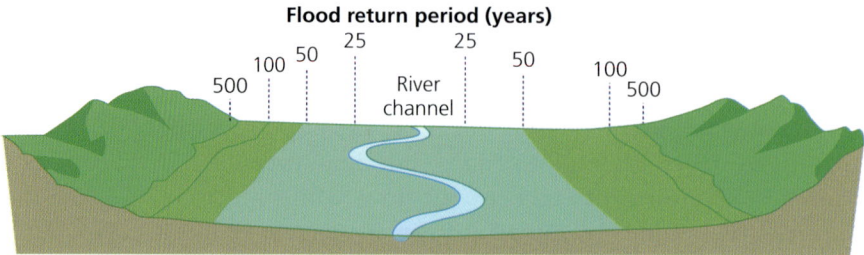
>
> **Figure 2.14** Understanding flood risk and flood return periods

The impacts of flooding

Floods are a common environmental hazard due to the widespread distribution of river flood plains and other low-lying areas. In the period between 1990 and 2010, the Emergency Events Database (EM-DAT) recorded over 3000 flood disasters worldwide, and an upward trend has continued. These were collectively responsible for 200,000 deaths and 3 billion people (around half of the world's population) being adversely affected – note this total includes many who were affected several times. Around 900 million people live in flood-prone areas and, of these, on average up to 75 million are exposed to flooding every year.

About 90 per cent of all flood deaths and 50 per cent of the economic damage occurs in Asia, notably in China, India, Bangladesh, Pakistan and Viet Nam.

However, damaging floods are not confined to emerging nations. Floods are the most frequent environmental disaster in Europe, and while fatalities are usually under 30 for each event, damage costs are very high. For instance, a cost of £2.5 billion has been quoted for the most recent of the series of flood events in the UK in 2019–20.

Socio-economic impacts

The degree of threat posed by a flood event depends on the depth and velocity of the water, the duration of the flood, and the quality of water (sediment load, presence of raw sewage, toxic chemicals, pollution from oil, etc.). Research on the impact of river depth and velocity suggests that water 0.5 m deep can wash cars away, and that foundations of buildings start to collapse at velocities of 2 m per sec. Stresses on structures such as bridges are also very closely related to depth and velocity.

Flood depth also has a very clear link to mortality. In many developing countries, people have not learned to swim, and can also be killed by poisonous snakes in the flood waters. Children and old people are particularly vulnerable. Post-flood **morbidity** is also extremely likely in low-income countries, mainly from water-borne diseases, which are secondary flood hazards. In developed countries, psychological stress is very common among flood victims.

> **Key term**
>
> **Morbidity:** A state of ill health.

Floods can affect people's livelihoods in many ways. Direct structural damage to property in countries at all stages of development is the major cause of tangible flood losses. There are all sorts of concerns post-flooding about getting flood insurance against future events, and also coping with reduced property values when a flood-prone property is resold.

Crops, livestock and agricultural infrastructure suffer major damage in intensively farmed rural areas. Where farming is subsistence, there is a direct loss of food supplies and famine can occur. In more developed countries (MDCs) floods can lead to escalating food prices as shortages of key products occur, as in the Big Dry in the Murray–Darling Basin of Australia in 2006.

In Cockermouth, Cumbria, the destruction of a key bridge connecting different parts of the town made communication and transport very difficult. In addition, flooded electricity substations meant that many people endured power shortages for up to three days.

Infrastructural losses are often extremely high in megacities such as Mumbai, as growth has outstripped flood defence systems.

A further loss of livelihood can occur when services and businesses are flooded – often putting them out of action for up to six months. In Carlisle, the McVitie's biscuit factory was flooded, leading to the temporary loss of over 1000 jobs.

Many areas that flood earn substantial income from tourism. Negative images of flood-affected areas lead to many cancellations in the short term – except for keen geography teachers and students out taking pictures of floods! Cumbria experienced a drop in tourism for up to a year after the floods of 2015.

Environmental impacts of flooding

In contrast to the horrendous tales of death and destruction from the socio-economic impacts of floods, there are some positive environmental impacts. In many natural ecosystems floods play an important role in maintaining key ecosystem functions and biodiversity, by linking the river with its land surroundings.

The floods can recharge groundwater systems, fill wetlands, increase connectivity between aquatic habitats, and move sediment and nutrients around the landscape and into marine environments. For many species, flood events trigger breeding, migration and dispersal.

Many natural ecosystems are resilient to the effects of moderate flooding, which can lead to increased productivity and maintenance of recreational environments.

However, in environments degraded by human activities, the impacts of flooding become more negative. Intense flooding, caused by excessive overland flow, can lead to oversupplies of sediment and nutrients, with possible **eutrophication** and the destruction of aquatic plants, as well as introducing pollution from nitrates, chemicals and heavy metals, which can degrade aquatic habitats.

Several research reports comment on the impacts on wildlife living in the soil, such as earthworms, moles, voles, hedgehogs and badgers, which can be poisoned by polluted waters.

In developing countries, many subsistence farmers have developed agricultural practices that rely on the annual inundation, which brings sediment and nutrients

> **Key term**
>
> **Eutrophication:** Excessive richness of nutrients in a lake or other body of water, frequently due to run-off from farming land, which causes a dense growth of plant life and death of animal life from lack of oxygen.

to the fields, so working with nature. When the Aswan Dam was built in the Nile Basin, one of the aims of this multipurpose scheme was to control flooding in the Nile. This had a negative impact on subsistence farmers, and also on sardine fishermen, as the sardines migrated away from the Nile Delta because of the loss of nutrient supply.

Flooding in the UK, 2019–20

From October 2019 to February 2020 the UK experienced some of its worst floods ever, based on the length of the time period and the widespread spatial coverage of the UK with no region unaffected within the four-month period.

The first wave of flooding occurred in November 2019, mainly affecting Yorkshire and Humberside and the West and East Midlands. In all these areas, the ground was saturated, especially in the West Midlands, as a result of record levels of October rainfall in mid Wales which feeds the Severn and Wye catchments.

The first wave of the floods resulted from heavy and prolonged rainfall from a series of depressions (a cause of winter flooding) in the first two weeks of November with Yorkshire and Humberside being particularly badly hit (Fishlake) as well as the West Midlands where many existing flood defences on the Severn were overtopped.

Flooding was also reported across southern England from 18–22 December, caused by repeated intense cold fronts moving across the area, with more than 50 mm of rainfall in less than 36 hours in many areas.

The second wave of flooding resulted in February from two very powerful European wide windstorms (see Table 2.2). These storms were extremely damaging as torrential rain leading to flooding was combined with very strong winds.

Another flooding event occurred in January 2021, again widespread across the UK.

Table 2.2 Storms that caused the second wave of flooding

	Type	Conditions	Damage	Impacts
Storm Ciara 8–11 February 2020	Extra tropical cyclone European windstorm	Highest wind 136 mph Lowest pressure 943 mb	£1.6 billion 13 fatalities	Disruption across UK but especially in Yorkshire Pennines and Lake District and also in mainland Europe
Storm Dennis 15–19 February 2020	Extra tropical bomb cyclone European windstorm	Highest wind 140 mph Lowest pressure 920 mb	Well over £1.2 billion 12 fatalities	Major flooding especially in southern Wales but widely across UK including all of southern England, Northern Ireland, Scotland and northern England

The impact of flooding in England and Wales, summer 2007

Met Office figures for the three months to 29 July 2007 showed that 387 mm of rain fell in England and Wales, more than double the average rainfall of 186 mm. The north of England was badly hit by floods in June, while western and southern areas were most affected in July. This was the wettest July on record (Figure 2.15), with 129 mm of rain in England and Wales.

Flooding in July was focused along the rivers Severn and Thames.

- Rainfall along the course of the River Severn had already reached record levels on 20 July 2007 when a powerful storm sent the river into flood. In the days that followed, this had a devastating effect on many of the towns and villages downstream.
- A dozen people were killed and the financial costs amounted to £6 million.

- On the upper Severn at Shrewsbury, the scene of much flooding in the past, the flow was six times the normal level, but mobile flood defences begun in 2003 largely did their job.
- On the mid-Severn around Worcester, river levels rose to 4.5 m above normal. The village of Upton-upon-Severn was the first serious casualty. Flooding occurred here six times in 2007.
- The greatest impacts were felt on the lower Severn, in Tewkesbury where the River Avon joins the Severn, and in Gloucester with its waterfront developments (Figure 2.16). Homes were flooded, power supplies damaged and water supplies cut off.

Over the same period, Oxford, Abingdon and other towns along the upper Thames were flooded. Additional localised, unpredictable flash floods occurred (for example, in Tenbury Wells).

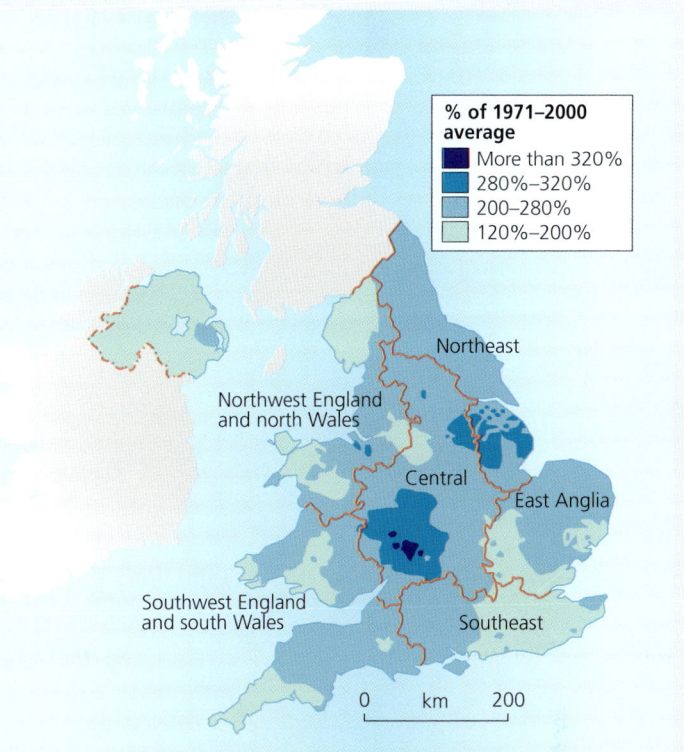

Figure 2.15 Rainfall patterns in England and Wales, summer 2007

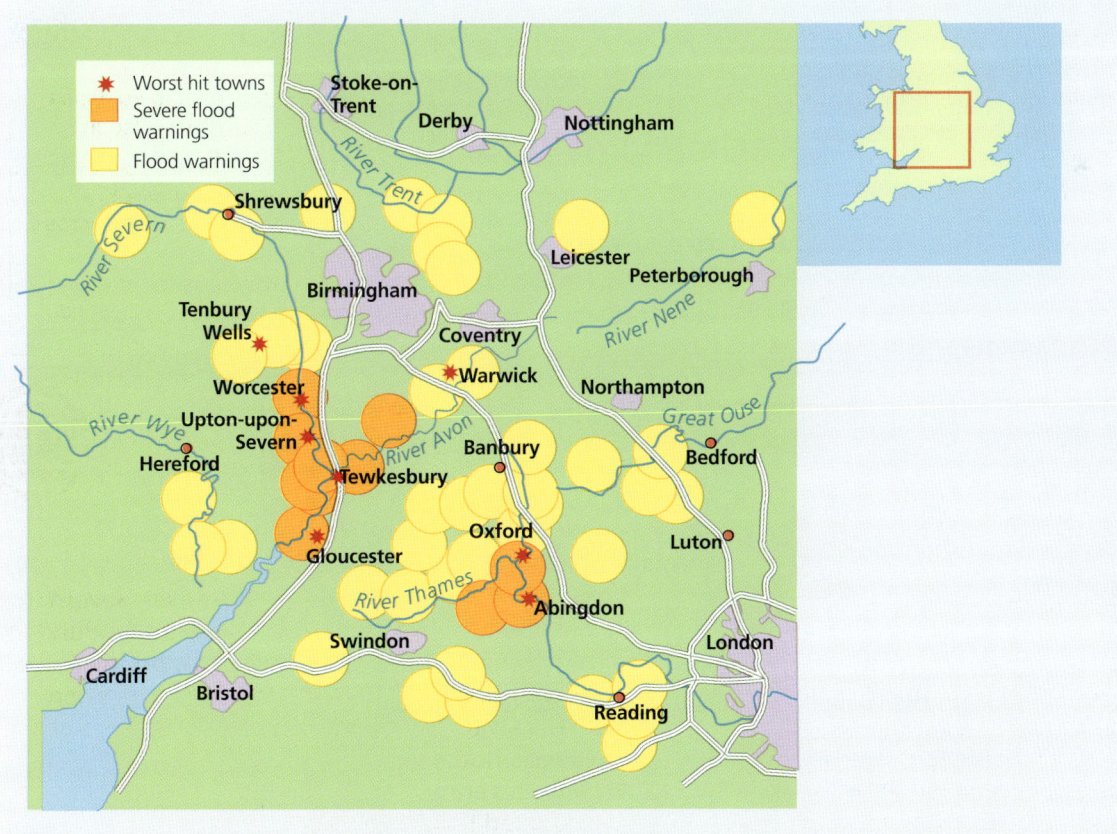

Figure 2.16 The areas of the UK worst affected by floods, July 2007

2 Short- and long-term variations in the hydrological cycle

2.3 The impacts of climate change on the hydrological cycle

Most scientists agree that climate change (global warming and oscillations such as ENSO) will result in an intensification, acceleration or enhancement of the global hydrological cycle.

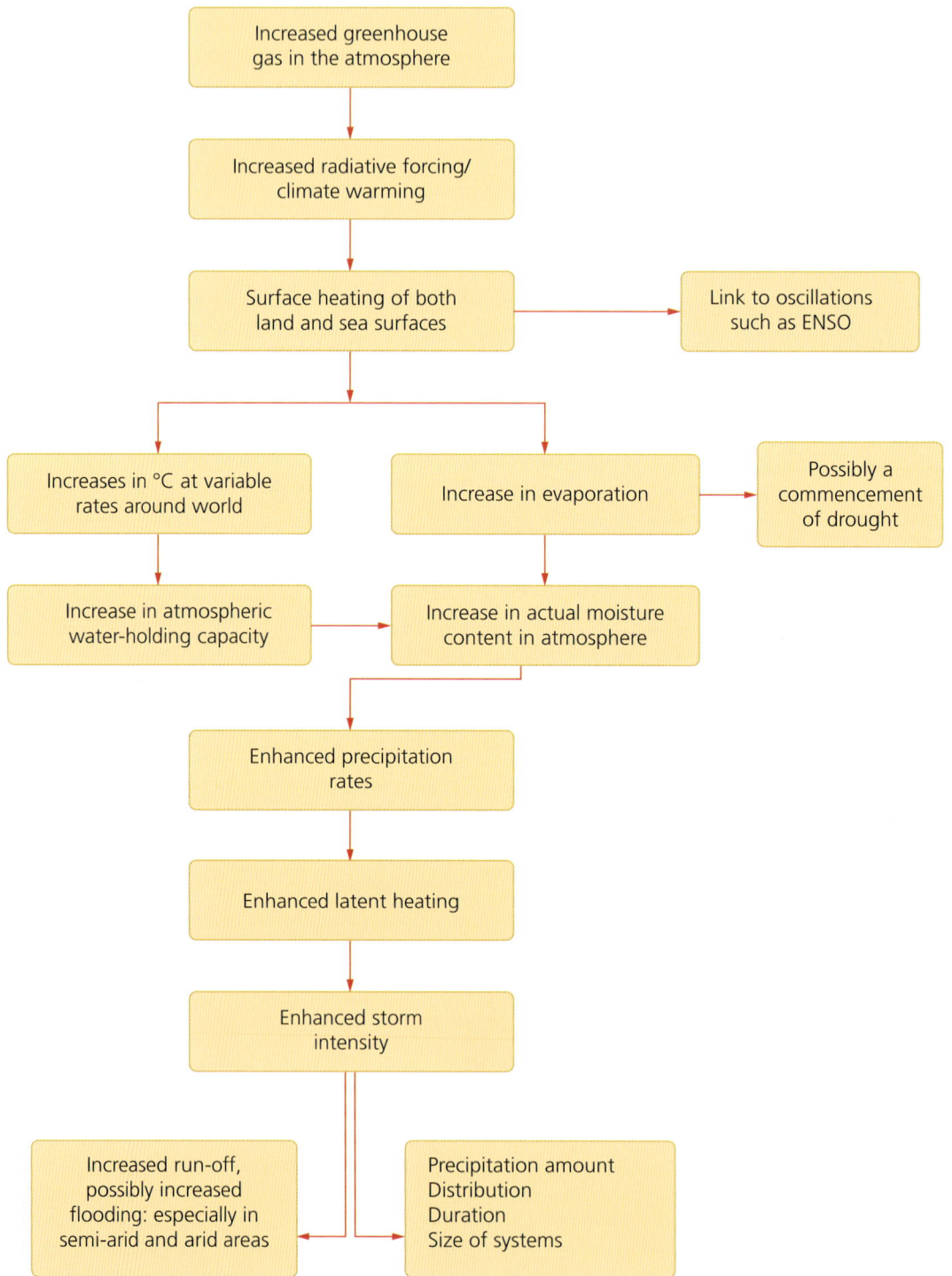

Figure 2.17 How climate warming modifies the hydrological cycle

As the impacts of climate change vary around the world, with differential amounts of temperature increase and varying changes to totals and distribution of rainfall, there will be differential changes in the way the hydrological cycle (open system) operates within the world's drainage basins. There will be different impacts on drainage basins in different climate zones of the world, for example, caused by differential precipitation and run-off predictions.

Therefore, decision makers in various countries will have to cope with changes in water budgets, and this will impact on the way climate change is managed to secure water resources for the future.

Modelling climate change trends tends to be complex for several reasons:

- Climate dynamics – the way the atmosphere, ocean, terrestrial, cryosphere and biosphere systems all interact with each other – is still only a partially understood science.
- As a result of teleconnections, in some instances it is difficult to distinguish between the impacts of oscillations such as ENSO and climate warming.
- Global records are very incomplete; in many parts of the world there is insufficient depth or detail of evidence to establish reliable trends for the impact of climate change or to make firm predictions about the future.

Figure 2.17 shows some of the possible changes brought about by climate change.

Table 2.3 summarises the key findings by experts about changes in inputs, outputs and stores brought about by climate change. Another complexity to resolve is that sometimes changes are brought about largely by human actions, which intensify the impact on the hydrological cycle (a positive feedback loop).

Table 2.3 Summary of key findings relative to trends in water cycle components

Hydrologic variable	Key findings
Precipitation input	The mode of the precipitation may be more important than the mean precipitation in determining hydrologic impacts.
	Widespread increases in intense rainfall events have occurred although overall amounts remained steady or even decreased.
	Areas of precipitation increase include the tropics and high latitudes, with decreases of 10–30° north and south of the equator. At the same time, length, frequency and intensity of heat waves has increased widely, especially in southern Europe and southern Africa. This has led to an increase in drought occurrence. With climate warming more precipitation now falls as rain, not snow, in northern regions. All this is consistent with a warmer atmosphere with a greater water-holding capacity.
Evaporation and evapotranspiration	Some research suggested that in large areas of Asia and North America actual evaporation is increasing, although increased cloud cover from increased water vapour may work against this. Transpiration is linked to any vegetation changes, which are linked to any changes in soil moisture and precipitation as well as increasing transpiration, which makes vegetation more productive.
Soil moisture	Results are ambiguous here – the amount of soil moisture is related to many factors, of which climate change is only one. Where precipitation increases, it is likely that soil moisture will also increase.
Run-off and stream flow (a 1°C rise in temperature could increase global run-off by 40%)	Evidence is developing to suggest that, along with more climate extremes, there will be an increase in hydrologic extremes, with more low flows (droughts) and high flows (floods). An accelerated cycle with more intense rainfall will increase run-off rates and reduce infiltration. There are marked decreases in the continental interiors of the Mediterranean, Africa and the US southwest.

Groundwater flow	Evidence is again limited, with no definitive link between groundwater amounts and climate change, as human abstraction is the dominant influence on supplies, especially for agriculture.
Reservoir, lake and wetland storage	Regional variations in lakes and reservoirs have been linked to regional changes in climate, for example, in Lake Chad. Changes in wetland storage are occurring, but they cannot be conclusively linked to climate change. Wetlands are affected where there are decreasing water volumes and higher temperatures.
Permafrost	Changes in the physical climate at high latitudes, primarily increasing air and ocean temperatures, are leading to permafrost degradation in northern areas. With the deepening of the active layer this has an impact on groundwater supplies and also releases methane from thaw lakes, which leads to positive feedback and accelerating change.
Snow	Most studies suggest that the length of the snow-cover season has decreased, especially in the northern hemisphere and, in the last 50 to 100 years, spring melt has occurred earlier, possibly accelerating in the last decade (with corresponding changes in river regimes).
Ice	There is strong evidence that glaciers have retreated globally since the end of the 'Little Ice Age', with downwasting (thinning of a glacier due to the melting of ice) accelerating in most areas since the 1970s. This is the result of rapid temperature increase and changes in the precipitation type (more rain/less snow). Tropical high-altitude glaciers, e.g. in the Andes, have shown the most rapid changes, leading to low flow from a dwindling cryosphere supply.
Oceans	Work on measuring sea surface temperatures has lagged behind land-based research, but in areas of ocean warming increased evaporation will occur, and there is limited evidence that more cyclones are generated.

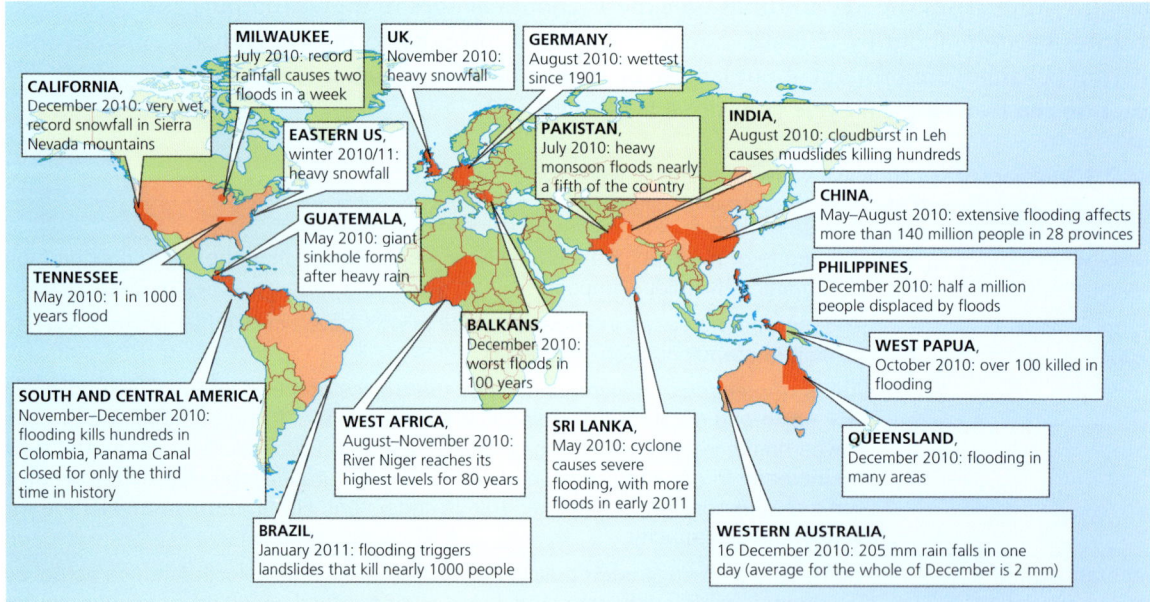

Figure 2.18 Major flooding events in 2010, the wettest year ever recorded (Adapted from *New Scientist*, March 2011)

Future trends – more global deluges and global drought

Floods and the future

Scientists generally agree that the hydrological cycle will intensify and that extremes will become more common. The moisture-holding capacity of the atmosphere has been increasing at a rate of about 7 per cent per degree Celsius of climate warming, creating the potential for heavier precipitation. There have *likely* been increases in the number of heavy precipitation events in many land regions.

Figure 2.18 shows the situation globally for 2010–11, with 2010 being the wettest year ever recorded (Figure 2.19). Questions are inevitably asked after such freakish weather: is climate change to blame, and is there worse to come?

Heavy precipitation events have certainly led to spectacular flooding, with economic losses rising ten-fold between 1990 and 2010. The fact that hydrological disaster losses have grown much more rapidly than precipitation or economic growth suggests that a climate change factor *may* be involved, although socio-economic factors such as land-use changes and greater use of vulnerable areas also play a huge part.

However, documented flood figures show no clear evidence of trends in either increasing frequency or magnitude of flood events globally. One study did show increasing frequency of 'large' floods in 16 large basins across the globe during the twentieth century, with more 1-in-100-year events, but another study on global change in river flows only showed positive trends in Europe, and only for late autumn and winter floods. Some floods from melting snow had actually decreased as the snow has disappeared.

> **Synoptic themes:**
> **Futures and uncertainties**
> Projections into the future occurrences of droughts and floods are very complex and hard to predict, both spatially and temporally, for a number of reasons:
> - multiple causes of climate change
> - insufficient understanding of teleconnections within Earth systems
> - lack of availability of long-term big data sets.

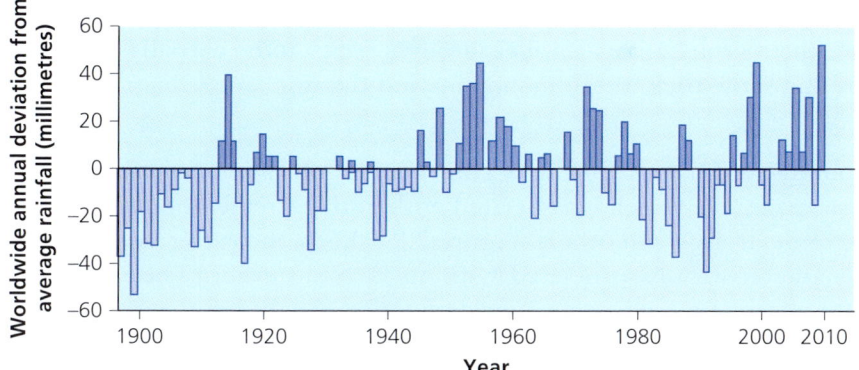

Figure 2.19 2010 was the wettest year ever recorded

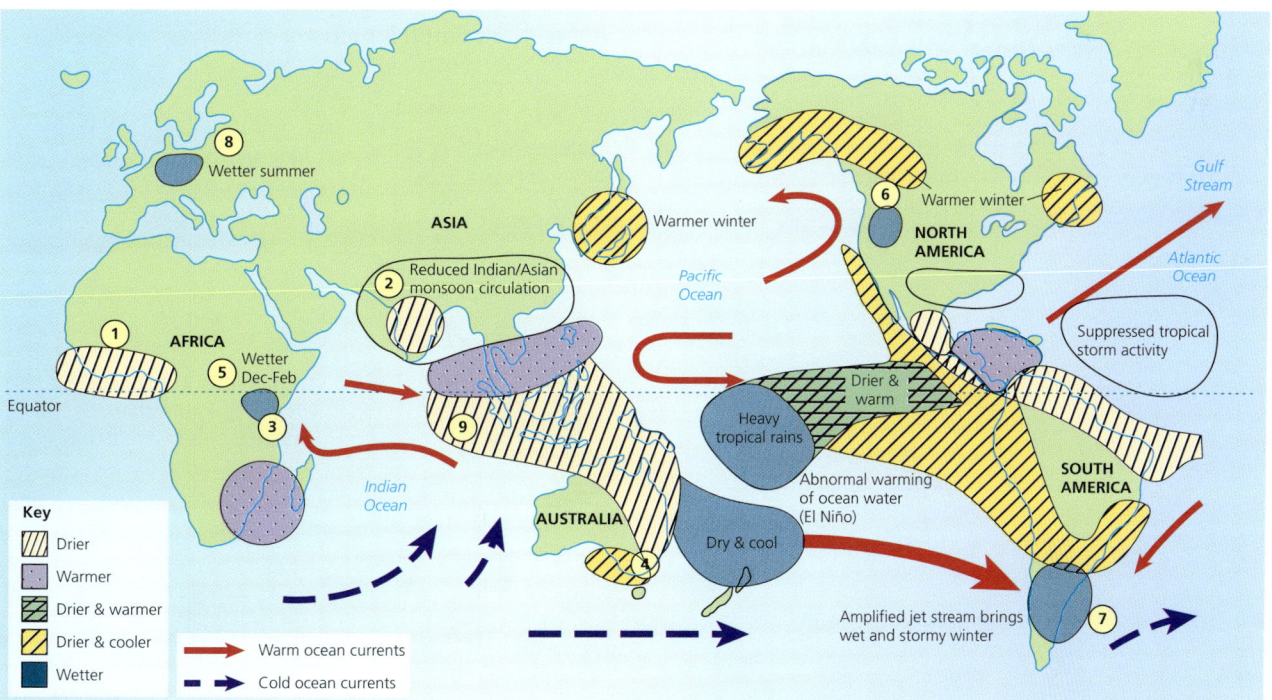

Figure 2.20 The pattern of droughts and floods associated with an El Niño event; 1–4 are areas of drought, 5–8 are where above-average rain leads to flooding, especially in Argentina and Kenya

2 Short- and long-term variations in the hydrological cycle

Low flows and drought in the future

Climate change is expected to influence precipitation, temperature and potential evapotranspiration and, through their combined effects, the occurrence and severity of droughts. In the past 30 years droughts have become more widespread, more intense and more persistent. The problem is that it is difficult to disentangle the possible impacts of climate change from those of human influences, such as land-use changes, and usual multi-decadal climate variability (see the Sahel case study on page 27).

However, more intense droughts have affected more people, and links to higher temperature and decreased precipitation have occurred (see the Australia case study on page 28, and the South Africa case study on page 29).

Research studies have proved inconclusive globally but some regions, such as southern Europe, southwestern USA and the Sahel, do seem to have a higher incidence and intensity of drought.

A further issue is that the occurrence of the droughts seems to be determined largely by changes in sea surface temperatures, especially in the tropics, through associated changes in atmosphere circulation and precipitation amounts. In the western USA, the recent Californian megadrought seems to be partially associated with diminishing snow pack in the mountains and the resultant reductions in soil moisture.

A further complicating factor for the occurrence of both floods and droughts is the ENSO (see page 26), which is associated with both extreme flooding in some areas and extreme droughts in others. Figure 2.20 shows the pattern of droughts and floods characteristically associated with an El Niño event. The impacts of the ENSO therefore also have to be considered when analysing future trends in short-term climate change.

Other oscillations, such as the North Atlantic Oscillation (NAO) or the Indian Ocean Dipole, may also contribute to the complex picture.

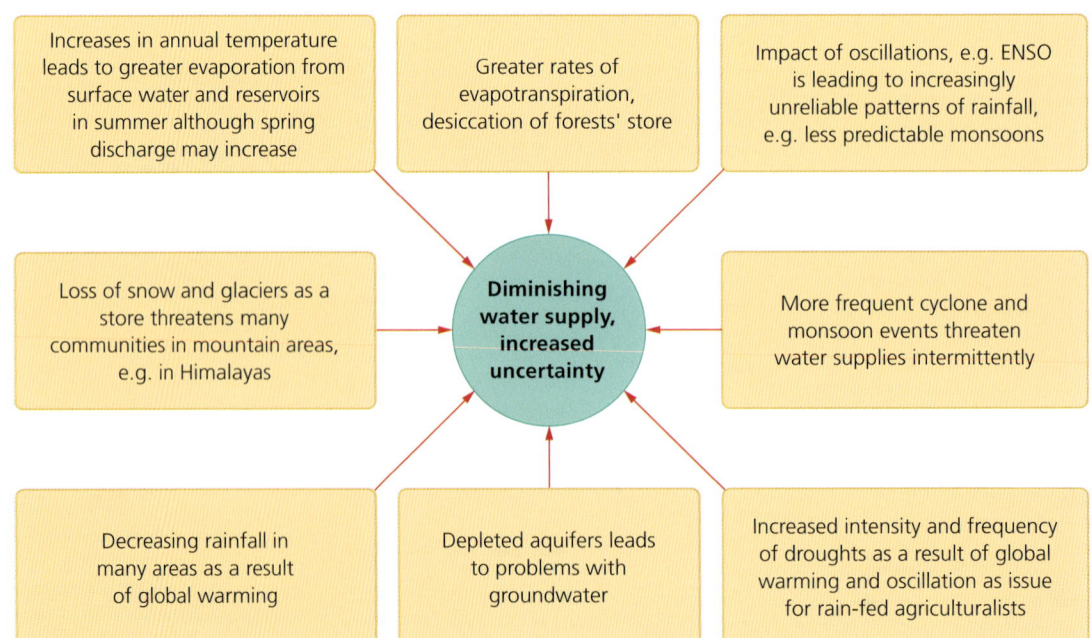

Figure 2.21 Summary of the impacts of short-term climate change on water supply

Figure 2.21 summarises the potential impacts of short-term climate change on water supply.

In conclusion, decision makers concerned with water planning and disaster management have to factor in the possibility of more extreme weather events – but there is no *definite* link *yet* made to climate warming as sufficient research has not yet been done. Countries developing their water-management strategies for the future need to consider a number of drivers of pressures on water, of which climate change is just one (Figure 2.22, page 46).

Alongside natural forces affecting water resources are many new human forces that affect the planet's water system, often related to human activities and economic growth. Water is unique among natural resources as it is vital for both ecosystem and human well-being.

> **Key concept: Climate change hotspots**
>
> There are seven areas in the world where the impacts of climate change will have a disproportionate impact for a variety of reasons:
> 1 **Murcia, Spain** which has experienced a 1.4°C (1°C globally) temperature rise combined with prolonged drought, and a growing demand for water has led to significant water security problems, and semi-arid conditions.
> 2 **Dhaka, Bangladesh** which will experience increased coastal flooding from rising sea levels leading to environmental refugees.
> 3 **Mphampha, Malawi** which is experiencing extreme temperatures and long-term regional drought, combined with flash flooding leading to frequent food deficits.
> 4 **Longyearbyen, Svalbard, Norway** which is experiencing dramatic climate warming of the Arctic, with widespread avalanches and melting permafrost.
> 5 **Manaus, Brazil** which is experiencing drying of the Amazon rainforest, with increased temperatures and decreasing humidity – with global impacts on the carbon cycle.
> 6 **New York, USA** – rising sea levels (76 cm by the 2050s) combined with storm surges and widespread flooding of this world city.
> 7 **Manila, Philippines** is a country experiencing multiple climate change impacts, rising sea levels, hotter temperatures, more variable precipitation and more extreme weather events such as cyclones.
>
> Whether it's faster than average warming (4), more severe than average drought (1), floods and storms (2) and (6) or more extreme weather where there are more vulnerable populations (3) and (7), it's clear that some places, i.e. hotspots, are being hit harder than others by climate change, which for them has become a climate emergency.

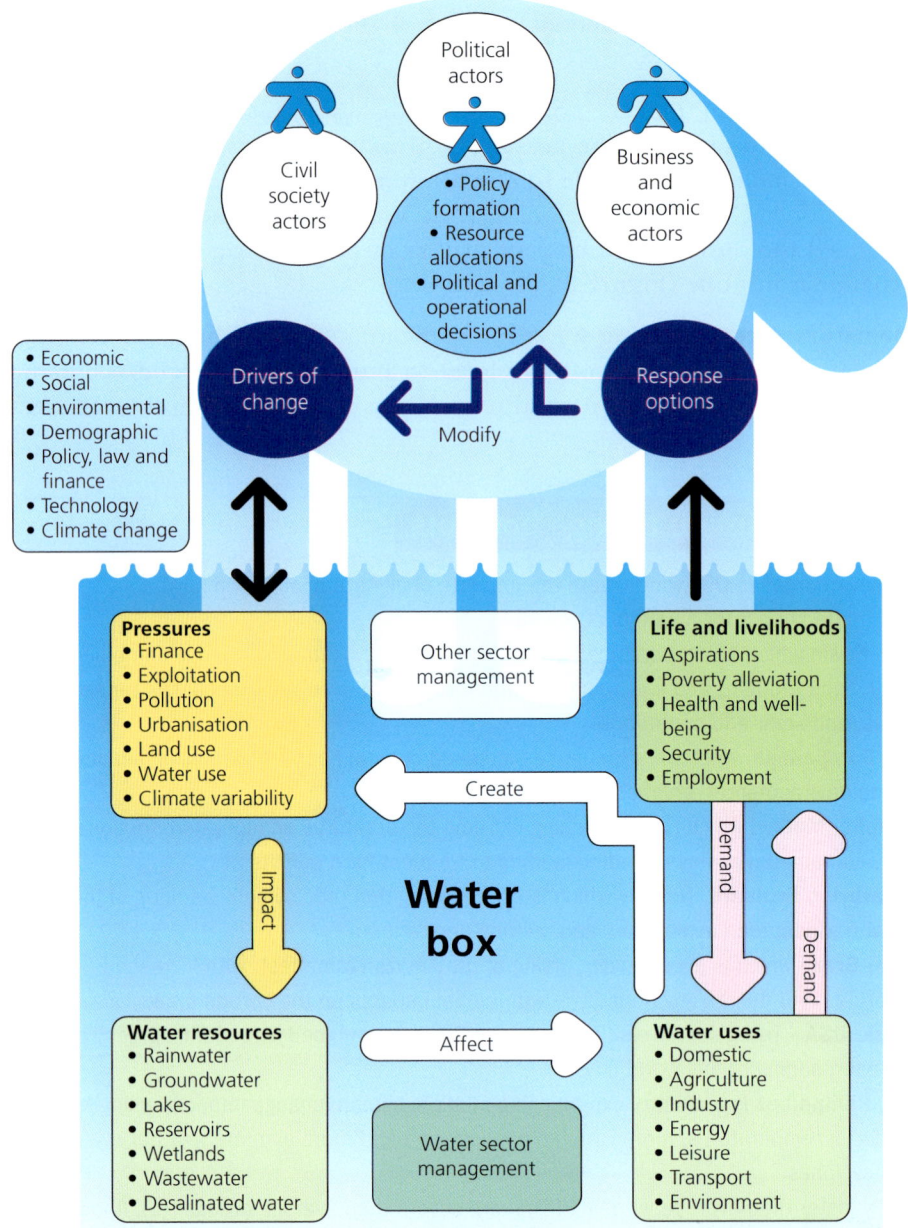

Figure 2.22 Decision-making affecting water management (Source: WWAP(2009))

Review questions

1 Using Figure 2.2 on page 26, write a short scientific article on the workings of the ENSO. Explain the differences between the normal circulation, El Niño and La Niña, and the impacts that the oscillation has both locally in South America and globally via teleconnections.

2 Use the internet to research the very strong El Niño of 2014–16: there are numerous newspaper and scientific sources. Make a map of possible climate extremes similar to Figure 2.20 on page 43 and compare the two.

3 Use a large database such as **http://floodobservatory.colorado.edu**, or EM-DAT (**https://public.emdat.be/**) to draw a map or table of large flood events around the world in a recent year. Use a technique to correlate social and economic impacts with the flood magnitude. Is the maxim 'more deaths in developing nations, more damage in developed nations' true?

The Water Cycle and Water Insecurity

4 Using newspaper articles and publications such as Geofile and Geo Factsheets, research two contrasting flood events (different causes, countries at different levels of development) and prepare a quadrant-form diagram to compare the environmental, social and economic impacts both immediately after the event and then short term and long term.

5 For either the Sahel *or* Australian drought (pages 27 and 28, respectively), carry out research to update the occurrence of the drought and chart its progress over recent years.

6 Study Figure 2.23.

   a Describe and suggest possible reasons for the distribution of predictions in changes in run-off for 2090.

   b What advice would you give to water planners in Regions A and B? Before you answer, think about the current state of these regions.

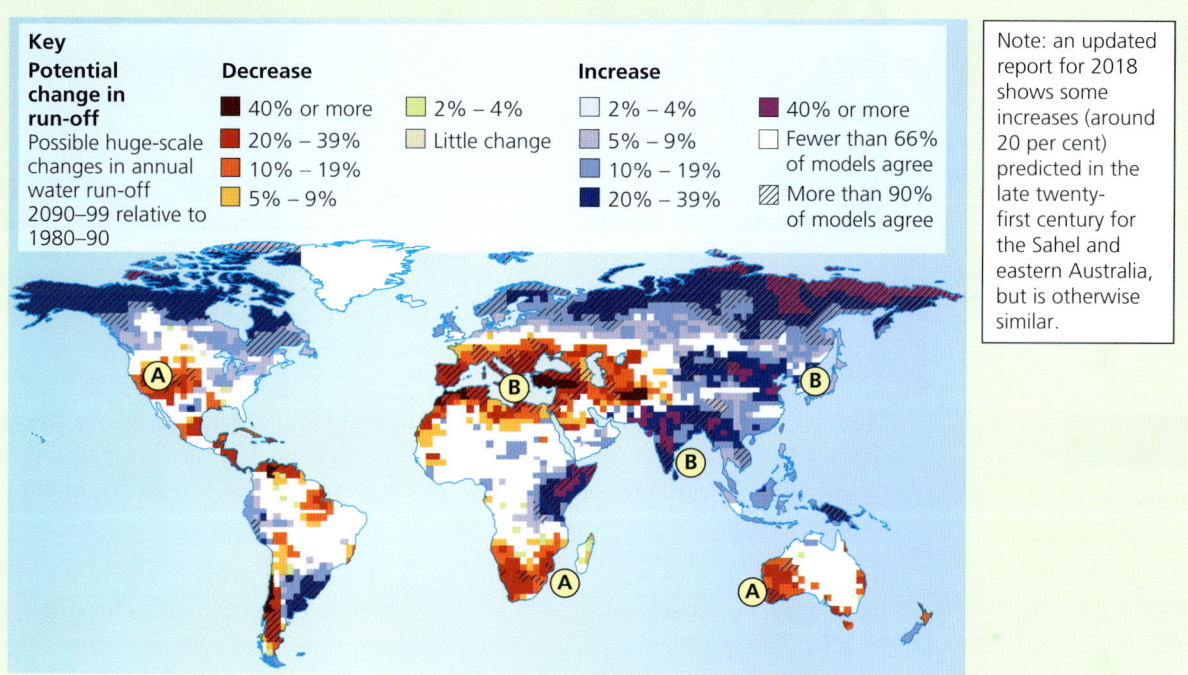

Figure 2.23 Potential changes in run-off (Source: IPCC Fourth Assessment Report: Climate Change, 2007)

Further research

National Drought Mitigation Center: http://drought.unl.edu

US Drought Monitor: http://droughtmonitor.unl.edu

Australian Bureau of Meteorology: www.bom.gov.au

UK Environment Agency (for all UK flood events): https://www.gov.uk/government/organisations/environment-agency

US National Flood Programme: www.fema.gov and https://www.floodsmart.gov

Flood Hazard Research Centre, Middlesex University: https://www.mdx.ac.uk/research/research-centres-and-groups/flood-hazard-research-centre-fhrc/

Data for gauging stations in UK: https://www.ceh.ac.uk/

The Geographical Association's pages on flooding: www.geography.org.uk/resources/flooding

The following websites are useful for researching the potential impact of climate change on the hydrological cycle: https://www.rmets.org, https://scied.ucar.edu/learning-zone/climate-change-impacts/water-cycle-climate-change

Search for relevant UNESCO case studies: https://en.unesco.org

The 'Reports' section of the IPCC website and the Climate Signals website provide useful future data on the impact of climate change: https://www.ipcc.ch and https://www.climatesignals.org

2 Short- and long-term variations in the hydrological cycle

3 Water security – is there a crisis?

> **How does water insecurity occur and why is it becoming such a global issue for the twenty-first century?**
> By the end of this chapter you will be able to:
> - understand the physical and human causes of water insecurity
> - appreciate that in terms of water insecurity it is an unequal world, both physically and economically
> - assess the consequences and risks associated with water insecurity at a variety of scales
> - evaluate the different approaches to managing water supply and usage
> - understand the need to plan for the future with more sustainable approaches to the management of this finite resource.

3.1 Introducing the issue – an unequal water world

Key term

Players: Individuals, groups or organisations with an involvement or interest in a particular issue.

Water, like energy and food, is a fundamental human need. Many **players** argue that all basic needs are best provided by market mechanisms, but others see access to available, affordable, safe supplies of clean water as a human right for all of the world's people.

Improving access to water and sanitation underpinned many of the UN's Millennium Development Goals (MDGs) and the subsequent Sustainable Development Goals (SDGs). The problem is that when a final assessment of progress was made in 2015, some 15 per cent of the world's people still did not have reliable access to safe water, and around 25 per cent still lacked clean sanitation.

Figure 3.1 re-emphasises how, of the volume of water in the global water pot, only 2.5 per cent is available as freshwater for humans to use, and only around 1 per cent is available as easily accessible surface water.

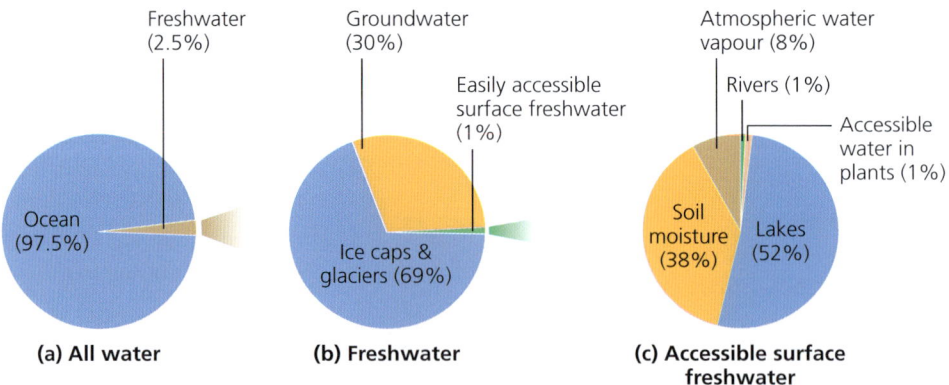

Figure 3.1 The global 'water pot'

Table 3.1 Water resources per capita across a spectrum of countries

Highest (m³ per year)		Lowest (m³ per year)	
Greenland (ice)	10,767,857	UK	2,465
Guyana (rainforest)	316,689	India	1,880
Canada	94,353	Denmark	1,128
Australia	25,708	Rwanda (LDC)	683
Mongolia	13,739	Yemen (desert)	223
Greece	6,998	Maldives (coral reefs, atolls)	103
Mexico	4,624	Kuwait (desert)	10

In theory, this should not be a problem as, according to the UN, our *basic* needs can be met by 1000 cubic metres per year. In 2020 it was estimated that nearly 63 per cent of this accessible freshwater – contained in rivers, lakes and groundwater aquifers – was being used, leaving some 37 per cent untapped so, in theory, there is more than enough to go round. So, what's the problem?

As you will see, the combination of rising demand and the diminishing availability of finite supplies could create a 'perfect storm' of resource shortages in combination with food and energy, for which water is a vital part of production. The phrase 'peak water' is being used increasingly to describe the state of growing constraints on quantity and quality of accessible water.

The fundamental problem lies with an unequal water world, as opposed to the generally satisfactory global situation. There are three facets to this state of affairs: physical distribution, the gap between rising demand and diminishing supplies, and the water availability gap.

Physical distribution

In terms of physical distribution there is a mismatch between where the water supplies are and where the demand is. Water supplies are spread very unevenly across the world: 60 per cent of the world's supplies are contained in just ten countries.

Table 3.1 suggests that both physical factors, such as location of precipitation belts/temperature, *and* level of development are important.

In conclusion, 66 per cent of the world's population live in areas receiving only 25 per cent of the world's annual rainfall. Clearly, there are areas of supply shortage such as most of the Middle East where there are potential sources of conflict over shared basin usage/dams and pollution.

Gap between rising demand and diminishing supplies

There is a global gap between rising demand (Table 3.2) and diminishing supplies.

Table 3.2 Projections for increasing global water usage

Year	Total annual water withdrawal (km³)
1900	579
1950	1,382
2000	3,973
2025	5,235 (projected)

Rising demand

This is driven by a number of factors:

- Population growth, possibly fuelled by an additional 2 billion people by 2050, especially in large urban areas and megacities in the developing world.
- Rising standards of living as countries such as China adopt dairy and meat-rich diets, which lead to higher consumption of water for agricultural purposes. There is also increased domestic use – for drinking, bathing and cleaning – as people become more affluent. Equally, the demand for consumer goods such as white goods and electronics encourages more use of water in manufacturing (i.e. embedded water). The combination of rising numbers and changing lifestyles, often in rapidly urbanising environments with high costs of providing water infrastructure, puts pressure on water supplies.
- Economic growth increases demand for water in all economic sectors (agriculture, industry, energy and services). The mining of unconventional energy sources, for example, **fracking**, puts huge demands on water.
- Irrigated farming places a particular strain on resources. Countries such as Israel or areas such as the Murray–Darling Basin in Australia (page 28) are experiencing increasing droughts as a result. The countries bordering the Aral Sea – Turkmenistan, Kazakhstan and Uzbekistan – have the highest water use per capita in the world, with around 99 per cent being used for irrigated crops. This overuse led to the environmental degradation of the ecosystems surrounding the Aral Sea (see page 62).

> **Key term**
>
> **Fracking:** Hydraulic fracking or oil/gas well stimulation is a technique in which rock is fractured by a pressurised liquid.

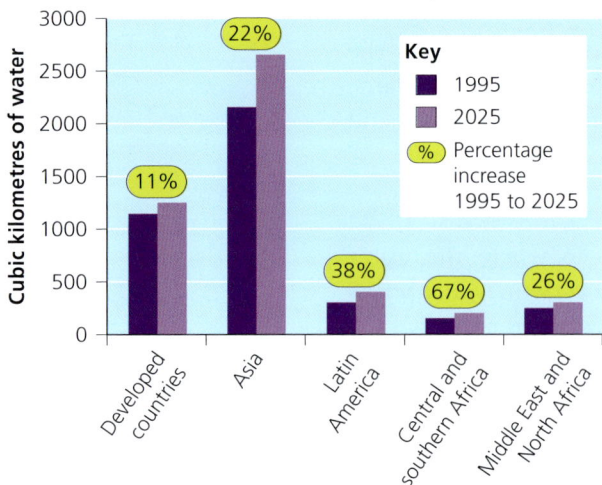

Figure 3.2 Total freshwater withdrawals by region, 1995 and 2025 (projected)

Figure 3.2 shows that total freshwater withdrawals are predicted to rise by 2025 in all regions of the world.

Dwindling supplies

The most serious manifestation of dwindling supplies concerns the diminishing supplies available from groundwater aquifers. The main reason is for irrigation, which is a voracious consumer of water. Comparatively cheap pumping technology, minimal legislation to regulate its use and threats from climate change-induced drought have combined to put pressure on supplies, leading to a falling water table as the groundwater supplies are being extracted faster than they can be replenished. Excessive withdrawals lead to land subsidence (as in Mexico City) and intrusion of salt water in coastal districts (as in coastal North Africa).

The conclusion is that groundwater can no longer be regarded as an unlimited supplement to surface water supplies, which are themselves being diminished by overuse.

Table 3.3 lists some problem areas for over-extraction for you to research.

Table 3.3 Regions experiencing problems due to over-extraction of water

USA	Ogallala Aquifer (Great Plains), Navajo Aquifer (Arizona)
Mexico	Mexico City, Ciudad Juárez–El Paso (US border)
Middle East	Yemen, Iran
Asia	North China Plain, Punjab (India, Pakistan), Gujarat (India)

The result of this imbalance between demand and supply has resulted in a number of pressure points where, nationally, regionally or locally, water supplies are under threat. Figure 3.3 summarises the reasons for pressure points (also called pinch points or water hotspots).

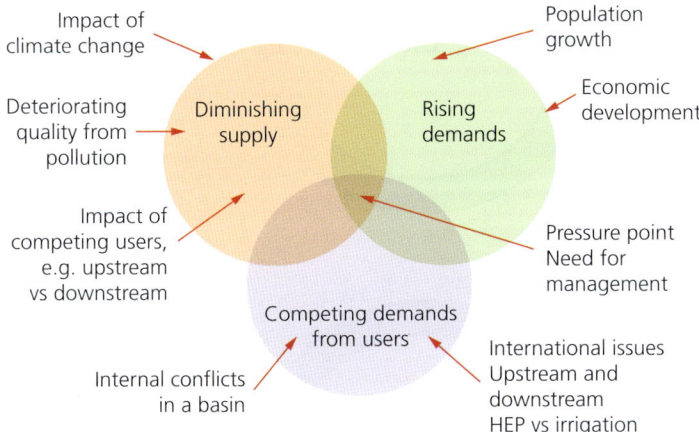

Figure 3.3 Water pressure points

Water availability gap

The underlying concept is that of a water availability gap between the 'have-nots', largely in developing nations (for example, Central African Republic, Chad and Ethiopia), and the 'haves', largely in developed nations (for example, Sweden and Finland). There is an imbalance of usage, with richer countries using up to ten times more water per head: they have a water profile that includes large percentages of embedded water as well as direct water use. Embedded water is known as **virtual water**, which comes embedded in all the farm products, food and manufactured goods that are imported.

> **Key term**
>
> **Virtual water:** The hidden flow of water when food or other commodities are traded.

Figure 3.4 (page 52) compares the water profiles of contrasting countries. Many countries will experience water stress (under 1700 m^3 per person per year), especially in some parts of western Asia, such as Pakistan, South Africa and Ethiopia and, in recent years, California in the USA. With the onset of climate change and the associated desertification of ecosystems, by 2050 some 4 billion people could be experiencing water stress.

By 2025, it is estimated that nearly half of the world's population will be water vulnerable (under 2500 m^3 per person per year). A state of vulnerability means that there is insufficient water and risks to supplies, especially when unusually hot or dry conditions result from short-term climate change. The list of vulnerable countries includes: Spain, Belgium, the UK, Bulgaria and Poland in Europe, and countries such as India, Ghana, Nigeria and most parts of China.

If there is around 3000 m^3 per person available, supplies are declared to be sufficient. This includes virtually the whole of the Americas (if complete countries are considered), Russia, Scandinavia and many countries in equatorial regions. Australia is a surprising example of a country which overall has sufficient water on a per capita basis but, with many drought-prone areas, has regional problems (for example, in the Murray–Darling Basin, see page 28).

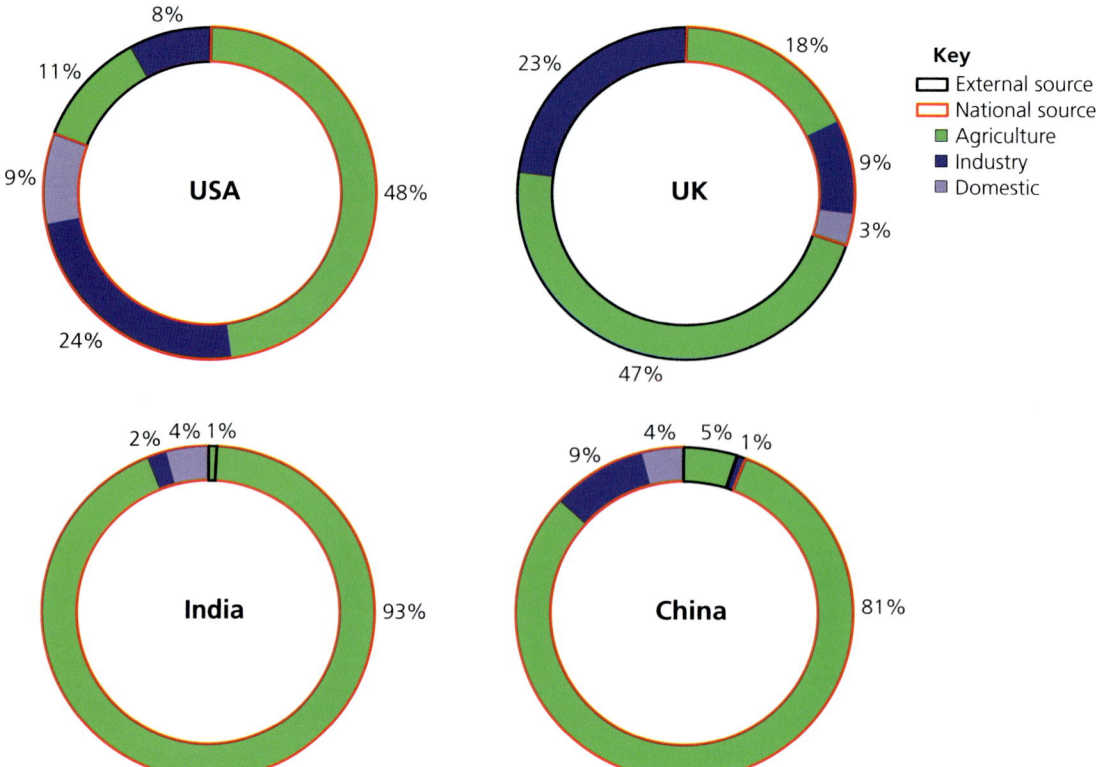

Figure 3.4 Four contrasting water profiles: compare the proportion of embedded water by source

Causes of water insecurity

Physical factors determining the supply of water

At a macro scale, climate determines the global distribution of water supply by means of annual and seasonal distribution of precipitation (rain and snow). Precipitation varies globally as a result of atmospheric pressure systems, with the low-pressure zone of mid-latitudes and equatorial regions having the highest totals and, therefore, being generally water secure. Also important is the seasonal distribution of rainfall, its reliability and its availability for use as water supply. As the study of the Sahel shows (page 27), lower annual totals of rainfall often have greater variation and therefore poorer reliability of supply. Short-term climate change (the ENSO and climate warming) are exacerbating the **water security** situation.

On a more regional scale, topography and distance from the sea have significant impacts. High relief promotes increased precipitation and rapid run-off, but may also provide greater opportunities for surface water storage in natural lakes and artificial reservoirs, especially where it is combined with impermeable geology. Snowfall and glaciers can be extremely important locally, as in the Bolivian Andes where climate warming has led to widespread melting, diminishing the cryosphere storage and threatening water supplies for La Paz–El Alto. The same issue is also affecting Nepal's water supply.

> **Synoptic themes:**
>
> **Futures and uncertainties**
>
> A bleak picture of a future with increasing water scarcity for many countries and people, driven by rising demand from all sectors and dwindling supplies of a finite resource.

> **Key concept: Defining water security levels**
>
> There are a number of measures used to define water shortages: if there is less than 1000 m³ per capita of available water, a state of water scarcity occurs. There are two types:
>
> - *Physical scarcity* occurs when more than 75 per cent of a country's or a region's blue water flows are being used – this currently applies to around 25 per cent of the world's population (water-scarce countries are clustered in the Middle East and North Africa, and regionally in some larger countries, such as northeast China and parts of the Great Plains of the USA). Some Middle Eastern countries, such as the desert Kingdom of Saudi Arabia, are using up to 4 per cent more water than their supplies and therefore have to rely on desalination (see page 75).
> - *Economic scarcity* occurs when the development of blue water sources is limited by lack of capital, technology and good governance. Around 1 billion people currently have satisfactory physical availability but can only access some 25 per cent of the water supplies because of the high levels of poverty prevalent in these developing countries. Solutions may be reliant on privatisation (research Tanzania or Ghana in Africa, or Bolivia in South America).
>
> By 2050, 1.5 billion people will be experiencing water scarcity, especially in the Middle East and parts of Africa.

The world's major river systems store large quantities of water and transfer it across continents. The Amazon, for example, has an average annual discharge of 175,000 cubic metres per second from its catchment areas of 6,915,000 square kilometres shared by Brazil and six other South American countries. Recent severe droughts in 2005 and 2010, with a dry period in between (a 1-in-100-years event), covered an area twice the size of California and had a huge impact on Brazil's water supply. Flows in the main river were at an all-time low, with several tributaries completely dry, along with record sea temperatures off the northeastern coast of Brazil. Many experts argued that deforestation was a contributor to the drought affecting the Amazon's hydrological pump.

Geology controls the distribution of aquifers (water-bearing rocks) that provide the groundwater storage. Permeable chalk and porous sandstones can store vast quantities of water underground, which is valuable as it is not subject to evaporation loss. The water supply comes from springs and can also be accessed by wells, giving an even supply throughout the year, despite the uneven distribution and variability of rainfall – *provided* they are not overused by demand rising at a faster rate than they can be replenished by natural recharge. Currently, there is a crisis caused by uncontrolled drilling of tube wells, leading to excessive abstraction and a falling water table, combined in many places with a less predictable pattern of rains, for example, in the monsoon areas of India and Pakistan.

Figure 3.5 (page 54) shows how these physical factors can combine to affect the water supply of India, a country that is vulnerable to water insecurity, especially in the Indo-Gangetic Plains, the backbone of the water intensive **Green Revolution**.

Human factors influencing the security of water supplies

Human activities can lead both to diminishing supply and rising demands. Humans can also impact on both the quantity of available water and its quality.

Quality

Human actions can pollute both surface water and groundwater supplies, so diminishing the quality of both sources and having a knock-on effect on the security of supplies. Pollution is widespread throughout the world, although its impact is felt

> **Key term**
>
> **Green Revolution:** The use of high yield varieties (HYVs) of crops along with the use of agrochemicals and irrigation to increase yields and improve food supplies; begun in the 1960s.

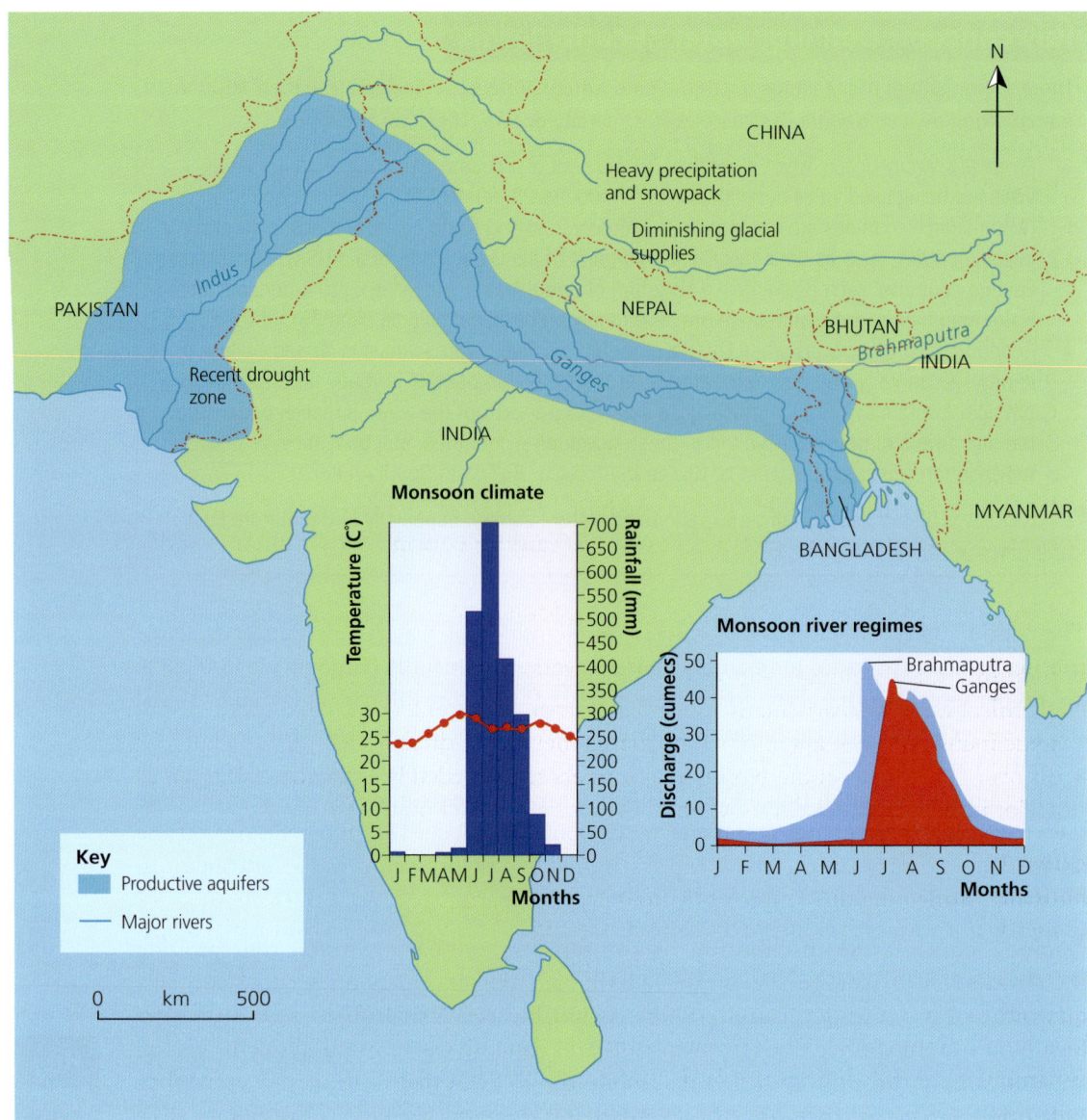

Figure 3.5 Factors influencing water supply in India

especially in developing countries (1 billion people are without safe water and 2.3 billion lack adequate sanitation). The difference of impact is related to the ability of developed countries to do something about it, either by prevention or remediation of supplies.

The pollution of surface water in rivers, streams and lakes is a cause for concern, for example, in China where 300 million people use contaminated water daily and 190 million suffer from water-related illnesses *annually*. In China, one-third of all rivers, 75 per cent of major lakes and 25 per cent of coastal zones are currently classified as highly polluted. It has been reported that longer-term 2 million Chinese people may suffer from water-related diseases, including those in the 'cancer cluster' villages in Guandong province where liver and digestive cancers were responsible for 80 per cent of recent deaths (heavy metal toxins from the Dabaoshan mine had washed into the Hengshi River). Contaminants usually enter waterways via run-off or sewage.

However, groundwater contamination is potentially even more serious if important aquifers are irreversibly damaged by the high levels of toxicity. Nearly 20 per cent of all the tube wells sunk in Bangladesh, often concentrated in particular villages, were found to be unsafe because of a high concentration of arsenic. This led to major health problems, with correlated social impacts, as the victims developed arsenicosis with skin lesions. Worldwide, 137 million people in over 70 countries have some signs of arsenic poisoning from drinking water.

Some common types of pollution include:

- Untreated sewage disposal, especially in countries where sanitation is poorer. This causes water-borne diseases such as typhoid, cholera and hepatitis. As many people are forced to use unsafe water, it was estimated by WHO that, in 2020, 135 million people could have died unnecessarily from these water-borne diseases. In India, only 20 per cent of sewage is treated before being discharged in rivers.
- Chemical fertilisers, used increasingly by farmers (part of the Green Revolution) contaminate groundwater as well as rivers, causing eutrophication in lakes and rivers. This leads to hypoxia and the formation of dead zones in coastal waters. Many of the pesticides used are banned in developed countries because of the health hazards.
- Industrial waste is dumped into rivers and, subsequently, oceans. Heavy metals and chemical waste are particularly toxic. The Ganges is a useful example to study; many toxic industries, such as tanneries, discharge their waste directly into the holy river.
- As over 60 per cent of the world's major rivers are impeded by large dams, this has a major impact on sediment movement, which can impact on river ecology.

Quantity

Humans can over-abstract from both rivers and lakes, and groundwater sources, for domestic purposes (drinking water), agriculture (largely irrigation) and industrial usage.

By 2025, total projected water withdrawals are predicted to reach over 5000 cubic kilometres per year, of which agricultural use will be two-thirds. Regionally and locally, a combination of a number of drivers (population growth, migration and urbanisation, rising living standards, economic development and industrialisation) will have increased water demand to unsustainable levels, hence the occurrence of Day Zero for some cities (see page 29).

The removal of freshwater from aquifers on coastal locations can upset the natural balance of saline and freshwater, and can lead to salt water incursion and salinisation of wells, boreholes and wetlands. Coastal storm surges and rising sea levels compound the problem, for example in many South Seas atolls.

Until recently agriculture absorbed over 70 per cent of extractions globally, but industrial usage is rising, especially in developed countries and emerging economies where the proportion of use can rise up to 60 per cent, especially in paper and metal industries.

The energy industry also requires increasing amounts of water for new energy developments such as biofuels and fracking. A number of technological developments are available to cut water usage in all sectors of the economy ('smart water' is the watch word) but, with a finite source, the damage has been done.

Figure 3.6 (page 56) summarises the human impact on water supply and quality, both of which impact on water security.

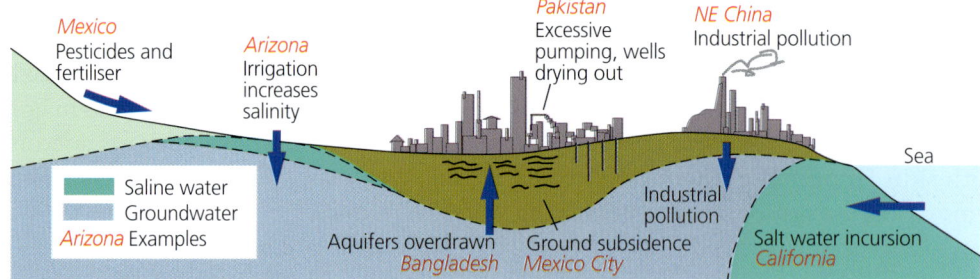

Figure 3.6 Human impacts on water supply and quality

Key concept: The water poverty index

In 2002, the Centre for Ecology and Hydrology published the first water poverty index (WPI). It is an assessment of the degree of water shortage and the subsequent water insecurity problems. Scores can be generally correlated with GNP per capita, with Canada having the highest score (78) and Ethiopia one of the lowest (48).

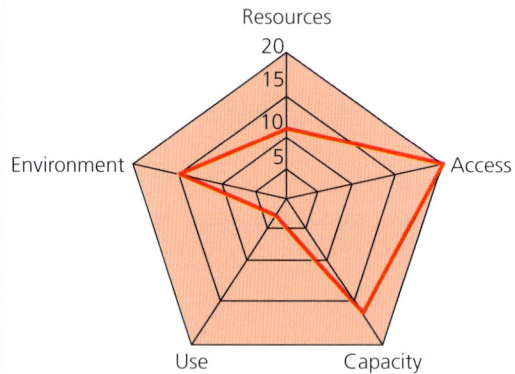

Figure 3.7 The water poverty index

The index uses five parameters:

- Resources: the quantity of surface and groundwater per person and its quality.
- Access: the time and distance involved in obtaining sufficient safe water.
- Capacity: how well the community manages its water (and health).
- Use: how economically water is used in the home and by agriculture and industry.
- Environment: ecological sustainability (green water).

Each of the parameters is scored out of 20 to give a maximum possible score of 100. Figure 3.7 shows the water poverty index. The red line illustrates how a country's score would be presented.

3.2 The consequences and risks associated with water insecurity

Water shortages

Access to water

Water insecurity means not having access to sufficient safe/clean water. Despite global efforts to improve water supply and sanitation (US$35 billion is spent each year worldwide by a number of players including international agencies, national and regional governments, private water companies and NGOs), around 1 billion people are still without access to clean water. Many of these people live in 30 or so developing countries where the root cause is poverty; others live in areas of physical scarcity where only technology and capital investment can overcome the shortage or unreliability of supply.

The problem of water insecurity is therefore related to:

- availability – having not only a water supply but a water distribution network
- access – freedom to use, or income to buy, water in a particular location
- usage – entitlement to, and understanding of, water use and health issues.

Physical scarcity

Physical scarcity is largely determined by climate (the balance between precipitation inputs and evapotranspiration outputs) with concentrations, in general terms, in high-pressure latitudinal bands between 23.5°N and S and 35°N and S. However, factors such as continentality (for example, the interior of Asia) and topography (the Murray–Darling Basin is in the rainshadow of the Great Divide and rain-bearing southeasterly trade winds) are significant regionally.

A number of factors may be significant at a more local scale, such as geology. The situation is not static, as temperate areas such as South Africa, northeastern Brazil or California can be affected by drought-related climate change. Climate change can lead to physical scarcity, as the case study of the Central Asian Highlands shows.

An interesting area to research is the Pacific Islands, where there are also acute water shortages for physical reasons, notably sea water encroachment on supplies.

Economic scarcity

Economic scarcity has a very different global distribution. Above all it is associated with developing countries that lack capital and technology and good governance to fully exploit their often adequate supplies of blue water. Central and southern Africa stands out as the key concentration of countries experiencing economic water scarcity, although there are one or two other countries, such as Haiti, the country with the lowest income in the western hemisphere, and Laos in Southeast Asia.

The Central Asian Highlands

Glaciers in the 'high heart' of Asia feed its greatest rivers – a lifeline for 1 billion people – including the Mekong, Yangtse, Huáng Hé (Yellow River) and Ganges. Glacial melt plays a vital role in maintaining river discharge before and after the summer monsoon rainy season, providing an abundance of water that needs to be captured for agricultural and domestic use otherwise it is lost forever.

As is the case in many mountain ranges (including the Andes, Rockies and Eastern Rift, as well as the Himalayas and Tibetan plateau) there is direct photographic evidence that the glaciers have dramatically retreated as a result of climate warming. Of 680 glaciers monitored by scientists, 95 per cent are shedding more ice than they are adding, leading to the deterioration of mountain pastures. Further issues are associated with the building of huge HEP dams by China and India in the upper reaches of the Himalayan rivers, which will have an impact on discharge downstream.

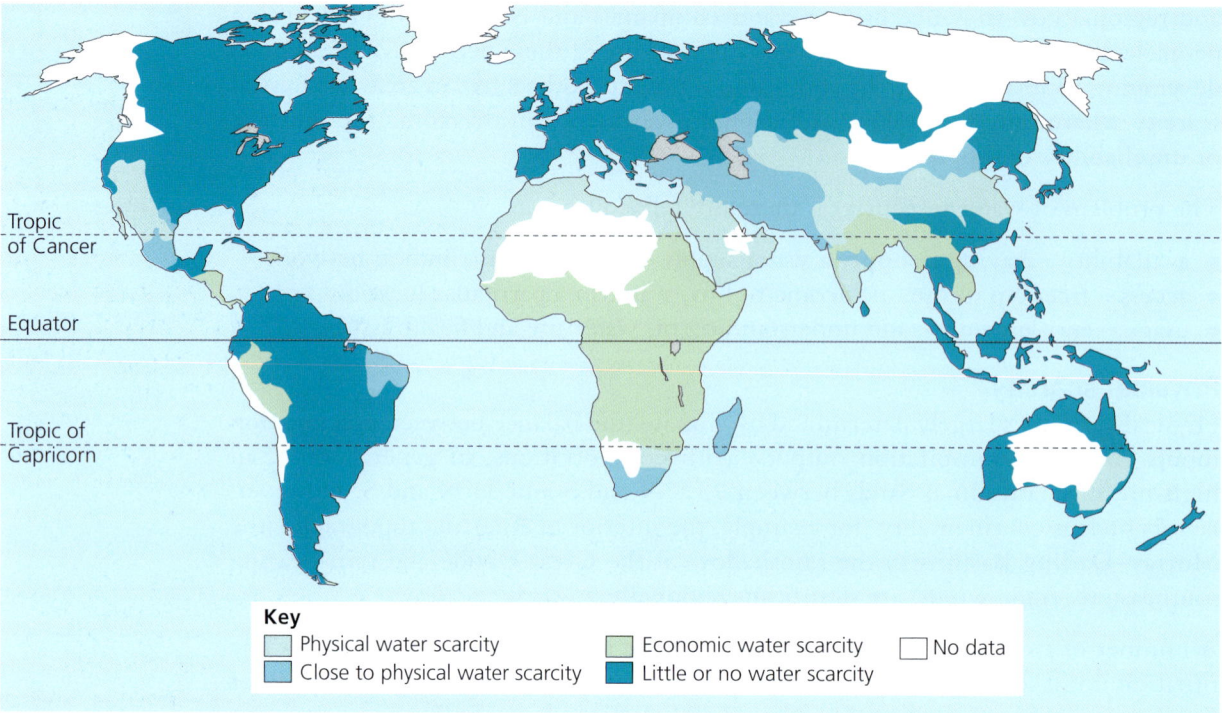

Figure 3.8 The global distribution of water scarcity

> ### Skills focus: Describing and explaining distributions
> Figure 3.8 shows the global distribution of water scarcity. When asked to describe or explain a distribution map like this, consider the following points:
> - Plan your answer to develop a solid structure; a useful tip is to use the structure provided by the key of the map.
> - Always support your description with precise locational detail, either by naming countries or regions, or by using latitudinal details.
> - When explaining, use the structure provided by the question. In the case of Figure 3.8 this could be to assess the *relative* importance of physical and human factors in explaining the distribution of water scarcity (see page 48–49 for factors).
> - Make sure that you refer to areas of non-scarcity too, to give the full picture.

The price of water

Price comparisons of water are a 'minefield'. There is very little correlation between prices paid and GNI per capita. As Figure 3.9 shows, a poor person in the developing world with no access to safe water at home pays a very high proportion of their income for 50 litres a day (the minimum threshold to maintain health, hygiene and for all domestic uses).

The price of water is determined by a number of factors:

- The physical costs of obtaining the supply. In some cities, the water has to be piped for many kilometres from mountain reservoirs (for example, the Californian coastal city of Los Angeles gets its water from Colorado through a very long pipeline).
- The degree of demand for the water. If water is scarce, as in the 2015 Californian drought, the price increases to manage demand (inevitably

The Water Cycle and Water Insecurity

the poor miss out). Even in cities in developed countries, such as Detroit and New York in the USA, there are considerable numbers of very poor people who do not have a direct supply to their homes, and so are supplied by water tankers.
- In developing world megacities such as Accra in Ghana, there is insufficient infrastructure. Poor people living in self-organised districts have to rely on water tankers, stand pipes and bottled water. The costs of water from informal vendors are nearly always twice that of standard tap connections; in Manila costs are four times higher.
- Who supplies the water is also an important influencing factor. In many areas in developing countries water is free, but usually it is not treated in any way and therefore is not clean. People (usually women and children) often have to spend many hours of the day walking up to 10 km to the supplies, carrying heavy containers (Figure 3.10, page 60). In many urban areas, the water is supplied by private water companies that charge the market price for it, as in Barranquilla, Colombia, with poor people, again, losing out. In some countries, such as Cuba, the government subsidises the price of water to ensure supplies are available for all.

If people are to have taps, safe drinking supplies and flushing toilets, there has to be an 'industry' to build and manage the infrastructure, and the water it delivers has to be paid for by someone. However, what the price, if any at all, should be to different consumers — farmers, industrialists, rich householders, residents of self-organised settlements — and who, if anyone, should benefit from its sale, remains a matter of considerable controversy.

Despite its status as a vital human need, in the twenty-first century water is increasingly seen as a commodity for which a realistic price should be paid. In the late twentieth century, politicians, financiers and other decision makers promoted the neo-liberal view in favour of privatisation of public utilities such as water, on the

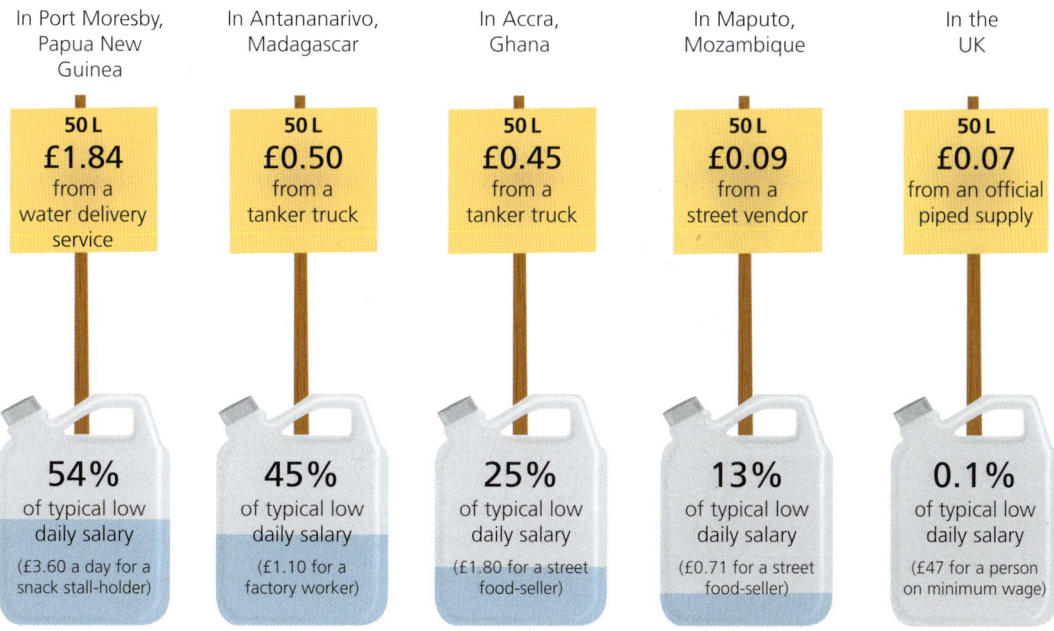

Figure 3.9 Water prices in five different world locations (Source: Adapted from a 2016 WaterAid report)

> **Key term**
>
> **Structural adjustment programmes (SAPs):** Policies promoted by the World Bank and IMF to help developing countries overcome their debt problems. These are now superseded by poverty reduction strategy papers (PRSPs) as for many countries SAPs resulted in unacceptable hardship and little progress with solutions to debts.

assumption that market mechanisms would simultaneously conserve water, improve efficiency and increase service quality and coverage. Subsidies would end, so all consumers would be charged for water at the price it costs to capture, treat and deliver it. With private companies, water is seen as a commodity from which profits could, and should, be made, so there are inevitable issues.

As part of the controversial neo-liberal policies of the 1970s and 1980s, the World Bank, in tandem with the International Monetary Fund (IMF), developed **structural adjustment programmes (SAPs)**, which it claimed would help developing countries to overcome their debt issues. The privatisation of utilities, including water, was seen as essential, as existing systems were inefficient, corrupt and failed to provide water to poorer citizens.

The provision of contracts by developing world governments to international water companies such as Veolia or Suez, both European transnational corporations (TNCs), proved a disaster in some cases, not only for the developing countries and their citizens (especially impoverished ones) but also for the water companies themselves, who had hoped for huge profits from these opportunities.

The cost of providing the water (often under very difficult conditions) meant huge price increases, which meant that the poor could not pay.

A seminal case of protest against privatisation took place in Cochabamba, Bolivia, in 1999–2000, where a local company (Aguas del Tunari, a subsidiary of US TNC Bechtel) was given a monopoly to collect water charges and actually took over water co-operatives run by the householders and tried to make *them* pay very high prices. Months of simmering protests culminated in the occupation of the city square by 80,000 people. After street battles, the company fled …

Other unsuccessful privatisation schemes to research include those in Dar es Salaam (Tanzania) in 2003 and in Jakarta, Indonesia.

In recent years, some Western TNCs have retreated from managing privatised water in developing countries, often in some disarray with governments abruptly cancelling their contracts, defeated by the many complexities and insufficient profit margins. Their place has been taken by Chinese and Indian companies as part of their policy of foreign direct investment (FDI) in developing countries, often using local companies to help with the work.

Privatised or not, the challenge of developing affordable water services in developing countries remains, as does the need to conserve water yet sell it at equitable prices. In some cities, such as Paris, there is a move to take water back into public ownership in order to do this.

Figure 3.10 People pumping and collecting potable water in the village centre, Ghana, West Africa

Water supply and economic development

Water plays a central role in all economic productivity, either directly as an input or as part of the context in which the economic activity takes place (for example, recreational tourism).

As shown in Figure 3.11, agriculture will continue to absorb around two-thirds of water extractions globally – but industrial usage is growing, especially in emerging countries such as China and India. The energy industry also requires water and, as with industrial usage, there are major concerns about the environmental impacts of these activities, from the destruction of ecosystems to uncontrolled discharge of polluted effluents.

Aquaculture too (fish farming) has expanded rapidly in recent years as wild fish stocks have declined. Again, there is an environmental downside: although it has provided both employment and vital food supplies, especially for Southeast Asian markets, the lack of regulation and the degradation of ecosystems, such as coastal mangroves, is a major concern in countries such as Thailand.

Agricultural use

Figure 3.12 shows the spectrum of agricultural practices from producing crops under entirely rain-fed conditions, using green water in the soil, to producing under fully irrigated conditions. In rain-fed agriculture, fields and grazing lands are entirely dependent on rainwater. Farmers focus on storing water (rainwater harvesting) to conserve supplies. Moving along the spectrum, more surface water or groundwater (blue water) is added to enhance crop production, as well as providing opportunities for multiple use.

Around one-fifth of the world's land is under full irrigation. In water-short and monsoon areas, traditional practices such as basin irrigation (for example, along the River Nile) have always been used. Industrial scale irrigation, which began in the 1960s using high-yield variety seeds combined with fertilisers and pest control, has greatly increased the pressure. Although the Green Revolution has improved food security enormously, it is causing environmental concerns:

- Around 30 per cent of this irrigation is provided using dams from which systems of irrigation canals radiate. Much irrigated land becomes waterlogged, leading to salination of the soils.
- The majority of irrigation is pumped up electrically from aquifers, leading to significant groundwater depletion, especially in India, the USA, China and Pakistan.

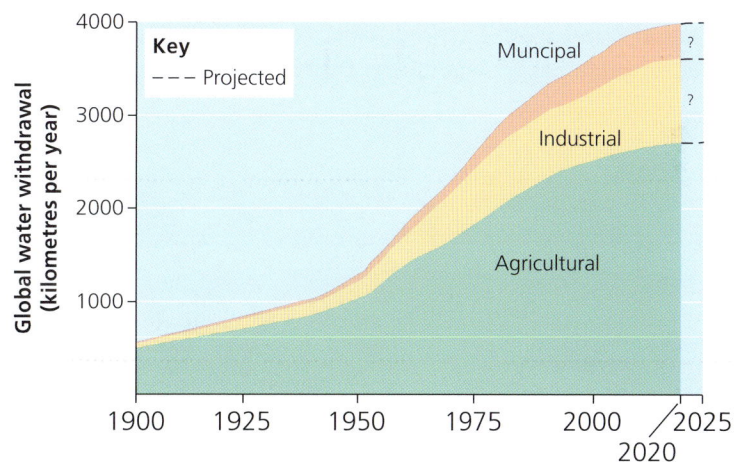

Figure 3.11 Trends in sectoral water use

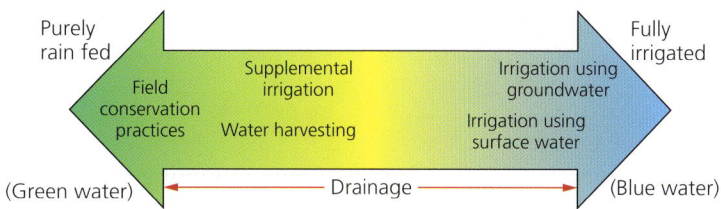

Figure 3.12 The spectrum of agriculture showing intensity of irrigation use

3 Water security – is there a crisis?

A study of the Aral Sea emphasises the environmental downside of large-scale irrigation schemes. It also emphasises how different stakeholders have been affected by the environmental and ecological catastrophe, and therefore have different opinions depending on whether they were winners or losers.

A further source of pressure on water supplies for agriculture is the dietary revolution in countries such as China, where there has been a huge rise in the consumption of dairy products and meat. While it may take around 2975 litres of water to produce 1 kg of rice, it takes nearly six times as much water to produce 1 kg of beef.

Clearly, managing agricultural demands is of paramount importance in managing overall water security, as it is by far the greatest water user. The maxim for irrigated systems is 'more crop per drop', using modern automated spray technology and more advanced drip irrigation.

Even more exciting developments are associated with phase 3 of the Green Revolution, which recognises how food security is closely interlinked with water security in drought-prone areas. It is focused on drought-restraint and salt-tolerant crop strains (the Gene Revolution).

Intermediate technology solutions for water conservation play a role too, such as the 'magic stones' system practised in the Sahel (page 27).

The Aral Sea

Once the world's fourth-largest inland sea (68,000 km²), the Aral Sea (Figure 3.13) has been steadily shrinking since the 1960s. In the late 1950s the Soviet government diverted much of the water from the Amu Darya and Syr Darya, which fed into the Aral Sea, for irrigation of agriculture. By 2007, the sea had declined to just 10 per cent of its original size and had split into separate lakes; its level had fallen to 40 m (Figure 3.14). This is an environmental catastrophe. An interactive map and satellite photographs can be seen at: https://earthobservatory.nasa.gov/world-of-change/AralSea

Figure 3.13 The location of the Aral Sea

Figure 3.14 Contrasting satellite images showing how the Aral Sea has decreased in size between 2000 and 2015

The Aral Sea crisis has involved several stakeholders:

- *The former Soviet government*. Communist leaders began an ambitious irrigation scheme to develop fruit and cotton farming in what had been an unproductive region and create jobs for millions of farm workers.
- *The fishing community*. A once-prosperous industry that employed 60,000 people in villages around the lakeshore has collapsed. Unemployment and economic hardship are everywhere. Ships lie useless on the exposed seabed.
- *Local residents*. Health problems are caused by the windblown salt and dust from the dried-out seabed. Drinking water and parts of the remaining sea have become heavily polluted as a result of weapons testing, industrial projects, and fertiliser and pesticide run-off. Infant mortality rates are among the highest in the world, with 10 per cent of children dying in their first year, mainly of kidney and heart failure.
- *The Uzbekistan government*. The irrigation schemes based on the Aral Sea allowed this poor country, with few resources, to become one of the world's largest exporters of cotton. It also hopes to discover oil beneath the dry seabed.
- *Scientists*. Only 160 of the 310 bird species, 32 of the 70 mammal species and very few of the 24 fish species remain. The climate has changed too, making the area even more arid and prone to greater extremes of temperature.
- *Kazakhstan farmers*. Irrigation has brought the water table to the surface, making drinking water and food crops salty and polluted.
- *International economists*. People in the region may no longer be able to feed themselves because the land has become so infertile. Up to 10 million people may be forced to migrate and become environmental refugees.
- *Water engineers*. Inspections have revealed that many of the irrigation canals were poorly built, allowing water to leak out or evaporate. The main Karakum Canal, the largest in Central Asia, allows perhaps 30–75 per cent of its water to go to waste.

Since 2007, Kazakhstan has secured World Bank loans to save the northern part of the Aral Sea – an extremely ambitious project aimed at reversing one of the world's worst environmental disasters, which has brought new wealth to the fishing villages of Kazakhstan in contrast to the south Aral Sea, which in 2018 remained completely dessicated.

Use by industry and energy

Just over 20 per cent of all freshwater withdrawal worldwide is for energy production and industry. While in developed countries this percentage is currently around 40–45 per cent of all water used, especially for the chemical, petroleum, paper and electronics industries, there has been a fall in use as heavy manufacturing industries, such as steel, have declined. A major concern is the global shift in industrial production towards emerging nations such as China and South Korea. This rapid industrialisation, particularly in developing countries, has contaminated both rivers and groundwater, affecting the quality of water. Considerable progress has been made by many TNCs, such as Coca-Cola India, to reduce their consumption by efficient recycling and also to control effluents.

Energy use is a very mixed picture. Over half of the water used is either for generating hydroelectric power (HEP) or for the cooling of thermal and nuclear power stations, so is returned to its source virtually unchanged, although its warmth can impact on river ecosystems. Countries that rely heavily on HEP for the production of electricity, such as New Zealand, are affected by changing patterns of rainfall – especially the decreasing amounts of rainfall associated with short-term climate change.

A further area of concern, for a variety of reasons, is the growth of biofuels. The crops grown to produce bioethanol and biodiesel are very thirsty: up to 10,000 litres of water is needed to produce 1 litre of bioethanol, and 20,000 litres for 1 litre of biodiesel.

Water supply and human well-being

Although huge improvements have been made via the thrust of the MDGs, and subsequent SDGs, around 15 per cent of the world's population still rely on unimproved water (unprotected wells, springs or rivers and other untreated surface water), and around 2.5 billion people have no access to improved sanitation facilities. Water and disease interact in two ways: unsafe drinking water can spread disease, but water used for personal and domestic hygiene (washing hands, etc.) can prevent disease transmission.

The fundamental source of much water-related disease is a lack of sanitation, which is estimated to contribute 10 per cent of the 'global disease burden'. This lack of sanitation (on-site dry systems as well as water-borne sewage) is a cause of low standards of personal hygiene. A major issue is indiscriminate or open defecation, which has been steadily decreasing across the world as household incomes increase, but remains notably high in rural South Asia. The diarrhoeal diseases – cholera, typhoid and dysentery – are examples of diseases transmitted by faecal–oral routes.

Figure 3.15 shows the interrelationships between water, sanitation and disease.

In many districts in developing countries, institutional indifference to improving programmes of community hygiene is the key issue, as opposed to acute water shortage. Diseases related to a lack of clean water or lack of improved sanitation also lead to high levels of morbidity, as they affect people's ability to work and look after their family and, therefore, their ability to escape poverty.

Water is also a breeding ground for many vectors of diseases, such as malarial mosquitoes, snails and parasitic worms. These lead to debilitating diseases such as malaria, dengue fever, zika and bilharzia.

Infections can be contracted from washing in surface water polluted with human faeces. New reservoirs behind dams expand the breeding ground for insects and snails. For some projects, however, increased prosperity, improved nutrition and access to medical facilities more than outweigh the additional risks of infection (for example, in rural Burkina Faso). Drinking from infected water sources can also increase disease risk, for example, from the Guinea worm parasite. Although major progress has been made in the eradication of the Guinea Worm disease, many of what the World Health Organization (WHO) called the 'neglected tropical diseases' in 2007 continue to exist because of marginalisation and isolation. The last communities affected by Guinea worm disease are often the most distrustful of government programmes.

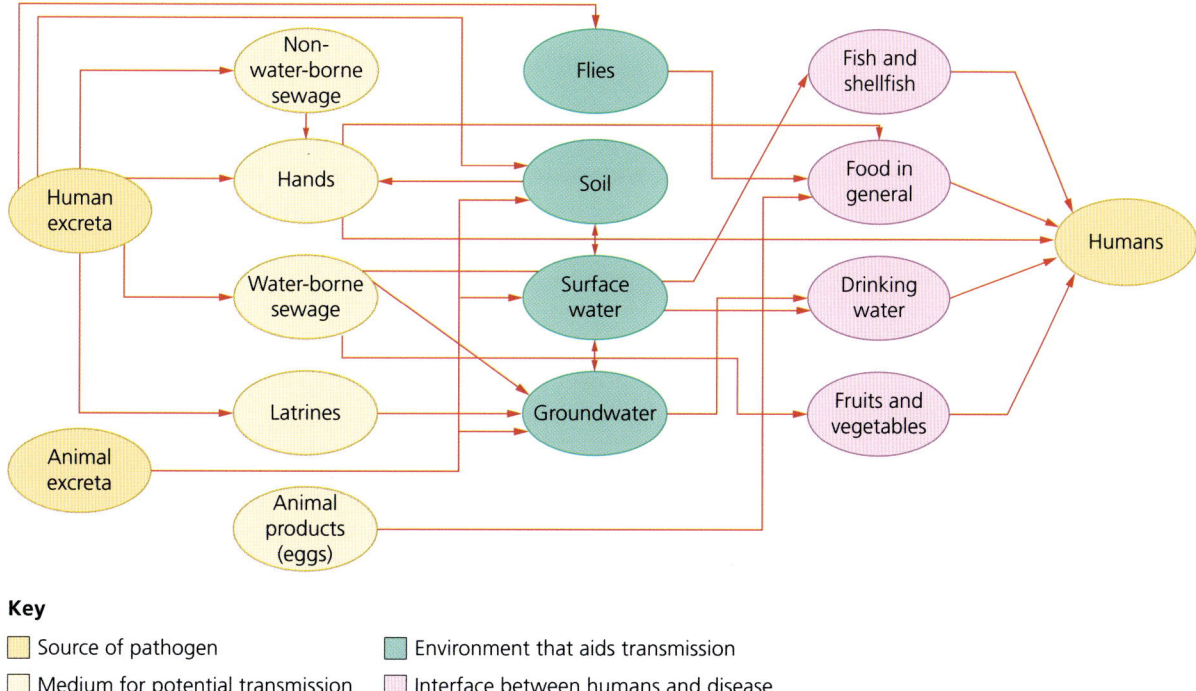

Key

☐ Source of pathogen   ☐ Environment that aids transmission
☐ Medium for potential transmission   ☐ Interface between humans and disease

Figure 3.15 Water, sanitation and disease

While there have been some spectacular developments and progress in eradicating water-related diseases around the world by NGOs (for example, the Gates Foundation programme for malaria, and WaterAid's projects to tackle the root causes of the problem by providing clean water and improved sanitation in rural districts), significant problems remain. Poverty of both countries and the people within them is the single most important factor retarding improvement.

Water insecurity and the potential for conflicts

When demand for water overtakes the available supply, and a number of stakeholders (players) wish to use the same diminished resources, there is potential for conflict at all scales. Competing demands for diminishing water supplies for irrigation, energy, industry, domestic use, recreation and ecosystem conservation can lead to tension both within countries and between countries.

Figure 3.3 (page 51) identifies how water pressure points (pinch points or water hotspots) can occur. While some observers suggest these tensions could escalate into conflicts, with water taking over from oil as the most contested resource of the twenty-first century, the evidence would suggest that this is by no means a certainty.

Between 1948 and 2008, out of nearly 2000 international 'events', only 25 per cent led to any form of conflict, and only 1.5 per cent caused serious wars. Of these conflicts, nearly two-thirds were about the quantity of water available, especially where upstream users had diverted or planned to divert water in a river basin at the expense of lower basin users (as is occurring in the Nile Basin). The other common source of conflict was about the building of dams and diversion canals, and their ecological impact (as in the Mekong). While there have been many military and terrorist actions threatening to destroy dams, cut off water supplies, or deliberately pollute water sources, the reality is that most action has been in the form of political campaigns and protests.

The most likely scenarios for conflicts to develop into wars occur where the river basins are transnational and where disputes over water are just one item on the agenda for wider wars, such as the longstanding conflict between Israel and Palestine. Water is a major issue there as Israel does not accept that this scarce resource should be shared equally with Palestine and extracts over two-thirds of supplies. Further examples include India and Pakistan, where there is a long history of boundary disputes. There are also a number of disputes in the Middle East. In each of the three most disputed basins, military power is not evenly balanced: one country is strong enough to get its own way most of the time.

- In the Jordan Basin, downstream Israel can threaten upstream Jordan with military power, so it leaves the flow of water alone.
- In the Tigris–Euphrates system, Türkiye is upstream and strong enough to do what it wants, despite protests downstream from both Iraq and Syria.

As well as surface water conflicts, groundwater conflicts often occur in similar areas to surface water ones – for example, between Israel and Palestine over the use of mountain aquifers. Israel has very advanced abstraction technology and so has an advantage over water-insecure Gaza (part of Palestine).

Many subterranean aquifers straddle international boundaries. The issues of shared groundwater usage are highly complex for the following reasons:

- Supplies are underground; it is difficult to understand the problem as it takes years for an effect to show.
- The boundaries are very unclear underground; it is difficult to negotiate an equitable share for each nation to exploit as nobody knows who owns what.
- UN legislation to sort out water sharing of aquifers between nations is only just being written.

Conflicts within a country

A number of quite simple conflicts can occur within a country, for example, over the building of a dam and water reservoir, as in Kielder, Northumberland, where there was concern over the flooding of a farming valley and villages.

Some reservoirs, such as the badly needed reservoirs for Southeast England, are so long disputed that they have not been built because of **nimbyism**. In a very crowded area, there is always concern about environmental and socio-economic impacts.

Further examples to research include:

- The Colorado Basin, where disputes occur between the states of California and Arizona in the USA, and internationally with Mexico (see Geofile 648 and Geo Factsheet 254).
- In India, where Karnataka and Tamil Nadu states have disputed the sharing of Kaveri River waters. Tamil Nadu rice farmers rely on waters from upstream dams for their summer crops, but in a dry year with a 'weak monsoon' the farmers of Karnataka strenuously oppose the loss of water.

> **Key term**
>
> **Nimbyism:** 'Not in my backyard' – people protesting about developments which they see as detrimental to their own neighbourhood.

The Great Ruaha River, Tanzania

An example of a more complex conflict between a number of players occurs in the basin of the Great Ruaha River, a semi-arid area in the Southern Highlands of Tanzania. Economically, this river basin is important to Tanzania as it provides water for rice growth and the generation of HEP, maintains a Ramsar-status wetland, and is important for wildlife tourism in the Ruaha National Park.

The Great Ruaha River has ceased flowing in the dry season because water levels in the large wetland in the upper course have dropped below a critical level, which is a major problem for lower river users.

Table 3.4 lists the viewpoints of the players (stakeholders) – essentially they all blamed each other.

The national and local concerns about the issue were:

- National power shortages resulting from low flows through the HEP scheme.
- Desiccation in the Ruaha National Park, with the wetland diminishing in size and causing problems for wildlife.
- Increased competition for water causing disputes as supplies kept being turned off for domestic users.

Table 3.4 summarises the opinions that stakeholders initially had about the causes of the changes. A programme of scientific research projects was developed to test the theories of the stakeholders, looking at the reasons for the reduction in the size and flows from the wetland, and the impact of the HEP developments, as well as the role of upper and lower basin agriculture.

The conclusions suggested that different parts of the system were affected by low summer flow for different reasons. Overgrazing and deforestation in the watershed area were ruled out. Mismanagement of releases from reservoirs to maximise HEP generation and overuse of water for rice irrigation in the dry season were also ruled in as contributing factors.

The results of the scientific research did impact on the views of some stakeholders to an extent, but the emphasis has now moved to developing integrated water management schemes to manage the problems.

Table 3.4 Viewpoints on hydrological changes in the Usanu basin and Ruaha River

Stakeholder	Initial viewpoint, mid-1990s	Perceptions after scientific research, 2003
General view	Shrinking wetland, drying river and low reservoir levels were all closely related	Shrinking wetland, drying river and low reservoir levels were separate issues
Investigators (SMUWC/RIPARWIN)	Various hypotheses were tested: combination of cattle, deforestation, climate change, irrigation, abstraction of water and total flows into Mtera/Kidatu	Dry season abstraction and environmental losses, which led to Ruaha River flows ceasing. Miscalculation of drawdown of stored water led to low reservoir levels
Ministry of Agriculture	Inefficient smallholder schemes required funding for improvement, which would allow more water to flow downstream	Smallholders competed over water and therefore were quite efficient in their management
Ministry of Natural Resources	Cattle and overgrazing were degrading the wetland, reducing its ability to hold and release water. Deforestation in the upper catchment was reducing base flows in rivers	Cattle and overgrazing in the wetland remained the cause. Deforestation remained a problem
Ministry of Water	Inefficient smallholder schemes. Deforestation in the upper catchment	Inefficient smallholder irrigation
Mbarali District	Cattle and overgrazing in the wetland. Deforestation in the upper catchment	Cattle and overgrazing in the wetland. Deforestation still a cause
Friends of Ruaha and WWF	Large-scale irrigation schemes were abstracting water during the dry season. Damaged wetland from overgrazing	Dry-season abstraction into all irrigation schemes. Damaged wetland
Electricity Supply Corporation	Scale and inefficiency of irrigation led to lack of water for power generation	Scale and inefficiency of irrigation

International conflicts

Table 3.5 summarises some of the current potential major conflicts over water. There is, however, no suggestion that these conflicts will develop into actual ones.

Table 3.5 Some potential international water conflicts

Location	Reason for pressure point
Tigris–Euphrates Basin	Concerns from Iraq and Syria that the GAP project in Türkiye will divert much of the water via a series of irrigation dams. Syria has also developed dams which, in the 1990s, initially led to conflict with Iraq. There is uncertainty as to how events will unfold with a newly elected Syrian government.
River Jordan	Use of the Jordan, largely by Israel but also Syria, Lebanon and Jordan, has reduced the flow of the river to a mere trickle. This also affects supplies to Palestine's West Bank. It is one of the most disputed and militarised waterways in the world. Current ongoing wars.
Ganges–Brahmaputra	India has built dams, such as the Farakka Barrage which has reduced the flow of the river into Bangladesh. Currently not at war.
Syr Darya and Amu Darya, Central Asia (Aral Sea)	Turkmenistan, Uzbekistan and Kazakhstan need more summer water for irrigation, but water has been diverted by Tajikistan/Kyrgyzstan. Restoration policy by Kazakhstan of the north Aral Sea (see page 62).
Colorado Basin	States in the USA dispute their allocation of water from the Colorado, which is so great that the quantity and quality reaching Mexico does not reach the standard agreement.
Nile Basin	While agreements exist, schemes developed in Ethiopia and Sudan may threaten supplies to Egypt. Currently both Egypt and Sudan are weakened by other crises and poor governance.
Asian Water Tower	China's move to utilise the headwaters of large rivers flowing out from the Tibetan plateau has caused deep concern to all downstream countries that rely on water for irrigation – this includes India, which has its own ambitions.
Bolivia–Chile	Bolivia plans to industrialise the Silala River headwaters, reducing the flow of water to the copper mines in Antofagasta, Chile. Old enemies from past wars.

Troubled waters on the River Nile

The 6700-km-long Nile is the world's largest river, with two main sources: the White Nile, whose source is Lake Luvironzo, near Lake Tanganyika, which subsequently flows into Lake Victoria through Uganda and into Sudan, and the Blue Nile, which rises in the Ethiopian Highlands. Their confluence is at Khartoum, from where the river continues to flow northwards into the desert state of Egypt and on to its delta in the Mediterranean Sea. The Nile Basin covers about 10 per cent of the African continent (Figure 3.16, page 69). There are both physical and human factors which could trigger conflict over the Nile's waters.

Hydrology of the River Nile

Eleven countries compete for the Nile's water and yet, with a measured flow of 84 billion cubic metres, the Nile has a very modest discharge compared to other great rivers of the world, for instance the Amazon and many other African rivers such as the River Congo.

There are three key features about the pattern of discharge that could increase the potential to cause disputes. These are linked to the south–north direction of flow through contrasting climate zones.

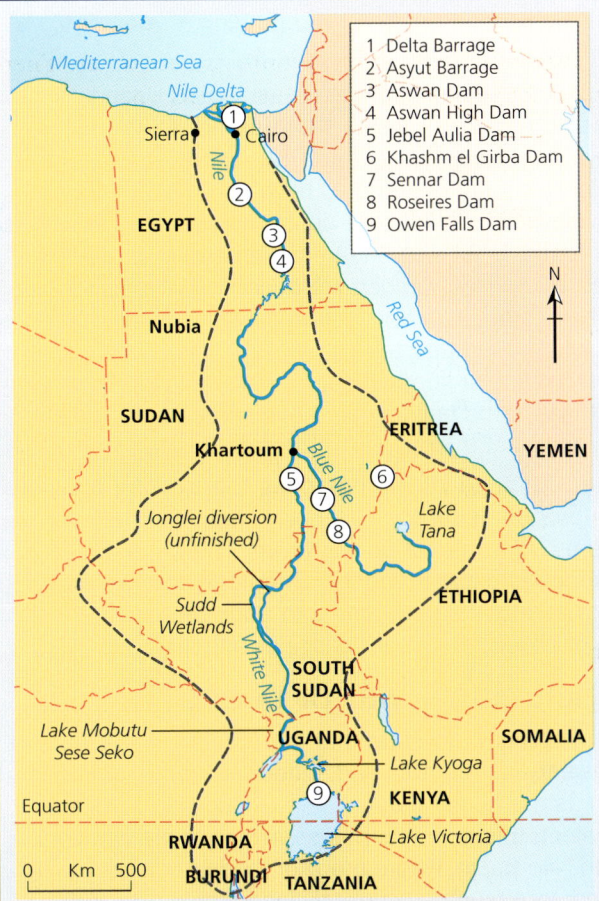

Figure 3.16 Map of the Nile Basin and existing dams

1. The White Nile provides a mere 30 per cent of the flows measured at Aswan, Egypt. While the catchment of the Blue Nile is small relative to that of the White Nile, the heavy monsoonal rainfall from July to September means that it is by far the greatest contributor to Lower Nile flows. The difference in the two major river regimes is very marked (Figure 3.17).

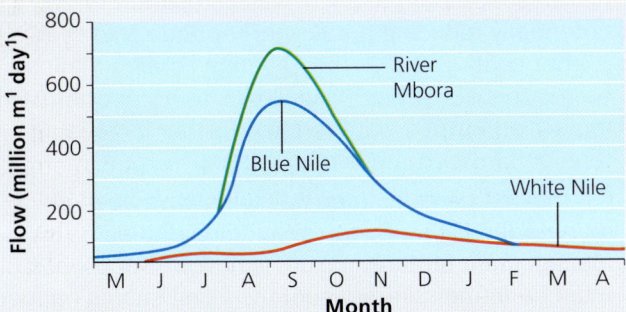

Figure 3.17 The discharge pattern of the Nile

2. The seasonal variation in flow poses a key challenge to river basin planners and agriculturists, especially as it is susceptible to major inter-annual and decadal fluctuations brought about by El Niño–La Niña cycles and, in the future, the possible impact of climate change. These drought and flood cycles are a particular problem for Ethiopia and Sudan, and in the past for Egypt before the building of the Aswan High Dam.

3. Much of the river system is located in hot, arid areas where evaporation losses are high, especially in the Sudd area of southern Sudan. Between entry to and exit from the Sudd swamp, the White Nile loses up to 50 per cent of its flow – leading to early summer water shortfalls in Egypt and Sudan when flows from the Blue Nile are at their pre-monsoonal low point. Egypt was very keen to sponsor the Jonglei Canal Scheme to speed up the flow through the Sudd, so reducing evaporation losses. The scheme was begun in 1979 but remains unfinished to this day due to deteriorating relations between Egypt and Sudan. It is being reviewed as part of the latest Nile Basin Initiative (NBI).

Geopolitical issues

One of the key geopolitical features of the Nile Basin is the large number of national borders that traverse it. Largely as a result of the European colonial era, these boundaries pay scant regard to the physical and human geography of the Nile Basin.

Currently, over 300 million people live within the Nile Basin, but this is expected to double to 600 million by 2030, placing further pressure on water supplies for domestic and agricultural use. Egypt is dependent on the Nile for 95 per cent of its water needs, but other states, such as Rwanda and Ethiopia, need large supplies of water to develop crop irrigation, HEP production and industrial processing in order to lift their nations and their people out of poverty.

For water conflicts to develop, there have to be underlying water scarcity issues (i.e. a threshold figure of below 1000 m³ per person per year). While many African nations do not currently have physical water scarcity issues, many, such as Tanzania, have economic water scarcity issues whereby they lack the capital and technology to exploit supplies. Demands from growing populations and development combined with the impact of climate change means that of the Nile Basin countries, Burundi, Rwanda, Egypt, Ethiopia, Kenya, and possibly Sudan and Tanzania, will all be potentially water scarce by 2025. This water scarcity will ultimately impact on food security, especially in countries such as Egypt, which already have high levels of virtual water use because of the need to import substantial quantities of food. (Source: Adapted from/ Geo Factsheet 282.)

The impact of history

The original agreements for sharing the Nile's waters were bilateral between Egypt and Sudan (the two 'Arab' nations), agreed by their colonial masters – Britain.

- In 1929, the first Nile Waters Agreement was signed, giving 48 million cubic metres to Egypt and 4 billion to Sudan, with only 14 per cent going to the other African countries. The big issue was that Egypt (as the downstream nation) was given the right of veto on any modifications in the use of the Nile's water in the other nine nations. Egypt was favoured as an important agricultural asset and the Egyptian-run Suez Canal was vital for British imperial ambitions. The British riparian colonies – Sudan, Uganda, Kenya and Tanganyika (now Tanzania) – as well as Ethiopia, had no say.
- In 1959, a second Nile Waters Agreement was signed, giving 55.5 billion cubic metres to Egypt and 18.5 billion to Sudan – in effect giving *all* the water to these two countries as the rest of the discharge is lost to evaporation. The increased allocation to Sudan represented the increased needs of the country for irrigation (for example, for the Gezira Scheme). This agreement was again signed by the British colonial powers on behalf of upstream countries, who felt that all these countries had plenty of water from other sources. Ethiopia refused to recognise the legitimacy of the agreement.

The acquired rights from these historic agreements have resulted in an unfair allocation of the Nile's waters. For example, Ethiopia has a major production of water but very low capture of the resources, in contrast to Egypt and Sudan, which have low internal renewable resources but high capture of the Nile's water.

Will there be water conflicts among the Nile nations?

Sudan and especially Egypt have ever-increasing needs for more Nile water. Nevertheless, a riparian-led process of joint decision-making called the Nile Basin Initiative (NBI) emerged during the 1990s.

Since 2005, nine of the ten Nile Basin countries (with Eritrea observing) have been exploring the development of the NBI in partnership with key external agencies such as the World Bank to establish a common vision. Two subsidiary action plans, the Eastern Nile Program and the Nile Equatorial Lakes Program, have been established.

The good news is that Egypt and Sudan are involved in the NBI, but the bad news is that so little was really achieved in spite of the funding of many worthwhile development projects.

In 2010, Ethiopia, Rwanda, Tanzania and Uganda signed a new water treaty, with Burundi, DRC and Kenya promising to sign later. This treaty stated that *all* riparian countries should have equal rights to use the Nile waters. While the upstream countries have now urged Egypt and Sudan to sign and eventually agree to the treaty, much will depend on the politics of the Nile, with Egypt and Sudan now politically weaker and less well supported by powerful allies. Indeed, in current unipolar geopolitics, it is the neo-colonisation of China that is building all the schemes and dams, especially in Ethiopia.

Table 3.6 (page 71) summarises the existing and new dam proposals for the River Nile, which will clearly have a major impact downstream. The big shift in influence is exploitation by Ethiopia, a country with major economic ambitions and an impressive growth rate of around 8–10 per cent per annum.

Conclusion – war or peace?

It is all too easy to state that there will be water wars in the twenty-first century and, indeed, in some locations such as the Middle East, where existing conflicts are prevalent, this does seem increasingly likely. However, in the Nile Basin, where many of the ingredients are present (such as currently inequitable use and increasing scarcity) the emphasis so far has been on co-operation. Some would argue that Egypt and Sudan are so politically weak with other wars that they have no spare capacity to fight a Nile war.

It is possible that the common vision of the Nile Basin Initiative, which seeks to achieve sustainable socio-economic development through the equitable usage of, and benefit from, the common Nile Basin resources, and peace, will prevail.

Ethiopia's mega-dam project, the Grand Ethiopian Renaissance Dam (GERD), was completed in 2024, potentially reducing the flow of the Blue Nile by 25 per cent. In 2023, the Egyptian Foreign Ministry stated the GERD was 'illegal' due to its downstream impact on Egypt's water supply, further increasing tension between the two countries.

For the last decade, three-way talks have been going on between Egypt, Sudan and Ethiopia to agree a management strategy including how fast and when (it must be in the rainy season) the filling should happen. The Blue Nile in Sudan has recently experienced several years of climate change-induced drought, but Sudan sees an opportunity to get HEP from Ethiopia's dam as part of the deal and so is in the middle of the two protagonists in spite of concerns of water loss.

Egypt is totally opposed to Ethiopia's intent to exercise 'hydro hegemony' and has been seeking support from all Arab league countries and the USA who chaired the negotiations to 'drum up' opposition to the dam. The talks have concluded with no solution and considerable acrimony.

Table 3.6 New plans for the Nile waters. All these countries are developing countries

Country	Population (million)	GDP (US$ billion)	Schemes
Egypt	74	75.1	Hydropower potential almost fully tapped; Nag Hammadi barrage under construction
			New Toshka irrigation scheme to be completed in 2017 to irrigate 220,000 hectares
			New Salam Canal to divert water to northern Sinai
			Many new desert settlements needed to house growing population
Sudan	36	19	Many dams raised in height, e.g. Roseires Dam
			Huge new Merowe Dam, as well as Kajbar Dam
			Plans to restart the Jonglei Canal as part of Sudan–Egypt co-operation
Ethiopia	77	9.8	Many new dams on the Blue Nile including Tis Abay, Tekezé and the GERD
			In all, proposals for 33 irrigation and HEP plants on the Blue Nile, which will make Ethiopia a food and power exporter, but will block sediment transfer and cut the flow of water considerably
Uganda	29	6.8	Construction of the Bujagali Falls Dams
Tanzania	52	48	Plans for a new pipeline from Lake Victoria to provide drinking water

3.3 Approaches to managing water supplies

The use and subsequent abuse of water is one of the most controversial and complex issues facing the world. Table 3.7 (page 72) shows that there is a wide range of players involved in any issue relating to water resources.

Controversy 1: social versus political players

Social players see access to clean, safe water as a human right, whereas political players see water as a human need which, like food, shelter and energy, can be provided in a number of ways through market mechanisms (private), public services (government) or public–private partnerships (via governments and NGOs). The infrastructure of water provision is extremely costly, so governments have to find a mechanism to pay for it in order to satisfy the needs of their people. It was estimated in 2010 that meeting the UN's MDGs of halving the proportion of the population without access to safe water supply and improved sanitation would cost over US$200 billion.

Controversy 2: economic versus environmental players

In order to keep pace with rising demand, business players favour hard engineering schemes such as mega dams, water transfer projects and clusters of desalination plants. Inevitably, these schemes have very high social and environmental costs, and are opposed by social and environmental players, who favour more sustainable approaches.

Table 3.7 Players involved in issues relating to water resources

Category	Players
Political	International organisations (e.g. the UN) responsible for the MDGs and SDGs; government departments (e.g. Defra); regional and local churches; lobbyists and pressure groups that form to fight against particular issues such as the building of mega dams
Economic (business)	The World Bank and IMF, which fund mega projects and ensure legislation is in place for trans-boundary schemes; developers of mega schemes; transnational water companies (utilities) that run the supply business; TNCs and businesses that are large users (agriculture, industry, energy and recreation)
Social (human welfare)	Individuals, residents, consumers, land owners and farmers, who feel access to water is a human right; health officials who try to ensure safe water; NGOs such as WaterAid or Practical Action, which develop sustainable schemes for the poor in LDCs
Environmental (sustainable development)	Conservationists who fight hard-engineering schemes or seek to save wetlands; scientists and planners who develop new schemes; NGOs such as the WWF, which try to influence world water policy; UNESCO/FAO/IUCN, which operate globally

> **Key terms**
>
> **Top down:** Large-scale capital intensive development schemes, usually developed by government.
>
> **Bottom up:** Small-scale development schemes.

There are therefore a number of responses that can be made to manage future water supplies, with action at a variety of levels, ranging from large-scale projects funded by international governmental organisations (IGOs) such as the World Bank, and by governments, down to changing individual consumers' attitudes to water use to encourage conservation at a local scale. The large projects usually employ **top down** approaches to ensure efficient delivery, which can disregard people's wishes, whereas more localised projects, such as sustainable WaterAid projects, tend to be **bottom up**, involving local people in their management, but sometimes with scaling-up difficulties.

Hard-engineering projects

In all cases, high levels of capital and technology are needed to carry out these projects. Economic costs are inevitably very high (multi-billion) and, while there are often economic benefits across a large area, questions have to be asked about the environmental and social costs.

Water transfer schemes

Water transfer schemes involve the diversion of water from one drainage basin to another (inter-basin transfer), either by diverting the river itself or by constructing a large canal to carry water from an area of surplus to an area of deficit.

In the UK, with its generally wet northwest and much drier southeast – a factor potentially exacerbated by projected climate change – a water grid has long been planned but ruled out because of the costs of the infrastructure, the energy-intensive need for pumping and, more recently, the difficulty of achieving co-operation between privatised water companies. With the population expected to rise in the South East by 1 million by 2025, an already water-stressed area, transfers may become essential.

Nevertheless, there are numerous inter-regional pipeline transfers – from Welsh reservoirs to Liverpool and Birmingham, from the Lake District to Manchester, and from Kielder Water to the River Tees and subsequently to the Yorkshire Ouse.

The controversy lies in large-scale, high-tech transfer schemes. The engineering itself and the actual water transfer schemes have been successful, but there are many environmental and social disadvantages. Additionally, continued use of

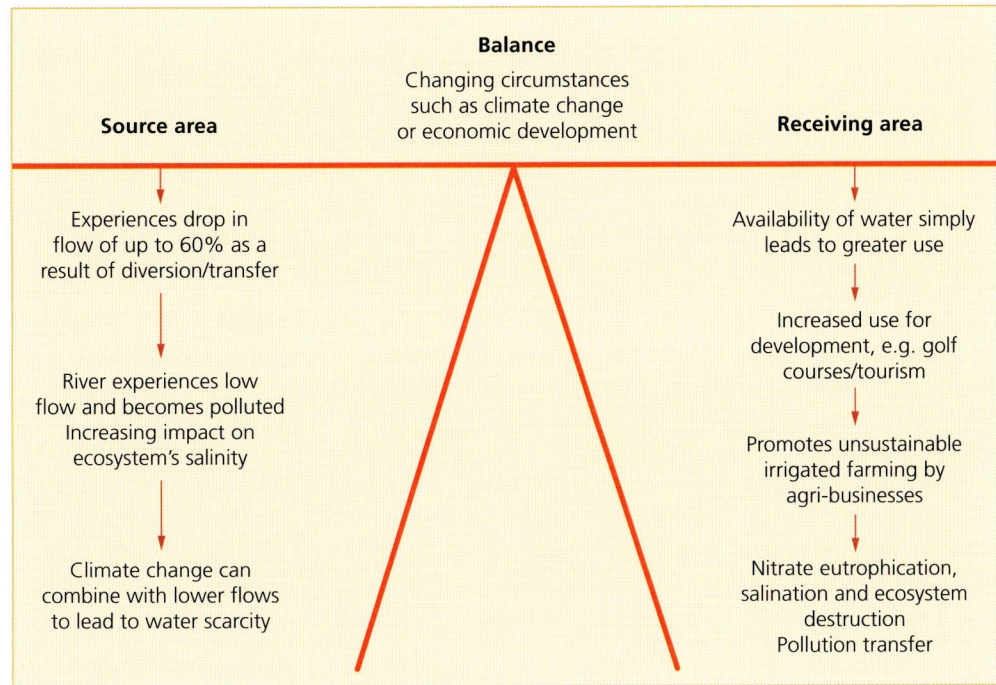

Figure 3.18 Water transfer issues

transferred water may lead to long-term changes to local and regional hydrological conditions, perhaps increasing flood risk, damaging fish stocks, spreading diseases and pollution, and acting as a pathway for introducing alien species into new river environments.

Figure 3.18 shows the balance of water transfer issues. Events such as increasing risks from climate change or inability to pay for the enormous costs due to a declining economic situation, can swing the balance for or against.

Table 3.8 (page 74) outlines three major existing schemes and some proposals for future schemes. China's South–North Transfer Project, which is in progress, dwarfs all others. (See place context on page 75.) Inevitably, with such a controversial large scheme, the risks are enormous. However, such is the severity of northeastern China's water problems (with all provinces using over 200 per cent of their annual renewable resources) that bold decisions had to be made. Half the population of the northeast rely on only 15 per cent of China's water supply.

Mega dams

Nearly 60 per cent of the world's major rivers are impeded by large dams. Rivers such as the Colorado have been impeded, stored, rechanneled and re-engineered in a technological exercise to redesign natural flows for the benefit of humankind.

In 2020 there were over 845,000 dams in the world, of which over 5000 could be considered to be mega dams. These mega dams have the facility to store 16 per cent of the annual global run-off, which is theoretically a huge addition to the blue water component. However, evaporation losses are very high as many dams are located in semi-arid areas. In the 1980s, mega dams were seen as the panacea for the world's water shortages and, through irrigation, for the food security issue – irrigated areas contribute up to 16 per cent of world-food production. Many dams were large

Table 3.8 Existing schemes and details

Schemes	Details
Existing schemes	
Snowy Mountains scheme in southeast Australia	Water is transferred from the storage lake of Eucumbene westwards by the Snowy tunnel to the headwaters of the Murray River to irrigate farms and provide water to an increasingly drought-stricken area
Melamchi Project in Nepal	Water is diverted from the Melamchi River via a 26 km tunnel to water-stressed areas in the Kathmandu Basin; in return, the residents of Melamchi are provided with improved health and education services
South–North Transfer Project in China	This project began in 2003 and will take 50 years to complete, costing up to US$100 billion. It will transfer 44.8 billion m³ of water per year from the relatively water-secure south to the drought stricken north via 1300 km of canals linking the Yangtze to the Yellow, Huai and Hai rivers
Proposed schemes	
Ebro scheme in Spain	Following on from the Tagus scheme, 828 km of canals will be built to divert the waters of the Ebro to southern Spain; now aborted and replaced by 20 desalination plants
Israel – transfer from any neighbours who would agree	Israel has a huge water deficit and plans for several schemes, such as transferring water from the Red Sea to top up the Dead Sea
Projected water transfer systems in Russia	Russia plans a whole series of schemes, diverting rivers such as the Ob to the drought stricken area of the Aral Sea; the diversions could have major implications for the Arctic Ocean as it would affect salinity; still at planning stage
Projected water transfer systems in India	India plans to develop a national water network to ensure better distribution of supplies to water-deficit areas such as the Deccan Plateau; still at planning stage because of interstate conflicts
Projected transfer projects in North America	Canada is a water-surplus country. NAWAPA is a scheme to take water from Alaska and northwest Canada to southern California and Mexico. A further scheme (Grand Canal) could take water from the Hudson Bay to the Great Lakes

multipurpose schemes with multiple benefits, providing irrigation, HEP and flood control, as well as domestic water supply. Well-known examples include the dams along the Colorado, the Three Gorges Dam in China, and the Aswan High Dam.

But what are the real costs of those mega dams? The World Commission on Dams report in 2000 (found at www.internationalrivers.org/campaigns/the-world-commission-on-dams) evaluated the key issues by looking at the economic, environmental and social costs.

Recently, the mega dam is 'back in fashion', with many new and ever-bigger dams planned for developing countries. There are several reasons for this:

- China is the world's leading dam builder (46 per cent of the world's total) and, as part of its FDI programme in Africa, is building many huge dams.
- With rising global concerns about global warming from greenhouse gas emissions, there is a move to produce more clean energy from renewables.
- In developed countries, most of the most technologically attractive sites have been used, whereas along rivers such as the Congo (Africa) or the Panama Basin in South America, there are plentiful sites to construct dams, largely for HEP development to fuel economic growth.

China's South–North Transfer Project

The south of China is rich in water resources but the north is not. To redistribute these resources and to even out the availability of water, a gigantic south–north water diversion project was begun in 2003. It is expected to take 50 years to complete and will cost close to US$100 billion. The project involves building three canals that run 1300 km across the eastern, middle and western parts of China and link the country's four major rivers: the Yangtze, Yellow, Huai and Han (Figure 3.19).

The scale of engineering involved in this scheme is awesome. It will transfer a total of 44.8 billion cubic metres of water per year. Central government will provide 60 per cent of the cost of the scheme, with the rest coming from local authorities, which, in turn, will charge domestic and industrial users. Water conservation, improved irrigation, pollution treatment and environmental protection are included in the plans.

Critics are concerned about the uncertainties and risks associated with the project. These include the likelihood of significant ecological and environmental impacts along the waterways, resettlement issues and worsening water quality. The Yangtze River is already severely polluted, and the water of the Yellow River is undrinkable. Some experts fear an ecological disaster.

Figure 3.19 China's South–North Transfer Project

Desalination

There has been a global boom in desalination, which draws from supplies from the ocean as opposed to from the 1 per cent of freshwater supplies available for use on Earth. It is, therefore in one way, a sustainable process as it conserves supplies for future generations. Although people have been desalinating water for centuries, recent breakthroughs in technology (for example, the development of the reverse osmosis process) have made desalination far more cost effective (given that freshwater exploitation costs are rising), less energy intensive and easier to implement on a large scale. However, it is still a costly option and does have a major ecological impact on marine life.

TNCs such as Veolia (France), Salini Impregilo (Italy), Doosan (South Korea) and GE (USA) are building desalination plants around the world. The top six nations by desalination capacity are (greatest first) Saudi Arabia, the USA, UAE, Spain, Kuwait and China (Figure 3.20 (page 76) shows the growth of global desalination since 1990).

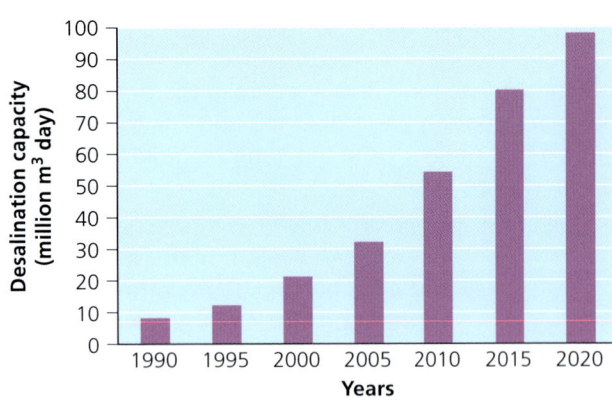

Figure 3.20 Global desalination capacity, 1990–2020

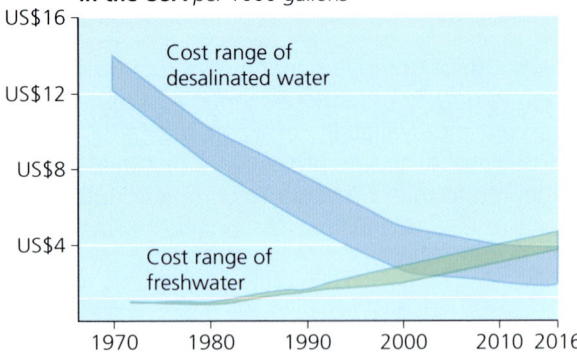

Figure 3.21 The converging costs of water provision

With the future advent of carbon nanotube membranes requiring less pressure and therefore greater energy efficiency, relative costs of desalination may decrease even further, making this option – in effect – the ultimate techno-fix. It is more viable than massive hard-engineering water transfers, but there are major concerns about its environmental impact. Left-over water returned from the desalination process has twice the salt concentration of sea water. Dumping it near the shoreline will have adverse consequences on coral reefs and their food webs.

> A number of US cities have outgrown or exhausted their local freshwater sources. To continue to deliver water to residents, their local governments have to pipe water in from further away or pump from greater depths. This infrastructure is expensive and has led to rising prices.
>
> The cost of desalinating water has come down by about 4 per cent each year. Improvements in technology mean that desalination is beginning to catch up with the increased cost of freshwater. However as fossil fuels are frequently used in the process, there is a large carbon footprint (Figure 3.21).

Many of the countries involved are relatively well developed technologically and increasingly water stressed. The OPEC countries of Saudi Arabia, UAE, Kuwait and Libya have the additional advantage of cheap energy to distil freshwater from the oceans; however, this involves the use of fossil fuels. Currently, there are 20,000 desalination plants in the world, including one in the Thames estuary.

In conclusion, technological innovations and adaptations have a role to play in meeting the mounting threats to freshwater supplies that should not be overlooked; it should not be overstated either, as these hard solutions cannot be regarded as fully sustainable.

The role of more sustainable solutions to managing water supply

Defining sustainable water management

Figure 3.22 uses the **sustainability** quadrant to focus on water management. Increasingly, the concept of water sustainability is enshrined and embedded in the notion of water security. For all the world's nations, communities and peoples, clean, safe water should be available, accessible *and* affordable (the three As).

Futurity Energy efficiency of development and economy of use (conservation) to manage demand yet at the same time ensuring security of supplies for the future.	**Environment** Achieving high standards of environmental protection. Restoration of damaged water supplies.
Public participation Involvement of communities. Decentralised decision-making to ensure bottom up appropriate technology solutions, e.g. from NGO WaterAid.	**Equity and social justice** Equitable allocation between users to ensure secure supplies at affordable prices delivered by good governance and management.

Figure 3.22 The water sustainability quadrant

Figure 3.22 will help you to consider what this actually means, and whether sustainable development, management and use can be achieved, to safeguard the resource for future generations.

> **Key concept: Sustainability of water**
>
> Environmental sustainability is a major issue. Many of the world's rivers are ecologically threatened as a result of human actions, which have polluted and damaged water supplies. The water is of poor quality and acts as a vector for water-borne diseases. Nearly 25 per cent of the world's peoples lack access to safe water, so environmental sustainability protects water quality.
>
> Economic sustainability involves guaranteeing security of access to water for *all* groups at an affordable price. Interestingly, many schemes to manage rivers, such as mega dams, actually dispossess people of their land, homes and livelihoods. Privatised schemes to bring clean, safe water to millions often fail to deliver at affordable prices. Economic sustainability is also achieved by minimising wastage and maximising efficiency of usage, for example, in irrigation.
>
> Socio-cultural sustainability manages water supplies in such a way that it takes into account the views of *all* users, including poor and disadvantaged people, and leads to equitable distribution within and between countries.

Water conservation

Conservation of water supplies is one of the main thrusts of the sustainable use of water as it manages demand. This can be done in a number of ways in the various economic sectors. As agriculture is the main user of water supplies (70 per cent), especially in semi-arid areas for irrigation, there are many ways that the use of water can be made more efficient.

In agriculture the maxim has to be 'more crop per drop' where cash crops are grown. Sprinkler and surface flood irrigation systems are steadily being replaced by modern automated spray technology and more advanced drip irrigation systems, which use less water. Israel is a major pioneer of water conservation. There are also great savings to be made in repairing leaks in irrigation systems.

For many farmers operating in areas of water scarcity, such as northern China or western USA, there is a pressing need to make water go further. Recycling of city waste water for agricultural use is a feasible, relatively low-cost option as this **grey water** does not need to be of drinking-water quality. This recycling happens regularly on the North China Plains.

Empowering farming communities to make their own decisions concerning water use has also been successful. There are numerous intermediate-technology solutions

> **Key term**
>
> **Grey water:** Refers to waste bath, sink or washing water. It can be recycled, resulting in savings in water usage.

to water conservation, such as the 'magic stones' system practised widely across the semi-arid Sahel (lines of stones 5–10 cm high are laid along the contours of a hill to prevent soil erosion and conserve soil moisture) or the development of devices to store and recycle rain in areas reliant on rains (for rain-fed agriculture). Rainwater harvesting experiments in Uzbekistan put farmers in control of the irrigation network and allowed them to decide how much water they needed, as opposed to giving them a fixed allocation. This cut consumption by 30 per cent.

Specialised NGOs such as Farm Africa and WaterAid have helped farming communities develop a whole range of low technology strategies to combat climate change-induced water scarcity. Farmers are trained in minimising tilling so that water is conserved in a layer of mulch on the field's surface, which absorbs the rainwater and limits evaporation. Agriculture advisers give guidance on types of crops that will generate good profits yet use less water, for example, substituting dry crops such as olives for thirsty citrus fruits. One current controversy is the thirstiness of crops used to produce supposedly sustainable biofuels such as bioethanol and biodiesel.

High technology also has a role to play. Second-generation genetically-modified (GM) crops are being bred that are not only tolerant of diseases but also of drought and salty conditions – these include strains of maize, millet and wheat, which are vital food crops.

Agronomists are also beginning to devise tools to help monitor the efficiency of water use. Some have designed algorithms that use satellite data about surface temperatures to calculate the rate at which plants are absorbing and transpiring water – this means development agencies can concentrate their efforts for improvement on the most thirsty crops.

> **Key term**
>
> **Hydroponics:** A method of growing plants using mineral nutrient solutions without soil.

Systems have also been devised to grow crops using little water. **Hydroponics** involves growing crops in huge greenhouses that are carbon dioxide and temperature controlled. The crops are grown in shallow trays where they are drip-fed nutrients and water (there is no soil). The only issue is that while it may be a sustainable system water-wise for supplying food to arid lands, it is very energy intensive.

For businesses, water is not discretionary as without water, industry and the global economy falter. Water is an essential ingredient for many food and beverage products, such as beer and soft drinks. It is also used in a huge range of other industries such as making silicon chips and for cooling thermal power stations.

Rapid industrialisation, particularly in developing countries, has contaminated both rivers and aquifers, and for many industries it is not so much the quantity of the water but the quality of the water that is important. Many large TNCs have reduced their consumption of water; for example, Coca-Cola bottling plants around the world committed to clean all their waste water by 2010 and then to recycle some of it for use as grey water in their plants for cleaning bottles and machinery. Coca-Cola is currently the largest beverage company in the world and has been the subject of adverse publicity over its intensive use of water (283 billion litres worldwide).

Many companies have improved their recycling of water as a response to legislation prohibiting the use of groundwater or due to rising costs in the price of water. In Beijing, in water-stressed northern China, zero liquid discharge rules ban companies from dumping waste water into the environment; this forces companies to recycle all their waste water by purifying it for reuse as grey water.

Conservation has also been very effective in reducing demand for domestic use. Here there is more of an attitudinal fix than a techno-fix, to persuade consumers to use less water, for example, when megacities face Day Zero.

Domestic water conservation includes reducing consumption by the installation of smart meters, which can monitor use and make higher charges in stress periods such as dry summers. Rain harvesting using a system of water butts is a further conservation measure in the garden. Strategies such as sharing a bath, putting a brick in the toilet cistern or using an eco-kettle can also cut down on consumption, often inspired by the threat of rising costs of metered water. In times of drought, water conservation can be enforced by hosepipe and sprinkler bans. The use of recycled water can be encouraged for flushing the toilet or for garden use such as watering plants with leftover washing-up water. Another development concerns the construction of climate-proofed gardens filled with drought-resistant species that can survive periods of water stress.

Filtration technology now means that there is very little dirty water that cannot physically be purified and recycled. Faced with the loss of cheap imports of water from Malaysia, Singapore has followed a path to water self-sufficiency – artificial rain catchments combined with treating sewage water. Water cleaned by a combination of dual membrane technology (microfiltration and reverse osmosis) and ultraviolet disinfection produces water that exceeds WHO quality thresholds. It is marketed as Newater and is now a key source of supply for the densely populated island (in 2016 some 40 per cent of Singapore's drinking water was Newater). However, a psychological barrier has to be overcome on drinking water from toilets – another attitudinal fix.

Technology can be useful in a number of ways, such as water companies carrying out projects to cut down on leakage from broken pipes and burst water mains or treating and reusing industrial and waste water at their waterworks.

The main strategies for reducing demands for water – recycling, grey water use and reducing consumption – can clearly make a major contribution to the sustainable management of water. Every little helps!

Restoration

A second thrust towards the more sustainable management of water is the restoration of damaged rivers, lakes and wetlands. A number of management strategies are being used to return water environments to their natural state.

At a local scale, this can involve restoring meanders, replanting vegetation and using sustainable methods to manage water courses to provide an alternative environment for all users. The website of the River Restoration Centre in the UK (www.therrc.co.uk/river-restoration) provides details of a number of schemes, for example, for the River Skerne and the River Cole.

On a larger scale, the US Army Corps of Engineers have finished the restoration of the Kissimmee River in Florida. More than 100 square kilometres of river channel have been restored. Equally successful has been the partial restoration of the marsh area in South Iraq deliberately drained by Saddam Hussein (see page 31).

Currently there are major EU-funded projects in the lower Danube Basin, while the largest project of all is the restoration of the northern part of the Aral Sea in Kazakhstan (see page 62), which is already showing some successful outcomes.

> **Synoptic themes:**
>
> ### Attitudes and actions
>
> Attitudes to water usage vary. Some social player providers such as NGOs and campaigners see provision as a human right whereas politicians see it as a human need which they have to supply.
>
> Businesses rely on secure water supplies at all costs for development, whereas environmentalists are concerned that provision should be sustainable.

> **Key term**
>
> **Integrated water resource management (IWRM):** A process which promotes the co-ordinated development and management of water, land and related resources in order to maximise economic and social welfare in an equitable manner without compromising the sustainability of vital ecosystems.

These schemes are clearly environmentally sustainable and have many socio-cultural benefits to the communities living there, but the question that has to be asked is: how economically sustainable are they?

Co-operation – the way ahead to water security?

Co-operation, as opposed to competition, can take place on two main fronts:

- the management of water demands in an integrated way, known as **integrated water resource management (IWRM)**
- the management of the political aspects of the river basin to ensure that all those competing for the water mutually co-operate over its usage in order to avoid 'water wars'.

Both are key to managing the risks from water insecurity, both present and future, and ensuring equity between users. There are many players operating in any river basin and this is a very complex task.

Integrated water resource management

Many researchers see the real water crisis facing the world as more one of water management. Pressure on future resources requires an efficient and equitable allocation between the rising demands of different types of users and their usage.

First advocated in the late 1990s, IWRM emphasises the river basin as the logical geographical unit for strategic planning, making co-operation between basin users and players absolutely central. The river basin is treated holistically – to protect the environmental quality of the rivers and the catchment, and also to ensure maximum efficiency of usage, and equitable distribution. IWRM encompasses many facets of sustainable water management. The scope of IWRM is shown in Figure 3.23.

Satellite images and water accounting are used to determine how much water there is (at various times of the year), how productively it is currently used, and how this could be improved. Of paramount importance is the way that water is managed

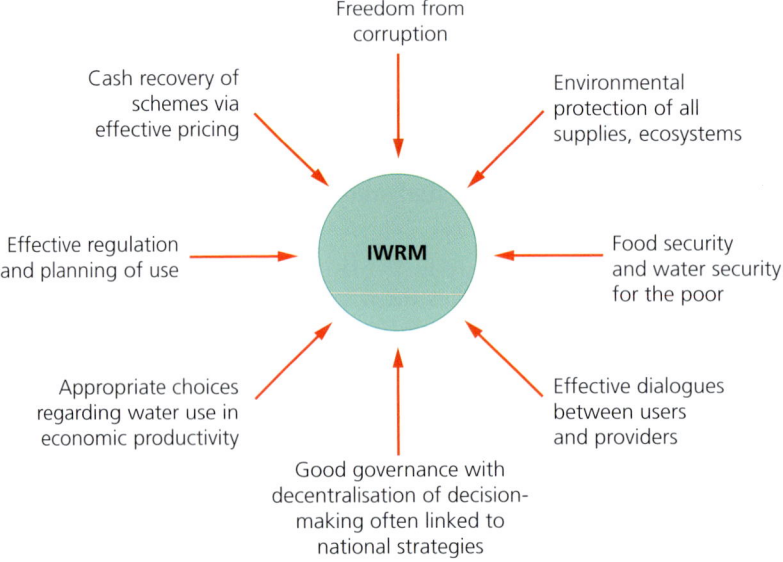

Figure 3.23 The main aims of IWRM

within a basin community and how usage impacts on the basin environment. Figure 3.24 explains the IWRM process.

IWRM is often very practicable at the community level but the larger the basin, the more complex it becomes, especially if a transnational basin is involved.

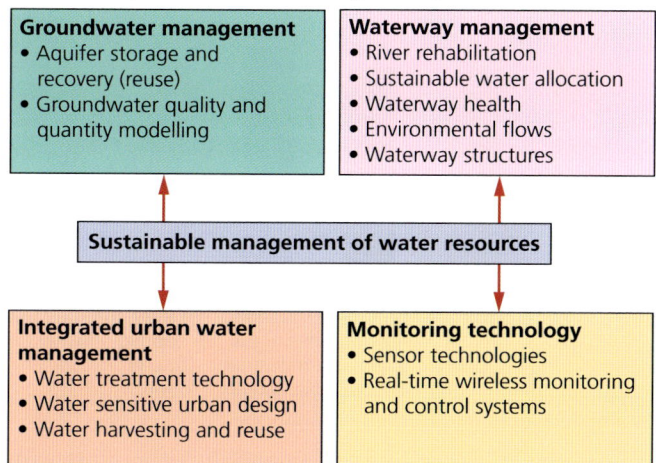

Figure 3.24 The process of IWRM

There are numerous successful local schemes such as those in Gujarat, India, with women taking the lead in inspiring action from local communities:

> 'The village of Thunthi Kankasiya in Gujarat, India, is an example of a locally successful IWRM where the bottom-up approach is designed to overcome water poverty. In the village appropriate technology schemes now provide a year-round water supply, enabling triple cropping and the quadrupling of production per hectare. This raised household incomes five-fold, well above the Indian average.'

The Atlas of Water: Mapping the World's Most Critical Resource, 2009

Some countries, such as Uganda, have developed a national blueprint but used a decentralised system for each catchment involving the regulation of water use through a permits system.

Water-sharing treaties and frameworks

In spite of the potential for hostilities over shared waters, particularly in retaliation against 'greedy' upstream behaviour, international co-operation is far more often the rule than the exception, with actual military conflict occurring in only 0.5 per cent of all drainage basin disagreements over river use in the last 60 years (Figure 3.25).

This is a more optimistic scenario than the water wars predicted by many students of hydropolitics. Even nations that are traditionally enemies are sharing river basins, for example, India and Pakistan. The Indus Waters Treaty, agreed in 1960, has been honoured in spite of the two countries going to war twice since its signing – so there are signs for optimism.

As a continent Africa has the most politically dispersed rivers and lakes due in some part to historical reasons with borders being drawn up by European colonial powers to enable trade routes (90 per cent of all surface water in this continent is in transboundary basins). Even the Nile Basin (see page 68), which is the hardest to sort out, is moving very slowly towards the Nile Basin Initiative to share the waters

---

Colorado in crisis – could IWRM work at a large scale?

A useful case study to research at a larger scale concerns the Colorado River Basin in the USA. This river drains 7 per cent of the USA. Throughout the twentieth century, numerous treaties and agreements were needed to allocate 'fair shares' of its water to the seven surrounding US states, plus Mexico.

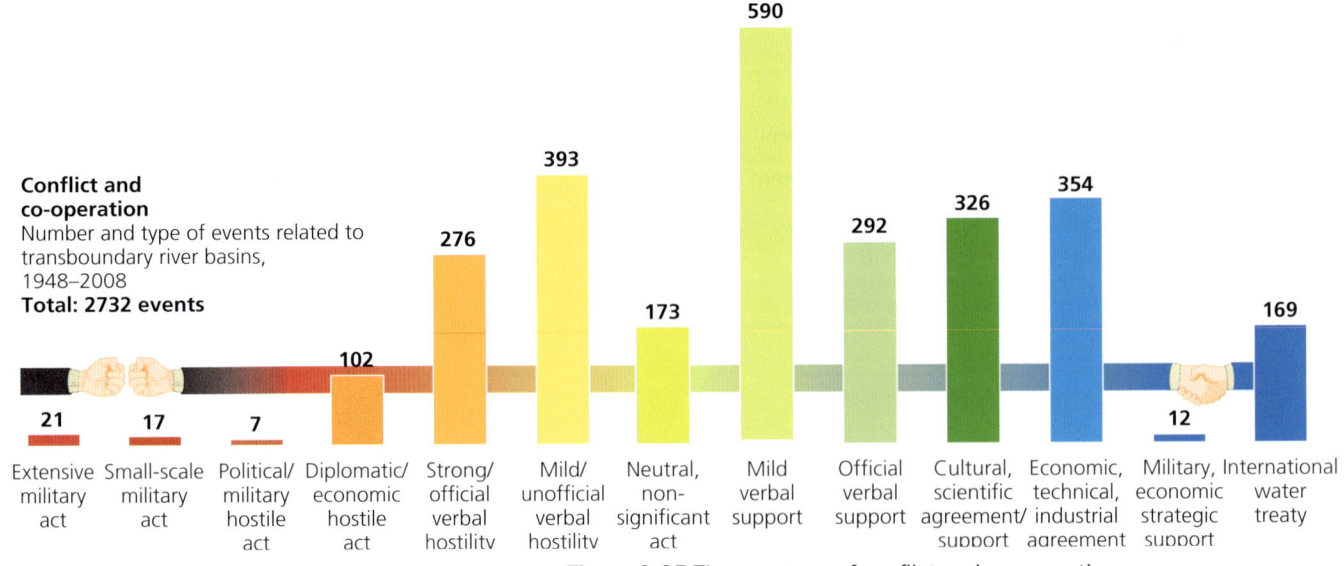

Figure 3.25 The spectrum of conflict and co-operation

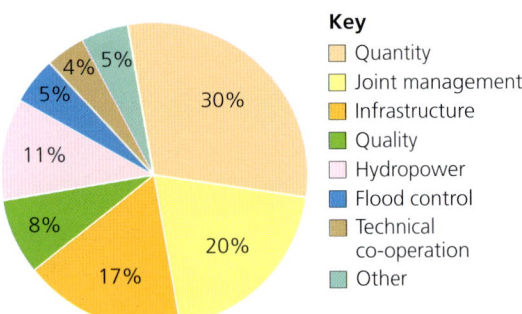

Figure 3.26 Water co-operation and conflict by type of issue

Synoptic themes:

Players

Players can reduce conflict risk at all scales. At a global scale, the UN sets rules such as the Helsinki or Berlin frameworks. However, this concerns NGOs such as the WWF as these frameworks are not regulated. Locally, a range of players are involved in IWRM including planners, environmentalists, water companies and water users, to minimise conflicts and develop co-operation.

more equitably between Egypt and Sudan and the other upstream countries, such as Ethiopia. Although the dispute over GERD is unresolved. Globally, co-operation has occurred across a whole range of issues (Figure 3.26). A number of semi-regulatory frameworks have been developed to share waters.

Under the Helsinki Rules (1966) there is general agreement that international treaties *must* include concepts such as 'equitable use' and 'equitable shares', and be applied to whole drainage basins, not single countries. The criteria for water sharing should be based on:

- natural factors – rainfall amounts, discharge along water sources, share of drainage basin, impact of climate change
- social and economic needs – population size, welfare of people, development plans
- downstream impacts – restructuring flow, lowering water tables, pollution
- dependency – availability of alternative sources
- prior use – the tricky question of (past) existing historic rights or potential future use
- efficiency – avoiding waste and mismanagement of water.

However, these guidelines are not fully backed by compulsory regulation.

The UNECE Water Convention promotes joint management and conservation of shared freshwater ecosystems in Europe and neighbouring regions. It was followed by various EU water frameworks and directives on issues of pollution and hydropower.

The UN Water Courses Convention offers guidelines on the protection and use of transboundary rivers. From 2005 to 2015 was designated the International Decade for Action 'Water for Life' by the UN.

However, the WWF, a leading environmental NGO, states that most of these agreements showed some gaps in coverage of issues and, above all, lacked appropriate enforcement mechanisms or monitoring provision. Moreover, they only apply to 40 per cent of the world's rivers with some, such as the Amazon and Zambezi, being outside any frameworks. Inevitably, climate warming is placing great strain on many agreements as the availability and reliability of supply of water is decreasing in many places.

Conclusion

Table 3.9 summarises four approaches to managing the rising risks of water insecurity. The question is: which of these approaches should be approved and why?

So, is there a water crisis? Do you think water will be the 'new oil' in terms of wars in the twenty-first century?

Table 3.9 Approaches to managing the rising risks of water insecurity

Water crisis	Business as usual	Sustainable management	Radical action
Increased use across agriculture, industry and domestic water with BRICS and NICs driving water consumption 'sky high'	Overall consumption rises up to 50% as exactly the same pattern of use continues	Global water consumption stabilises; no major food security issues; increasing costs control supplies	Strict control of water allocations, especially for agriculture (see northern China), averts water crisis backed up by enforcement
Food production will decline dramatically as surface and groundwater supplies run out; rising food prices	Some reduction in food security in developing countries; strategies in southwestern USA, China, India, North Africa and Egypt	Use of sustainable techniques of water management, e.g. Israel	Emphasis on water conservation
Dramatic decline of water supplies in China, India, parts of Central Africa, Latin America, Spain and even southern England	Issues of weak management persist; prevents efficient and equitable allocation between users and consumers	Increasing involvement of players in research of crop types and improved smart management	Water supplies keep pace with increasing demand and the impacts of climate change and variability
Conflicts possible, even wars	Potential conflicts	Most disputes solved by negotiation	Enforced sharing by legislation

Review questions

1. Using Figure 3.2 on page 50, explain the reasons for the trends in freshwater withdrawals shown. Hint: look at both totals and percentage increases.
2. Using Figure 3.4 on page 52, suggest reasons for the contrasting water profiles.
3. Using Figure 3.7 on page 56, calculate the water poverty index for the country shown. What does the profile suggest about the country?
4. Research the global bottled water industry and the high price paid for it. What are your feelings on the bottled water issue?
5. Read through the text on hard-engineering solutions (page 72). Using the sustainability quadrant (Figure 3.22, page 77), assess the extent to which any of the three solutions could be considered sustainable.
6. Study Table 3.9 and evaluate the approaches to managing the world's impending water security crisis. Which one would you choose and why?

Further research

World Wide Fund for Nature (WWF): www.panda.org
World Resources Institute: www.wri.org
United Nations Environmental Programme: https://www.unep.org
AQUASTAT, the FAO's global water information system: www.fao.org/nr/water/aquastat/main/index.stm
Website of International Rivers: www.internationalrivers.org
WaterAid: www.wateraid.org
WHO, a good resource for information on water-related diseases: www.who.int
Special Report on Thirsty planet, April 2019, from *The Economist*

Exam-style questions

1. **a** Label three features (from A–F) of the soil moisture budget graph, Figure 3.27. [3]

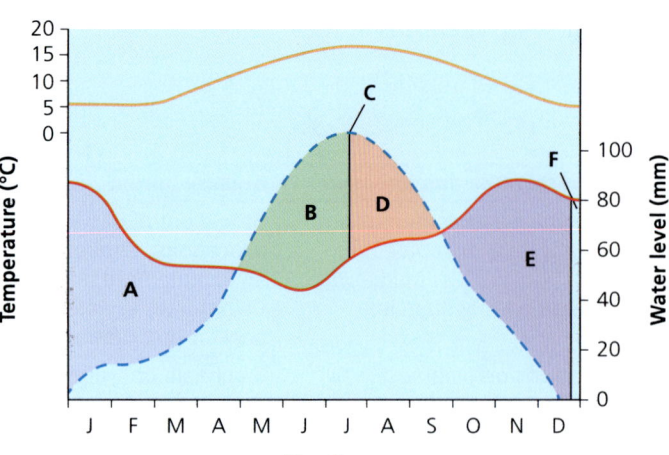

Figure 3.27 A soil moisture budget graph

   b Explain the impact climate type can have on soil water availability. [6]

2. Explain how climate change might have significant impacts on the operation of the water cycle. [8]

3. Assess the extent to which some approaches to future water supply management are more sustainable than others. [12]

4. Evaluate the extent to which conflicts might occur between users within a country, and internationally, over the use of water and energy. [20]

Topic 6
The Carbon Cycle and Energy Security

Chapter 4: The carbon cycle and planetary health

Chapter 5: Consequences of the increasing demand for energy

Chapter 6: Human threats to the global climate system

4 The carbon cycle and planetary health

> **How does the carbon cycle operate to maintain planetary health?**
> By the end of this chapter you should:
> - understand that the carbon cycle operates as a system, at a range of spatial scales and timescales
> - understand the geological and biological processes that control carbon movements between its stores
> - be aware that humans are having an increasing impact on natural carbon cycle functioning, especially through fossil fuel consumption
> - be aware that a balanced carbon cycle is important in maintaining planetary health.

Synoptic themes:

Players

Measuring the stores and fluxes of the carbon cycle, especially oceanic ones, is difficult for the players involved. Local ecosystem studies are 'scaled up' to give wider estimates of stores and fluxes. Technology is improving: supercomputers can process remote sensing and ocean drone data.

Key term

Carbon cycle: The biogeochemical cycle by which carbon moves from one sphere to another. It acts as a closed system made up of linked subsystems that have inputs, throughputs and outputs. Carbon stores function as sources (adding carbon to the atmosphere) and sinks (removing carbon from the atmosphere).

4.1 Terrestrial carbon stores

This section considers the role of land-based processes in the carbon cycle, focusing on slow movements and longer-term stores of carbon.

Cycles, stores and fluxes

Carbon is called the main 'building block of life'. It is present in the stores of:

- the atmosphere, as carbon dioxide (CO_2) and compounds such as methane (CH_4)
- the hydrosphere, as dissolved CO_2
- the lithosphere, as carbonates in limestone and fossil fuels such as coal, oil and gas
- the biosphere, in living and dead organisms.

Carbon moves from one sphere to another by linked processes known as the biogeochemical **carbon cycle**. This includes every microbe, leaf, puddle, grain of rock, dead being and volcanic eruption. Complete decomposition of organic matter results in carbon returning to inorganic forms such as CO_2 and carbonates contained in rock and seawater. Processes including photosynthesis and diffusion drive the flows or fluxes between the stores, operating at local and global scales.

If sources equal the sinks, the carbon cycle is balanced, or in equilibrium, with no change in the size of the stores. Changes in the system may result in negative or positive **system feedback**.

Key concept: System feedback

Earth systems normally operate by negative (stabilising) feedbacks, maintaining a stable state by preventing the system moving beyond certain thresholds. Any change is cancelled out, maintaining equilibrium. Positive (amplifying) feedback loops occur when a small change in one component causes changes in other components. This shifts the system away from its previous state and toward a new one.

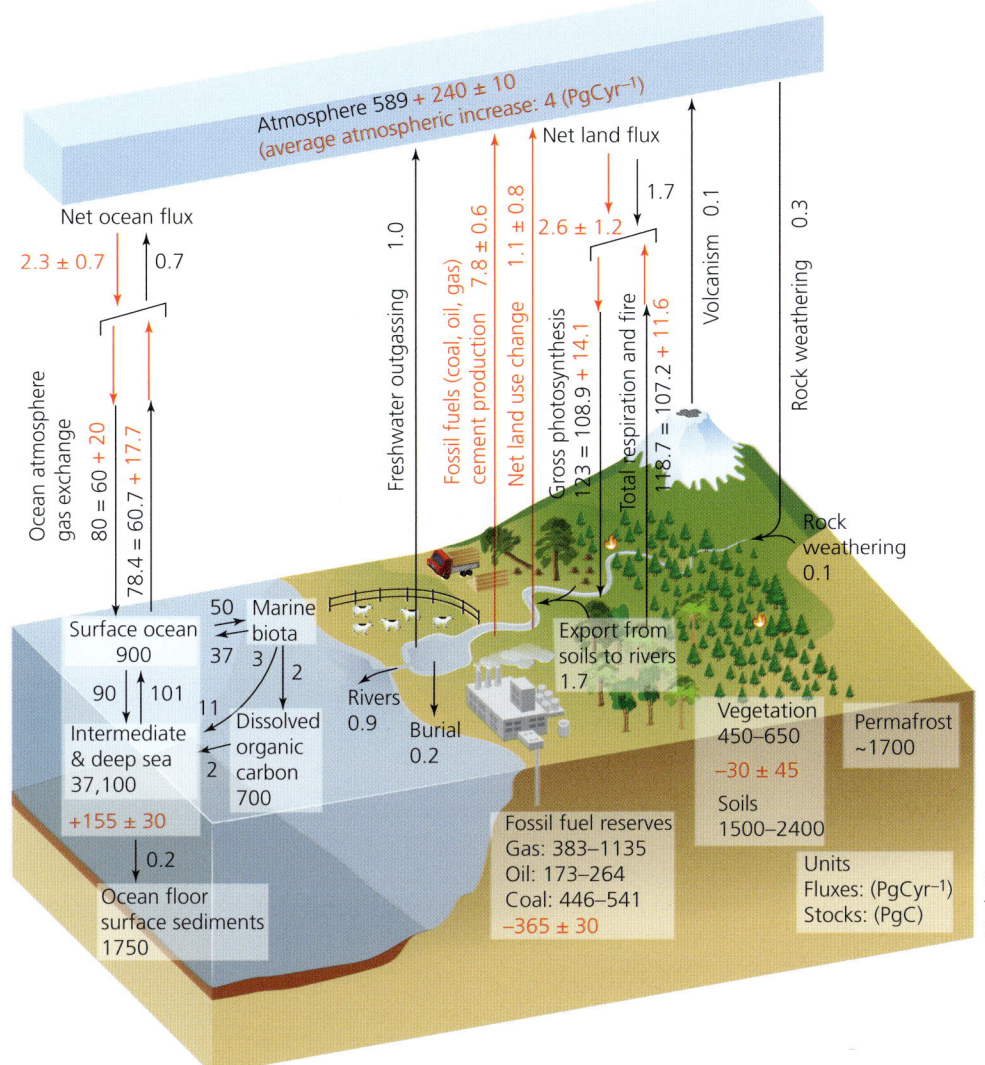

Figure 4.1 The carbon cycle as modelled by the IPCC

Figure 4.1 shows the main stores and fluxes of the carbon cycle before (in black) and after (in red) major **anthropogenic** (human) influences. The numbers represent estimated carbon pool sizes in PgC (**petagrams** of carbon) and the magnitude of the different exchange fluxes in PgC yr^{-1} (petagrams of carbon per year), averaged between 2000 and 2009.

One of the most important drivers of the carbon cycle is the water cycle (outlined in Chapter 6); for example, run-off and rivers transport eroded rock and soil into oceans.

Single carbon stores of the larger cycle can often have several fluxes, adding and removing carbon at the same time.

There are two main components of the carbon cycle.

Key terms

Intergovernmental Panel on Climate Change (IPCC): The leading international organisation for the scientific assessment of climate change.

Anthropogenic: Processes and actions associated with human activity.

Petagrams (Pg) or gigatonnes (Gt): The units used to measure carbon; one petagram (Pg), also known as a gigatonne (Gt), is equal to a trillion kilograms, or 1 billion tonnes.

4 The carbon cycle and planetary health

> **Key terms**
>
> **Reservoir turnover:** The rate at which carbon enters and leaves a store is measured by the mass of carbon in any store divided by the exchange flux.
>
> **Sequestering:** The natural storage of carbon by physical or biological processes such as photosynthesis.

The geological carbon cycle (slow carbon cycle)

This slow part of the cycle is centred on the huge carbon stores in rocks and sediments, with **reservoir turnover** rates of at least 100,000 years. Organic matter that is buried in deep sediments, protected from decay, takes millions of years to turn into fossil fuels. Carbon is exchanged with the fast component through volcanic emissions of CO_2, chemical weathering, erosion and sediment formation on the sea floor.

The biological or physical carbon cycle (fast carbon cycle)

This fast component of the carbon cycle has relatively large exchange fluxes and 'rapid' reservoir turnovers of a few years up to millennia. Carbon is **sequestered** in, and flows between, the atmosphere, oceans, ocean sediments and on land in vegetation, soils and freshwater. Table 4.1 summarises carbon stores in rank order of petagrams.

Fluxes are measurements of the rate of flow of material between the stores. Because fluxes are a rate, the units are mass per unit time. At a global scale they are expressed as Pg per year: $PgC\ yr^{-1}$, or a $GtC\ yr^{-1}$. You will need to be able to construct proportional arrows to show these varying fluxes.

Table 4.1 Carbon stores

Carbon storage times	Store type (before anthropogenic influences)	PgC (average)
Long-term stores: hundreds of years to millennia		
Crustal/terrestrial geological	Sedimentary rocks, very slow cycling over millennia	100,000,000 Fossil fuels store an extra 4,000
Oceanic (deep)	Most carbon is dissolved inorganic carbon stored at great depths, very slowly cycled	38,000
Short-term stores: seconds to decades		
Terrestrial soil	From plant materials (biomass); micro-organisms break most organic matter down to CO_2 in a process that can take days in hot, humid climates to decades in colder climates	1,500
Oceanic (surface)	Exchanges are rapid with the atmosphere through: • physical processes (CO_2 gas dissolving into the water) • biological processes (plankton) Some of this carbon sinks to the deeper ocean pool	1,000
Atmospheric	CO_2 and CH_4 store carbon as greenhouse gases with a lifetime of up to 100 years	560
Terrestrial ecosystems	CO_2 is taken from the atmosphere by plant photosynthesis; carbon is stored organically, especially in trees; rapid interchange with atmosphere over seconds/minutes	560

Skills focus: Use of proportional flow diagrams

Using a copy of Figure 4.2, showing carbon fluxes and stores, convert the flux data into proportional flow lines. Use a scale where the thickness of the line represents the amount of flux.

Speed of flux in years	Flux PgC/year	
Very fast less than 1	1. Photosynthesis 2. Respiration 3. Gases from volcanic eruptions	103 50 50
Fast 1–10	4. Surface water and atmosphere fluxes diffusion into ocean 5. Diffusion out of ocean 6. Vegetation to soils decomposition	92 90.3 50
Slow 10–100	7. Weathering and erosion	0.8
Very slow over 100	8. Sedimentation/fossilisation	0.2

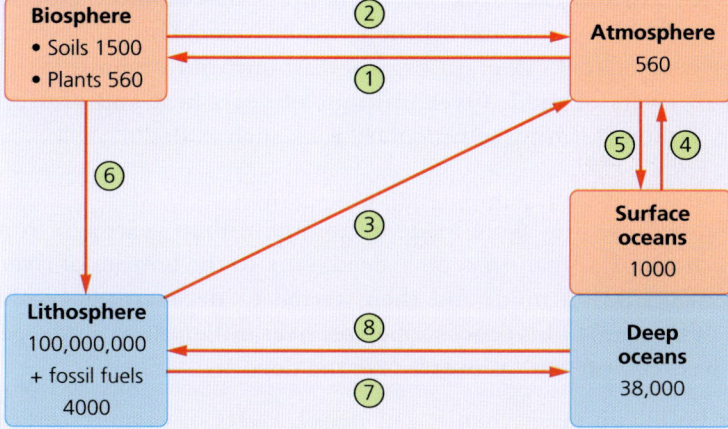

Figure 4.2 Carbon fluxes and stores

Geological origins

Most of the Earth's carbon is geological, resulting from the formation of sedimentary **carbonate rocks** in the oceans, and biologically derived carbon in rocks like shale and coal. Slow geological **processes** release carbon into the atmosphere through chemical weathering of rocks, shown in Table 4.2, and volcanic outgassing at ocean ridges, hotspot volcanoes and subduction zones.

Limestone, shale and fossil fuels are important carbon stores.

Key concept: Geological fluxes

These are small on an annual basis, but without them the carbon stored in rocks would accumulate and remain there forever, eventually depleting the sources of CO_2 that are vital to life forms.

Carbon in limestone and shale

One of Earth's largest carbon stores is the Himalayas, which started off as oceanic sediments rich in calcium carbonate. Folded up by mountain building, this carbon is being actively weathered, eroded and transported back to the oceans.

Key terms

Carbonate rocks: Carbonate rocks contain a high proportion of carbonate ions (CO_2^{3-}) within minerals such as calcite, aragonite and siderite. All limestones (including chalk) and dolomites are carbonate rocks.

Processes: The physical mechanisms that drive the flux of material between stores.

Table 4.2 Key processes in the geological carbon cycle

Processes	Results		
1 Mechanical, chemical and biological weathering of rocks on land *in situ* (on the spot)	**Mechanical weathering** The **breakup** of rocks by frost; shattering and exfoliation produces small, easy-to-transport particles	**Chemical weathering** The **breakdown** of rocks by **carbonic acid** in rain, which dissolves carbonate-based rocks	**Biological weathering** Burrowing animals and the roots of plants can break rocks up
2 Decomposition	Plant and animal particles that result from decomposition after death and surface erosion store carbon		
3 Transportation	Rivers carry particles (ions) to the ocean, where they are deposited		
4 Sedimentation	Over millennia these sediments accumulate, burying older sediments below, such as shale and limestone		
5 Metamorphosis	Deep burial of sedimentary rocks combined with compression due to plate tectonics causes sedimentary rocks to be altered by heat and pressures into metamorphic ones such as slate and marble: shale becomes slate and limestone becomes marble		

In the oceans today, 80 per cent of carbon-containing rock is from shell-building (calcifying) organisms (corals) and plankton. These are precipitated onto the ocean floor, form layers, are cemented together and lithified (turned to rock) into limestone. The remaining 20 per cent of rocks contain organic carbon from organisms that have been embedded in layers of mud. Over millions of years heat and pressure compress the mud and carbon, forming sedimentary rock such as shale.

Carbon fossil fuels

Fossil fuels are so called because they were made up to 300 million years ago from the remains of organic material. Organisms, once dead, sank to the bottom of rivers and seas, were covered in silt and mud, and then started to decay anaerobically (without the presence of oxygen). This process operates over millennia. The deeper the deposit, the more heat and pressure is exerted on the deposits.

When organic matter builds up faster than it can decay, layers of organic carbon become oil, coal or natural gas instead of shale. Table 4.3 summarises the processes involved.

Table 4.3 The formation of fossil fuels

Oil and natural gas	Coal
Formed from the remains of tiny aquatic animals and plants Gas and oil occur in porous rocks, migrating up through the crust until meeting **caprocks** Natural gas, such as methane, is made up of the **fractions** of oil molecules, so small they are in gas form not liquid, and usually found with crude oil Other hydrocarbon deposits include oil shales, tar sands and gas hydrates	Formed from the remains of trees, ferns and other plants There are four main types of coal: • **anthracite** is the hardest coal; it has the most carbon and, hence, a higher energy content • **bituminous coals** are next in hardness and carbon content • soft coals such as **lignite** and **brown coal** are lower in carbon (25–35%) and energy potential; these are the major global source of energy supplies but emit more CO_2 than hard coals • **peat** is the stage before coal; it is an important carbon and energy source

Geological processes

Table 4.1 (page 88) showed that slow geological processes are an important control on the carbon cycle. Through a series of chemical reactions and tectonic activity, carbon takes between 100 and 200 million years to move between rocks, soil, ocean and atmosphere. On average, 10^{13} to 10^{14} grams (10–100 million metric tonnes) of carbon move through this slow carbon cycle annually. This compares with the faster carbon cycles of ecosystems (10^{16} to 10^{17} grams annually) and anthropogenic cycles (10^{15} grams annually).

The specification focuses on two specific processes: the chemical weathering of rocks, and volcanic outgassing at ocean ridges and subduction zones, as shown in Figure 4.3.

Chemical weathering

The geological part of the carbon cycle interacts with the rock cycle, a series of constant processes through which Earth's materials change from one form to another over millennia. These processes can be broken down into five phases:

1. In the atmosphere, water reacts with CO_2 to form weak carbonic acid. When this precipitation reaches the surface it reacts with minerals in rocks, slowly dissolving them. Chemical weathering processes release dissolved carbonate **ions** from carbonate rocks like limestone (solution), and ions such as calcium and sodium from **silicate rocks** like granite (hydrolysis).
2. The ions are transported by rivers to the oceans where marine organisms combine calcium ions (Ca^{2+}) and carbonate ions (CO_3^{2-}) into solid calcium carbonate ($CaCO_3$, or calcite) to make their shells. Calcium carbonate also precipitates directly out of ocean water in tropical oceans.
3. Sedimentation of dead shelled organisms and precipitates onto the ocean floor eventually turns calcium carbonate sediment into limestone rock, as sediment is buried and compressed.
4. At subduction zones, plate tectonics moves some carbonate rock into deep geological stores.
5. Some of this carbon rises back up to the surface within magma, then is 'outgassed' as CO_2 and returned to the atmosphere during volcanism, i.e. the eruption of volcanoes at the surface.

Volcanic outgassing

CO_2 exists in the Earth's crust. Disturbance by volcanic eruptions or earthquake activity may allow pulses or more diffuse fluxes into the atmosphere. Outgassing occurs at:

- active or passive volcanic zones associated with tectonic plate boundaries, including subduction zones and ocean ridges
- hotspot volcanoes such as on the Hawaiian islands and Yellowstone in the USA (which has active hot springs and geysers)
- direct emissions from fractures in the Earth's crust.

Volcanoes currently emit 0.15–0.26 Gt CO_2 annually. In comparison, humans emit about 35 Gt, mainly from fossil fuel use, so volcanic degassing is relatively insignificant.

About 70 surface volcanoes are currently active. An example of a major degassing as a pulse is the 1991 eruption of Mt Pinatubo in the Philippines, part of an island arc created by a subduction zone. Such eruptions not only return CO_2 to the atmosphere; the fresh silicate rock erupted starts the carbon cycle again (Figure 4.3).

Volcanically active mid-ocean ridges, such as in Iceland, are found on the growing edges of tectonic plates. Constructive plate boundaries zigzag nearly 60,000 km across

> **Key terms**
>
> **Ions:** Ions are atoms or molecules with a positive or negative charge. Chemical weathering of carbonate rocks (limestone) by solution produces carbonate ions (HCO_3^-) whereas hydrolysis of silicate minerals such as feldspar produces calcium ions (Ca^{2+}) and carbonate ions (CO_3^{2-}). These ions are transported in solution through the hydrological cycle to the oceans.
>
> **Silicate rocks:** Igneous, metamorphic and sedimentary rocks that contain large quantities of silicate minerals such as feldspar, mica, clays and quartz can be called silicate rocks. Examples include granite (igneous), schist (metamorphic) and sandstone (sedimentary). Chemical weathering of silicate minerals frequently produced calcium ions as well as other weathering products.

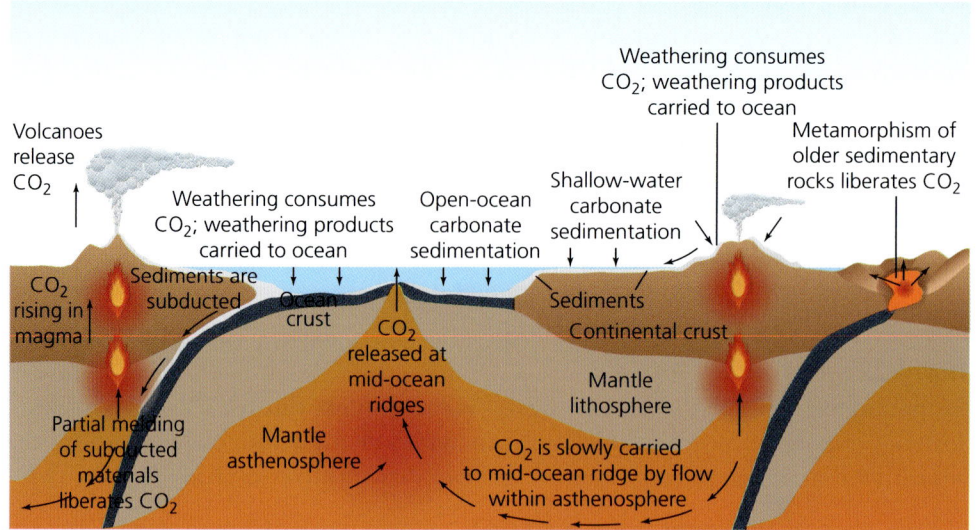

Figure 4.3 The slow geological carbon cycle (Source: Adapted from D. Bice, 'The long-term carbon cycle', 2001)

the sea floor. Currently they seem to be in a fairly languid state, despite producing more lava annually than land volcanoes. Their magmas are basic and low in silicate. Hence, although they produce much lava, the CO_2 emitted is about the same as the smaller number of alkaline magma land volcanoes, about 88 million metric tonnes a year. CO_2 can drive explosive eruptions in normally effusive lava flows.

Two other locations contribute CO_2 to the atmosphere: isolated volcanic hotspots such as Kilauea in Hawaii, and tectonic collision zone volcanoes such as Etna in Sicily.

An interesting negative feedback mechanism regulates the natural geological carbon cycle, shown in Figure 4.4.

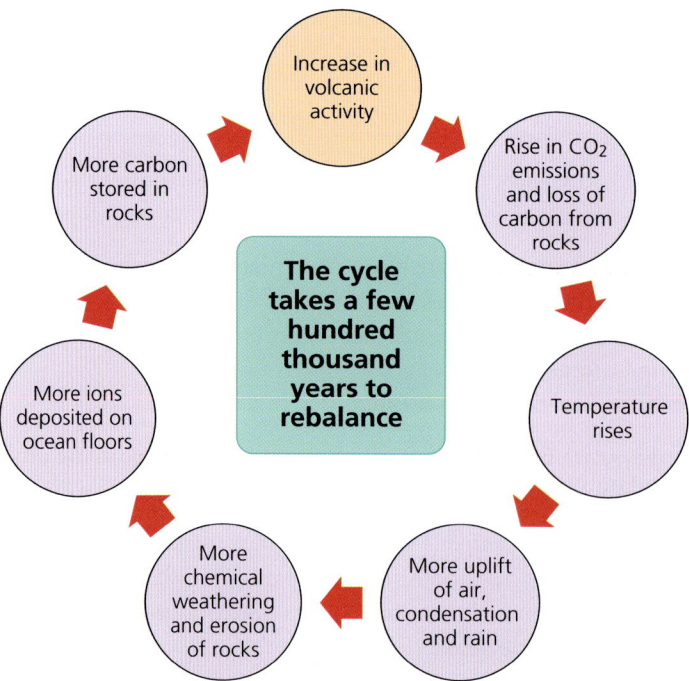

Figure 4.4 Negative feedback regulating the geological carbon cycle

The Carbon Cycle and Energy Security

4.2 Biological processes sequestering carbon

This section outlines the processes involved in the fast carbon cycle, linking the atmosphere, ocean and ecosystems.

Oceanic sequestration

The oceans are the Earth's largest carbon store, being 50 times greater than that of the atmosphere; 93 per cent of carbon is stored in undersea algae, plants and coral, with the remainder in a dissolved form. Small changes in oceanic carbon cycling can have significant global impacts.

The CO_2 gas exchange flux between oceans and atmosphere operates on a timescale of several hundred years. There is also a significant input of both organic carbon and carbonate ions from continental river run-off. Only a small proportion of this carbon is eventually buried in ocean sediments, but these are important long-term carbon stores with fluxes operating over millennia, unlike most terrestrial systems.

Carbon cycle pumps

The specification requires you to understand two key processes in depth when looking at the **carbon cycle pump**: the **biological pump**, and the actions of the linked **carbonate pump** involving **thermohaline circulation**. This circulation is part of a third important process called the **physical pump**. These pumps flux surface ocean CO_2 to the deep ocean, as illustrated in Figure 4.5 and summarised in Table 4.4 (page 94).

> The average depth of the ocean is about 3688 m, with the deepest part in the Mariana Trench (10,994 m) in the western Pacific Ocean. Far less is known about this store than about terrestrial stores.

> **Key terms**
>
> **Carbon cycle pumps:** The processes operating in oceans to circulate and store carbon. There are three sorts: biological, carbonate and physical.
>
> **Thermohaline circulation:** The global system of surface and deep water ocean currents is driven by temperature (thermo) and salinity (haline) differences between areas of oceans.

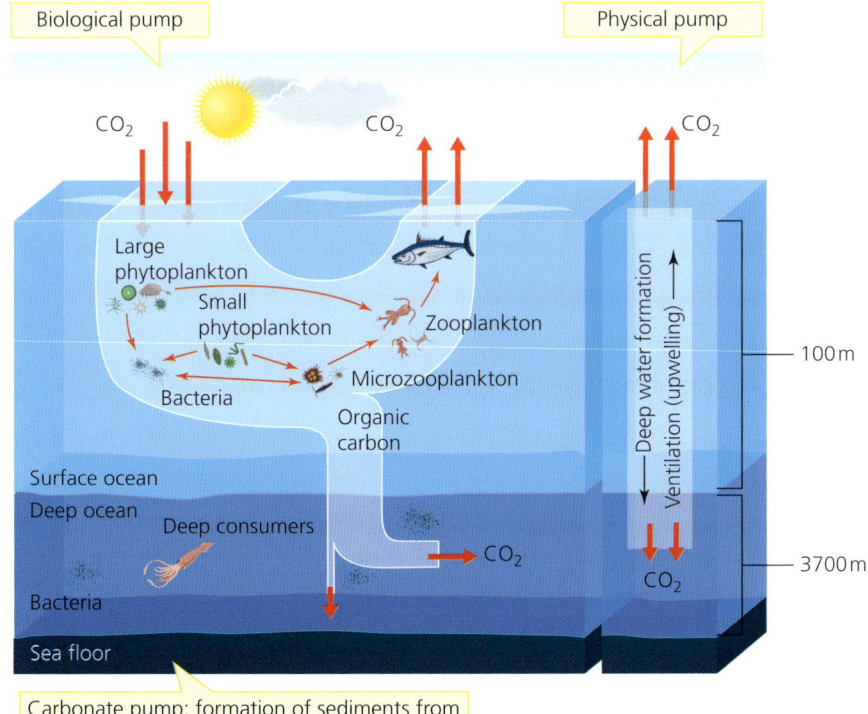

Figure 4.5 Oceanic carbon pumps

Table 4.4 Fact file on oceanic carbon pumps

Biological pump	This is the organic sequestration of CO_2 to oceans by phytoplankton. These microscopic, usually single-celled, marine plants float near the ocean surface to access sunlight to photosynthesise. They are the base of the marine food web. Although minute, their huge numbers make up half of the planet's biomass. Phytoplankton have rapid growth rates, called net primary productivity (NPP), especially in the shallow waters of continental shelves, where rivers carry nutrients far out to sea, and in nutrient upwelling locations. The Arctic and Southern Oceans are very productive areas. Carbon is then passed up the food chain by consumer fish and zooplankton, which in turn release CO_2 back into the water and atmosphere. Most is recycled in surface waters. Only 0.1% reaches the sea floor after the dead phytoplankton sink, where they either decompose or are turned into sediment. Decomposition is faster than on land because of the lack of woody plant structures. Phytoplankton sequester over 2 billion metric tonnes of CO_2 annually to the deep ocean.
Carbonate pump	This relies on inorganic carbon sedimentation. Marine organisms utilise calcium carbonate ($CaCO_3$) to make hard outer shells and inner skeletons, such as some plankton species, coral, oysters and lobsters. When organisms die and sink, many shells dissolve before reaching the sea floor sediments. This carbon becomes part of deep ocean currents. Shells that do not dissolve build up slowly on the sea floor, forming limestone sediments such as those in the 'White Cliffs' of Dover.
Physical pump	This is based on the oceanic circulation of water including upwelling, downwelling and the thermohaline current (Figure 4.6). CO_2 in the oceans is mixed much more slowly than in the atmosphere, so there are large spatial differences in CO_2 concentration. The colder the water, the more potential for CO_2 to be absorbed. CO_2 concentration is 10 per cent higher in the deep ocean than at the surface, and polar oceans store more CO_2 than tropical oceans. Warm tropical waters release CO_2 to the atmosphere, whereas colder high-latitude oceans take in CO_2 from the atmosphere. More than twice as much CO_2 can dissolve into cold polar waters than in warm equatorial waters. As major ocean currents such as the North Atlantic Drift (Gulf Stream) move waters from the tropics to the poles, the water cools and can absorb more atmospheric CO_2. High latitude and Arctic zones with deep oceans have cooler water, which sinks because of its higher density, taking CO_2 accumulated at the surface downwards.

Thermohaline circulation

Thermohaline circulation is a vital component of the global ocean nutrient and carbon dioxide cycles. Ocean currents circulate carbon, with water flows equivalent to over 100 times that of the Amazon River. It takes 1000 years for any given cubic metre of water to travel around the system. Warm surface waters are depleted of nutrients and CO_2, but they are enriched again as they travel through the conveyor belt as deep or bottom layers. The foundation of the planet's food chain depends on the cool, nutrient–rich waters that support algae and seaweed growth. The

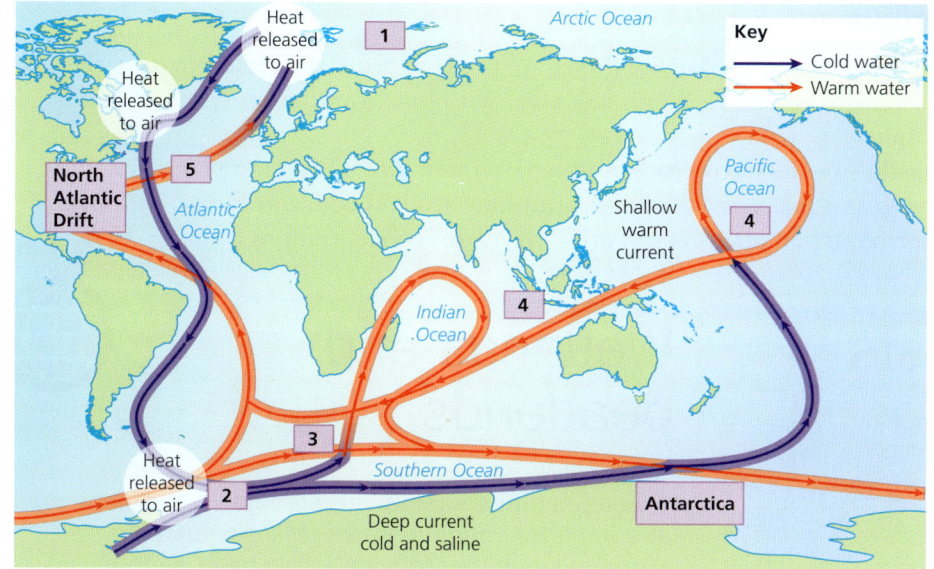

Figure 4.6 The thermohaline conveyor belt

1. The main current begins in polar oceans where the water gets very cold; sea ice forms; surrounding seawater gets saltier, increases in density and sinks.
2. The current is recharged as it passes Antarctica by extra cold salty, dense water.
3. Division of the main current: northward into the Indian Ocean and into the western Pacific.
4. The two branches warm and rise as they travel northward, then loop back southward and westward.
5. The now-warmed surface waters continue circulating around the globe. On their eventual return to the North Atlantic they cool and the cycle begins again.

circulation also helps shift carbon in the carbonate pump cycle from upper to deeper waters. Figure 4.6 shows five distinct phases in this circulation.

The *balance* of total carbon uptake (92 PgC) and carbon loss (90 PgC) from the ocean is therefore dependent on organic and inorganic processes acting at both surface and deep ocean locations. Until the start of the twenty-first century oceans were able to sequester increased CO_2 emissions, but evidence now suggests a slowing of this storage. Increased oceanic acidification, due to increased CO_2, reduces the capacity for extra CO_2 storage. This is discussed further in Chapter 6.

Terrestrial sequestration

This part of the carbon cycle, based on organic carbon, has the shortest temporal (time) scale – only seconds, minutes or years – as shown in Table 4.1 (page 88).

- Primary producers – plants – take carbon out of the atmosphere through photosynthesis and release CO_2 back into the atmosphere through respiration.
- When consumer animals eat plants, carbon from the plant becomes part of its fats and proteins.
- Micro-organisms and detritus feeders such as beetles feed on waste material from animals, and this becomes part of these micro-organisms.
- After plant and animal death, tissues such as leaves decay faster than more resistant structures, such as wood. Decomposition is fastest in tropical climates with high rainfall, temperatures and oxygen levels; it is very slow in cold, dry conditions or where there is a shortage of oxygen. In Arctic biomes, ecosystems are 'locked down' by extreme cold for substantial time periods.

Globally, the most productive biomes are tropical forests, savannah and grassland, which together account for half of global NPP. Figure 4.11 on page 102 shows typical values for different biomes of productivity and carbon storage capacity. Storage is mainly in plants and soils, with smaller amounts in animals, and micro-organisms (bacteria and fungi). The largest store is in trees, which can live tens, hundreds and even thousands of years.

Carbon fluxes vary:

- *diurnally* – during the day the fluxes are positive, from the atmosphere to the ecosystem; at night the flux is negative, with loss from the ecosystem to the atmosphere
- *seasonally* – in the northern hemisphere winter, when few land plants are growing and many are decaying, atmospheric CO_2 concentrations rise; during the spring, when plants begin growing again, concentrations drop.

> Carbon sinks can become carbon sources with anthropogenic influence, for example, by forest burning.

Tropical rainforests

These are one of the largest organic stores of carbon on Earth. The Amazon rainforest alone, covering 5.3 million km², sequesters 17 per cent of all terrestrial carbon, more than any other land-based biome. Some species, such as Brazil nut trees, dominate this process. One per cent of the Amazon's 16,000 tree species store 50 per cent of its carbon, removing CO_2 from the atmosphere for centuries.

Wetlands and peatlands

Wetlands that contain peat, an organic sediment, are important carbon stores. Many peatlands formed during the Holocene have been a store for thousands of years; with climate change and overuse, however, they are becoming net carbon sources.

Biological carbon

Soils store 20–30 per cent of global carbon, sequestering about twice the quantity of carbon as the atmosphere and three times that of terrestrial vegetation. However, whether it sequesters or actually emits CO_2 depends on local conditions.

There are two sources of carbon in soils. Arid and semi-arid soils, and those developed on limestone, contain inorganic carbon. However, the most important store is from organic sources through plant photosynthesis and subsequent decomposition both above and below ground. Living organisms represent about 5 per cent of the total soil organic matter. They have seasonal as well as daily patterns, and not all are active at the same time.

Since all parts of plants are made of carbon, any loss to the ground means a transfer or flux from the plant to the soil. Litterfall and branch litter includes whole plants, leaves and branches shed during any year. Roots may be shed as well. Carbon is stored as dead organic matter in soils for years, decades or even centuries in colder climates or wetland environments, before being broken down by soil microbes and released back to the atmosphere.

After death there are thousands of compounds in the soil to be decomposed. The most long-term process is the formation of humus. This is seen easily in soils as it has a dark, rich colour. Humus soils are 60 per cent carbon and are important for sequestration as well as for water storage.

In general, carbon cycling and formation is most active in topsoil horizons. Stabilised carbon, with longer turnover times, is located in deeper soil layers. In permafrost regions, over 61 per cent of carbon is stored deeper than 30 cm. An additional long-term carbon store in many soils is pyrogenic carbonaceous matter, formed from biomass burned and carbonised during wildfires. This resists microbial decomposition and can remain in soils for long periods.

The capacity of soil to store organic carbon is determined by:

- **Climate**, which dictates plant growth and microbial and detritivore activity. Rapid decomposition occurs at higher temperatures or under waterlogged conditions. Places with high rainfall have an increased potential carbon storage than the same soil type in lower rainfall places. Arid soils store only 30 tonnes per hectare compared with 800 tonnes per hectare in cold regions.
- **Soil type**: clay-rich soils have a higher carbon content than sandy soils. Clay protects carbon from decomposition.
- **Management and use of soils**: since 1850, soils globally have lost 40–90 billion tonnes (Gt) of carbon through cultivation and disturbance. Current rates of carbon loss due to land-use change are 1.6 ± 0.8 billion tones (Gt) of carbon per year.

Rising atmospheric CO_2 has indirect effects on the carbon dynamics and stability of soils, by affecting vegetation and litter stores and flows. For example, the Arctic biome contains one-third of the Earth's soil but with a rapidly warming climate and rising CO_2, so its net storage function may have already 'flipped' from a store to a source.

> **Key concept: Carbon balance**
>
> The carbon stores of the atmosphere, ecosystems and soils are in constant exchange. The carbon balance in soils is regulated by plant productivity, microbial activity, geology, erosion, climate and the amount of upward and downward (leaching) water movement in the soil.

4.3 Human interference

A balanced carbon cycle is important in sustaining other systems. It plays a key role in regulating the Earth's global temperature and climate by controlling the amount of CO_2 in the atmosphere, which then affects the hydrological cycle. Ecosystem development and agriculture depend on the carbon cycle. Carbon stores and fluxes involve natural processes that have helped regulate the carbon cycle and atmospheric CO_2 levels for millions of years. However, as Figure 4.1 (page 87) and Table 4.5 illustrate, the system is being increasingly altered by anthropogenic actions.

> **Synoptic themes:**
>
> **Players, futures and certainties**
>
> Humans have not created more carbon on Earth, but have depleted or enhanced some stores, and speeded up some fluxes. Atmospheric carbon has become a major focus for decision makers because of the role of CO_2 and CH_4 as greenhouse gases. Human interference has consequences for the future climate, ecosystems and food supply.

Table 4.5 Estimates of anthropogenic influence on the carbon cycle; pink = increases, purple = decreases

Stores	Changes to stores by humans from pre-Industrial Revolution, PgC yr^{-1}	Direction of fluxes	Changes to fluxes by humans, PgC yr^{-1}
Atmosphere	240 (average atmospheric increase: 4 PgC yr^{-1})	Ocean to atmosphere	17.7
Vegetation	−30	Atmosphere to ocean	20
Fossil fuels	−365	Land to atmosphere from fossil fuels	7.8
Land	2.6	Land to atmosphere by land use changes	1.1
Ocean	155		

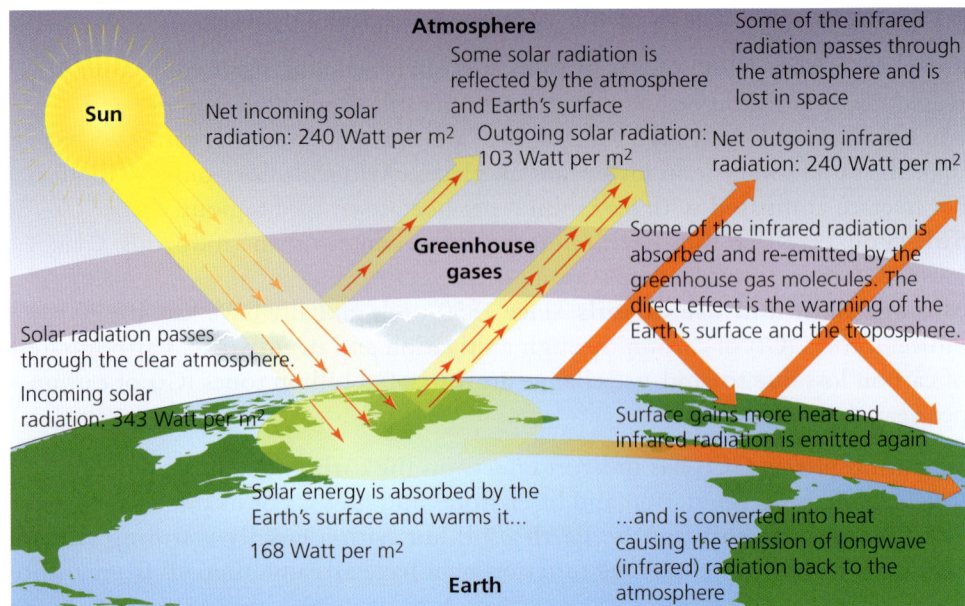

Figure 4.7 The greenhouse effect

The greenhouse effect

The Earth has a natural temperature-control system that relies on greenhouse gases. The concentration of atmospheric carbon (carbon dioxide and methane) strongly influences the natural greenhouse effect (Figure 4.7 and Table 4.6).

The Earth's climate is driven by incoming shortwave solar radiation:

- approximately 31 per cent is reflected by clouds, aerosols and gases in the atmosphere and by the land surface
- the remaining 69 per cent is absorbed; almost 50 per cent is absorbed at the Earth's surface, especially by oceans
- 69 per cent of this surface absorption is re-radiated to space as longwave radiation
- however, a large proportion of this longwave radiation emitted by the surface is re-radiated back to the surface by clouds and greenhouse gases (Figure 4.7); this 'trapping' of longwave radiation in the atmosphere is what gives a life-supporting average of 15°C, the 'natural greenhouse effect'.

In the Earth's past, the carbon cycle has responded to natural climate change driven by variations in the Earth's orbit affecting solar energy. In the Pleistocene era, the northern hemisphere summers cooled and the last Ice Age slowed down the carbon cycle. Increased phytoplankton growth increased the amount of carbon that the ocean took out of the atmosphere. As an example of positive feedback, the drop in atmospheric carbon then caused additional cooling. At the end of the last Ice Age, temperatures rose as did atmospheric CO_2.

The Anthropocene

The current geological era is called the Holocene, which began 12,000 years ago. Some scientists have proposed a new geological era, the Anthropocene, beginning about 8000 years ago when humans first began farming. 'Anthropocene' reflects a view that humans, not natural processes, are now the main driver of many Earth processes.

Table 4.6 Selected greenhouse gases and their relative global warming potential (GWP)

Greenhouse gas	Atmospheric concentration, ppm, ppb and ppt (parts per million/billion/trillion)		Atmospheric lifespan, years	100-year global warming potential (GWP)
	Pre-Industrial Revolution 1000–1750	2018		
Carbon dioxide (CO_2)	280 ppm	408 ppm	50–200	1
Methane (CH_4)	0.7 ppm	1870 ppb	12	23
Nitrous oxide (N_2)	0.270 ppm	331 ppb	114	296
Tetrafluoromethane (CF_4)	40 ppt	80 ppt	Over 50,000	5,700
Sulfurhexafluoride (SF_6)	0	10 ppt	3,200	22,200

Constant levels of atmospheric CO_2 help to maintain stable global average temperatures.

- Fast carbon cycling is thought to have been relatively balanced before the Industrial Revolution, which started in the eighteenth century. It functioned in a 'steady state system'.
- The slow carbon cycle, volcanism and sedimentation, have been fairly constant over the last few centuries, although erosion and river fluxes have been modified by changes in land use.
- Natural exchange fluxes between the slow and fast domains of the carbon cycle were relatively small, at under 0.3 PgC yr^{-1}. Evidence from ice cores shows relatively small variations of atmospheric CO_2 until the late nineteenth century, despite small emissions over the last millennia from land-use changes caused by human activity.

Greenhouse gas increases raise temperatures, which in turn affect precipitation patterns. The temperature at any place depends on the input of solar radiation. Average figures may hide important seasonal differences and also changes over longer climatic periods. Maps and graphs showing anomalies from the average may help.

Atmosphere, plants and soils

The carbon cycle relies on ocean and terrestrial photosynthesis. This section focuses on the role of photosynthesis in regulating the composition of the atmosphere, and how soil health and ecosystem productivity is influenced by stored carbon.

Photosynthesis and the atmosphere

The atmosphere is stratified into a number of different layers. Greenhouse gases absorb radiation from the Sun and maintain the temperature of the Earth. Photosynthetic organisms play an essential role in helping to keep CO_2 levels relatively constant, thereby helping to regulate Earth's average temperature. There are distinct spatial patterns in plant productivity and carbon density (carbon storage), as shown in Table 4.7 (page 101).

Skills focus: Interpreting maps

The specification requires you to practise the geographical skills of analysing maps. Use Figures 4.8 and 4.9, showing global temperature and average precipitation distribution between 1960 and 1990, to practise your skills. Use the acronym **PEA**:

- **Pattern**: describe the big patterns before any details.
- **Evidence**: refer to specific geographical areas and places.
- **Analysis**: suggest a range of reasons.

Focus on physical factors only: solar input, albedo, latitude, continentality, role of ocean currents and altitude.

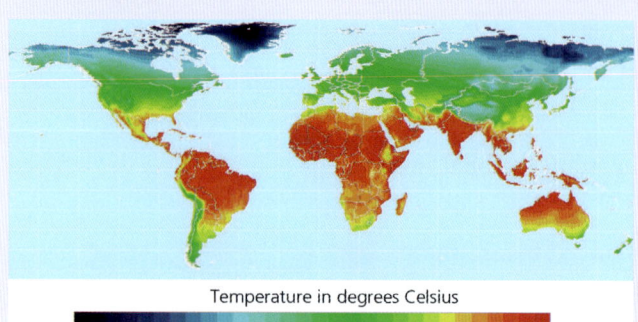

Figure 4.8 Interpreting maps: global temperature between 1960 and 1990

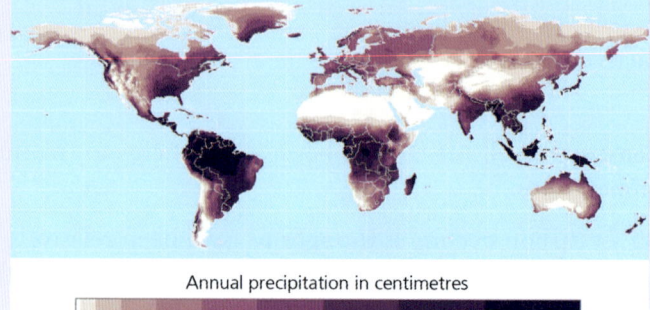

Figure 4.9 Interpreting maps: average precipitation distribution between 1960 and 1990

(Source: Figures 4.8 and 4.9 are used by permission of The Center for Sustainability and the Global Environment, Nelson Institute for Environmental Studies, University of Wisconsin-Madison. Data from Climatic Research Unit, University of East Anglia.)

Climate and nutrients are the main controls on NPP, which is a measure of the size of carbon sink. Highest productivity occurs:

- On land: in areas that are warm and wet. The amount of water available limits primary production; for example, deserts and dry shrub lands have little biomass above ground, although their huge extent nonetheless means a significant store. Forests store the largest amount of carbon collectively. Tundra has the least spatial extent but has the highest density of carbon storage in its permafrost.
- In the oceans: in shallower water, allowing higher photosynthesis, and in places receiving high nutrient inputs.

The rank order of rates of NPP per hectare is: estuaries, swamps and marshes, tropical rainforests, and temperate rainforests. However, when NPP is multiplied by ecosystem extent, the rank order changes to: open oceans, tropical rainforests, savannahs, and tropical seasonal forests.

Ecosystems have varied in their role as a sink or source of carbon, as summarised in Table 4.8 (page 101). Regrowth of forests from past land clearance, discussed in Chapter 6, can increase the carbon sink, but the result of anthropogenic activity on the land globally has increased net carbon fluxes to the atmosphere.

Key concept: CO_2 fertilisation

Anthropogenic rises in CO_2 should speed up the rate of photosynthesis, and hence NPP, by 63 per cent by 2100. However, plant growth is limited by nutrient and water availability (nitrogen and phosphorus), needed in order to utilise CO_2. As a result, the IPCC estimates extra growth rates of only 20 per cent in tropical rainforests, savannahs, boreal forests and tundra.

Table 4.7 Carbon density and ecosystem productivity

Natural biomes	Global NPP, PgC yr^{-1} pink = slow, orange = intermediate, red = fast	Ecosystem NPP, PgC ha yr^{-1}	Total carbon stored by biome, GtC	Where most carbon is stored	Main threats producing potential carbon emission or decreased storage
Tundra	0.5–1	80–130	155.4	Permafrost	Rising temperatures
Boreal forest	2.6–4.6	1773–2238	184.2	Soil	Fires, logging, mining
Temperate forest	4.6–9.1	465–741	160	Biomass above and underground (roots)	Historic losses but now mainly ceased
Temperate grasslands	3.4–7.0	129–342	183.7 Likely sink	Soil	Degradation by livestock and crops
Deserts and dry shrub land	0.5–3.5	28–151	178.0 Uncertain sink	Soil	Fire
Savannah and tropical grasslands	14.9–19.2	345–393	Not estimated	Soil	Fire followed by conversion to pasture/crops
Tropical forests	16.0–23.1	871–1086	547.8	Above ground biomass	Deforestation for logging, agriculture, minerals and hydroelectric power (HEP)
Wetlands/peatlands	Not estimated	Not estimated	550	Soil	Drainage, conversion to agriculture, fire, fuel and soil fertiliser use
Oceans	Not estimated	Not estimated, but plankton have fast NPP	Not estimated	Deep ocean	No emissions but decreasing uptake capacity. Degrading coastal zones becoming sources of carbon

Table 4.8 Changes in the carbon storage of ecosystems

Era	Plants as a net sink/source of CO_2	Data from IPCC, MEA and other sources
Before the nineteenth century	Sink	Until the eighteenth century human disturbance was localised
Nineteenth and early twentieth centuries	Source	Globalising scale of degradation and **destruction of ecosystems**: deforestation, desertification, soil erosion, resource extraction and urbanisation
Mid-twentieth century	Sink	Despite carbon loss from land-use changes, net carbon sequestration because of afforestation and reforestation in North America, Europe and China, as well as improved agricultural practices
2015 onwards	An increasing source?	Warmer temperatures trigger faster decomposition and recycling of carbon in dead plants and soils, fluxing more carbon back into the atmosphere

Soil health

Soil health depends on the amount of organic carbon stored in the soil. This depends on its inputs (plant and animal residues and nutrients) and outputs (decomposition, erosion and use in plant and animal productivity). Figure 4.10 illustrates the stores and fluxes in soil nutrient cycling, while Figure 4.11 shows how these depend on the climate.

> **Key term**
>
> **Millennium Ecosystem Assessment (MEA):** The UN Millennium Ecosystem Assessment was the first major global audit of the health of ecosystems in 2005, highlighting their degradation (the loss of natural productivity through overuse and destruction).

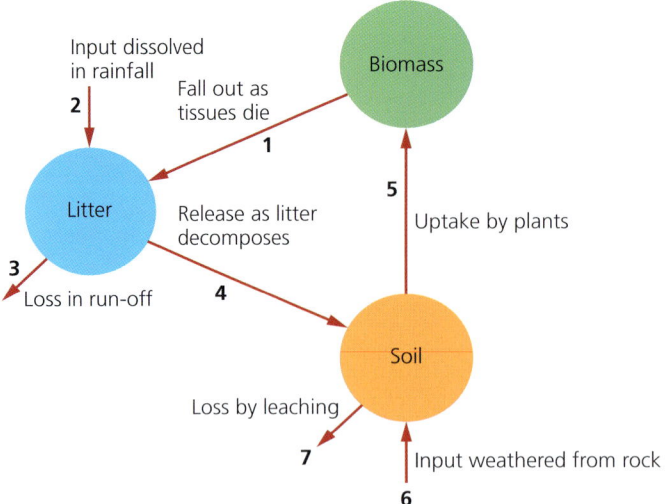

Figure 4.10 Nutrient and carbon cycling in soils

Key concept: Soil carbon balance

If plant residue is added to the soil at a faster rate than soil organisms convert it to CO_2, carbon will gradually be removed from the atmosphere and sequestered in the soil.

Carbon is the main component of soil organic matter and helps give soil its water-retention capacity, its structure and its fertility. This is in contrast to 'active' soil carbon found in topsoil. Organic carbon is concentrated in the surface soil layer as easily eroded small particles, so soil erosion is a major threat to carbon storage and soil health.

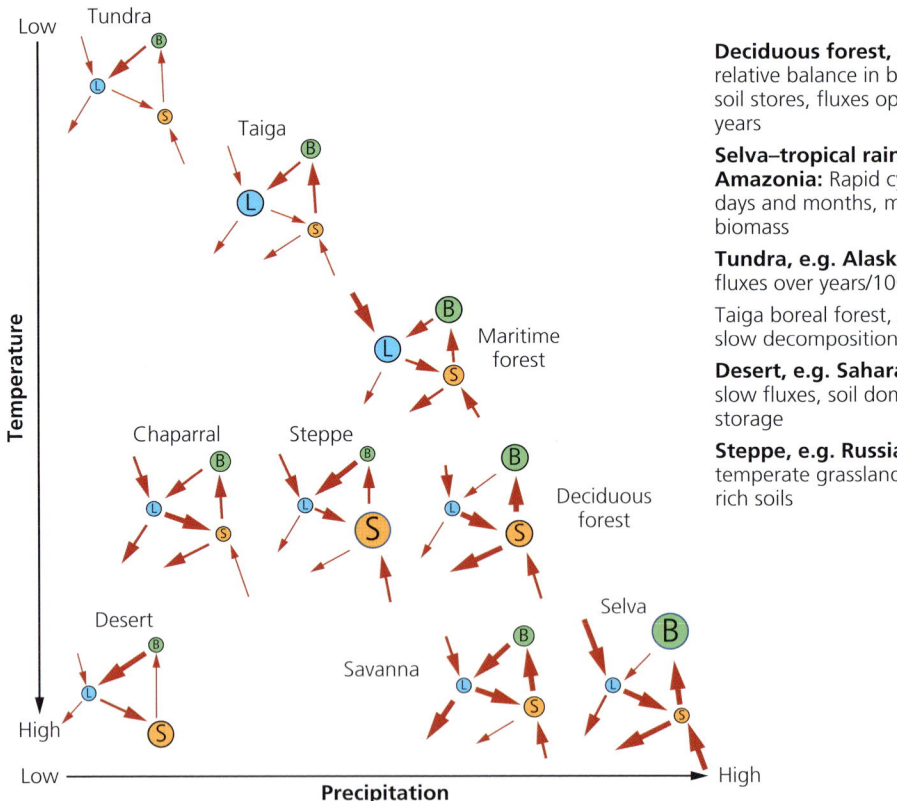

Deciduous forest, e.g. in UK: relative balance in biomass and soil stores, fluxes operate over years

Selva–tropical rainforest, e.g. Amazonia: Rapid cycling over days and months, main store in biomass

Tundra, e.g. Alaska: very slow fluxes over years/100s of years

Taiga boreal forest, e.g. Canada, slow decomposition of litter

Desert, e.g. Sahara: arid areas slow fluxes, soil dominates storage

Steppe, e.g. Russia and USA: temperate grasslands, nutrient-rich soils

Figure 4.11 Nutrient cycling in different biomes

The Carbon Cycle and Energy Security

Fossil fuel combustion

Fossil fuels have been burnt at increasing rates since the start of the Industrial Revolution. They continue to be the primary energy source driving modern civilisation. Without human interference, the carbon in fossil fuels would flux very slowly into the atmosphere through volcanic activity. Fossil fuel combustion shifts this flux from slow to fast carbon cycling.

About half of the extra emissions of CO_2 since 1750 have remained in the atmosphere. The rest has been fluxed from the atmosphere into the stores of oceans, ecosystems and soils shown in Figure 4.1 (page 87). IPCC modelling suggests:

- increased fluxes to the biological store
- increased soil storage in high latitudes, only limited by nitrogen availability
- loss of storage in unfreezing permafrost and in the Southern Ocean and North Atlantic because of warming.

This already changed balance of carbon pathways and stores has varying implications for the climate, ecosystems and hydrological cycle.

Implications for the climate

The impacts of fossil fuel combustion on climate are at global and regional scales. The 2014 IPCC report, aimed at policy makers, explicitly linked greenhouse gas concentrations to fossil fuel emissions, rising global temperatures and sea levels, as shown in Table 4.9.

> **Key concept: Climate forcing**
>
> Climate forcing means the causes, or drivers, of climate change. Currently the most important driver is fossil fuel combustion.

Table 4.9 Climate predictions

Global	Regional
On average, the Earth will become warmer, hence more evaporation and precipitation	Some regions will become warmer and drier, others wetter
Sudden shifts in weather patterns	Some regions will have less snow, more rain
More extreme, intense and frequent events: floods, droughts	Storm surges may increase
Rising mean sea level	

Global warming and the alteration of ocean temperatures and salinity levels could affect the thermohaline current by slowing or reversing the North Atlantic Drift (NAD), also called the Gulf Stream. This happened 20,000 years ago when temperatures dropped. The NAD keeps UK temperatures 5°C higher than they would otherwise be in winter.

> **Key concept: Positive feedback**
>
> Global warming creates ice melt, and permafrost thawing releases trapped methane. Drying forests and warming oceans emit CO_2. Increased greenhouse gases mean increased warming.

Implications for ecosystems

Ecosystems are valued for the services they provide for the planet as well as for humans, helping regulate carbon and hydrological cycles. By the end of the century, global warming and its impacts may be the dominant direct driver of changes in these services and in biodiversity. Already at risk are species with low population numbers, limited climatic ranges and restricted or patchy habitats. There is increasing evidence of changes in the distribution and geographical ranges of species, their population size and timings of reproduction and migration.

Marine organisms are threatened with progressively lower oxygen levels and high rates and magnitudes of ocean acidification, as well as rising temperatures, which may alter the foundation of the food chain: plankton growth. Impacts on coastal ecosystems and low-lying areas at risk from sea level rise will continue for centuries, even if the global mean temperature stabilises.

Although more species will be negatively affected by climate change, there may be some that benefit. Cool, moist regions such as the UK could provide habitats for additional species, while in hotter, arid regions species diversity may decline. The two biomes most at risk immediately are Arctic and coral ecosystems, as shown in Figure 4.12.

Figure 4.12 Coral disease and bleaching in the Red Sea

> ### Skills focus: The impacts of climate change
> You should be aware of the impacts on different biomes, especially marine ecosystems, including coral and tropical rainforests.

Implications for the hydrological cycle

Figure 4.13 shows how the hydrological cycle's flows and stores are vulnerable to global warming, which may:

- increase evaporation rates and hence trigger more moisture circulating throughout the cycle, rather than storage in oceans, and intense precipitation events
- change precipitation type, as in the northern hemisphere where spring snow cover has decreased in extent; earlier springs mean earlier peaks in snowmelt and resulting river flows and flows into oceanic stores
- increase surface permafrost temperatures, a trend recorded since the early 1980s
- reduce sea ice, ice cap and glacier storage, as in the Arctic already
- change the capacity of terrestrial ecosystems to sequester carbon and store water; an example of the importance of water storage is in the Amazon, where 60 per cent of precipitation originates from evapotranspiration by upwind ecosystems.

Lastly, the complex El Niño–Southern Oscillation (ENSO) is an important factor in the Earth's climate system and affects the hydrological cycle. Droughts and floods driven by ENSO may be more intense and increase in frequency because of a warming atmosphere and ocean surface.

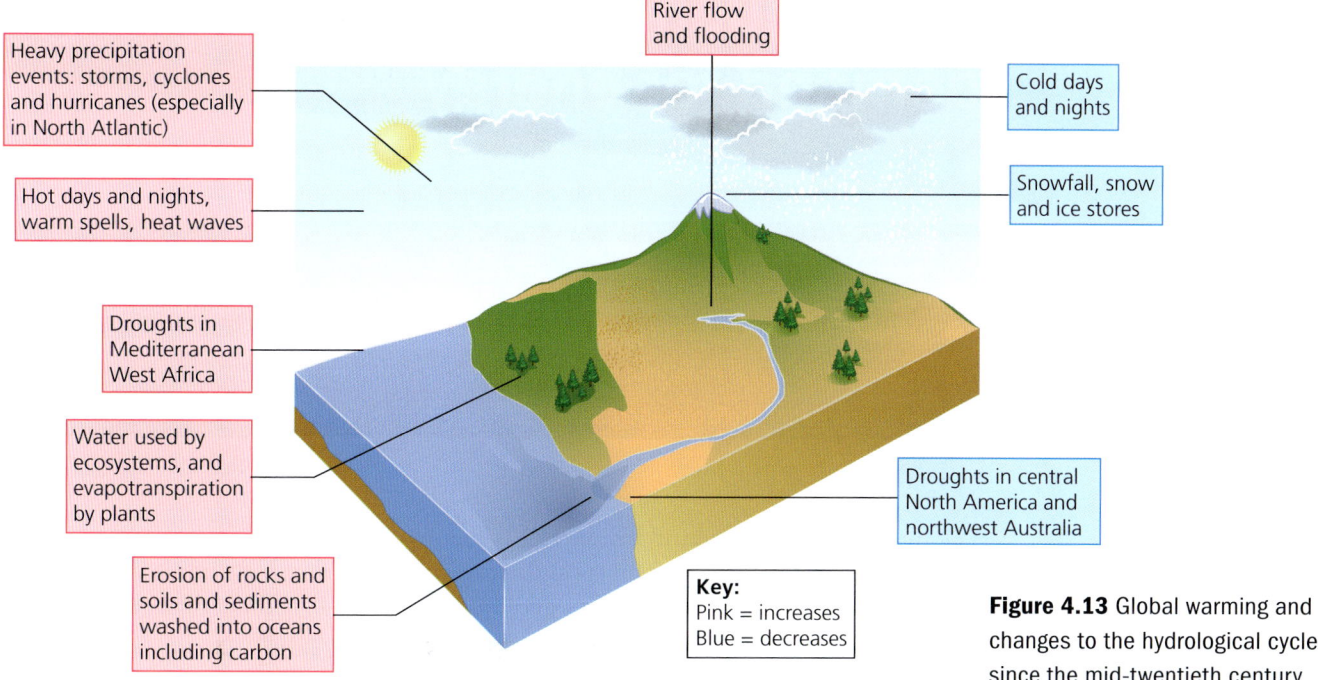

Figure 4.13 Global warming and changes to the hydrological cycle since the mid-twentieth century

Review questions

1. Give reasons for the different sizes of carbon stores.
2. What factors affect the rates of carbon flows between stores?
3. Explain the difference between positive and negative feedback in systems.
4. How do geological processes differ from biological processes in the carbon cycle?
5. Study Figure 4.1 on page 87 and Table 4.5 on page 97. Write a short account describing how humans are altering the natural carbon cycle.
6. Outline the stores and fluxes of the carbon cycle.
7. Summarise the role of the thermahaline circulation in the carbon cycle.
8. Evaluate the implications of anthropogenic climate change for ecosystems and biodiversity.

Further research

The IPCC's website has a wealth of data: http://ipcc.ch

The UNEP's *The Natural Fix?* has useful basic data on carbon cycling: http://www.grida.no/publications/151

Explore the Millennium Assessment Reports: www.millenniumassessment.org/en/index.html

This website maps and graphs IPCC data: www.globalcarbonatlas.org

NOAA is a US government research organisation with useful data and animations: www.esrl.noaa.gov/gmd/ccgg/trends/index.html

San José State University in the USA has a useful video of human effects on CO_2: https://youtu.be/0yD9sIzbRN8

The UK Meteorological Office has short videos and infographics on carbon cycling and climate change: www.metoffice.gov.uk/news/in-depth/climate-infographic

5 Consequences of the increasing demand for energy

> By the end of this chapter you should:
> - understand that all countries strive for energy security
> - be aware that energy security still relies heavily on fossil fuels
> - be aware that much economic development today continues to depend on fossil fuels
> - understand that there are alternatives to fossil fuels, but each has its costs and benefits.

As long as the rising demand for energy is met mainly by the burning of fossil fuels, changes to carbon stores and fluxes will continue to be much the same as they have for the last 200 years. In short, the geological store of carbon will continue to be depleted and the carbon flux to the atmosphere will continue.

5.1 Energy security

> **Key concept: Energy security**
>
> Energy security refers to the uninterrupted availability of energy sources at an affordable price. It has many aspects, for example:
> - long-term energy security mainly deals with timely investments to supply energy in line with economic developments and environmental needs
> - short-term energy security focuses on the ability of the energy system to react promptly to sudden changes in the balance between energy demand and energy supply.

At the start of this chapter, the following aspects of **energy security** need to be stressed:

- It is usually evaluated at a national level; that is, countries are either energy secure or they are not.
- There are four important aspects to the supply side: availability, accessibility, affordability and reliability (see Figure 5.1).
- It requires an accurate prediction of future energy demands.
- Those countries that are likely to be most energy secure will be those that are able to meet all or most of their energy needs from within their boundaries.

Energy security is vital to the functioning of a country, particularly its economy and the well-being of its people. Just pause for a moment and think about the various ways in which energy is essential:

- It powers most forms of transport.
- It lights our settlements.
- It warms or cools our homes and powers a whole host of domestic appliances.
- It is vital to modern communications.
- It drives most forms of manufacturing.

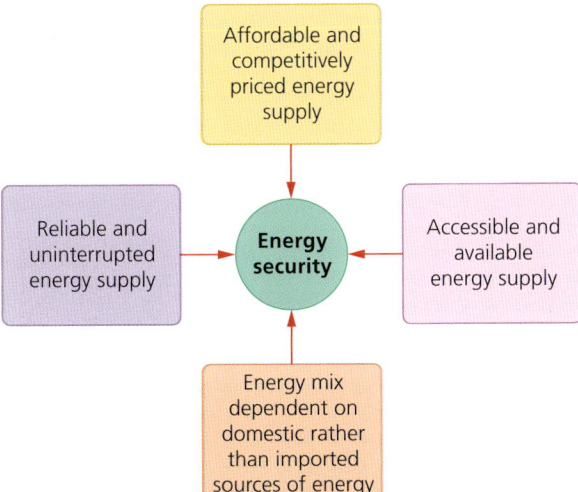

Figure 5.1 Energy security

The consumption of energy (demand) is constantly increasing as a result of development, rising living standards and population growth. Figure 5.2 shows how the global consumption of energy has increased since the mid-1990s.

Consumption and the energy mix

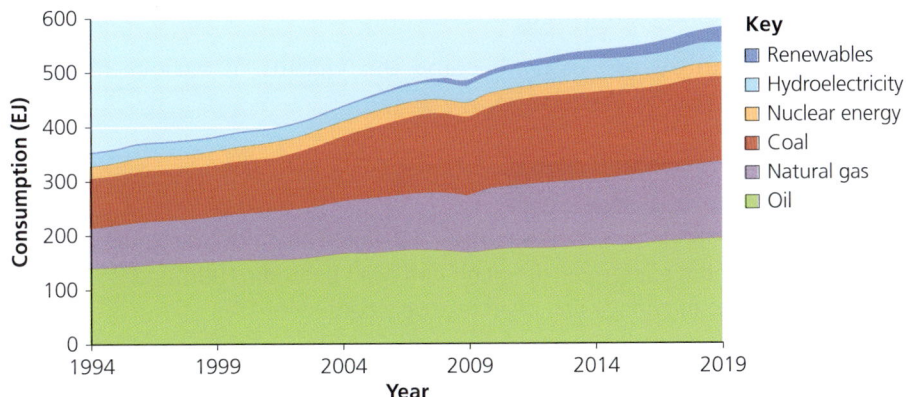

Figure 5.2 World energy consumption, 1994–2019

Energy consumption

The consumption of energy is usually expressed in per capita terms using one of the following measures:

- kilograms of oil equivalent per year (kgoe/yr)
- gigajoules (10^9 joules) per year (GJ/yr) or exajoules (10^{18} joules) per year (EJ/year)
- megawatt hours per year (MWh/yr).

It is also important to know how efficiently energy is being used. One possible measure here is **energy intensity**.

> ### Key concept: Energy intensity
>
> A measure of how efficiently a country is using its energy. It is calculated as units of energy used per unit of gross domestic product (GDP). A high energy intensity indicates a high price or cost of converting energy into GDP. It is generally recognised that energy intensity decreases with economic development: energy is used more efficiently and so the cost per unit of GDP becomes less, and more developed service economies are likely to use less energy per US$1000 of GDP than less developed manufacturing economies.

5 Consequences of the increasing demand for energy

> **Key term**
>
> **Energy mix:** The combination of different available energy sources used to meet a country's total energy demand. The exact proportions or mix vary from country to country. It is an important component of energy security.

Energy mix

Every country satisfies its energy needs in a particular way, referred to as its **energy mix**. A critical aspect is the mix of primary energy sources used to generate electricity (the form in which most energy is consumed). These sources include:

- non-renewable fossil or carbon fuels, such as oil, natural gas and coal
- recyclable fuels such as nuclear energy, general waste and biomass
- many types of renewable energy, such as wind, geothermal, water and solar.

While all countries have their own particular energy mix, globally fossil fuels account for over 80 per cent of the energy mix. The major challenge facing the world today is to lessen reliance on non-renewables. The future of the Earth depends on it, not so much because the non-renewables will run out one day, but rather due to the damage done to the global system by the continued burning of carbon fuels.

An important dimension of the energy mix is the balance between the amount of energy that comes from domestic sources and that which is imported. Energy security increases as dependence on imported sources of energy decreases. A high dependence on imported energy puts a country at risk from sudden threats, for example:

- artificial and abrupt hikes in energy prices
- supplies cut off by military campaigns or civil unrest.

Importing countries can all too easily become the victims of various forms of blackmail.

> **Skills focus: Graphical analysis**
>
> Simple line graphs are used to show how a particular variable changes over time. Figure 5.2 shows how global energy consumption has risen between 1994 and 2019. It also shows that the rate of change has been relatively consistent.
>
> Figure 5.2 is in fact a compound graph showing how the global energy mix has changed with the rise in energy consumption. Clearly, three fuels have dominated the energy scene since 1994: coal, oil and gas. A similar type of graph could be used to show a country's changing energy mix.
>
> Another widely used technique for illustrating and comparing the energy mixes of countries is the pie chart. Two pie charts showing the energy mix at different times can give a clear visual impression of how a country's mix might change (Figure 5.3).
>
>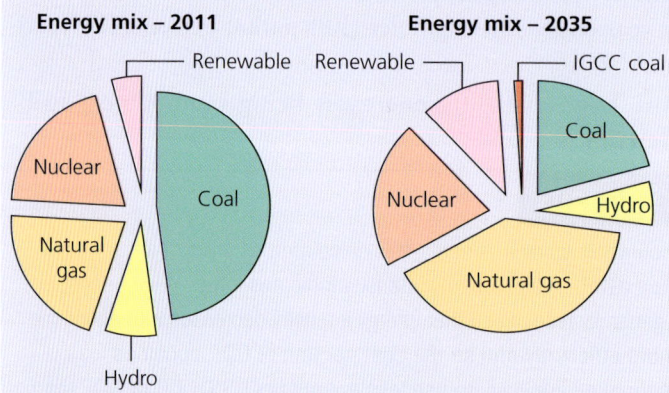
>
> **Figure 5.3** Predicted changes in the energy mix of the USA, 2011 and 2035

The consumption of energy

Remember that consumption is a function of demand: the greater the demand, the higher the consumption. But there are other factors at work, as illustrated in Figure 5.4.

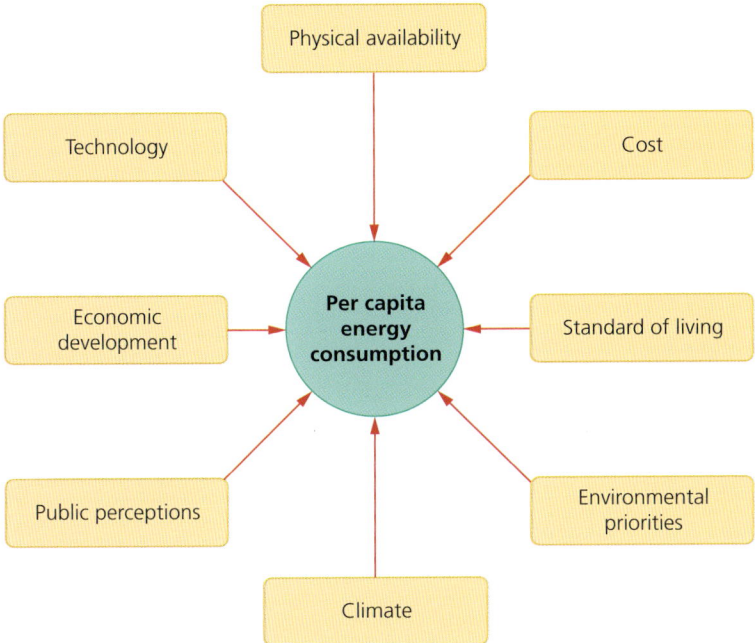

Figure 5.4 Some factors affecting per capita energy consumption

Physical availability

This is perhaps the most fundamental of all the factors. Critical questions here include: are the energy resources available within the country concerned, or do they have to be imported? If the latter, then transport costs are likely to add to the overall cost of energy to the consumer. Rising costs are likely to be a downward pressure on energy consumption.

Even if there are domestic energy resources, another potential issue is their accessibility. Is their exploitation going to be technically difficult and expensive? These questions lead to the next significant factor.

Technology

Modern technology can certainly help in the exploitation of energy resources that are not so readily accessible, for example, deposits of oil and gas that require deep drilling through a contorted geology. There are two sides to technology. It can help tap energy resources that are not so readily accessible. This in itself is likely to encourage energy consumption. At the same time, much of the modern technology that is now part of everyday living is energy thirsty.

Cost

The factors above are two of the three main determinants of access to energy resources. The third is cost. This includes a number of separate costs, for example:

- physical exploitation
- processing (converting a primary into a secondary resource)
- delivery to the consumer.

Clearly, relatively low energy costs may be expected to boost energy consumption.

Economic development

Costs are, of course, relative. The same energy costs may be perceived as expensive in one country and acceptable in another. The public perception will depend very much on the level of economic development and the standard of living. The higher these are, the less the sensitivity to energy costs.

Figure 5.5 clearly shows that developed countries have relatively high levels of energy consumption. A critical factor here is the energy needs of all the domestic appliances that help make everyday living more comfortable and those forms of transport that allow us to travel more easily. There is certainly a broad correlation between economic development (GDP per capita) and energy consumption.

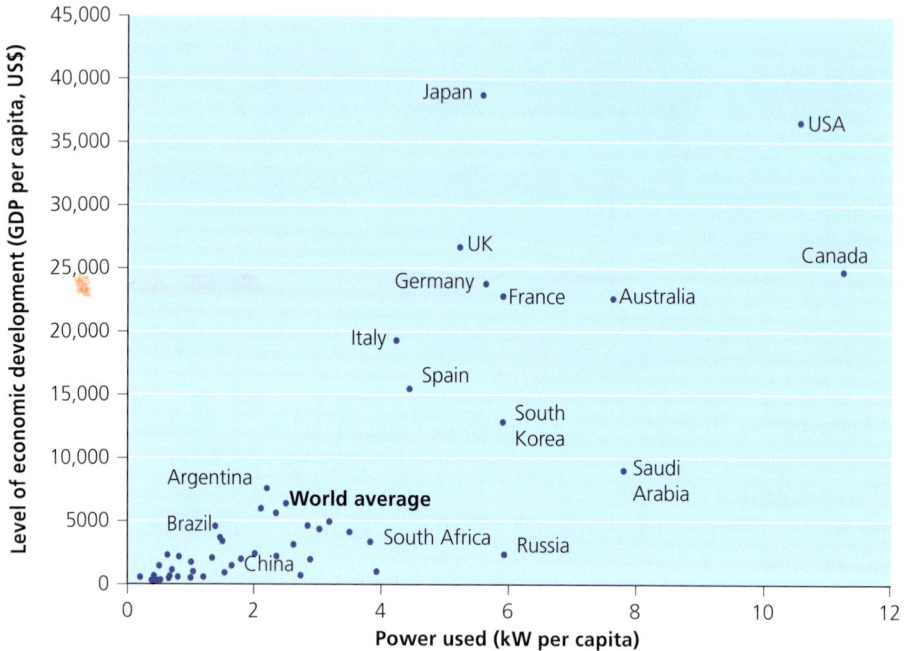

Figure 5.5 Energy consumption and economic development in a sample of countries

Climate

Climate is another factor affecting energy consumption. Very high levels of consumption in North America, the Middle East and Australia reflect the extra energy required to make the extremes of heat and cold more comfortable, not only in the home but also at work and in public places. The low energy consumption in much of Africa, a continent of considerable heat, reflects its relatively low levels of economic development.

Environmental priorities

It might be that, out of concern for the environment in general, and about carbon emissions in particular, a government does not take the cheapest route to meeting its energy needs. Remember, however, that renewables such as wind turbines and solar panels are not necessarily environmentally friendly. Also, the cost of a 'green' energy policy could have a slightly depressing impact on consumption, as would any government's drive to raise energy efficiency and energy saving.

Energy players

Meeting the demand for energy involves **energy pathways** from producer to consumer. At both ends of such pathways, there are influential players (organisations,

> **Key term**
>
> **Energy pathway:** The route taken by any form of energy from its source to its point of consumption. The routes involve different forms of transport, such as tanker ships, pipelines and electricity transmission grids.

Figure 5.6 The presence of Shell along the energy pathway: a) exploration, b) trans-shipment, c) delivery on the forecourt

groups or individuals) with particular involvements in the energy business. At the supply end, there are energy companies and the governments of energy-producing countries (Figure 5.7). There are governments at the demand end also, as well as a range of consumers. There are also players involved at various points along the pathways, such as the companies responsible for the movement and processing of energy (Figure 5.6). The five major players are described in Table 5.1.

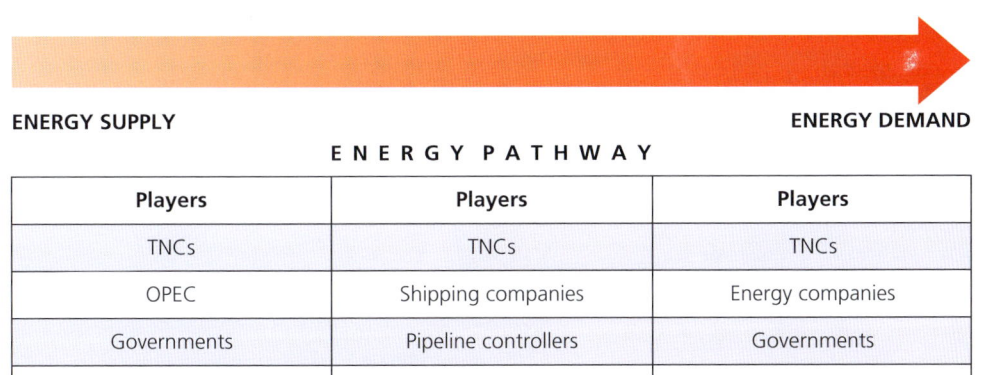

Figure 5.7 The energy pathways of oil and gas

Table 5.1 Major players in the world of energy

Player	Role
Transnational corporations (TNCs)	The big names in the oil and gas business are shown in Figure 5.9. Nearly half of these companies are state-owned (all or in part) and, therefore, very much under government control. Because of this, strictly speaking they are not TNCs. Most are involved in a range of operations: exploring, extracting, transporting, refining and producing petrochemicals.
Organization of the Petroleum Exporting Countries (OPEC)	OPEC has 13 member countries which between them control around two-thirds of the world's oil reserves. Because of this, it is in a position to control the amount of oil and gas entering the global market, as well as the prices of both commodities. OPEC has been accused of holding back production in order to drive up oil and gas prices.
Energy companies	Important here are the companies that convert primary energy (oil, gas, water and nuclear) into electricity and then distribute it. Most companies are involved in the distribution of both gas and electricity. They have considerable influence when it comes to setting consumer prices and tariffs.
Consumers	An all-embracing term but probably the most influential consumers are transport, industry and domestic users. Consumers are largely passive players when it comes to fixing energy prices.
Governments	They can play a number of different roles; they are the guardians of national energy security and can influence the sourcing of energy for geopolitical reasons.

5 Consequences of the increasing demand for energy

Energy portraits of the USA and France

The USA and France are two of the top ten energy consuming countries. The total consumption in France is seen to be only one-tenth of that of the USA. This big difference is explained mainly by differences in population: 331.2 million compared with 65.2 million. Climate is another contributory factor. The higher US consumption reflects the fact that quite large areas of this huge country experience great extremes of heat and cold. Counteracting these extremes requires large inputs of energy into heating and lighting on the one hand and air conditioning on the other.

Carbon fuels produce 82 per cent of the energy consumed in the USA (Figure 5.8). Only 11 per cent comes from renewables and 8 per cent from nuclear energy. The French energy mix is rather different, with 50 per cent of its energy coming from fossil fuels, nearly 10 per cent from renewables and 41 per cent from nuclear energy. France is, in fact, a major player in the nuclear power industry. It has over 50 nuclear reactors in operation currently.

When it comes to dependency on imported supplies, the figure for France is surprisingly high at 46 per cent. This reflects the fact that all of its natural gas and oil are imported, plus the uranium needed for its large number of nuclear power stations. In contrast, the USA imports only 15 per cent of its primary energy – a difference that makes the USA altogether more energy secure.

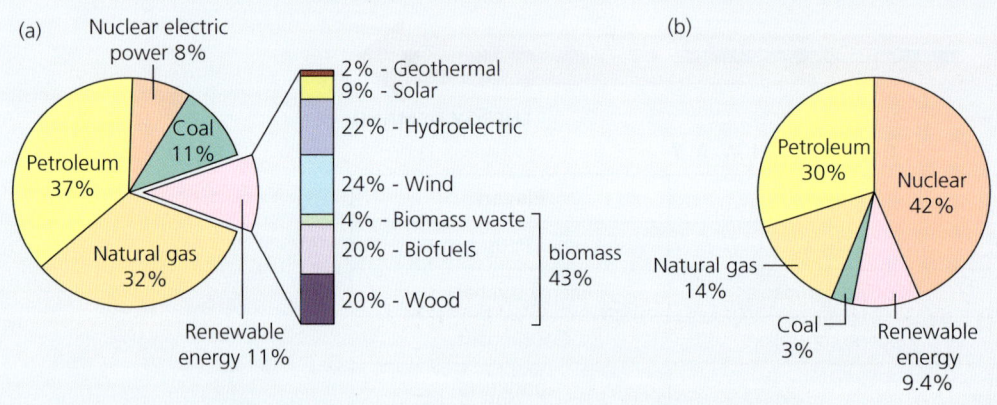

Note: Sum of components may not equal 100% because of independent rounding.

Figure 5.8 Energy consumption by source, 2019: (a) USA and (b) France

Synoptic themes:

Players

Five major players in the world of energy:

- TNCs
- OPEC
- Energy companies
- Consumers/ pressure groups
- Governments

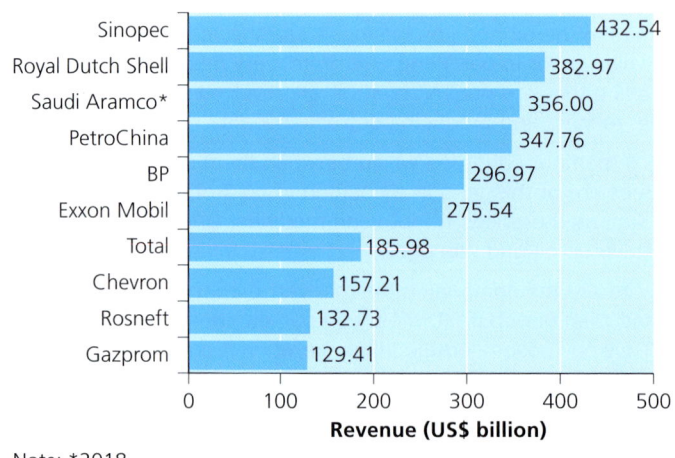

Note: *2018.

Figure 5.9 The world's biggest energy companies, 2019

112 The Carbon Cycle and Energy Security

5.2 Reliance on fossil fuels

The point has already been made that today's world continues to rely on fossil fuels for the greater part of its energy needs. The human use of carbon fuels has progressed through a long sequence, starting in prehistoric times with fuelwood and possibly peat. This was eventually followed by the coal era of the Industrial Revolution. During the twentieth century, oil overtook coal as the major carbon fuel, but today it is being challenged by natural gas as the dominant fuel. Figure 5.10 takes into account all three fossil fuels and tells us who today's major consumers are.

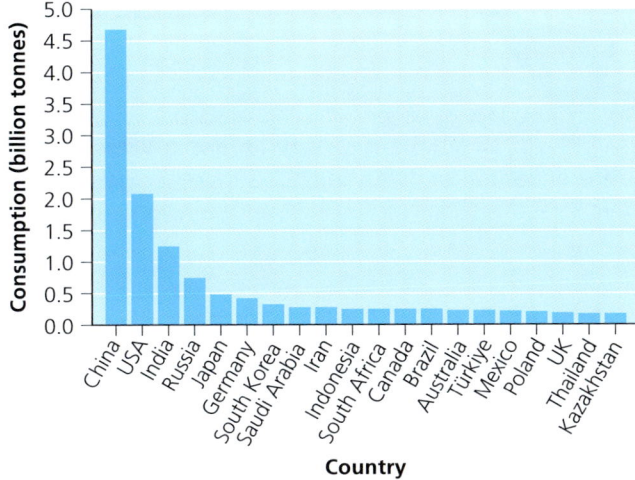

Figure 5.10 The leading consumers of fossil fuels, 2019

Mismatch between fossil fuel supply and demand

Any assessment of the degree of mismatch between fossil fuel supply and demand needs to look at the three fuels separately. Here the focus will be on two distributions for each fuel. In the following section, the focus will shift to the outcomes of any mismatches in terms of trade flows and energy pathways.

Coal

Although the consumption of coal is declining relative to the other two fossil fuels, production continues to increase. China is by far the largest producer. China's dominance over global coal production has been significant in the last decade. During the early 1970s, China accounted for approximately 13 per cent of global production but by 2016 it had increased to over 44 per cent. A long way behind come Australia, Indonesia and India. The mismatch between the distributions of coal production and consumption appears to be small, in that China and the USA are clearly the two largest consumers. It is interesting to note that a number of the remaining leading consumers are, in fact, also producers of coal. This reflects the fact that coal is characterised by high transport costs relative to a low **energy density**.

> **Key term**
>
> **Energy density:** The amount of energy stored in a given system, substance, or region of space per unit volume.

Oil

The pie chart in Table 5.2 highlights a significant feature of oil supply: namely that well over half comes from the two international groups of OPEC and North America (Canada, USA and Mexico). The relatively small amount of oil production in oil-thirsty Europe clearly signals one important mismatch. Interestingly, all

Table 5.2 The world's leading coal producers and consumers

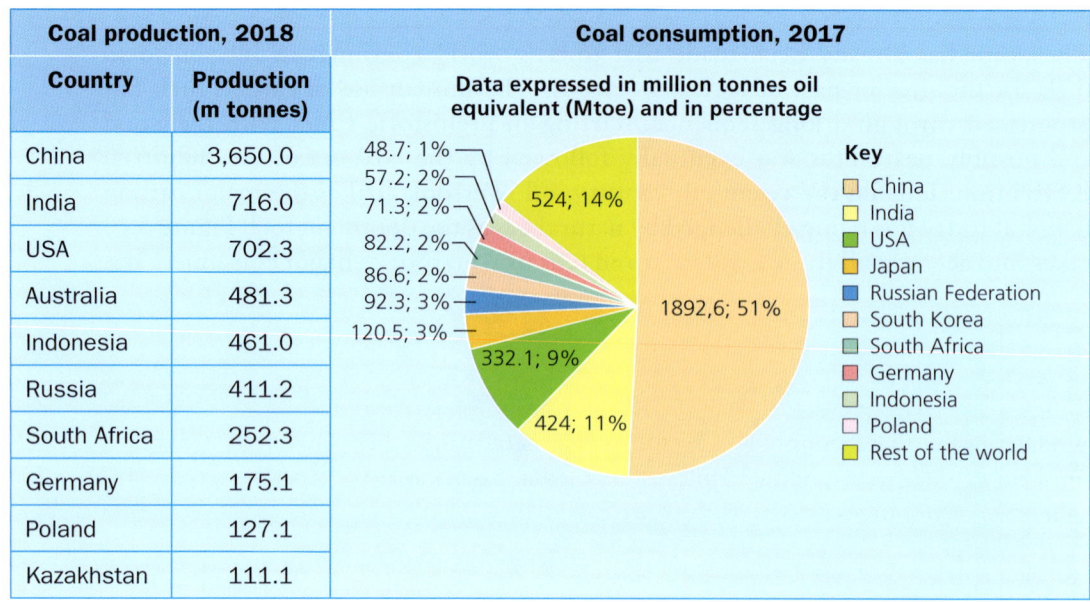

Coal production, 2018	
Country	Production (m tonnes)
China	3,650.0
India	716.0
USA	702.3
Australia	481.3
Indonesia	461.0
Russia	411.2
South Africa	252.3
Germany	175.1
Poland	127.1
Kazakhstan	111.1

four BRIC countries (Brazil, Russia, India, China) are shown to rank among the oil producers. Figure 5.11 shows that three of the oil producers (USA, China and India) are also major importers of oil, however. The other big oil consumers are all industrialised countries in either Asia or Europe.

Oil is the most interesting fossil fuel because of its demand as a transport fuel; there is no substitute, as there is with coal and gas (mostly electricity and industry), so there is a deeper global market and differences between consumers and producers.

Gas

Global gas production is dominated by the USA and Russia (Figure 5.12). Other noteworthy suppliers are to be found in the Middle East, Asia and Canada. As with oil,

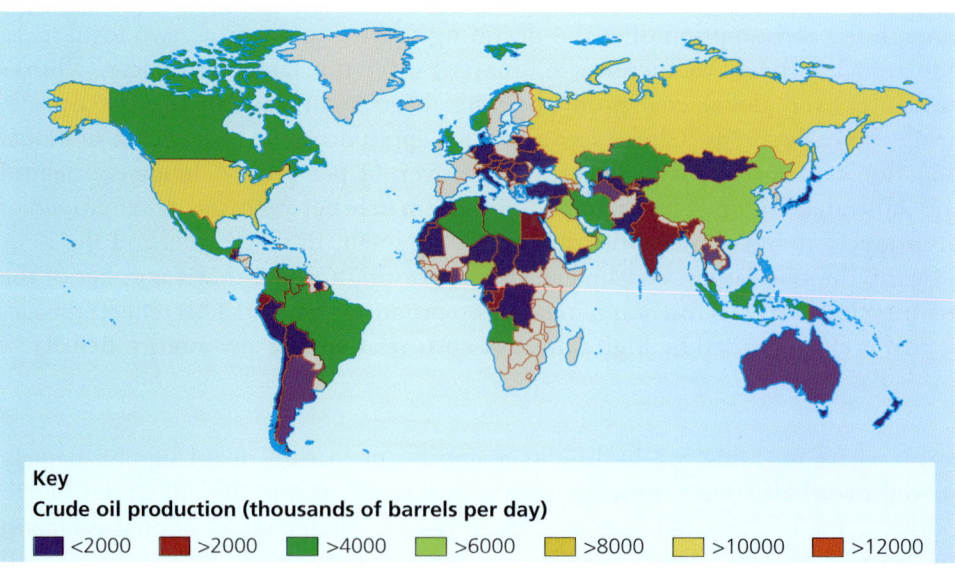

Figure 5.11 World's leading oil producers, 2020

114 The Carbon Cycle and Energy Security

gas production has a global but uneven spread. A particularly interesting aspect of Figure 5.12 is that some of the leading gas producers are also leading gas consumers. Of course, it makes basic sense for any country to make the fullest use possible of its domestic sources of energy.

Energy pathways

It should be clear from the above brief examination of the global production and consumption distributions of coal, oil and gas that there are mismatches. Such mismatches are resolved by the creation of energy pathways that allow transfers to take place between producers and consumers. The spatial tensions between supply and demand largely reflect a tension between physical and human geography. The former has determined the location of energy resources, while the latter has conditioned where those resources are needed.

Coal

Figure 5.13 shows that there is still a significant global trade in coal. Particularly notable is the fact that three of the largest coal producers – China, India and the USA – also import coal. Clearly, Australia and Indonesia export large quantities of coal to the first two countries, as well as to Japan, South Korea and Taiwan.

Oil

The energy pathways of oil are considerable (Figure 5.14). The Middle East is clearly the number one global producer; there are flows from here to Asia, Europe and even to the USA. The USA, despite being an oil producer, also draws oil flows from northern South America, West Africa and even Europe. There is only one energy pathway from Russia, and that is to Europe. In 2020, China bought significant amounts of Russian oil albeit delivered through Baltic ports.

Table 5.3 The world's leading oil consumers, 2017

Country	Million barrels per day	Share of world total
USA	19.96	20%
China	13.57	14%
India	4.34	4%
Japan	3.92	4%
Russia	3.69	4%
Saudi Arabia	3.33	3%
Brazil	3.03	3%
South Korea	2.63	3%
Germany	2.45	2%
Canada	2.42	2%
Total top 10	59.33	60%
World total	98.76	

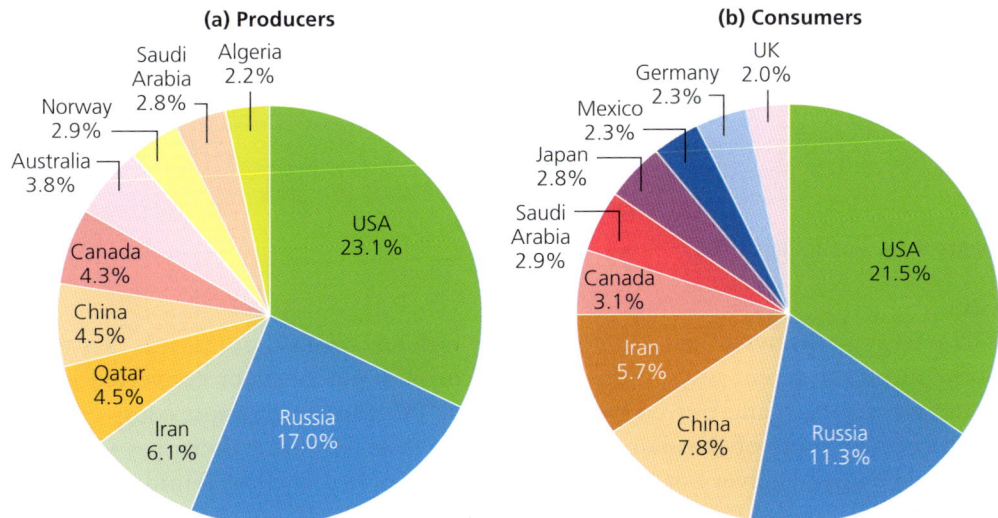

Figure 5.12 The world's leading gas (a) producers and (b) consumers, 2019

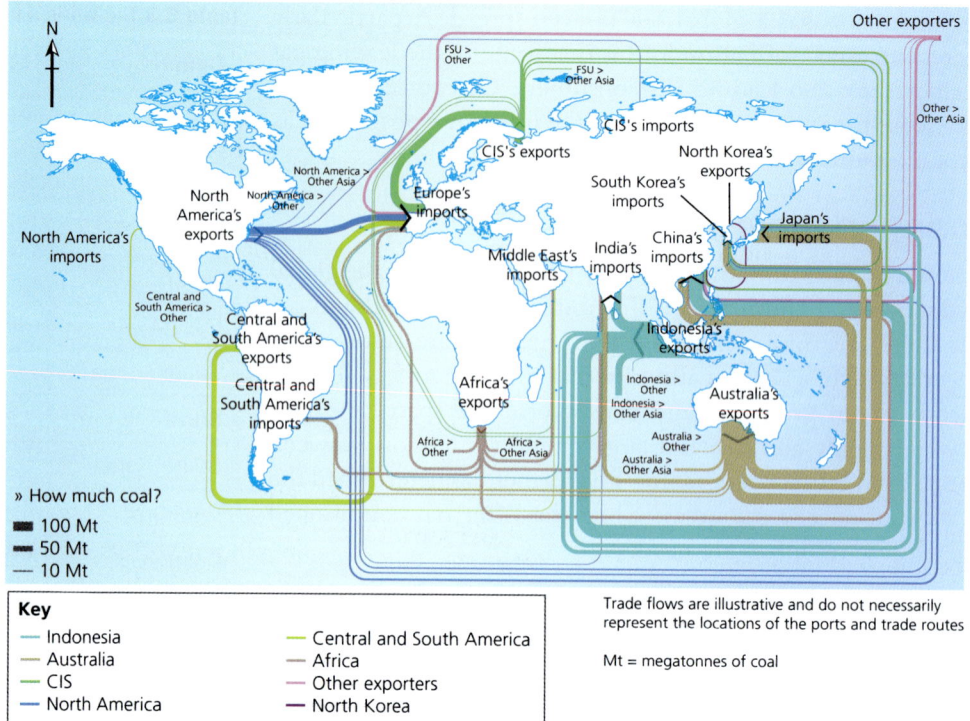

Figure 5.13 Global trade in coal

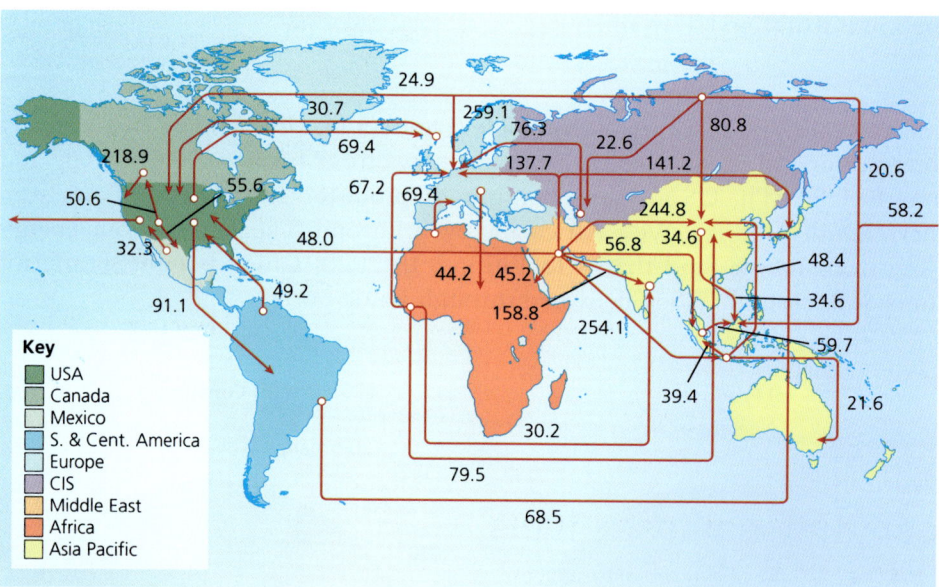

Figure 5.14 Global trade in oil

Gas

The energy pathways of gas are not so very different from those of oil. Gas flows occur in two different ways: either directly through pipelines, or converted into a liquid form (LNG) and moved by tanker ships. The former are most conspicuously used for the export of Russian gas to Europe, while Middle Eastern gas is largely delivered by sea. Delivery of gas from other supply nodes, such as Trinidad, Nigeria and Indonesia, is also mainly in the form of LNG.

One inescapable fact to be noted is that all the energy pathways are themselves consumers of energy.

Russian gas to Europe

It has already been shown that:
- Russia is the world's second largest producer of gas (see Figure 5.12, page 115)
- it also imports a significant amount of gas, presumably to help meet its own needs.

In recent years, the quantity of natural gas exported to western Europe has increased by 40 per cent. In 2018, Russia exported 243 billion m³ of natural gas.

Figure 5.15 looks in more detail at the delivery of Russian gas to the EU in 2017. The delivery is through four pipelines, and a planned fifth pipeline, Nordstream 2.

Europe's vulnerability to Russian oil and gas supplies became evident as a consequence of the war between Russia and Ukraine which began in 2022.

The war has had a number of impacts on energy markets and caused a global energy crisis.

Oil and gas prices on world markets spiked, causing inflation for consumers and businesses. Consumer energy prices were so high that governments began to provide state subsidies so people could afford electricity and gas.

European countries cut back imports of gas and oil from Russia as part of economic sanctions on the country. Existing pipelines were shut down and plans for Nordstream 2 were suspended. Russian gas exports through Ukraine to the EU finally ended in December 2024. This crisis showed how dangerously dependent Europe had become on a single source of gas, with serious consequences for energy security.

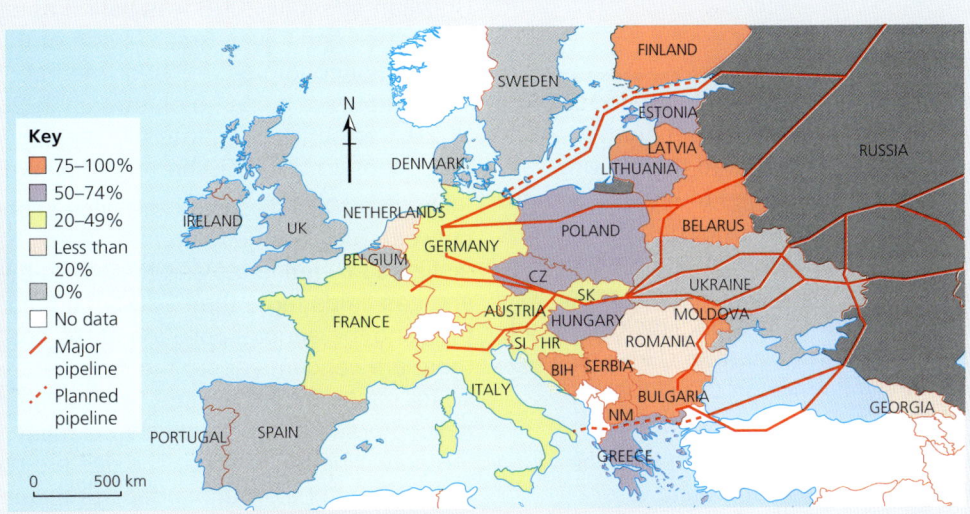

Figure 5.15 European dependency on Russian gas in 2017

Unconventional fossil fuels

Table 5.4 Unconventional sources of fossil fuel

Resource	Nature	Extraction
Tar sands	A mixture of clay, sand, water and bitumen (a heavy, viscous oil)	Tar sands have to be mined and then injected with steam to make the tar less viscous so that it can be pumped out
Oil shale	Oil-bearing rocks that are permeable enough to allow the oil to be pumped out directly	Either mined or shale is ignited so that the light oil fractions can be pumped out
Shale gas	Natural gas that is trapped in fine-grained sedimentary rocks	Fracking: pumping in water and chemicals forces out the gas
Deepwater oil	Oil and gas that is found well offshore and at considerable oceanic depths	Drilling takes place from ocean rigs; already underway in the Gulf of Mexico and off Brazil

Three examples of the exploitation of unconventional fossil fuels

Canadian tar sands

There are numerous deposits of tar sands in the world, but the biggest are in Canada and Venezuela. Exploiting the Canadian deposits on a commercial scale started in 1967 and has focused on the province of Alberta, most notably the Athabasca area. Currently, tar sands produce about 40 per cent of Canada's oil output. The 2015 fall in the global price of oil had a depressing impact on the tar sands industry because extracting bitumen is relatively expensive, largely because of the high energy input.

The exploitation of tar sands is not without its environmental and human costs, such as the scale of the strip mining, which requires the clearance of large areas of taiga.

US shale gas

In 2000, shale gas provided 1 per cent of the USA's gas supply. In 2015 it was nearly 25 per cent. Most of that increased production is due to the growing use of hydraulic fracturing (fracking) to release oil and gas from underground formations that are otherwise too difficult to drill. The most important shale gas fields have been found in the states of New York, Pennsylvania, Texas and West Virginia. Fracking for oil in the USA is now a key determinant of US oil security and has an increasing influence on the global oil price.

Environmental concerns associated with fracking include possible contamination of groundwater by the chemicals in the pumping fluid and surface subsidence. Fracking is known to produce airborne pollutants such as methane, benzene and sulphur dioxide. There are also reports of 'fraccidents', such as mysterious animal deaths and industrial explosions.

Figure 5.16 Three unconventional fossil fuels: a) Canadian tar sands, b) US shale gas, c) Brazilian deepwater oil

Brazilian deepwater oil

The discovery of huge oil deposits far off the Brazilian coast in 2006 was hailed as one of the biggest oil finds ever. Brazil is one of the leading emerging economies and badly needs oil and gas. The deepwater oil came on stream in 2009. By 2020, Petrobras (the state oil company) aims to raise production to 500,000 barrels of oil a day.

Since the 2006 discovery, the once scenic coast between Rio de Janeiro and São Paulo has been disfigured by refineries and the bases that serve the oil and gas fields more than 200 km offshore. Apart from this, and the serious pollution of the coastal waters, there are concerns about the risky nature of drilling so far offshore. The rigs are beyond the range of most helicopters, and access by ship is made hazardous by the prevailing rough seas. The rigs are drilling more than 2000 m below the surface of the sea and then many more thousands of metres below the seabed. The deposits lie below a thick layer of salt.

The oil and gas reservoirs contain huge amounts of toxic, flammable and explosive gases. Many observers are wondering what will happen if there is an accident, such as occurred at the Deepwater Horizon rig in 2010, when an explosion on the oil platform, killing eleven workers and causing nearly 5 million barrels of oil to be spilled into the Gulf of Mexico.

All of the resources shown in Table 5.4 have the potential to help meet future energy needs. Because of the fairly wide distribution of reserves, they offer some countries, currently dependent on imported energy, the prospect of greater energy security. However, offset against this are a number of serious costs.

- They are all fossil fuels, so their exploitation and use will continue to threaten the carbon cycle and contribute to global warming.
- Extraction is costly and requires a high input of complex technology, energy and water.
- Extraction threatens environmental damage (for example, the scars of opencast mines and possible ground subsidence) and environmental risks (for example, pollution of groundwater and oil spills). It is an unfortunate fact that a large proportion of the proven reserves of these fuels happen to occur in fragile environments, both on land and in the oceans.

There are four major players here (Table 5.5).

> **Synoptic themes:**
>
> **Players**
>
> Players in the harnessing of unconventional fossil fuels:
>
> - Exploration companies
> - Environmental groups
> - Affected communities
> - Governments

Table 5.5 Major players in the harnessing of unconventional fossil fuels

Player	Role
Exploration companies	These are not always the major players, such as Shell, Exxon or Petrobas. There is a large amount of subcontracting to companies specialising in exploration, such as Halliburton. Ultimately, though, the big energy companies have to bear the financial risks associated with finding and opening up new energy reserves. But their searches are in the context of oil and gas, rather than the wider search for completely new (i.e. renewable) sources of energy.
Environmental groups	It is very evident that exploitation of all unconventional fossil fuels has adverse impacts on the environment, which protest groups such as Greenpeace have done much to publicise. They are there to monitor the progress and their campaigns are well articulated.
Affected communities	The tendency is to focus on the negative impacts that energy production has on nearby communities, including various forms of pollution, disturbance of traditional ways of life, and so on. However, there are also benefits: job opportunities, inflows of investment and improved services. So, the role of such communities should not always be to object to each and every development.
Governments	The role of governments is a tricky one. Most governments wish to be seen as caring for the environment. Equally, for strategic reasons, they have a responsibility to ensure and improve energy security. Appealing to many governments is the fact that it is the private sector taking the financial risks associated with the search for new sources of energy.

5.3 Alternatives to fossil fuels

The global drive to reduce carbon emissions and to decouple fossil fuels from economic growth must involve increasing reliance on alternative sources of 'clean' energy. Renewable and recyclable energy sources are part of the multi-energy approach to energy security and protecting the carbon cycle.

Renewable and recyclable energy

The main forms of renewable energy being harnessed today are hydro, wind, solar (mainly via photovoltaic cells), geothermal and tidal. All are up and running, but their contributions to the energy budget vary from country to country. It is a simple fact of physical geography that not all countries have renewable energies to exploit. For example, not all countries have coasts or 'hot rocks'; not all countries have warm climates with long sunshine hours; not all countries have permanently flowing rivers or persistently strong winds. Furthermore, hydro and tidal power are the only sources that could provide base-load electricity.

It is frequently claimed that renewable sources of energy will be the saviour of the global energy challenge. However, some sobering facts are often overlooked.

- There are very few, if any, countries where renewables might completely replace all the energy currently derived from fossil fuels. The most likely are those with good hydro resources.
- As oil prices tumbled during 2015, renewables – with their slightly higher costs – became less attractive as an option.
- Upping the importance of renewables is likely to have significant impacts on the environment: more valleys would be drowned to create HEP reservoirs; large areas of land and the offshore zone would be covered by wind and solar farms.
- It is particularly frustrating that, while the majority of people believe that we must make greater use of renewable sources, most suddenly go off the idea when there is a proposal to construct a wind or solar farm close to where they live. They protest even when the wind farm is to be located well offshore!

Another unpalatable fact is that those countries with high levels of energy consumption will have no option but to look to nuclear energy to generate their electricity supply in a reasonably carbon-free manner. An added attraction is that nuclear waste can be reprocessed and reused, thereby making it a recyclable energy source. Nonetheless, nuclear power does have a downside. There are issues related to:

- safety, as exemplified by the incidents at Chernobyl (Ukraine) and Fukushima (Japan)
- the security of nuclear-powered stations in an era of international terrorism
- the disposal of highly toxic radioactive waste with an incredible long decay life
- the technology involved, which effectively means that the nuclear option is only open to the most developed countries
- costs – although operational costs are relatively low, the costs of building and decommissioning are high.

Biofuels

Of all the energy sources, fuelwood perhaps has the longest history. However, while fuelwood remains important in the energy mix of some parts of the world, **biomass** has recently come into prominence with the commercial use of a number of relatively new **biofuels**. They have now joined the ranks of recyclable energy, alongside nuclear energy.

Increasing attention is now being paid to the growing of biofuel crops as a way of decreasing the consumption of fossil fuels. These so-called energy crops include wheat, maize, grasses, soy beans and sugar cane. In the UK, the two main crops are oilseed rape and sugar beet. Most are converted into ethanol or biodiesel and used mainly as a vehicle fuel.

> **Key terms**
>
> **Biomass:** Organic matter used as a fuel, especially in power stations for the generation of electricity.
>
> **Biofuel:** A fuel derived immediately from living matter, such as agricultural crops, forestry or fishery products, and various forms of waste (municipal, food shops, catering, etc.). A distinction is made between primary and secondary biofuels:
> - Primary biofuels include fuelwood, wood chips and pellets, and other organic materials that are used in an unprocessed form, primarily for heating, cooking or electricity generation.
> - Secondary biofuels are derived from the processing of biomass and include liquid biofuels such as ethanol and biodiesel, which can be used by vehicles and in industrial processes.

The UK energy mix

The UK government is very mindful of the need to become energy secure and to play its part in reducing global carbon emissions. Figure 5.17 clearly shows that when it comes to primary energy consumption, while there has been a complete shift away from a direct use of coal, the reliance on oil and natural gas seems to have settled at a rather high level, providing close to 80 per cent of the UK's primary energy. Much of the petroleum is used by transport and most of the natural gas is used to generate electricity. Forecasts suggest that this is unlikely to change much in the near future. The electricity shown in Figure 5.17 is in fact 'primary electricity' generated by renewable (hydro, wind, solar, photovoltaic and geothermal) and recyclable (nuclear) energy.

Figure 5.18 shows that some electricity is still generated by coal-fired power stations, while oil has virtually ceased to be used in this way. Natural gas produces more electricity than nuclear energy. The contribution by renewables remains disappointingly small.

One piece of good news about the UK's energy budget is that today we consume less energy than we did in 1970, despite a population increase of some 6.5 million. The UK is now more efficient, both in producing energy and in using it. The rise of a less energy-intensive service sector at the expense of industry has also played a part. Households now use 12 per cent less energy while industry uses 60 per cent less. However, these savings have been offset by transport, particularly the big increase in the number of vehicles on the road and of flights. It now looks as if the UK will be using the same amount of energy in 2030 as it does today.

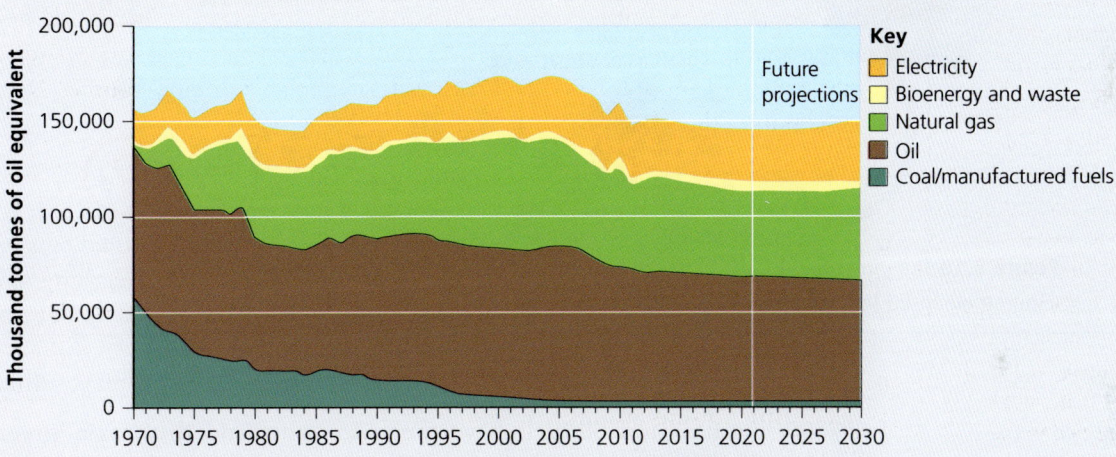

Figure 5.17 The UK's changing primary energy consumption, 1970–2030

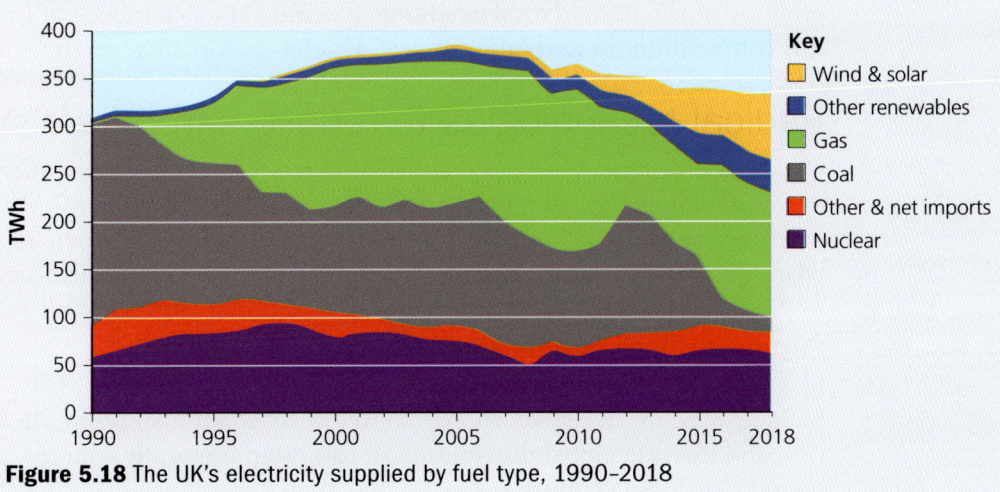

Figure 5.18 The UK's electricity supplied by fuel type, 1990–2018

5 Consequences of the increasing demand for energy

Biofuels in Brazil

Brazil took action in the 1970s to diversify its energy sources in order to combat concerns about its energy security. It has since invested in alternative energy sources, initially in hydroelectricity and more recently in biofuels. Today, 4 per cent of its energy comes from renewable sources, and approximately 90 per cent of new passenger vehicles sold in Brazil contain flex-fuel engines that work using any combination of petrol and sugar cane ethanol. This has led to a significant reduction in the country's carbon dioxide emissions.

Brazil is now the world's largest producer of sugar cane; in 2019/20, sugar cane production was estimated to be 643 million metric tonnes. It has also become the leading exporter of sugar and ethanol. Since 2003 the area planted with sugar cane has increased significantly. Sugar cane production is concentrated in the central southern region. The result has been the displacement of other types of agriculture, particularly cattle pasture. The knock-on effect has been to create a need for replacement pastures. This, in turn, has resulted in the large-scale clearance of the tropical rainforest (Figure 5.19). This deforestation is, in effect, now cancelling out the reduction in carbon dioxide emissions related to the increasing use of ethanol.

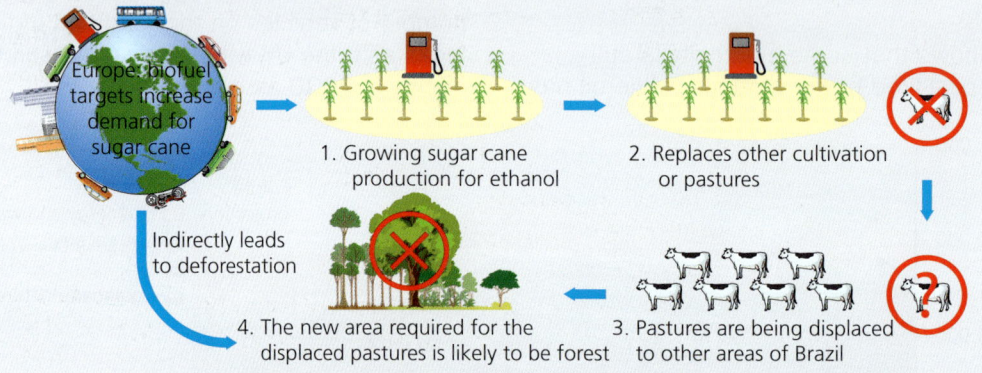

Figure 5.19 The direct and indirect consequences of an expanding sugar cane production (Source: Adapted from Friends of the Earth)

For some biomass and biofuels have much to commend them as a 'green' source of energy. Their increasing use is not without costs, however. The most notable of these is the fact that a hectare of space used to grow energy crops is a hectare less for growing much-needed food in a hungry world. The experience of Brazil shows that this competition for agricultural space also has unfortunate environmental costs.

Radical technologies to reduce carbon emissions

Greater use of renewable and recyclable sources of energy clearly offers one pathway to a more sustainable energy future. But are there any radical new technologies on the horizon that might play a part in reducing carbon emissions in the near future? Let's take a quick look at two: carbon capture and storage (CCS) and hydrogen fuel cells.

Carbon capture and storage (CCS)

It is widely accepted that coal will never cease to be a part of the global energy budget: it is an attractive energy source as it is abundant and cheap. Because of this, and its wide global distribution, it can often be locally sourced, particularly by poorer developing countries.

Skills focus: Emissions comparison

Possibly the fairest way of comparing the emissions of different fuels is to see how they all perform when they are used to generate electricity. Surprisingly, every fuel produces greenhouse gases (GHGs), but in varying quantities. This is because account needs to be taken of what are known as 'life cycle' emissions. These include not just the emissions that occur while electricity is being generated, but also those GHGs given off during the construction and decommissioning of generating plants. So, for example, coal-fired electricity releases large emissions of GHGs during the operational stage, whereas with wind, hydro, solar and nuclear energy, emissions occur mainly during construction and decommissioning. In short, there are no carbon-free forms of energy.

Figure 5.20 shows the emissions per gigawatt hour for nine fuels. Proportional bars are a good technique for representing such data as they make for an easy visual comparison. Another important point is that the data are derived from no less than 20 independent emissions studies, so the range of results is shown for each fuel. The figures show how relatively 'green' the renewable sources of energy are compared with fossil fuels.

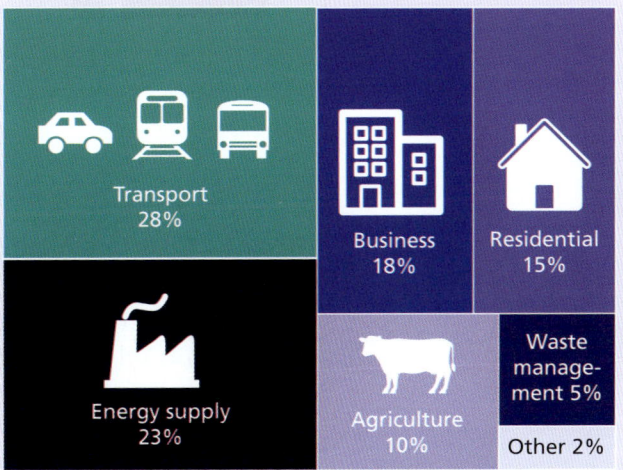

Figure 5.21 Greenhouse gas emissions, 2018

Figure 5.21 shows important messages about total greenhouse gas emissions:
1. Transport is the largest contributor because most transport uses oil.
2. Energy supply – largely electricity generation – is a major contributor.
3. Farming is a major emitter, despite being a small component of the economy of many countries.

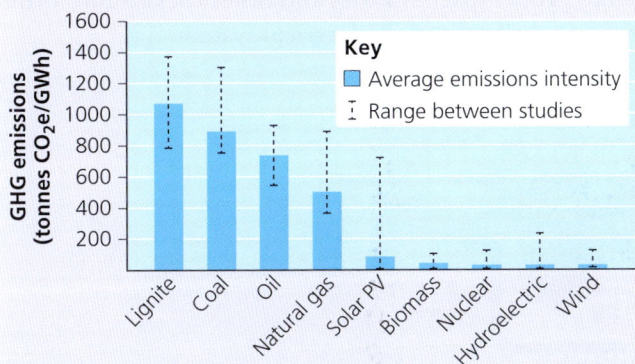

Figure 5.20 Life cycle greenhouse gas emissions of nine electricity-generating fuels

CCS involves 'capturing' the carbon dioxide released by the burning of fossil fuels and burying it deep underground (Figure 5.22). This technique promises the greatest savings in emissions where coal is being used to generate electricity. A slightly different technique, which 'scrubs' some of the carbon dioxide out of natural gas is already used quite widely, either at the point of production or at energy facilities from which gas is distributed to consumers.

It is frustrating that the implementation of this apparently simple idea is throwing up considerable challenges:

- It is expensive because of the complex technology involved.
- No one can be sure that the carbon dioxide will stay trapped underground and that it will not gradually leak to the surface and enter the atmosphere.

Hydrogen fuel cells

Although it is chemically simple and an abundant element, hydrogen does not occur naturally as a gas on Earth. It is always combined with other elements, for example,

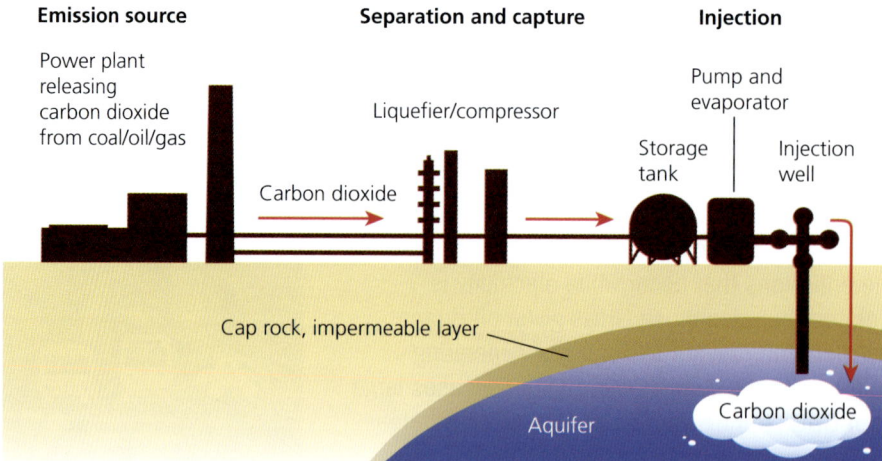

Figure 5.22 How carbon capture works

with oxygen in water. Currently, most hydrogen is extracted from other forms of fuel, such as oil and natural gas. Hydrogen is high in energy, and an engine that burns pure hydrogen produces almost no pollution (Figure 5.23). Since the 1970s, NASA has used liquid hydrogen to propel space shuttles and other rockets into orbit. Hydrogen fuel cells have also powered the shuttles' electrical systems.

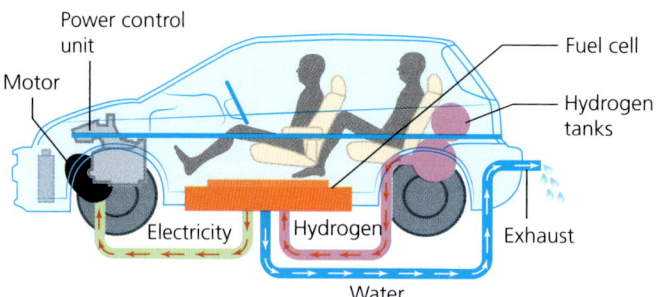

Figure 5.23 A hydrogen-powered car promises lower carbon emissions

A fuel cell combines hydrogen and oxygen to produce electricity, heat and water. It will produce electricity as long as fuel (hydrogen) is supplied, and it will never lose its charge.

Fuel cells are a promising technology for use as:

- a source of heat and electricity for buildings
- a power source for electric vehicles.

Hydrogen can be produced in two ways. Natural gas reforming uses fossil fuel natural gas to produce hydrogen, but this is not carbon neutral. Electrolysis splits water molecules into hydrogen and oxygen, using an electrical current. This process could be carbon neutral if renewable electricity (wind or solar power) is used.

Given where we are today, it is the second of these radical technologies that offers the brightest prospect of reducing carbon emissions. There seems to be a fair measure of certainty about the role it is just beginning to play in the context of transport. It does seem to be a very promising way of meeting future energy needs in an environmentally safe manner.

A world with no need to burn any fossil fuels is highly improbable. However, a world deriving much of its energy from renewable and recyclable sources, and making full use of the hydrogen cell, does promise relatively little disturbance of the carbon cycle and its stores and fluxes. It would also promise a longer human survival on Earth.

> **Review questions**
>
> 1 Explain each of the following key terms:
>   - energy security
>   - energy mix
>   - energy intensity.
> 2 Study Figure 5.3 on page 108. Write a short account describing the USA's predicted energy mix for 2035.
> 3 Explain why energy consumption varies between countries.
> 4 Identify the major players in the world of energy.
> 5 Explain what energy pathways are and why they exist.
> 6 Which of the following unconventional fossil fuels do you think is best?
>   - Canadian tar sands
>   - US shale gas
>   - Brazilian deepwater oil
>
>   Justify your choice.
> 7 How far do you agree that for some countries there is no energy alternative to the nuclear option?
> 8 Evaluate the UK's level of energy security.
> 9 'There is no such thing as a carbon-free source of energy.' Explain why this might be so.
> 10 Assess how far the burning of fossil fuels unbalances natural carbon stores and fluxes.
> 11 Study Figures 5.13 and 5.14 (page 116). Compare the patterns of global trade in coal and oil.
> 12 Assess the direct and indirect costs of biofuels.
> 13 Assess the value of alternatives to fossil fuels in terms of improving energy security.
> 14 Evaluate the role of radical technologies in reducing carbon emissions.

> **Further research**
>
> Explore the Global Carbon Atlas:
> http://www.globalcarbonatlas.org/en/content/welcome-carbon-atlas
>
> Find out more about the particular advantages of the hydrogen fuel cell:
> www.fuelcelltoday.com
>
> Read a good introductory piece on fracking: www.bbc.co.uk/news/uk-14432401

6 Human threats to the global climate system

> **How are the carbon and water cycles linked to the global climate system?**
> By the end of this chapter you should be able to:
> - understand the anthropogenic threats to the interlinked carbon and water cycles
> - be aware of how degradation of the carbon and water cycles affects human well-being
> - evaluate the pros and cons of climate change adaptation and mitigation
> - explain the roles of different players in reducing the risks of enhanced carbon emissions.

6.1 Cycles threatened by human activity

Growing resource demands

The terrestrial biosphere sequesters about a quarter of fossil fuel CO_2 emissions annually, directly slowing down global warming.

Growing demands for food, fuel and other resources have led to contrasting regional trends from **land conversion**.

Deforestation

Forests cover 31 per cent of the Earth's land area, although only 15 per cent are 'natural' primary forests (found in Canada, Alaska, Russia and the northwestern Amazon basin). Forests absorb rainfall and increase groundwater storage. Even small forest losses can disrupt natural weather patterns and the longer-term climate, enhancing or even creating destructive flood and drought cycles (Figure 6.1).

> **Key term**
>
> **Land conversion:** Any change from natural ecosystems to an alternative use; it usually reduces carbon and water stores and soil health.

Impact on water cycle
- Reduced intercepted rainfall storage by plants; infiltration to soil and groundwater changes.
- Increased raindrop erosion and surface run-off, with more sediment eroded and transported into rivers.
- Increased local 'downwind' aridity from loss of ecosystem input into water cycle through evapotranspiration.

The common method of either large tract or small plot deforestation is by burning trees by two methods: **Clear cutting** removes all primary forest, while **slash-and-burn agriculture** eventually allows growth of secondary forest.

Impact on carbon cycle
- Reduction in storage in soil and biomass, especially above ground.
- Reduction of CO_2 intake through photosynthesis flux.
- Increased carbon flux to atmosphere by burning and decomposing vegetation.

Figure 6.1 The impacts of deforestation on water and carbon cycling

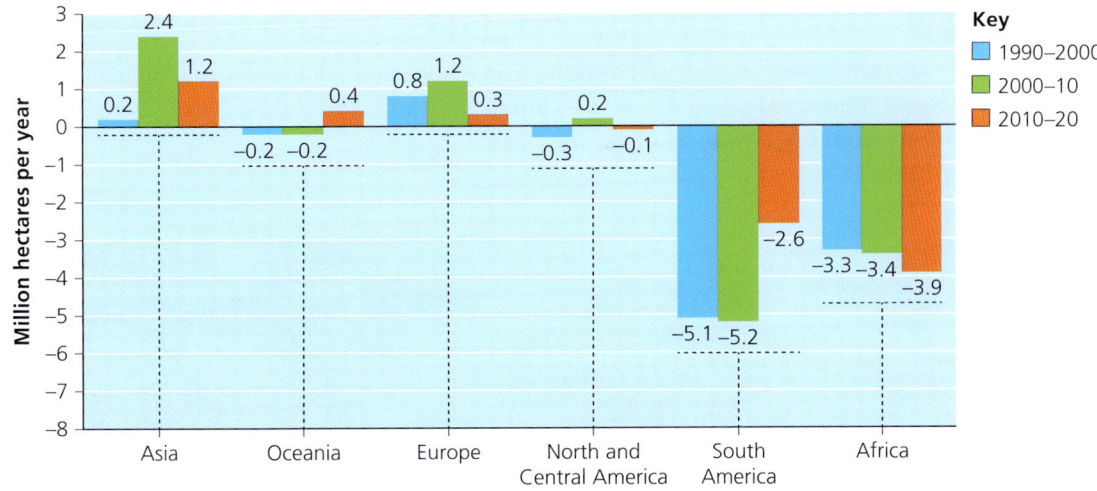

Figure 6.2 Annual forest area net change, by decade and region, 1990–2020 (Source: FAO)

The main driver of deforestation, shown in Figure 6.2, is the increasing demand for commodity production. Half of all current deforestation is for soy, palm oil, beef and paper production. Other causes of land conversion are: dams and reservoirs, infrastructure and opencast mining. By 2015, 30 per cent of all global forest cover had been completely cleared, 20 per cent degraded and the rest fragmented. Approximately 13 million hectares are deforested annually, equivalent to 36 football fields of trees lost per minute.

Regional trends include:

- Temperate forests such as in the UK and the USA have a long history of exploitation; 90 per cent was deforested by the nineteenth century.
- Boreal forests have been increasingly threatened since the mid-twentieth century, for example, by oil and tar sands production in Russia and Canada.
- Tropical forests have lost half their area since the 1960s, especially in Africa and South America. However, **remote sensing** shows that Indonesia has recently overtaken Brazil in the rate of deforestation, mainly for palm oil production and logging, as shown in Figure 6.2. Around 25 per cent of the rainforest has been clear-felled or burnt in 25 years, with Borneo (Kalimantan) most affected. Much of Southeast Asia suffers from the 'brown haze' created.

Afforestation

In 2014, the New York Declaration on Forests set a global target to restore 350 million hectares of deforested and degraded forest landscapes by 2030. By 2019, there was limited evidence to suggest that the original goals are on track to achieve the expected targets. **Afforestation** and **reforestation** is beneficial for CO_2 sequestration but can also be controversial in its impacts on landscape character as well as on carbon, water and soil systems. Monocultures of commercial trees, such as in palm oil plantations and non-indigenous species, often store less carbon, use more water and are disease prone. China's Three-North Shelterbelt Project – a 4500 km green wall of trees designed to reduce desertification – demonstrates many of these issues.

> **Key terms**
>
> **Remote sensing:** Surveillance by satellites such as Landsat generates data that can authenticate, or refute, official government data.
>
> **Afforestation:** The planting of trees in an area that has not been forested in recent times.
>
> **Reforestation:** Planting trees in places with recent tree cover, replacing lost primary forests.

Skills focus: The use of GIS

Figure 6.3 shows an example of a **GIS** map from Global Forest Watch. Use its interactive website to choose a location and create your own map (**www.globalforestwatch.org/map**).

Start with place names and administrative boundaries, then add layers of information to show forest cover/loss.

Figure 6.3 Measuring forest loss over time, 2001 and 2019

Key terms

Geographical Information System (GIS): Digital maps with 'layers' of information are an important tool in analysing place characteristics.

pH: A logarithmic measure of acidity or alkalinity. A value of 7 means neutral; above this the pH is alkaline, below this it is more acidic.

Ocean acidification: The decrease in the pH of the Earth's oceans caused by the uptake of carbon dioxide from the atmosphere.

Ecosystem resilience: The level of disturbance that ecosystems can cope with while keeping their original state.

Grassland conversion

There are two main types of grassland covering 26 per cent of global land area:

- Temperate grasslands have no trees and a seasonal growth pattern related to a wide annual temperature range. Those with fertile chernozem soils, an important carbon store, are prized for agriculture and hence suffer most degradation. Only 2 per cent of North America's prairies remain from land conversion, with similar issues in Russia's steppe biome.
- Tropical grassland or savannah have scattered trees, such as Africa's Serengeti and Brazil's Cerrado. Land conversion is increasing despite often infertile soils.

The carbon and water cycles are disrupted in grasslands that are used too intensively for animals or when ploughed up. Rapid increases in population and changes from nomadic to sedentary farming, coupled with the effects of climate change and poor management, are the drivers of change. Soil and ecosystem degradation is now a worldwide issue, resulting in carbon store loss.

Ocean acidification

Oceans are an important carbon sink, but their function as a fossil fuel greenhouse gas sink is increasingly changing their overall **pH** and **acidifying** them. Ecosystems are being affected, such as the complex food web based on coral.

For the past 300 million years, up to the early nineteenth century, the average oceanic pH was 8.2, dropping to 8.1 by 2019. Since the Industrial Revolution began, the pH of surface seawater has decreased by 0.1, a 30 per cent drop. By the end of the twenty-first century, the additional decrease in surface ocean pH may be between 0.06 and 0.32. Reefs, the foundation stone of coral ecosystems, stop growing if the pH is less than 7.8.

When **ecosystem resilience** is reduced, the potential for crossing a threshold is increased. Acidification increases the risk of marine ecosystems reaching a

The Carbon Cycle and Energy Security

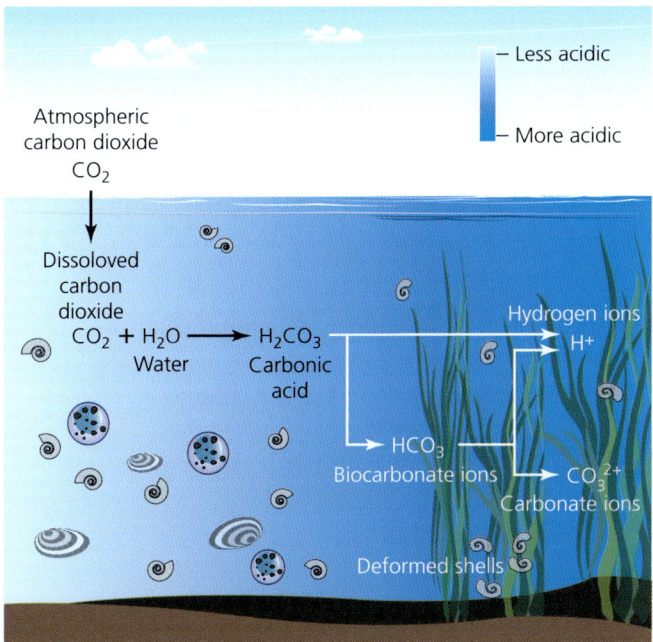

Figure 6.4 Ocean acidification

critical threshold of permanent damage. Ocean acidification impacts will be amplified because of other stressors, such as warming temperatures, destructive cyclones and pollution. One key factor is the speed of acidification: organisms may be able to adapt and be more resilient if changes are slow enough.

The Arctic Ocean is likely to be first affected because of its low pH, threatening its vulnerable 'cold water' corals. Coral reefs globally already show bleaching from warming temperatures (see Figure 4.12, page 104). Acidification affects shell-building marine organisms because carbonic acid reacts with carbonate ions in the water to form bicarbonate. Reduced carbonate ions mean animals like coral expend more energy building their shells, resulting in thinner, more fragile shells. The more acid the water is, the more it dissolves carbonate shells, weakening them and allowing attack from bio-erosion by molluscs, worms and sponges. Major functions of the ecosystem, especially reef building, may collapse. This could lead to an irreversible changed state and net reef loss by the mid-twenty-first century if current levels of carbon emissions continue.

Climate change

There is now unequivocal evidence of humans triggering the **enhanced greenhouse effect** with resulting climate changes.

The year 2015 was exceptional with respect to climate change: it was the first full year to exceed the key benchmark for global warming of 1°C above pre-industrial levels. Many parts of the world experienced unusual weather patterns associated with global warming: severe droughts in parts of Africa, India and Pakistan; flooding in Europe and the USA; and very warm temperatures in Siberia, northern Russia, and North America's east coast. The UK had the wettest and warmest December since 1910, resulting in severe flooding.

Researchers blame a combination of long-term anthropogenic influences combined with the 'spike' of the strongest El Niño in a generation.

> **Key terms**
>
> **Critical threshold:** An abrupt change in an ecological state. Small environmental changes can trigger significant responses. Negative and positive feedback loops reinforce or undermine changes once an alternative stable state has become established.
>
> **Enhanced greenhouse effect:** The intensification of the natural greenhouse effect by human activities, primarily through fossil fuel combustion and deforestation, causing global warming.

> **Oceanic processes**
>
> Research on oceanic processes is generally at a 'frontier' stage, hence the uncertainty about the consequences of warming and acidification for both marine ecosystems and the people relying on them.

Climate change may increase the frequency of drought due to shifting climate belts, such as in the Amazon. This may have an impact on the role of plants, especially forests, as carbon stores.

There are more than 30 different climate zones on Earth, simplified as equatorial, tropical, temperate and polar. However, they are not static, and warming by 2°C could lead to 5 per cent of the Earth's land area shifting to a new climate zone. There is already evidence of an expansion of subtropical deserts, and a poleward movement of stormy wet weather in the mid-latitudes.

- Northern middle and high latitudes will undergo more changes than the tropics.
- In the tropics, mountainous regions will experience bigger changes than low-altitude areas.
- The coldest climate zones will largely reduce in size, while dry regions will increase.
- Cool summers will change to hot summers in many places.

Key term

Inter-tropical convergence zone (ITCZ): A concentration of warm air that produces rainfall as part of a global circulation system (the Hadley cell). It moves north and south across the equator seasonally. Small shifts in its location can cause drought.

Key concept: The role of forests in climate regulation

All forests help control the climate at local, regional and global scales. They absorb and store rainfall, then add to atmospheric humidity through transpiration. Positive feedback operates: deforestation decreases rainfall locally and contributes to global warming, which in turn dries out the rainforest and causes it to die back.

The Amazon's changing climate

The forest acts as a global and regional regulator, pumping 20 billion metric tonnes of water into the atmosphere daily, 3 billion more than the River Amazon discharges into the ocean. The forest's uniform humidity lowers atmospheric pressure, allowing moisture from the Atlantic Ocean to reach further inland than areas without forest cover. Rain-bearing winds travel west until deflected by the Andes and normally transport moisture south to Buenos Aires and east to São Paulo.

However, since 1990 a more extreme cycle of drought and flood has developed in Amazonia, with a wetter rainy season, linked to shifts in the ITCZ. Rainfall has appreciably decreased downwind of deforested areas, with São Paulo suffering a water crisis. Severe droughts in 2005, 2010, 2015 and 2016 have increased stress already high due to decades of deforestation. A drier Amazon leads to the forest becoming a net carbon emitter rather than, as at present, a major global store.

6.2 Implications for human well-being

Impacts of forest loss

The UN has described the world's forests as 'fundamental' to human well-being and survival. Over 1.6 billion people depend on forests and over 90 per cent of these are the poorest in societies. Forests, like other ecosystems, are essential for human well-being through their 'services', as summarised in Table 6.2, as well as being the source of 80 per cent of global biodiversity.

Table 6.1 Annual rate of forest area change, 1990–2020

Period	Net change (million hectares/year)	Net change rate (%/year)
1990–2000	−7.84	−0.19
2000–10	−5.17	−0.13
2010–20	−4.74	−0.12

(Source: FAO State of the World's Forest Report 2020)

Table 6.2 Forest ecosystem services and human well-being

Type of ecosystem service		Forest functions and threats
Supporting functions • Nutrient cycling • Soil formation • Primary production	**Provision of goods** • Food • Freshwater • Wood and fibre • Fuel	• 1.1% of the global economy income • 13.2 million 'formal' and 41 million 'informal' jobs • Improve food and nutrition security • Source of livestock fodder in arid and semi-arid areas • Fuelwood source for one in three people globally for cooking and boiling drinking water • A genetic pool: a source for improving plant strains and medicines
	Regulation of Earth systems • Earth's 'green lungs' regulating climate, floods, disease • Water purification	Deforestation creates: • water-related risks (landslides, local floods and droughts) • an ongoing loss of biodiversity
	Cultural value • Aesthetic • Spiritual • Educational • Recreational	• Direct reliance by many indigenous peoples • Some cultures and religions see forests as sacred • Leisure and tourism

Data: FAO and UN State of the World's Forests Report 2020

In 2020, the FAO's State of the World's Forest Report indicated global forest area had decreased from 32.5 per cent to 30.8 per cent between 1990 and 2020. This amounted to a loss of 178 million hectares of forest which is equivalent to roughly the size of Libya. However, net loss was reduced because of afforestation. But there is evidence to suggest that the amount of forest loss is declining by 40 per cent between 1990–2000 and 2010–20 with the annual loss of forests reducing from an average of 7.84 million hectares per year to 4.74 hectares per year. This is leading to some countries having a forest gain. Figure 6.5 depicts a model often used in geographical studies: the environmental Kuznets curve. This suggests some societies reach a tipping point where exploitation changes to more protection. Factors affecting the timing of this attitudinal change are:

- the wealth of countries
- the rising knowledge of the role the environment plays in human well-being
- the aid given to some nations to help choices over exploitation
- the political systems and enforcement of environmental laws
- the participation of local communities
- the power of TNCs.

Increased **sustainable management** will hopefully result in greater reductions in carbon emissions from forests.

> **Key term**
>
> **Sustainable management:** The environmentally appropriate, socially beneficial and economically viable use of ecosystems for present and future generations.

6 Human threats to the global climate system

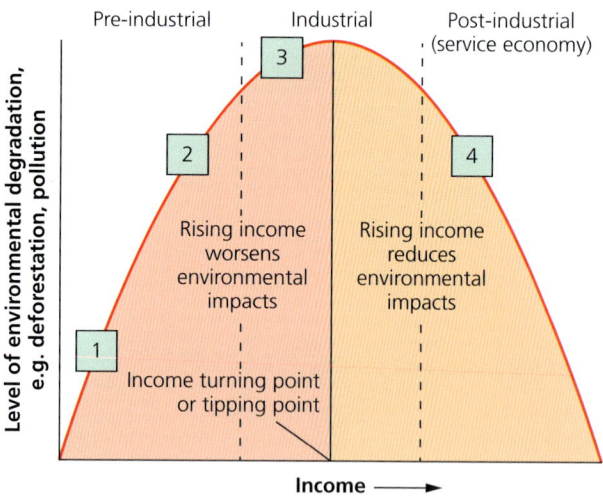

The environmental Kuznets curve is a generalisation and there are exceptions:
- Some developing and emerging countries (Costa Rica, Kenya, Viet Nam, Ghana, India) have reduced their deforestation rates and have active reforestation programmes and expanding areas of protected forests.
- Some developed countries (Canada, Sweden) have high deforestation rates, even of primary forest, although they also reforest cut areas of forest.
- It implies that environmental degradation is an unavoidavle and even necessary result of income growth, placing too much focus on specific income levels rather than environmental policies and education.

Key
1. UK pre-Industrial Revolution, remote Amazonia today, Indonesia pre-1970s
2. Indonesia today, China in the twentieth century
3. China today
4. UK today

Figure 6.5 The environmental Kuznets curve and tipping points

UK forests

By the end of the nineteenth century, forest cover in the UK had declined to under 10 per cent after centuries of exploitation. The Forestry Commission planted fast-growing exotic conifers such as Sitka Spruce on the moors of Wales, the Scottish Highlands and Lake District. Forest cover increased by 25 per cent between 1870 and 1947, and by 50 per cent between 1948 and 1995. By 2020, 13 per cent of the UK was forested, with increasing numbers of indigenous species planted.

Synoptic themes:

Attitudes and players

Human well-being can be enhanced through a more sustainable interaction with ecosystems. The support of different players, especially governments and NGOs, is important. However, players may have different attitudes: economic, social, political and environmental.

21 per cent of forests worldwide have some form of legal protection (national parks, forest reserves, conservation areas). In Costa Rica, the figure is 56 per cent. A Payments for Ecosystem Services (PES) scheme – giving financial rewards to landowners to protect and replant forests – has led to increased forest cover, from 22 per cent in the late 1980s to over 50 per cent by 2024. Between 1990 and 2020, Viet Nam's forest cover expanded from 28 per cent to 42 per cent. This was driven by government schemes to reduce flooding and landslides during typhoons. Forests and carbon stores have increased, but Viet Nam's approach has been criticised for promoting single- and non-native species plantations rather than natural forests.

Impacts of rising temperatures

Global warming is increasing temperatures globally, and since warmer air affects evaporation and stores more water, the amount of water in the atmosphere will increase. As a result, rain dropped during individual storms can increase, resulting in flash floods. This specification requires you to know about the uncertainty of global projections and three specific implications for:

- precipitation patterns
- river regimes
- cryosphere and drainage basin water stores.

Figure 6.6 Changes to Arctic sea ice extent

You should refresh your knowledge of the water cycle in Chapter 3. This section will focus on the patterns, regimes and stores in the compulsory case study of the Arctic.

> **Key concept: The Arctic water cycle**
>
> The Arctic plays a large role in global climate as its sea ice regulates evaporation and precipitation. The Arctic carbon cycle is very sensitive to climate change, making future projections difficult.

Earth's cryosphere has already been affected by global warming. Over the last 20 years the Antarctic and Greenland ice sheets have been losing mass; most glaciers have continued to shrink, and Arctic sea ice and northern hemisphere spring snow cover have continued to decrease in extent and thickness (Figure 6.6).

The Arctic is an early warning system for the rest of the planet, acting as a **barometer** of the environmental impacts resulting from fossil fuel climate forcing. In the past few decades average Arctic temperatures have risen twice as fast as global averages: 3–4°C in Alaska and northwest Canada. They could soon rise another 3–4°C over land, and up to 7°C over the oceans. There are huge implications for ocean currents, air circulation, sea level rise and flooding beyond the region.

As a net sink, the Arctic stores far more carbon than any other region, with 5–14 per cent of total oceanic stores, but it is also more vulnerable to global warming. Amplified atmospheric warming over the Arctic Ocean in autumn is already evident, with varying predicted effects, as shown in Table 6.3.

In the short term an increase in CO_2 uptake is predicted, but with further sea-ice loss, increases in marine plants such as phytoplankton may cause a limited net increase in the uptake of CO_2 by Arctic surface waters. In the long term, a net outward flux of carbon is expected because of rivers bringing carbon from melted permafrost stores, and loss of methane hydrate from destabilised sea floor deposits stored for thousands of years. Carbon uptake by terrestrial plants is increasing because of longer growing seasons and also the slow northward migration of boreal forests.

> **Key term**
>
> **Arctic barometer:** The idea that the sensitive Arctic region provides an early warning of the pressures and environmental changes caused by global warming.

6 Human threats to the global climate system

Table 6.3 Effects of global warming on the Arctic and beyond

Effects on the water cycle	Effects on the carbon cycle
• Warm water flowing into the Arctic from the Pacific and Atlantic • Rising local air temperatures • Shrinkage of sea ice; Arctic sea ice averages only 3 m thick and melting is increasing faster than anticipated • Run-off of fresh, cold water, which will alter marine ecosystems and the food chains dependent on the saline waters; this is predicted to affect areas outside the Arctic Ocean by 2100 • Funnelling of more cold water into the oceanic conveyer belt	Increased or new emissions of: • CH_4 from destabilisation of wetlands and sea floor deposits containing methane hydrate, stored for thousands of years • mainly CH_4 and some CO_2 from thawing permafrost • CO_2 from increased forest fires as boreal forests dry out; they may also absorb more CO_2 and CH_4 from the atmosphere

There is a high risk of irreversible feedback, called runaway global warming. Figure 6.10 on page 139 shows such feedback loops. There are two types of positive climate feedbacks in the Arctic from rising temperatures: sea-ice loss and carbon feedbacks.

> **Key concept: Modelling**
>
> Scientists have to use models because of limited first-hand data. Supercomputers are used to fill in the gaps. Modelling suggests parts of the Arctic and Europe may experience greater precipitation as the Arctic transitions toward a seasonally ice-free state. Figure 6.7 shows the wide variations predicted by modelling in 2015.
>
>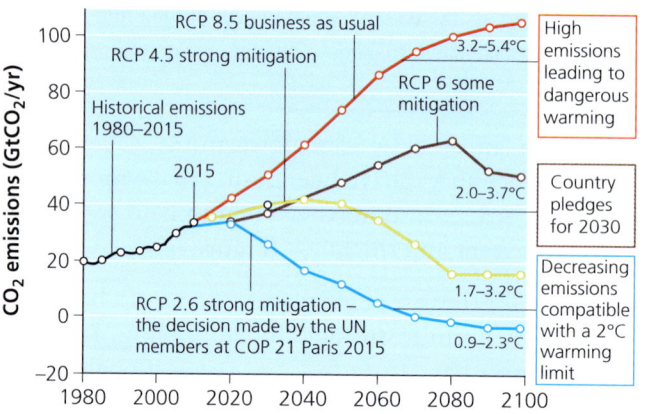
>
> **Figure 6.7** Future pathways modelled by the IPCC (RCPs = representative concentration pathways)

> **Key term**
>
> **Albedo flip:** When the sunlight reflected by white ice is suddenly absorbed as ice melts, creating a dark surface of open water.

Loss of Arctic albedo

Albedo is a measure of how sunlight is reflected away from the Earth's surface. Ice has a high reflectivity index, so a reduction in the amount of sea ice may create a positive feedback loop: melting allows more heat absorption, causing more melting. Loss of reflective (cooling) albedo is from:

- less summer sea ice, replacing reflective ice with (darker) ocean that absorbs heat.
- the replacement of (lighter) tundra with (darker) forests as they 'advance' north with improving temperatures from climate shift.

Black carbon (soot) pollution on snow adds to heat absorption and melting.

Carbon CO₂ feedbacks

Carbon emissions will outpace uptake as warming continues:

- Increased CO_2 emissions from tundra soils.
- Forest growth will absorb more of the Sun's energy, accelerating climate change.
- Methane hydrates are found in permafrost and ocean sediments in shallow water. They store more carbon than all of the proven reserves of coal, oil and natural gas together. However, they destabilise after thawing and will add to greenhouse gasses; CH_4 is 25 times more powerful a greenhouse gas than CO_2.

Table 6.4 IPCC projections for the future: two extreme RCP scenarios

RCP 2.6 Strong mitigation (the 2 degree future)	RCP 8.5 Business as usual
Agreed at COP21 Paris 2015 by all countries	We continue to use fossil fuels with no mitigation (runaway global warming)
Mean temperature rise projected to be between 1.5°C and 2°C by the end of the twenty-first century	Emissions continue to rise, so global temperature is predicted to reach 5.6°C above the pre-industrial level by 2100
The average warming for land regions is 2.3°C, compared to global average of 1.8°C	Some regions are projected to warm by more than 15°C (e.g. the Arctic)
Many regions will experience much greater (or lesser) increases in temperature; the Arctic is likely to have increases of about 8°C by 2100	The impacts of such a scenario are likely to be large and costly

Skills focus: Analysing climate model maps

You need to be able to analyse climate model maps to identify areas at most risk from water shortages and floods in the future. Start with the big picture, and try to name specific large-scale areas shown.

Figure 6.8 shows the two extreme **RCPs** out of the four scenarios put forward by the IPCC in 2014. Table 6.4 gives the background on these.

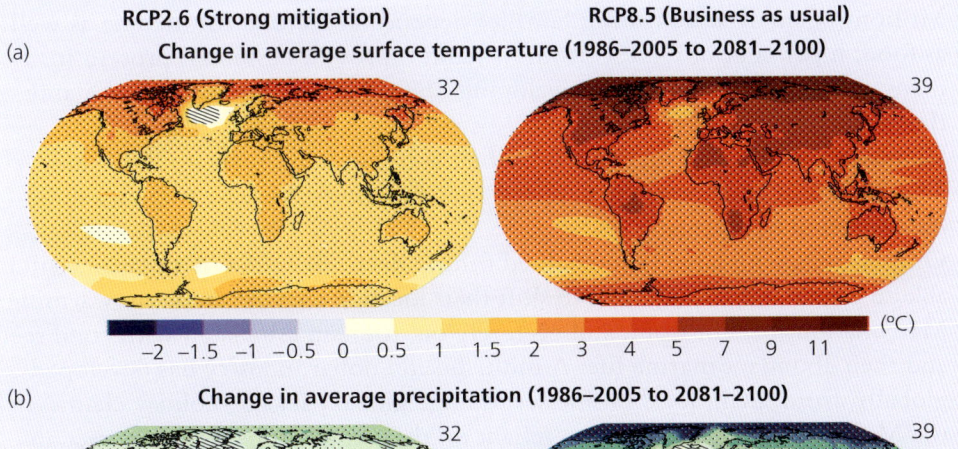

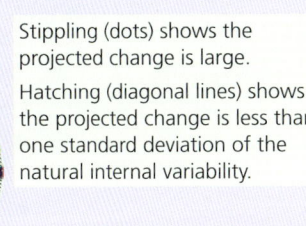

The number of models used to calculate the average multi-model is shown in the upper right corner of each panel.

Stippling (dots) shows the projected change is large.

Hatching (diagonal lines) shows the projected change is less than one standard deviation of the natural internal variability.

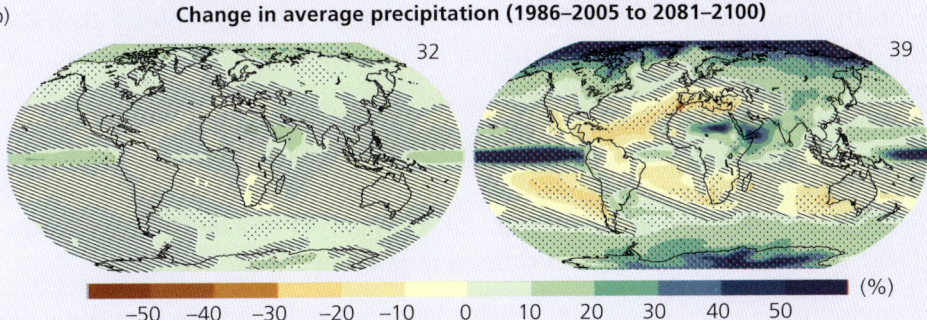

Figure 6.8 Analysing climate model maps: changes in average surface temperature and precipitation for 2081–2100 relative to 1986–2005; RCP2.6 (left) and RCP8.5 (right) (Source: IPCC Fifth Assessment Report, 2014)

6 Human threats to the global climate system

> **Key terms**
>
> **Representative Concentration Pathway (RCP):** The IPCC has a range of very different views or scenarios, called RCPs, of how the world may look in 2100, based on the level of CO_2 in the atmosphere. Their numbers show different radiative forcing, measured in watts per square metre, by 2100. This means the difference in atmospheric energy inputs and outputs since the Industrial Revolution.
>
> **Aquaculture:** The farming of aquatic organisms such as fish, crustaceans, molluscs and aquatic plants.

Ocean health

The World Wide Fund for Nature (WWF), an international pressure group, warns that climate change is affecting ocean temperatures, the supply of nutrients, ocean chemistry, food chains, wind systems, ocean currents and extreme events such as cyclones. The changes may be categorised under: bleaching, acidification, rising sea levels and loss of sea ice. These changes then affect the distribution, abundance, breeding cycles and migrations of the marine plants and animals that millions of people directly or indirectly rely on for food and income.

Research, such as that carried out in the North Sea, suggests that marine organisms may be responding faster to climate change than terrestrial plants and animals, with some shifts of animals and some plants towards the poles to compensate for a warming environment.

The resulting changes to the marine food web from global warming pose threats for a large proportion of the planet one way or another. This applies to both natural oceanic ecosystems and **aquaculture**. The concept of forest ecosystem services, shown in Table 6.2 on page 131, may also be applied to oceans. The following fact file shows the importance of oceanic 'health':

- All countries, even landlocked places, eat and either sell or buy fish and shellfish. The marine fishing industry is now globalised with a high level of trade.
- The FAO estimates that fishing supports 500 million people, 90 per cent of whom are in developing countries.
- Fish is the cultural choice of many wealthier countries such as Iceland and Japan, but an absolute necessity for well-being in poorer countries such as Namibia, Ghana and Senegal, and poorer communities within these. Fish provides 16 per cent of the annual protein consumption for 3 billion people, and is the main source of cheap protein for over a billion people.
- Millions of small-scale fishing families depend on seafood for income as well as food; 6 per cent of GDP is from fish and it provides essential protein in many of the 49 Small Island Developing States (SIDS), such as St Lucia and the Maldives.
- Countries that depend on exports of their fish resources, such as China and Thailand, will be affected by depleted and stunted stocks. The tropics will be most affected by warming.
- Only nations with large industrialised fishing fleets, like the UK or Japan, will be able to follow fish that are able to shift their location to adapt to ocean warming.
- Many nations benefit from, and even rely on, tourism associated with coral reefs and their abundant marine life. A multi-billion dollar industry has developed globally, from small, local businesses to huge global package-holiday chains. In the Maldives 220,000 people are reliant on their coral atolls, which attract an annual influx of 1 million tourists.
- Coral reefs also offer another 'service': over 200 million people live in coastal areas protected from waves by fringing reefs, for example, Hawaii.
- Environmental groups campaign to increase Marine Conservation Zones, established around the UK, St Lucia and Australia, to protect overfished areas. These may help give ecosystems time to respond to climate changes.

The environmental pressure group Oceana has produced a composite ranking of the vulnerability of coastal countries to increased threats, including acidification. They identify the top five as being the Maldives, Togo, Comoros, Iran and Libya. Rising sea levels are already a stressor, with warming and acidification posing more threats. Oceania's risk equation is:

Exposure + dependence + lack of adaptive capacity = vulnerability

Poorer, less diverse economies with rising populations are less able to cope with changes to ocean health, and may be unable to import alternatives. Paradoxically, they are often the least responsible for historic emissions of carbon dioxide.

6.3 Responses to the risk of further global warming

An uncertain future

Further planetary warming risks large-scale releases of stored carbon. This requires responses from different players from local to global scales. However, uncertainty about the type and rate of changes from a warming climate has been an excuse for procrastination by decision makers. Before the mid-twentieth century, scientists were unsure whether the human increases in CO_2 would stay in the atmosphere or be absorbed by forests and oceans.

Since 1958, the atmospheric concentration of carbon dioxide has been continually measured at the Mauna Loa Observatory, Hawaii (Table 6.5). Technology has moved on dramatically since the early twentieth century: CO_2 levels are now sampled weekly at 100 sites around the world. Higher atmosphere readings are taken by aircraft and sensors in space and processed by supercomputers, although the ocean remains a relatively dark area in our knowledge of Earth systems.

Skills focus: Plotting graphs

The specification expects you to be able to plot graphs, calculate means and rates of change. The data in Table 6.5 is taken from NOAA using the CO_2 measured in ppm above Mauna Loa, Hawaii.
- Plot a line graph to show the data; put time on the x axis, rate on the y axis.
- Calculate the overall mean between 1959 and 2019 and draw it as a horizontal line on the graph.
- Calculate the overall change from 1959 to 2019.
- What evidence is there of a steady increase in CO_2 emissions?

Table 6.5 Five-yearly changes in CO_2 rates, 1959–2019

Year	Mean annual atmospheric CO_2 (ppm)
1959	315.97
1960	316.91
1965	320.04
1970	325.68
1975	331.08
1980	338.68
1985	346.04
1990	354.35
1995	360.80
2000	369.52
2005	379.80
2010	389.85
2015	400.83
2019	411.43

Previous chapters have set the background to our rather uncertain future regarding the use of fossil fuels, changes to the carbon cycle and resulting climate change. This section summarises and categorises the many natural and human factors, and considers the role of feedback mechanisms.

Natural factors and the role of carbon sinks

Chapter 4 detailed the current storage rates and recent changes. Table 6.6 outlines current research on whether carbon sinks will be permanent, or will increase in strength, saturate or disappear.

Table 6.6 Future changes to carbon stores

Terrestrial sinks	Oceanic sink
Modelled to increase generally until 2050. When saturation is reached they begin to act as sources: • thawing tundra permafrost in the Arctic • shift of boreal forests to the north as tundra thaws; these may be able to store more CO_2 if more nutrients (nitrogen) available • tropical rainforests are already at their carbon capacity and may reduce their storage, especially after drought	Increased store in sea grasses and algae, but overall reduction as sink because: • tropical oceans have decreased CO_2 solubility because they are warming, so less uptake of CO_2 • decreased efficiency and slowing down of the biological pump taking nutrients and dissolved inorganic carbon from the surface to ocean floor sediment sink

Human factors

These factors include economic growth, population growth and energy sources used. The IPCC has identified some key factors driving anthropogenic GHG emissions, shown in Figure 6.9.

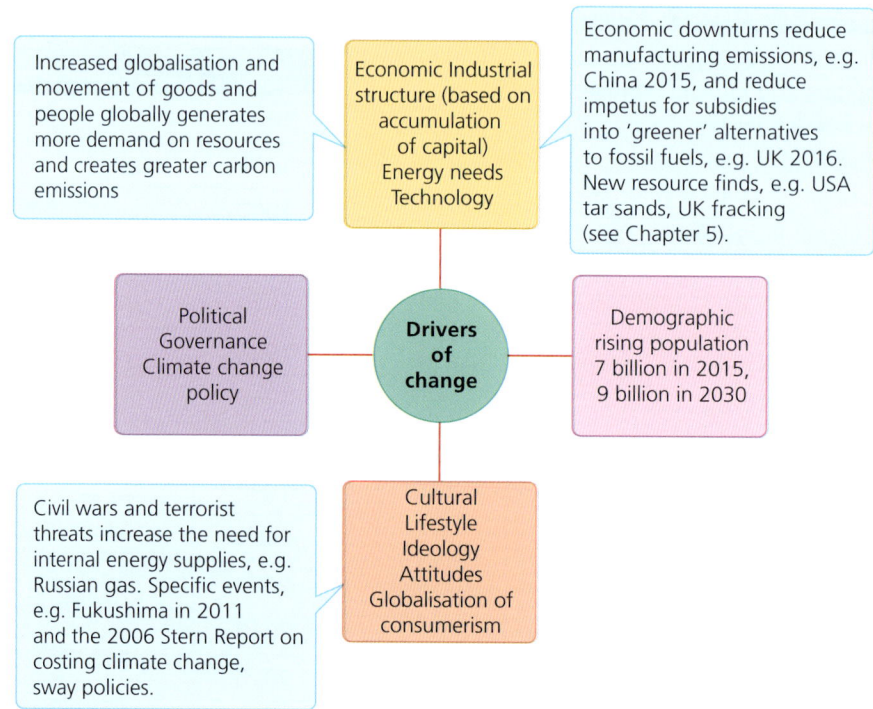

Figure 6.9 Key factors driving anthropogenic GHG emissions

By 2014 the three largest CO_2 emitters were China, the USA and India. In 2006, China overtook the USA because of a global shift in manufacturing production and its rapid urbanisation and industrialisation. In terms of CO_2 emissions per capita, in 2019, China accounted for 27.8 per cent of the global CO_2, emitting 9.43 billion metric tonnes. In 2nd place, the USA contributed 15.7 per cent of the global share, emitting 5.15 billion metric tonnes. This reflects the level of economic development, with a positive relationship between economic growth and CO_2 emissions, although the environmental Kuznets curve suggests it is not quite so simple (Figure 6.5, page 132). Figure 6.1 on page 126 showed that land use is the main driver of global warming; however, modelling suggests that land use emissions in the twenty-first century will be less than half of those from 1850 to 2016.

Positive feedback mechanisms

Chapter 4 introduced the concept of feedback and tipping points in systems, which was revisited in this chapter in the sections on forests and oceans. Figure 6.10 shows the various feedbacks in global systems and their interchanges. You should refresh your understanding of the **positive feedback** mechanisms behind carbon release from peatlands and permafrost, forest die back, thermohaline circulation and the concept of tipping points.

> **Synoptic themes:**
> ### Futures
> Under a low-emissions scenario RCP, the global average surface temperature is predicted to increase by 1–3°C by 2100. This compares with a high-emissions scenario RCP with warming of 2–6°C. Predictions are essential to help governments make decisions about policies on future emissions.

> **Key term**
> **Positive feedback:** This occurs when a small change to a system imbalances that system and this leads to further changes which are amplified over time. In terms of climate change, small increases in temperature lead to further changes to the climate system, and even higher temperatures.

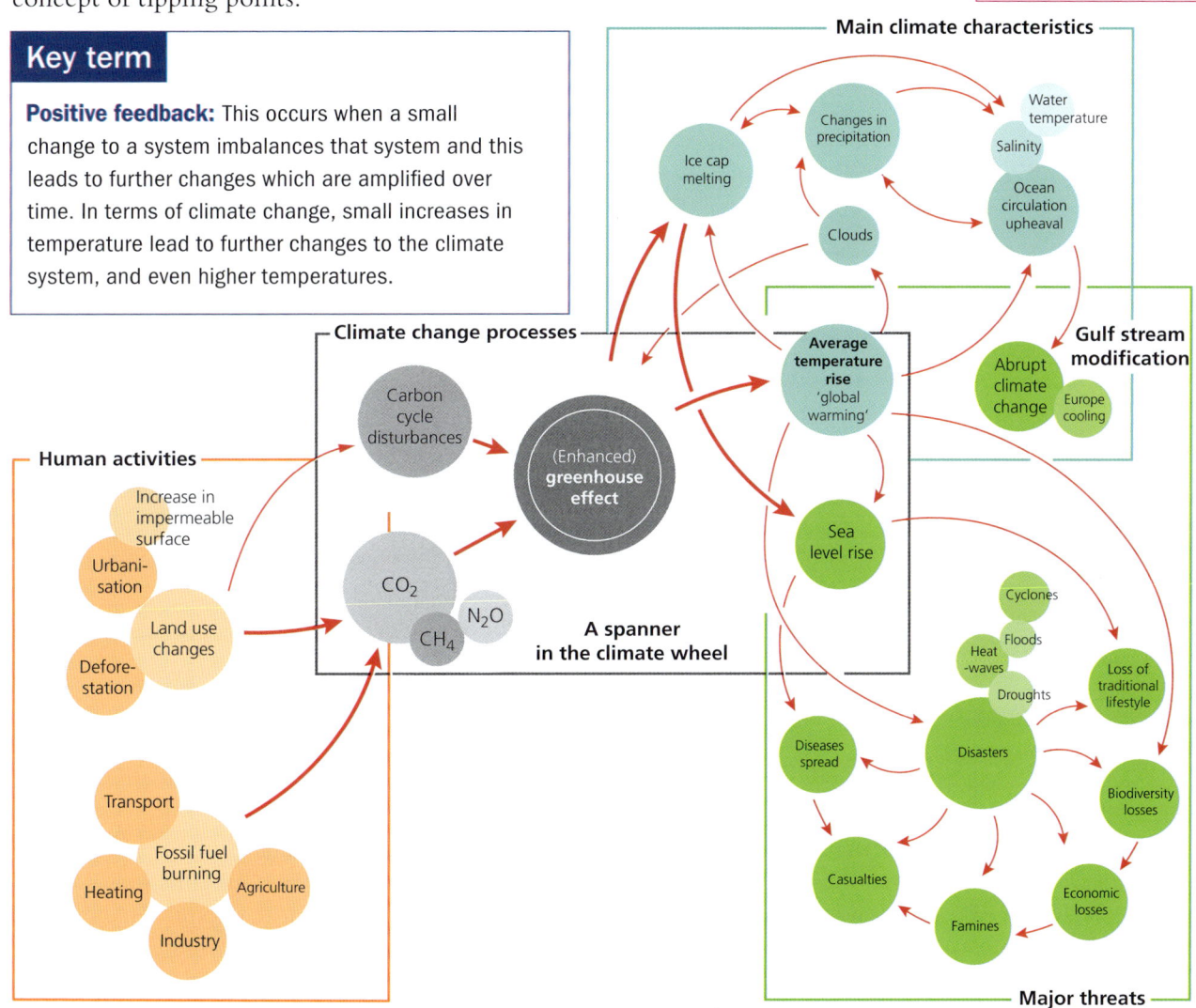

Figure 6.10 Global systems and feedbacks

6 Human threats to the global climate system

Key concept: Climate forcing

Predictions are uncertain because changes to the climate have several causes and feedback mechanisms. Increasing GHGs, changes in volcanic aerosols and ozone levels all cause similar circulation changes in models of the atmosphere.

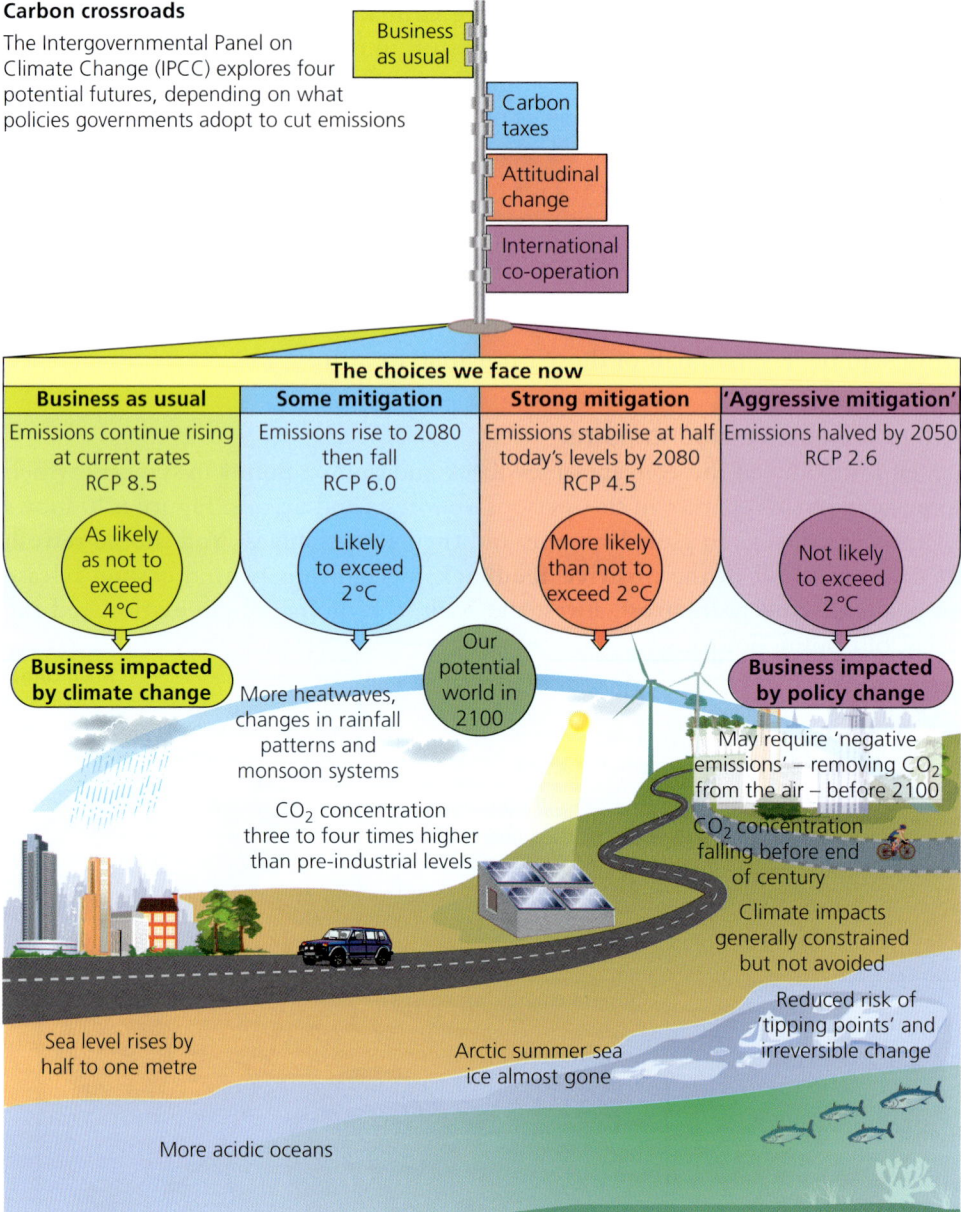

Figure 6.11 The carbon crossroads

There has never been a greater need for effective international agreements on all aspects of the carbon and water cycles. Figure 6.11 shows the 'carbon crossroads' we have reached.

Adaptation strategies for a changed climate

How much we will need to adapt to our changing climate depends on how much more the Earth's atmosphere warms. In turn, this depends on how much more carbon dioxide and other greenhouse gases we emit. Even if all emissions stopped today, we would still have enhanced global warming because of past emissions and the length of time it takes for CO_2 and other GHGs to disperse from the atmosphere.

There are a range of **hard and soft strategies** and **climate change adaptation** strategies. Some are low in technology and upfront costs, while conservation and

Key terms

Climate change adaptation: This includes any passive, reactive or anticipatory action taken to adjust to changing climatic conditions. There are two types:
- **hard strategies** require technology, for example, wind farms
- **soft strategies** involve legislation, such as land-use zoning.

efficiency also rely on changes in cultural practice. Most need a strong lead by governments.

This is possible in stable political regimes, although democracies may encounter more delays than more authoritarian-led countries such as China, because of public debate and changes in policy following elections. Most strategies depend on a partnership between the players involved to make them work. Times of economic austerity may make it more difficult to fund new schemes or impose emission limits on struggling industries.

> The more CO_2 there is in the atmosphere, the greater the effects on Earth's systems, and hence the more difficult adaptation will be.

Some of the costs and risks of specific adaptation strategies for a changed climate are outlined in Table 6.7.

Table 6.7 Costs and benefits of adaptation

Adaption strategies	Benefits	Costs and risks
Water conservation and management	Less resources used, less groundwater abstraction Attitudinal change operates on a long-term basis: use more grey water (recycled water)	Efficiency and conservation cannot match increased demands for water Changing cultural habits of a large water footprint needs promotion and enforcement by governments, e.g. smart meters
Resilient agricultural systems	Higher-tech, drought-tolerant species help resistance to climate change and increased diseases Low-tech measures and better practices generate healthier soils and may help CO_2 sequestration and water storage: selective irrigation, mulching, cover crops, crop rotation, reduced ploughing, agroforestry More 'indoor' intensive farming	More expensive technology, seeds and breeds unavailable to poor subsistence farmers without aid High energy costs from indoor and intensive farming Genetic modification is still debated but increasingly used to create resistant strains, e.g. rice and soya Growing food insecurity in many places adds pressure to find 'quick fixes'
Land-use planning	Soft management: land-use zoning, building restrictions in vulnerable flood plains and low-lying coasts Enforcing strict run-off controls and soakaways	Public antipathy Abandoning high-risk areas and land-use resettling is often unfeasible, as in megacities like Dhaka, Bangladesh, or Tokyo-Yokohama, Japan A political 'hot potato' Needs strong governance, enforcement and compensation
Flood-risk management	Hard management traditionally used: localised flood defences, river dredging Simple changes can reduce flood risk, e.g. permeable tarmac Reduced deforestation and more afforestation upstream to absorb water and reduce downstream flood risk	Debate over funding sources, especially in times of economic austerity Land owners may demand compensation for afforestation or 'sacrificial land' kept for flooding Constant maintenance is needed in hard management, e.g. dredging; lapses of management can increase risk Engrained culture of 'techno-centric fixes': a disbelief that technology cannot overcome natural processes

Rebalancing the carbon cycle

Rebalancing the carbon cycle is seen by scientists, and increasingly by decision makers, as an essential way of preserving Earth's life systems. The IPCC warned in 2014 that it was technically and economically possible to still keep within the target of no more than a 2 °C increase in average global temperatures, but that fossil fuel use needed significant reductions and total elimination by 2100. This

Key terms

Mitigation: Involves the reduction or prevention of GHG emissions by new technologies and low-carbon energies (renewables, nuclear), becoming more energy efficient, or changing attitudes and behaviour. Solar radiation management (SRM) is considered a type of mitigation.

Solar radiation management (SRM): Solar radiation management (SRM), often called geo-engineering, would involve planet-scale engineering to reflect more solar radiation (sunlight) back into outer space so less reaches Earth's surface. This could offset warming caused by greenhouse gas emissions. It is controversial and would require global co-operation. SRM does not reduce emissions, so works in a different way to other mitigation methods, but the outcome would be similar by slowing temperature rise or even reversing it. In contrast, adaptation methods try to cope with rising temperature and other changes to climate.

was to keep within what is widely considered to be the 'safe' limit for global warming.

Key to this goal is **mitigation**. Governments have to weigh up the costs and benefits of mitigation with those of adaptation to climate change. Long-term goals may involve controversial decisions in the short term. The power generated by burning fossil fuels is integral to our way of life, not just in more advanced economies like the UK but in the rapidly growing economies of countries like China and India, and increasingly in poorer economies. Making rapid emissions cuts will be very difficult.

Climate change mitigation can operate in a number of different ways, as shown in Table 6.8.

Table 6.8 Climate change mitigation methods

Method	Explanation
Carbon taxation	Taxes can be applied to carbon dioxide and other greenhouse gases emitted during production and consumption of goods and services. In the UK, car tax ranges from £0–£2745 per year depending on vehicle emissions.
	Many economists argue that taxing greenhouse gas emissions is an efficient way of reducing them because producers and consumers have an economic incentive to avoid the carbon tax.
Renewable switching	Renewable energy including wind, HEP, solar and biomass contributed only 2% of UK electricity generation in 1991 but 41% by 2023.
	Moving away from fossil fuels reduces emissions but renewable energy requires investment and space for solar and wind installations; however, the costs of renewable energy have declined sharply in the last 15 years.
Energy efficiency	Using energy efficient appliances such as LED lightbulbs and reducing energy wastage through building insulation can reduce energy demand and therefore emissions.
	About 15–20% of all global electricity is used for lighting, so switching to more efficient forms could dramatically reduce emissions.
Afforestation	Trees sequester carbon from the atmosphere so planting trees can capture carbon emissions from fossil fuels.
	However, this requires land which not all countries have, and the land might be needed for food production or other uses.
Carbon capture and storage (CCS)	CCS has the potential to capture CO_2 emissions from power stations and factories and then trap the CO_2 underground in rocks (geological stores), thus reducing emissions.
	However, it is a costly and immature technology: less than 1% of global CO_2 emissions were captured by CCS in 2022.
Solar radiation management (SRM)	SRM could involve using satellite mirrors to reflect some solar radiation back into space (like a giant sunshade) or emitting sulphur dioxide particles into the atmosphere to reflect back incoming solar radiation.
	These geoengineering methods are potentially very costly and may have unintended consequences or externalities. They also don't tackle the root cause of global warming, i.e. greenhouse gas emissions that are too high for the climate system to cope with.

The Carbon Cycle and Energy Security

Conservative party politics and austerity measures in the UK since 2011 have also played a role. Will it be possible to cut GHG emissions by 50 per cent by 2050? Technology ranging from nuclear power to smartphone applications will play a large role in replacing fossil fuels and conservation. It would be advisable to refer back to Chapter 4 at this point.

The Kyoto Protocol

The first major international effort to encourage both long- and short-term climate change mitigation was the 1997 Kyoto Protocol, an agreement to cut GHG emissions by 5 per cent on 1990 levels by 2012. It had mixed success, as summarised in Table 6.9.

Figure 6.12 Local players: WinACC members demonstrating before the Paris COP21 meeting in 2015

> ### Synoptic themes:
>
> ### Attitudes and players
>
> To be truly effective, mitigation needs agreement at global, national and individual scales. International efforts to reduce GHG emissions have increased. The UN sets goals and plans, called 'roadmaps'. Countries then use 'carrot and stick' measures to follow roadmaps, setting standards for TNCs and moulding public behaviour. The EU also sets emissions standards for its member states.
>
> TNCs such as Shell are powerful players, leading research and marketing in a globalised market. International pressure groups such as Greenpeace, and local ones such as the city of Winchester's WinACC, lobby for changes in attitude and actions (Figure 6.12).
>
> The public, as consumers, are critical. Online e-petitioning is growing, for example, the 38 Degrees campaigns for more renewable energy and to stop fracking.

Table 6.9 Kyoto's mixed results

Successes	Failures
Kyoto operates until 2020. It started a global approach to tackling anthropogenic climate change. It was the beginning of regular UN conferences on climate change (COP).	Slow ratification. The UK was one of the first but others struggled (Russia) or withdrew (USA, Canada and Japan), fearing economic impacts.
The Clean Development Mechanism supports 75 developing countries in developing less polluting technology.	Only industrialised countries were asked to sign, not developing nations. The top emitters – the USA and China – were left out of the agreement.
Kyoto paved the way for new rules and measurements on low carbon legislation, such as the UK's 2008 Climate Change Act. China is slowing emissions (although that may be due to an economic downturn as well as greener energy).	Complex trading systems were started allowing the trading of 'carbon credits', buying emission allowances from countries not needing them. Carbon sinks were allowed to 'offset' emissions. Both these are criticised for allowing polluters to pollute.
By 2012 emissions were 22.6 per cent lower than the 1990 levels, well beyond the 5 per cent goal. However, 2015 showed a 65 per cent increase above 1990 levels, mainly driven by India and China.	Emission reductions may be because of other factors, such as cheaper gas replacing coal, and a global shift of manufacturing from MEDCs to the 'global south'.

While Kyoto had limited results, the 2015 Paris COP21 climate change conference was touted as a new opportunity to reverse global warming. COP21 had a number of new features compared to Kyoto:

1 All countries were involved, not just developed countries.
2 Countries were asked to bring proposed emissions reduction to the meeting, rather than negotiate them at the meeting.
3 Global warming's potential threats had a much higher public profile in 2015 compared to 1997, and an expectation that 'something needed to be done'.
4 There was clarity in the overall aim, i.e. to limit warming to below 2°C.

Between 1997 and 2015 major changes in emissions had taken place so that China accounted for almost 30 per cent of CO_2 emissions and India 7 per cent. The EU's and USA's relative global contributions had fallen to 10 and 14 per cent, respectively. The 2015 Paris Agreement might be considered as a 'limited success' because:

- Emissions pledges should reduce emissions from about 60 gigatonnes of carbon per year by 2030, to around 55 gigatonnes.
- However, emissions by 2030 need to be about 40–45 gigatonnes to limit warming to 2°C and 30–35 gigatonnes to limit warming to 1.5°C.

Overall, the world needs to reduce emissions to 1990 levels (35 gigatonnes) in order to limit global warming to only 1.5°C – a time when world population was only 5.2 billion and global average income only US$5500. At best, Paris slowed the increase in emissions; it did not stop or even reverse emissions.

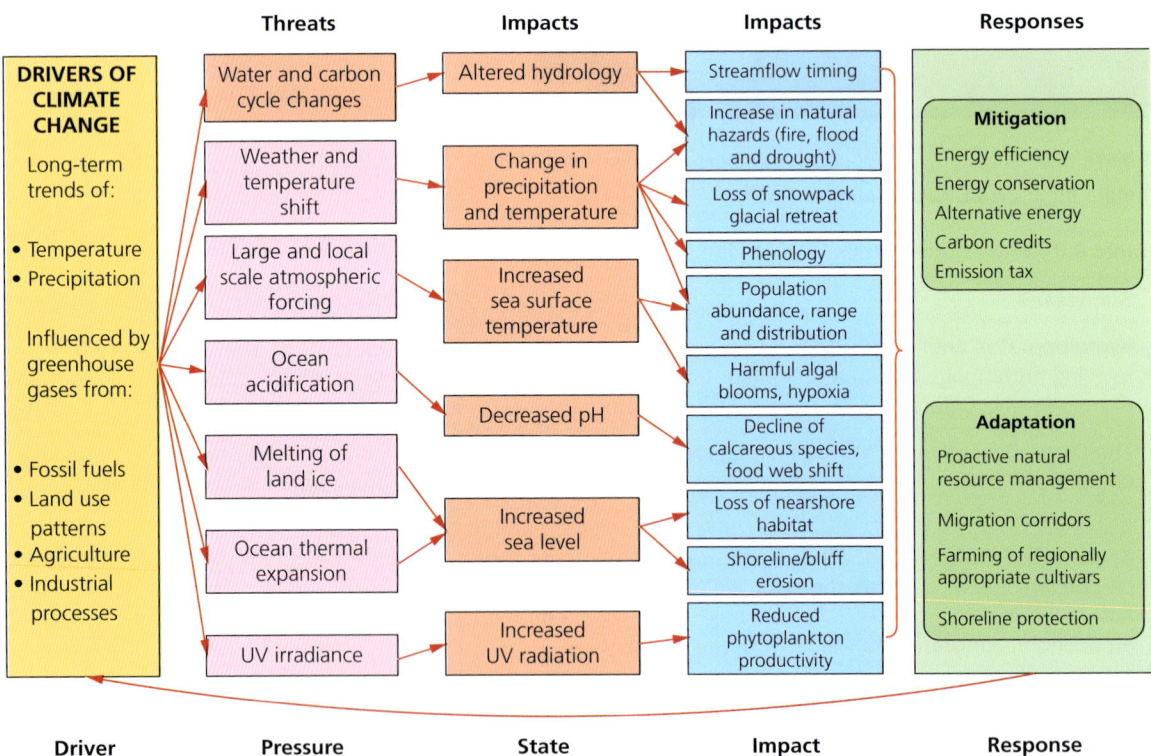

Figure 6.13 Summary diagram: Overview of climate change

Review questions

1. Give reasons why deforestation and grassland conversion are so important in the carbon cycle.
2. How are oceanic warming and acidification pushing ecosystems to a tipping point?
3. Study Figure 6.2 on page 127 and write a short account describing the patterns of forest changes.
4. Explain the links between the enhanced greenhouse effect and climate shift.
5. 'Ecosystems services are critical for human well-being.' Explain why this statement may be correct in relation to forests and oceans.
6. 'Feedback loops in the carbon cycle are complex.' Explain this statement with reference to the Arctic.
7. What role does technology have in helping us to cope with an uncertain future?
8. Assess the costs and benefits of adaptation strategies to cope with the impacts of global warming.
9. Evaluate the success of the 1997 Kyoto Protocol and 2015 Paris Agreement.
10. Categorise the role of the different players involved in balancing the carbon cycle according to their stance and importance.
11. Explain the key factors driving anthropogenic GHG emissions.

Further research

The original data source of most online articles on climate change and the carbon cycle is the IPCC: www.ipcc.ch

Explore the IPCC's archive: https://archive.ipcc.ch/

UK Met Office's summary of the IPCC findings:
https://www.metlink.org/resource/ipcc-updates-for-geography-teachers/

The UK government website has many links:
www.gov.uk/guidance/climate-change-explained

The Global Carbon Project provides good updates:
http://www.globalcarbonproject.org/carbonbudget/index.htm

The World Resources Institute has a wealth of data by country and global trends, including the Global Forest Watch interactive GIS: www.wri.org

The FAO: The State of the World's Forests 2020:
http://www.fao.org/3/ca8642en/CA8642EN.pdf

Exam-style questions

1. Explain why energy consumption varies between countries. [3]

2. Study Figure 6.14 below. This graph shows the change in global surface temperature relative to average temperatures from 1880 to 2020. Temperatures above the 0 °C line are called positive anomalies; those below are called negative anomalies.

   Explain how the data in this graph is useful in the study of climate change. [8]

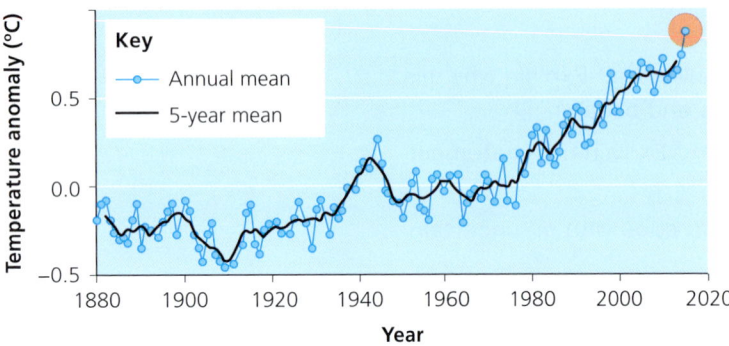

Figure 6.14 Global land ocean temperature index

3. Explain the costs and benefits of biofuels. [3]

4. Study Figure 6.7 on page 134, which shows the IPCC's range of CO_2 future carbon pathways (RCPs). Explain why there are differences between the scenarios projected. [3]

5. Explain how radical technologies could help reduce carbon emissions. [8]

6. Assess the role of increased resource consumption on the carbon cycle. [12]

7. Evaluate the role of different players involved in the carbon cycle. [20]

8. Evaluate the view that re-balancing the carbon cycle is more achievable though mitigation strategies compared to adaptation strategies. [20]

Topic 7
Superpowers

Chapter 7: What are superpowers?

Chapter 8: Superpower impacts

Chapter 9: Superpower spheres of influence

7 What are superpowers?

> **What are superpowers and how have they changed over time?**
> By the end of this chapter you should:
> - understand how different types of powerful country can be defined and characterised using a range of criteria
> - understand how and why patterns of power have changed over time, and how this creates stable or unstable geopolitical situations
> - understand the strengths and weaknesses of emerging powers, and how some are challenging the existing geopolitical order.

7.1 The power of superpowers

A **superpower** is a nation with the ability to project its influence anywhere in the world and be a dominant global force. The term 'superpower' dates from the late 1940s when it was used to describe the three dominant world powers at the time: the USA, USSR and British Empire.

A useful definition of a superpower is a nation that can '*conduct a global strategy including the possibility of destroying the world; to command vast economic potential and influence; and to present a universal ideology*' (Professor Paul Dukes). This definition stresses the *global* nature of superpower status: everyone is aware of a superpower's influence and potential to affect world affairs. In some cases, a lone superpower is referred to as a **hyperpower**. Emerging superpowers are those nations whose economic, military and political influence is already large and is growing. Regional powers are smaller. They influence other countries at a continental scale – a good example is South Africa within Africa.

Superpower status depends on what might be called pillars of power (Figure 7.1). Some nations, such as the USA, have all of these pillars of power, whereas other nations are strong in some areas but weaker in others.

- *Economic power*: this represents the 'base' of the temple in Figure 7.1 and is a prerequisite of power. A large and powerful economy gives nations the wealth to build and maintain a powerful military, exploit natural resources and develop human resources through education.
- *Military power*: this is used in two ways: firstly, the threat of military action is a powerful bargaining chip; secondly, military force can be used to achieve geopolitical goals. Some forms of military power, such as a **blue water navy**, drone, missile and satellite technology, can be deployed globally and reach distant places.
- *Political power*: the ability to influence others through **diplomacy** to 'get your way' is important and is exercised through

Key terms

Superpower: A nation with the ability to project its influence anywhere in the world and be a dominant global force.

Hyperpower: An unchallenged superpower that is dominant in all aspects of power (political, economic, cultural, military); examples include the USA from 1990 to 2010 and the UK from 1850 to 1910.

Blue water navy: One which can deploy into the open ocean, i.e. with large, ocean-going ships. Many smaller nations only have a green water navy designed to patrol littoral waters, i.e. those close to the nation's coastline.

Diplomacy: The negotiation and decision-making that takes place between nations as part of international relations, leading to international agreements and treaties.

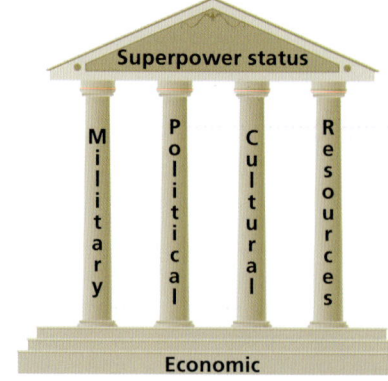

Figure 7.1 The pillars of superpower status

international organisations such as the UN and World Trade Organization, and through bilateral talks between countries.
- *Cultural power* includes how appealing a nation's way of life, values and **ideology** are to others, and is often exercised through film, the arts and food.
- *Resources* can be in the form of physical resources (fossil fuels, minerals, land) but also human resources. The latter includes the level of education and skills in a nation, but also the sheer numbers of people ('demographic weight').

> **Key term**
>
> **Ideology:** A set of beliefs, values and opinions held by many people in a society. These determine what is considered normal or acceptable behaviour. Superpowers project their ideology on others. In the case of the USA this includes 'Western values' of free speech, individual liberty, free-market economics and consumerism.

The extent to which nations enjoy the full range, or only some, of these aspects of power determines their status. Figure 7.2 shows countries by

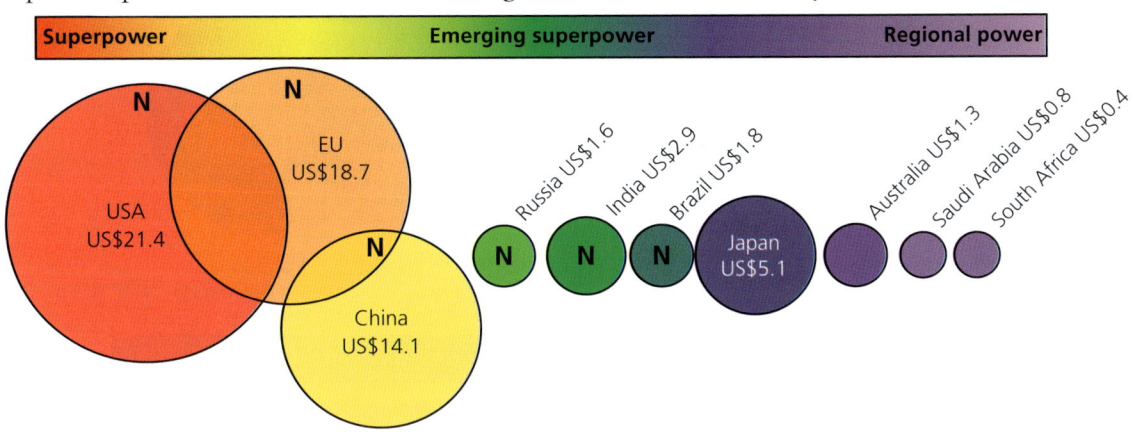

N = Nuclear weapons

Figure 7.2 The power spectrum of countries by GDP (trillions of dollars), 2019

Skills focus: Constructing a superpower index

Data on superpower characteristics can be used to quantify their power and influence. Single measures, such as population size or total GDP, are unlikely to be good indicators because they only quantify one aspect of power.

Table 7.1 shows a power index using four quantitative measures:
- total PPP GDP (economic)
- total population (resources/demographic)
- nuclear warheads (military)
- TNCs (economic/cultural).

Each measure has been ranked, with 1 being the highest score. Total PPP GDP and TNCs have been scaled (multiplied by three and two, respectively) to reflect their greater importance as measures of power.

Using this index, with the USA and China have equal scores of 11. This reflects China's 'catching up' with the USA in the last decade. Other countries are some distance behind these scores. This reflects the fact that they are powerful in some areas but not others. A wide range of data can be used to construct power indices, although the results are often surprisingly similar.

Table 7.1 A superpower index, 2019/20

	Total PPP GDP US$ trillions	Rank, scaled × 3	Total population millions	Rank	Nuclear warheads	Rank	Fortune Global 500 TNCs	Rank scaled × 2	Sum of Ranks
China	27.8	3	1,402	1	290	3	119	4	11
Germany	4.2	16.5	83	6	0	6.5	29	8	37
India	11.3	9	1,362	2	140	5	7	12	28
Japan	5.4	12	126	5	0	6.5	52	6	29.5
Russia	4.2	16.5	146	4	6,500	1	4	14	35.5
UK	3.0	21	67	7	215	4	18	10	42
USA	21.4	4	329	3	6,185	2	121	2	11

Data: World Bank, Wikipedia, https://fortune.com/global500/

Key concept: Hard versus soft power

The political scientist Joseph Nye coined the term 'soft power' in 1990. Soft power is the power of persuasion. Some countries are able to make others follow their lead by making policies attractive and appealing. This contrasts with 'hard power', which means getting your own way by force. Both forms of power have existed for centuries. Economic power can be thought of as sitting somewhere between hard and soft power (Figure 7.3). A third type of power, sharp power, refers to the power of manipulation such as attempts to interfere in other countries democratic elections or manipulate views on social media. This type of power is increasingly important. China and Russia are widely believed to use it.

Nye has argued that the most powerful countries utilise 'smart power': a combination of hard and soft mechanisms to get their own way. The use of different types of power is necessary because:

- Invasions, war and conflict are very blunt instruments. They often do not go as planned and fail to achieve the aims of those exercising hard power.
- Soft power alone may not persuade one nation to do as another says, especially if they are culturally and ideologically very different.

Today, perhaps true smart power comes from combining hard, soft and sharp power mechanisms (Figure 7.4).

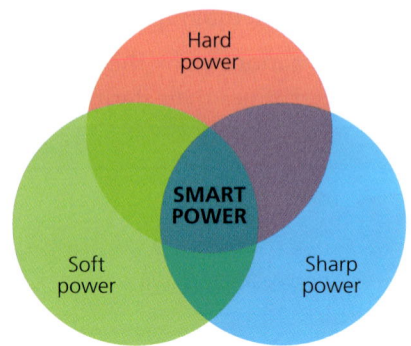

Figure 7.4 Types of power

Hard power	Economic power	Soft power
The spectrum of power		
• Military action or the threat of it • Economic and military alliances • Economic sanction	• Economic and development aid • Favourable trade agreements	• Cultural attractiveness • Appealing values and ideology • Moral authority of foreign policy

Figure 7.3 The power spectrum

---

total GDP. Only the USA, China and the EU could be considered true superpowers. China's case is weakened by its lack of cultural and political influence. The EU is a special case because its weakness lies in the fact that its 27 member states do not always act coherently so, as a whole, the bloc's economic size is not reflected in global influence. In 2020, the UK left the EU. Arguably, this will weaken the EU as the UK is the world's sixth largest economy and a nuclear power with a permanent seat on the United Nations Security Council. If GDP is measured on a PPP (purchasing power parity) basis reflecting the local cost of living, Chinese PPP GDP was larger than that of the USA in 2019 (PPP US$28 trillion versus US$21 trillion).

Very broadly, it is the case that hard power has become less important and soft power more important over time. During the colonial and imperial era, from 1600 to 1950, powerful countries conquered and controlled territory by military force. Armed forces were often positioned across the world in a way akin to a global game of chess.

In the twenty-first century, superpowers and emerging powers do not need to move military hardware about like pieces on a chessboard, and soft power diplomacy is proportionally more important. However, hard power does play a role even today, though most significant military attacks are carried out by the use of air force or drones:

- The Gulf War (1990–1), the invasion of Iraq in 2003 by US-led forces and the American-led war in Afghanistan (2001–14) all involved hard power to achieve aims by force.

China's Belt and Road Initiative (BRI) 1

China's BRI is an example of a **geo-strategic policy** designed to increase China's economic, geopolitical and military influence. It began in 2011 and it is due to be completed by 2049. The BRI involved 140 countries by 2023 with China having:

- invested in six new urban and industrial transport corridors to diversify its access to trade routes across Asia, the Middle East and Europe
- invested in road, rail, port and energy infrastructure in Africa to improve its access to raw materials (ores, minerals, fossil fuels)
- invested in digital, electricity and security infrastructure throughout much of Asia.

The BRI has the potential to tie many countries to China in economic terms and increase China's geopolitical influence. However, this influence is a concern for the USA which is allied to many other Asian countries including South Korea, Taiwan and Japan. There are additional concerns about the sustainability of China's investment in some emerging and developing countries in terms of its environmental impact, human rights and debt levels.

- The Russian annexation of Crimea in 2014 and invasion of Ukraine in 2022, and the subsequent economic sanctions imposed on Russia in response by the EU, the USA and other nations, are recent examples of hard power.

7.2 Changing patterns and polarity

Geopolitics can exhibit different types of polarity:

- A unipolar world is one dominated by one superpower, e.g. the British Empire.
- A bipolar world is one in which two superpowers, with opposing ideologies, vie for power, e.g. the USA and USSR during the **Cold War**.
- A multi-polar world is more complex: many superpowers and emerging powers compete for power in different regions.

The high point of superpower polarity, it could be argued, was the British Empire. The UK, a relatively small country, managed to maintain a global empire that, by 1920, ruled over 20 per cent of the world's population and 25 per cent of its land area (Figure 7.5). The Royal Navy dominated the world's oceans during this period, protecting the colonies and the trade routes between them and Britain. In 1914, Britain's navy was about twice

Key terms

Geo-strategic policies: Policies that attempt to meet the global and regional policy aims of a country by combining diplomacy with the movement and positioning of economic and military assets.

Cold War: A period of tension between ideologically rival superpowers, the capitalist USA and communist USSR, that lasted from 1945 to 1990. It was also the period when nuclear weapons, and systems to deliver them, were perfected, adding to the tension.

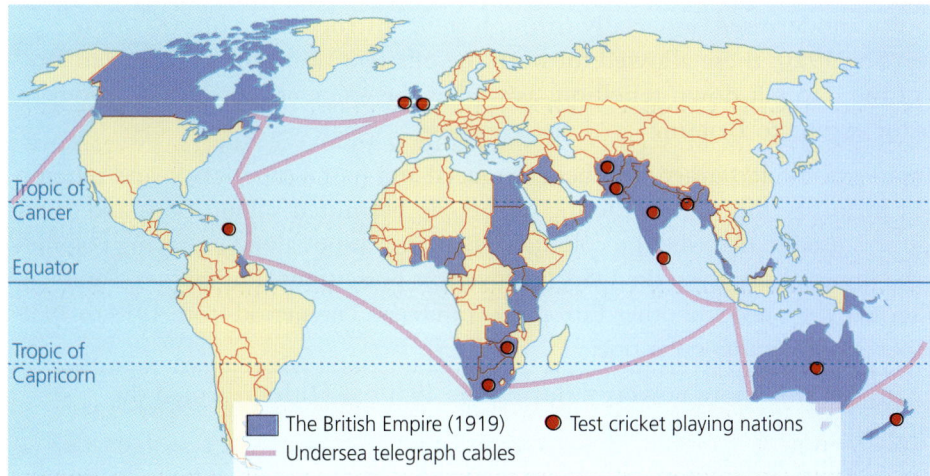

Figure 7.5 The British Empire in 1919

Table 7.2 The two phases of empire building

Mercantile phase, 1600–1850	Small colonies are conquered on coastal fringes and islands, e.g. New England (now the USA), Jamaica, Accra (Ghana) and Bombay (India), and defended by coastal forts.
	The forts, and navy, protect trade in raw materials (sugar, coffee, tea) and slaves.
	The economic interests of private trading companies such as the Royal African Company, Hudson's Bay Company and East India Company are defended by British armed forces. Some of these companies had vast complex operations. The East India Company had the authority to print its own money, set its own taxes and embark on wars on behalf of the national interest.
Imperial phase, 1850–1945	Coastal colonies extend inland, with the military conquest of vast territories.
	Religion, competitive sport (e.g. cricket) and the English language are introduced to colonies.
	Government institutions with British colonial administrators are set up to rule the colonial population.
	Complex trade develops, including the export of UK-manufactured goods to new colonial markets.
	Settlers from Britain set up farms and plantations in colonies.
	Technology, such as railways and telegraph, is used to connect distant parts of the empire.
	Rule was maintained through a combination of British colonial administrators and local elites benefiting from British largesse and protection.
	Uprisings against British Rule were quashed through massacres (e.g. at Jallianwala Barg, General Reginald Dyer ordered his troops to fire at the people assembled there – between 15,000 and 20,000 men, women and children).

> **Key terms**
>
> **Colonial control:** The direct militarised control exerted over territories conquered by mainly European powers in the period 1600 to 1900. They were ruled by force, with almost no power or influence being given to the original population.
>
> **Acculturation:** A process of cultural change that takes place when two different cultures meet and interact; it includes the transfer of a dominant culture's ideas on to a subordinate culture.

as large as the next largest, that of Germany. The British Empire can be seen as growing in two distinct phases (Table 7.2).

Colonial India is a good example of how colonies were controlled directly:

- British military personnel, civil servants and entrepreneurs emigrated to India to run the Raj.
- Educated Indians (speaking English, and wearing European dress) occupied many lower administrative positions.
- Symbols of imperial power, such as the residence of the governor-general in Delhi and the Howrah Bridge in Kolkata, demonstrated Britain's imperial wealth and technical prowess.
- A process of **acculturation** was undertaken as British traditions, such as cricket, afternoon tea and the English language, were introduced.
- A strict social order was maintained that differentiated the ruling white British and the elite who were often used by the British to maintain power from the Indians.
- India was modernised, especially through the construction of 61,000 km of railways by 1920; these allowed for both the efficient transport of troops to any part of India to put down rebellion, as well as the efficient transport of goods to ports for export to Britain.

The Empire was underpinned by a belief in racial and moral superiority, and religion was used to justify this. The 'othering' of people permitted subjugation, and racist language was used to describe the colonised. Britain was not alone in having colonies during the imperial era. Even the USA had overseas possessions (Philippines, Panama), as did Germany, France and other European countries. The period from 1919 to 1939 was an increasingly multi-polar one:

- Germany became more powerful during the 1930s as Hitler rearmed the country and prepared it for war.
- Imperial Japan began to be an increasing power in Asia.
- The USA became economically and militarily stronger, challenging Britain's traditional global leadership.
- European powers were still strong, but weakened by poor economic performance and the costs of maintaining empires.

The global geopolitical situation in the 1930s was messy to say the least, with emerging powers threatening the traditional geographical **spheres of influence** of established superpowers and regional powers. Military power was increasingly important and, as war approached in 1939, an arms race took place, with countries strengthening their naval power (Figure 7.6).

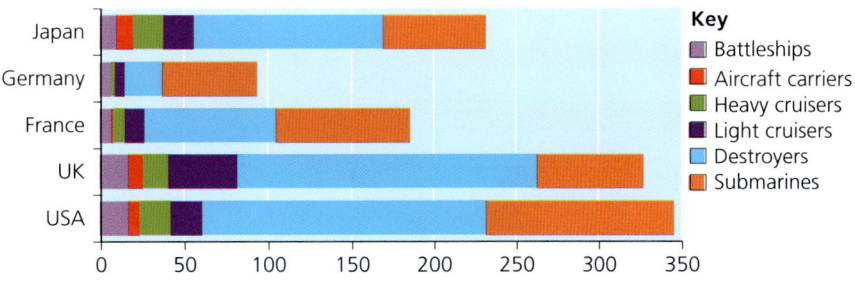

Figure 7.6 The relative size of navies in 1939

The post-colonial era

The colonial era came to an end relatively quickly after the end of the Second World War in 1945. Most colonial powers had lost their colonies by 1970. The reasons for this include:

- Post-war bankruptcy meaning there was no money to run, or defend, colonies.
- The focus on post-war reconstruction at home meant that colonies were viewed as less important.
- Anti-colonial movements, for example, in India, grew increasingly strong and demands for independence could not be ignored.

Much of the era from 1945 to 1990 was dominated by the Cold War. During this bipolar era the USA became an increasingly global superpower with worldwide military bases aimed at containing the USSR and preventing the spread of communism. The USSR built a contiguous core of countries it either allied with (Eastern Europe) or invaded (Afghanistan). As Table 7.3 shows, these two superpowers were very different.

> **Key term**
>
> **Sphere of influence:** A geographical area over which a country believes it has economic, military, cultural or political rights. Spheres of influence extend beyond the borders of the country and represent a region where the country believes it has a right to influence the policies of other countries.

Table 7.3 The USA and USSR Cold War superpowers compared

	USA	USSR
Human resources	Population of 287 million in 1989	Population of 291 million in 1991
Physical resources	Self-sufficient in most raw materials; oil importer	Self-sufficient in most raw materials; oil exporter
Economic system	Capitalist, free-market economy and global TNCs	Socialist, centrally planned economy; most businesses were state owned
Political system	Democracy with free elections held every four years	Single-party state with no free elections (dictatorship)
Allies	Western Europe through NATO. Strong economic and military ties to Japan and South Korea	Eastern Europe (the Warsaw Pact countries) and alliances with Cuba and other developing nations
Military power	World's largest navy and most powerful air force, with a 'ring' of bases surrounding the USSR. Large nuclear arsenal and global network of nuclear bases. Extensive global intelligence gathering through the CIA	Very large army, and large but often outdated naval and air force capability. Nuclear weapons. Troops stationed in Eastern Europe. Extensive global intelligence gathering through the KGB
Cultural influence	Film, radio, television and music industry proved a powerful vehicle for conveying a positive view of consumerism, family values, democracy and affluence to a global audience	Exported a 'high' culture message focused on ballet, classical music and art in contrast to the 'popular' culture of the USA. Strict censorship within the USSR

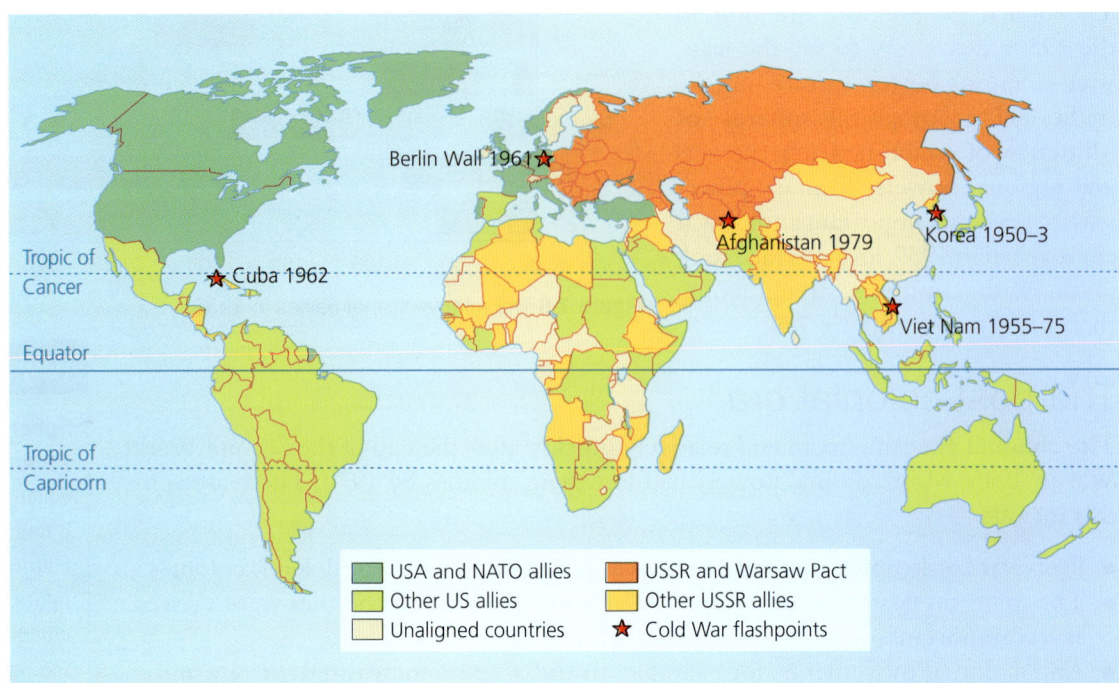

Figure 7.7 Cold War alliances circa 1980

The Cold War was so called because it never led to 'hot war', that is, direct conflict between the USA and USSR. It was a 45-year-long stand-off between what were, by far, the world's two most powerful countries at the time.

Superficially at least, it appears to have been a period of stable, if tense, geopolitics. However, there was conflict during this era. During the Cold War many countries allied themselves with the USA or with the USSR, either formally or informally (Figure 7.7). There were numerous flashpoints that came perilously close to direct conflict, or were proxy wars. Proxy wars occurred when the USA and/or USSR supported one side in a conflict but did not directly fight each other:

- the 1950–3 Korean War, which led to the division of Korea into US-backed South Korea and Chinese/Russian-backed North Korea
- the 1955–75 Viet Nam War, fought directly by the USA but indirectly by communist China with some weaponry from the USSR
- the Cuban Missile Crisis in 1962, the closest the USSR and USA got to direct conflict during the Cold War.

Neo-colonialism

A key issue during the Cold War was how developing countries could be influenced and controlled after they gained independence from their former colonial masters. The term 'neo-colonialism' was coined to refer to an indirect form of control that meant newly independent countries were not actually masters of their own destiny. Tellingly, the term was first used by Kwame Nkrumah, the first president of independent Ghana (see box to the left).

A key question is: how does a powerful country indirectly control decision-making and policy in another country? Several neo-colonial mechanisms have been suggested (Table 7.4).

> 'Neo-colonialism is … the worst form of imperialism. For those who practise it, it means power without responsibility, and for those who suffer from it, it means exploitation without redress. In the days of old-fashioned colonialism, the imperial power had at least to explain and justify at home the actions it was taking abroad. In the colony those who served the ruling imperial power could at least look to its protection against any violent move by their opponents. With neo-colonialism, neither is the case.'
>
> Neo-Colonialism: The Last Stage of Imperialism, 1965

Table 7.4 Possible mechanisms of neo-colonial control

Strategic alliances	Military alliances between developing nations and superpowers make the developing nation dependent on military aid and equipment from the superpower
Aid	Development aid comes with 'strings attached' (tied aid), forcing the recipient to agree to policies and spending priorities suggested by the aid donor
TNC investment	Investment from abroad may create jobs and wealth, but be dependent on the receiving country following 'friendly' policies
Terms of trade	Low commodity export prices contrast with high prices for imported goods from developed countries, inhibiting development
Debt	Developing countries borrow money from developed ones, and then end up in a debtor–creditor relationship
Control over institutional economic institutions	A number of international economic institutions like the World Bank, IFC and the IMF regulate and control the international economy. The developed states have a monopolistic control over these institutions.

During the Cold War, neo-colonial relationships provided both the USA and USSR with numerous allies throughout the developing world. However, these relationships often propped up corrupt, anti-democratic and violent regimes, such as those of President Mobutu in Zaire (now the Democratic Republic of the Congo) from 1965 to 1997 supported by the USA, and President Mengistu in Ethiopia from 1977 to 1991 supported by the USSR.

It is interesting to consider which type of geopolitical polarity (uni-, bi-, multi-polar) (Figure 7.8) is the most and the least stable. There is no simple answer.

- A unipolar world dominated by one hyperpower might appear stable, but the hyperpower is unlikely to be able to maintain control everywhere, all the time, which could lead to frequent challenges by rogue states not accepting of the hyperpower's hegemonic position.
- A bipolar world could be stable, as it is divided into two opposing blocs. Stability will depend on diplomatic channels of communication between the blocs remaining open and each superpower having the ability to control countries in its bloc; breakdown of control and/or communication could lead to disastrous conflict.
- Multi-polar systems are complex as there are numerous relationships between more or less equally powerful states; the opportunities to misjudge the intentions of others, or fears over alliances creating more powerful blocs, are high and may increase the risk of conflict.

It could be argued that the period between 1910 and 1945 was a multi-polar one, and that this complex geopolitical situation contributed to two world wars. Many observers believe that the twenty-first century could be multi-polar as countries such as India and China become increasingly powerful while the power of the USA and EU wanes.

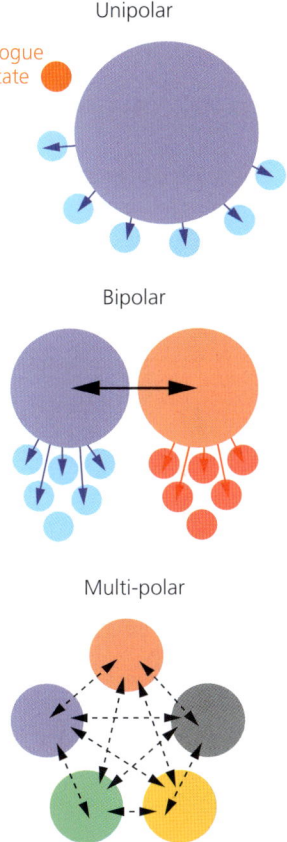

Figure 7.8 Uni-, bi- and multi-polar worlds

Key concept: Hegemony

Hegemony is a term used to describe the dominance of a superpower (the hegemonic power, or hegemon) over other countries. Hegemony can be exercised in several ways. The sheer size and multiple capabilities of US military forces give it dominance over world affairs, which deters others from acting against it. The USA spends more on its military (US$732 billion in 2019) than the next nine largest spenders combined (Figure 7.9).

Hegemony can be much more subtle. The Italian Marxist philosopher Antonio Gramsci described a type of power called cultural hegemony. Imprisoned by the Italian fascist dictator Mussolini, Gramsci was nonetheless impressed by the dictator's ability to influence opinion without resorting to direct force. Gramsci argued that people consented to Mussolini's power because the values of the powerful were accepted as 'common sense' by the majority of people. This 'common sense' view is projected and reinforced by:

- education systems that teach a particular ideology
- religion, which can subtly reinforce political ideology
- music, television and film, whose themes and storylines can reinforce certain values and demonise others
- news media, which controls the message people hear.

It is perhaps no surprise that the dominant superpower of today, the USA, has an enormous global reach in terms of culture and media. Think of CNN, Disney and Pixar, blockbuster action movies, *The Simpsons*, McDonald's and KFC — all of these spread an American view of the world and American values.

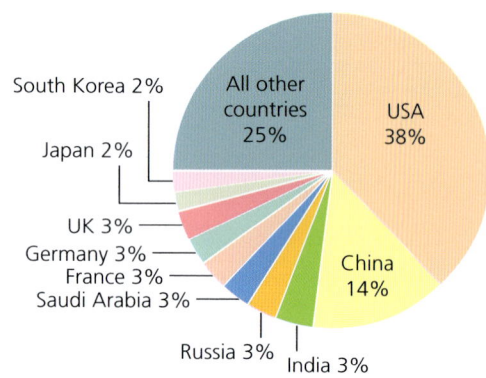

Figure 7.9 Military spending 2019 (percentage of global spending)

7.3 Emerging powers

In 2019, 46 per cent of global GDP was accounted for by the USA and the EU. China's economy in 2019, at US$14.1 trillion, represented 16.2 per cent of global GDP, and the next largest, Japan, only 6 per cent. However, in the future things are likely to change.

- The EU and Japan have ageing populations and are not likely to grow demographically; in turn, their economies will grow only slowly, if at all.
- The USA has less of an ageing problem, but economic and population growth are likely to be moderate rather than rapid.
- China's population is also ageing, but its manufacturing-based economy has huge potential as it shifts to growth based on services and consumerism; this is also true of Brazil which, like China, has a growing middle class.
- Some other demographically large countries will get much bigger, and potentially much richer – India, Indonesia and Nigeria are examples.

Future superpowers are likely to emerge from two groups of countries, which overlap:

- BRIC: Brazil, Russia, India and China were identified as a group of emerging powers in 2001 by Goldman Sachs banker Jim O'Neill. South Africa is sometimes included (BRICS); these five countries set up a formal association in 2009. In 2024, Iran, Egypt, Ethiopia, Saudi Arabia and the United Arab Emirates were also identified as part of this group of emerging powers.

- The G20 major economies: this group formed in 1999 and meets annually. It is made up of 19 countries plus the EU and includes some potential emerging powers, such as Mexico, Indonesia, South Korea, Saudi Arabia and Türkiye.

The G20 countries collectively account for 90 per cent of global GDP, 80 per cent of world trade and about 65 per cent of the world's population.

Emerging power strengths and weaknesses

Trying to spot the superpowers of tomorrow is very difficult. In terms of population and economy, it is possible to project growth into the near future (Table 7.5). Countries with both a large population and a large economy are likely candidates for future superpower status. By 2030, it is very likely that China and the USA will be more equal in terms of power, with India being a more significant global player. Despite being demographically large, Indonesia and Nigeria will have annual GDPs of US$2 trillion and US$1 trillion, respectively, by 2030, making them significant regional powers but no more than that.

All emerging powers have strengths and weaknesses, which are outlined in Table 7.6. It is worth mentioning the EU at this point. On the face of it the EU might be considered a superpower. The combined economy of the 27 EU nations is similar to that of the USA. However, the EU has a number of weaknesses:

- The 27 nations rarely agree easily, so many EU decisions are compromises that weaken its 'global message'.
- The EU economy has been weak since the global financial crisis of 2007–8.
- Demographically, it has an ageing population; the very high social costs associated with unemployment and pensions are a drag on economic growth.
- The UK's decision to leave in 2020 may weaken the EU, and the long process of 'Brexit' negotiations has distracted the bloc from its core economic and social purposes.

Table 7.5 The largest countries by population and GDP in 2030

Five largest countries by population in 2030		Five largest countries by GDP in 2030	
India	1,520 million	USA	US$25 trillion
China	1,390 million	China	US$22 trillion
USA	360 million	India	US$7 trillion
Indonesia	280 million	Japan	US$6 trillion
Nigeria	260 million	Germany	US$6 trillion

Table 7.6 Strengths and weaknesses of the emerging powers

	Strengths	Weaknesses
Brazil	▹ Regional leader in Latin America 💰 Strong agricultural economy and exporter 💰 Energy independent in oil and biofuels 💰 Growing middle class and maturing consumer economy 🎭 Culturally influential with 2014 World Cup and 2016 Olympics	✈ Small military with only a regional intervention capacity 💰 Economy suffers from 'boom and bust' phases 🌍 Needs to control the destruction of its forests 💰 Education levels lag behind competitors ▹ Populist politics often lead to unstable governments and corruption
Russia	✈ A nuclear power with very large military capacity 💰 Huge oil and gas reserves are a source of wealth ▹ Permanent seat on the UN Security Council	👤 An ageing and declining population, which is also unhealthy 💰 Extreme levels of inequality 💰 Economy is overly dependent on oil and gas ▹ Difficult diplomatic and geopolitical relationships with the EU and USA
India	👤 A youthful population with large economic potential 🎭 English is widely spoken and graduate education is widespread ✈ Nuclear armed, and has sophisticated space and missile technology	💰 Possible future resource shortages, especially water and energy 💰 Poor transport and energy infrastructure 💰 Very high levels of poverty ▹ Poor political relations with its neighbours, especially Pakistan
China	💰 A highly educated, technically innovative population 💰 Soon to be the world's largest economy and leads in fields such as renewable energy ✈ Military technology and reach is growing, and challenging the USA 💰 Modern infrastructure in terms of transport, e.g. high-speed rail	👤 Will soon have problems with an ageing population 🌍 Major pollution issues in terms of air and water quality ▹ Tense relationships with its neighbours in Southeast Asia 💰 Rising wages make its economy increasingly high cost for TNCs 💰 Relies on imported raw materials ▹ Plays a limited geopolitical role; not yet a leader on the global stage
Mexico	💰 Part of the United States–Mexico–Canada Agreement (USMCA) trade bloc and has an advanced economy ▹ Slowly becoming more democratic and 'open for business'	🎭 Crime, often drug related, gives it a poor global image 💰 Population is increasingly obese, leading to high healthcare costs 👤 Many well-educated, skilled workers migrate from Mexico
Indonesia	👤 A youthful, potentially dynamic population with economic growth potential 💰 Large untapped natural resources	🌍 Deforestation is a growing environmental disaster 💰 High levels of urban and rural poverty ▹ Internal political instability has been a recent problem
Türkiye	💰 Economy is increasingly integrated with the EU, could become an EU member ▹ Member of the NATO military alliance 👤 Youthful population with good education levels	▹ Internal political problems with its Kurdish minority ▹ External political instability in the Middle East on its borders
Key	💰 Economic ▹ Political ✈ Military 🎭 Cultural 👤 Demographic 🌍 Environmental	

Japan: a lesson from history

It may seem obvious that China is the next global superpower, but the case of Japan provides a warning against making such assumptions. In the 1980s it seemed obvious that Japan was, economically at least, destined for superpower status. Per capita GDP was higher than the UK and USA by 1980, and it was a world leader in exporting consumer electronics, ICT, cars and industrial machinery to the rest of the world.

Figure 7.10 shows the total GDP of Japan, the USA and China from 1980 to 2014. Back in the 1990s economists expected Japan's economy to continue to grow (shaded area in Figure 7.10) but this did not happen. In fact, Japan's economy has barely grown at all since 1995. This was being called the 'lost decade' by 2005, but is now two lost decades. This happened because:

- A property value bubble burst in 1989–90, which led to a collapse in the Japanese stock market.
- High interest rates of 4–6 per cent encouraged saving, not spending, so the economy slowed.
- Japan's ageing population quickly became a problem, slowing the economy further.
- More competitive Asian economies, such as South Korea and then China, stole Japan's lead on hi-tech consumer goods.

Japan warns us that predicting the economic power of countries in the future is really just educated guesswork.

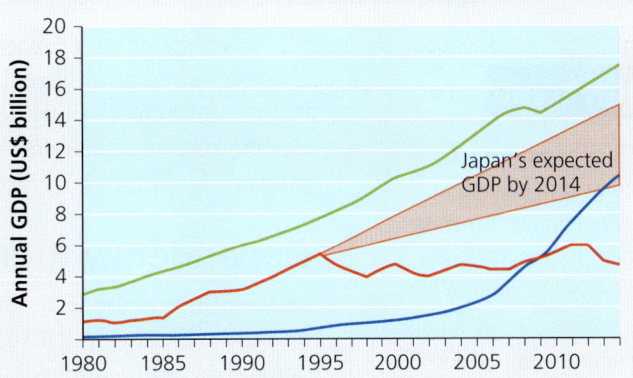

Figure 7.10 GDP of China, Japan and the USA, 1980–2014

Contrasting development theories

Modernisation theory

Modernisation theory is a useful framework within which to consider how some nations become wealthy and powerful. In 1960, the American economist WW Rostow outlined a theory that argued that countries develop in five stages. This has become known as the 'take-off model', or modernisation theory.

Rostow argued that pre-industrial societies would develop very slowly until certain preconditions for economic take-off were met:

- Exports of raw materials to generate income.
- Development of key infrastructure, e.g. roads, ports, electricity.
- Technology, e.g. telephones, radio, television, becomes more widespread.
- Education, leading to increased social mobility.
- Banking and financial systems, to allow places to take part in global trade.
- Governance and legal systems, to protect investors, property owners and trade transactions.

Once these were in place, industrialisation and the growth of secondary (manufacturing) industry would begin, along with increasing urbanisation. A country would rapidly become an industrial one and wealth would increase. Referring back to Figure 7.10, China's take-off point can be seen around 1995, with GDP growing rapidly from 2003 onwards.

Modernisation theory saw capitalism as the answer to poverty, and emphasis was laid upon cultural and institutional transformations if the developing nations were to modernise. The prevailing belief was developing countries needed to be integrated into the world market and establish free trade if they were to enjoy the benefits of economic growth.

7 What are superpowers?

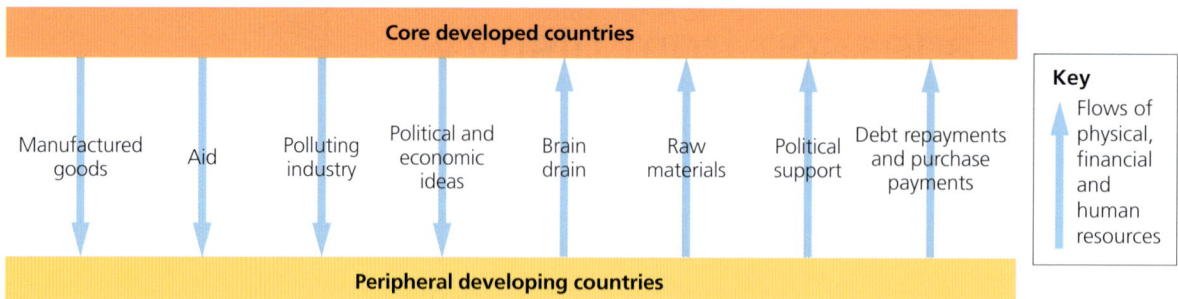

Figure 7.11 The mechanisms of dependency theory

Rostow's theory only really describes the process of economic change and growth. It does not help us to understand how some countries gain the political and cultural aspects of power needed to be a superpower. However, economic wealth and an industrial economy are precursors to obtaining military power.

Dependency theory

In the 1960s some left-wing academics began to argue that, despite independence, many developing African, Asian and Latin American nations existed in a state of dependency and underdevelopment. The political economist Andre Gunder Frank outlined his **dependency** theory in the mid-1960s. Frank stated that:

'contemporary underdevelopment is in large part the historical product of past and continuing economic and other relations between the satellite underdeveloped and the now developed metropolitan countries. Furthermore, these relations are an essential part of the capitalist system on a world scale as a whole.'

The Development of Underdevelopment, *1972*

Frank saw 'satellite' (periphery) countries providing a range of services to metropolitan (core) countries (Figure 7.11), for example:

- cheap commodities, such as oil, copper, coffee and cocoa
- labour in the form of migration, especially 'brain drain' migration of skilled workers
- markets for manufactured goods and locations for investment, such as mines and HEP dams.

For their part, the developed countries controlled the development of developing nations by setting the prices paid for commodities, interfering in economies via the World Bank and IMF, and using economic and military aid to 'buy' the loyalty of satellite states. Dependency theory is relevant to superpower status in a number of ways:

- Superpowers that control developing nations are gaining economic wealth and power by exploiting them.
- In turn, superpowers actively contribute to the 'underdevelopment' of these countries which reduces the number of potential emerging powers.
- Wealthy local elites, who own exporting/importing businesses and have political connections, benefit from the dependency relationship because they control the limited trade in goods and services, but the wider population does not benefit.
- Frank's experience researching in Latin America led him to believe that development was not possible while developing countries were locked in an asymmetric relationship with core developed countries.

> **Key term**
>
> **Dependency:** In the context of economic development it means that the progress of a developing country is influenced by economic, cultural and political forces that are controlled by developed countries.

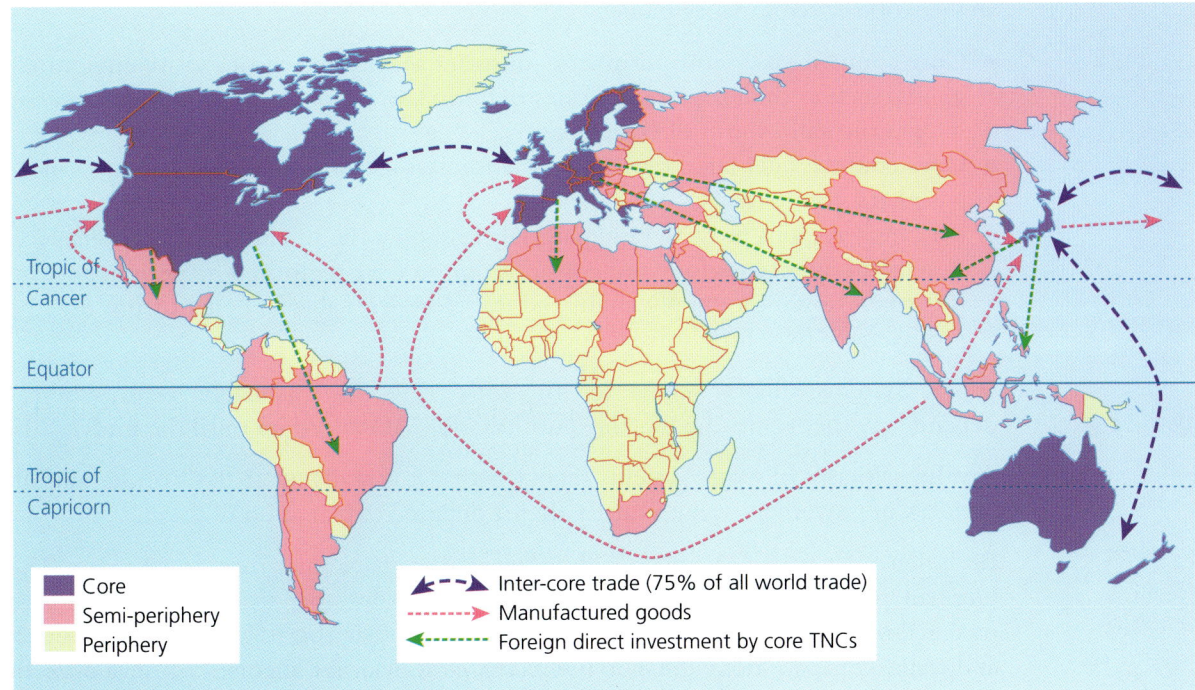

Figure 7.12 World systems theory

Against dependency theory is the fact that newly industrialised countries (NICs) have developed, for example, Taiwan, South Korea, Singapore and Malaysia. On the other hand, many Asian NICs received significant economic aid and political support from the USA to help them develop, creating a ring of strong capitalist economies to contain communist China.

World systems theory

A weakness of dependency theory is that it is static. The theory suggests that countries are stuck in a permanently underdeveloped state. The rise of NIC nations since 1980 suggests the world is more complex than a simple superpower core and undeveloped periphery. Immanuel Wallerstein developed his world systems theory in the 1970s (Figure 7.12). Wallerstein stressed that development should be viewed within a global economic context rather than focusing on individual countries. As the global capitalist economy expanded in the nineteenth and twentieth centuries some once peripheral regions have increasingly become part of the world economy, so that three broad economic development categories can be recognised:

- core regions – the Organisation for Economic Co-operation and Development (OECD) countries and the USA and EU superpowers
- semi-periphery regions – the NICs of Latin America and Asia, including emerging powers such as India and China
- periphery regions – the rest of the developing world.

Wallerstein's theory has the flexibility to recognise that countries may change group over time. In the future some areas, such as China, may move into the core world. A more classically Marxist theory, such as Frank's dependency theory, is more rigid, simply seeing the world as divided into 'owners' (rich countries) and 'workers' (poor countries).

7 What are superpowers?

In the world systems theory model:

- Core countries use semi-periphery countries as cheap locations to manufacture goods, such as the Free Trade Zones of China, or as locations for cheap services, such as the call centres of Bangalore.
- Core countries get large returns on the foreign investment they make in semi-periphery countries.
- Periphery regions provide raw materials to supply the manufacturing industry in semi-periphery regions and consumption in core regions. The periphery is furthest down the supply chain and therefore least able to benefit from the profits made by selling finished goods and services.

World systems theory seems to 'fit' today's world reasonably well, whereas dependency theory fits the 'north–south' world that existed up until the mid-1980s.

A valid criticism of world systems theory is that it is really an analysis of the world's patterns of power and wealth rather than a detailed explanation of them.

Both dependency theory and world systems theory have their roots in Marxism. The German philosopher Karl Marx argued that capitalist societies were divided into owners and workers, in other words a wealthy elite that controlled industry; business and trade set apart from the poor working class. Marxist theories of development work within this paradigm of rich and powerful versus poor and powerless. Rostow's modernisation theory ignores the Marxist class division and instead argues that countries develop in a linear way from 'poor' to wealthy and powerful. The very fact that few countries have attained the full modernisation envisaged by Rostow might suggest that it is at best a partial explanation.

Review questions

1. Define the term 'superpower'.
2. Explain what is meant by 'smart power'.
3. Explain how the 'polarity' of superpower patterns has changed over time.
4. The Cold War period could be described as 'stable but dangerous'. How far do you agree?
5. Explain the concept of hegemony.
6. Using Table 7.5 (page 157), and your own research, rank the emerging powers in terms of their strengths and weaknesses. Comment on which are most and least likely to threaten the USA by 2030.

Further research

Use data from the World Bank and CIA World Factbook to construct your own superpower index using economic, political, cultural and military data: https://data.worldbank.org and https://www.cia.gov/the-world-factbook/

Explore the military power of different countries: https://www.globalfirepower.com/countries-listing.asp

Explore a timeline of the Cold War era: https://coldwar.org/default.asp?pid=16879

8 Superpower impacts

> **What are the impacts of superpowers on the global economy, political system and the physical environment?**
> By the end of this chapter you should:
> - understand how superpowers attempt to control the global economy, and take advantage of it
> - recognise that some superpowers have significant cultural influence, which is a source of power
> - understand the role that superpowers play in global economic, political and environmental governance
> - understand that superpowers and emerging powers have a disproportionate impact on the global environment and global resource consumption.

8.1 The global economic system

Since the end of the Cold War in 1990 the world has been dominated by the free-market capitalism economic system. This economic system operates in most of the world with a few exceptions. For much of the twentieth century, capitalism operated alongside socialist centrally planned economies, an alternative economic system. The two systems are compared in Table 8.1.

Table 8.1 Capitalism versus centrally planned economic systems

	Free-market capitalism	**Centrally planned economy**
Features	Private ownership of property, e.g. homes and possessions	Government ownership of property and land
	Private ownership of businesses; wages based on supply versus demand and skill level	Most businesses state owned and wages determined centrally
	The right to make a profit and accumulate any amount of wealth	Profits taken by the government and used to provide public services
	The buying and selling of goods and services in a competitive free market, with limited restrictions	Prices controlled by the government, which also controls the supply of goods and services
Examples	USA, Canada, Japan, Western Europe	USSR, China, Eastern Europe, Cuba

Free-market capitalism has become increasingly dominant as alternative economic systems have weakened, notably:

- the collapse of socialist economies in the USSR and Eastern Europe after 1990
- China's movement away from a socialist economy towards what has been called 'state capitalism'
- reform in communist Cuba, allowing limited private ownership of business.

Free-market capitalism and **free trade** have been promoted by many global **inter-governmental organisations (IGOs)**, strengthening its grip on the world economy (Table 8.2, page 164). Superpowers, especially the USA and EU, attempt to control the global economy through free trade, IGOs, TNCs and their cultural influence.

> **Key terms**
>
> **Free trade:** The exchange of goods and services free of import/export taxes and tariffs or quotas on trade volume. Taxes, tariffs and quotas are forms of protectionism designed to make imports more expensive than locally produced goods (thus protecting local producers).
>
> **Inter-governmental organisations (IGOs):** Regional or global organisations whose members are nation states. They uphold treaties and international law, as well as allowing co-operation on issues such as trade, economic policy, human rights, conservation and military operations.

Table 8.2 Global organisations and capitalism

World Bank	Makes development loans to developing countries, but within a 'free-market' model that promotes exports, trade, industrialisation and private businesses, which benefits large developed-world TNCs.
International Monetary Fund (IMF)	Promotes global economic security and stability, and assists countries to reform their economies. Economic reforms often mean more open access to developing economies for TNCs.
World Economic Forum (WEF)	A Swiss non-profit organisation that promotes globalisation and free trade via its annual meeting at Davos, which brings together the global business and political elite.
World Trade Organization (WTO)	IGO that regulates global trade. Established in 1995, it has brokered many agreements aimed at promoting open trade and reducing protectionism. Previously known as the General Agreement on Tariffs and Trade.

Skills focus: World trade graphs

Trade between countries has grown enormously since the end of the Second World War. The main beneficiaries of this have been developed free-market capitalist economies and their TNCs. The growth of trade is challenging to show graphically, however, because of the huge differences in value from year to year. In 1960, the total value of exports from all countries was just US$123 billion but had risen to US$19 trillion by 2018.

Figure 8.1 shows world export data using a linear scale, whereas Figure 8.2 shows the same data using a logarithmic scale. Each has its advantages and disadvantages:

- The exponential nature of export trade growth can be seen in Figure 8.1.
- Very little can be made of trade growth up to about 1970 as the values are too small to see on this linear scale line graph.
- The impact of the 2008 global financial crisis – and rapid recovery of world exports by 2010 – can clearly by seen.
- Figure 8.2, using a logarithmic scale, picks out the acceleration of trade in the late 1960s better than Figure 8.1.
- Figure 8.2, however, makes the 2008 financial crisis seem much less significant.

Overall, the logarithmic scale (Figure 8.2) is perhaps a more accurate representation of world trade in exports because it clearly identifies periods of more rapid growth.

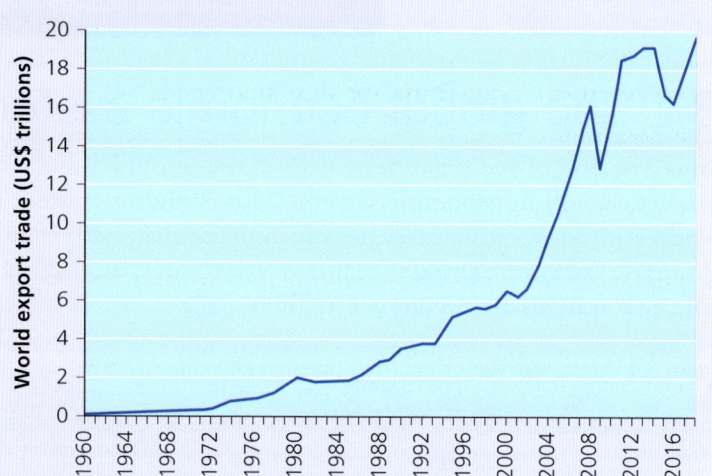

Figure 8.1 Growth in world export trade 1960–2018 using a linear scale

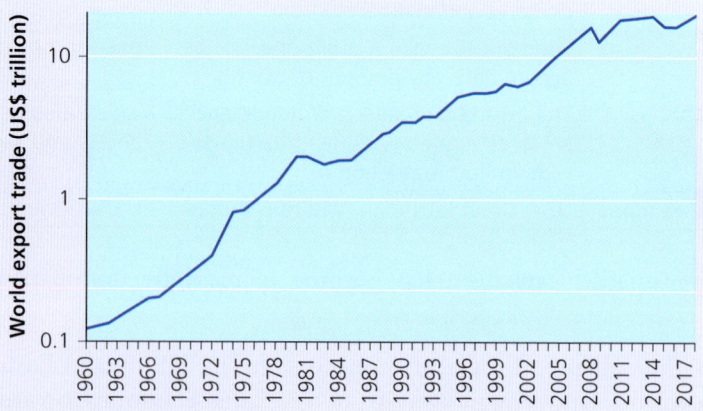

Figure 8.2 Growth in world export trade 1960–2018 using a logarithmic scale

Global TNCs and their role

It should not be a surprise to find that the world's largest TNCs originate from the world's superpowers and emerging powers. TNCs can be broadly categorised into two types (Table 8.3):

- publicly traded TNCs whose shares are owned by numerous shareholders (usually other TNCs, banks and large financial institutions such as pension funds) around the world
- state-owned TNCs that are majority or wholly owned by government.

In many cases state-owned TNCs are large but not well known as their brands are not global.

In 2018, the world's 500 largest companies had US$32.7 trillion in revenues, made US$2.2 trillion in profits and employed 65 million people. In emerging superpowers many of the largest companies are state owned. The 12 biggest companies in China are all state owned.

The dominance of TNCs in the global economy has been caused by a number of factors:

- Their economies of scale mean they can outcompete smaller companies and, in many cases, take them over.
- Their bank balances and ability to borrow money to invest has allowed them to take advantage of globalisation by investing in new technology.
- The move towards free-market capitalism and free trade has opened up new markets, allowing them to expand.

TNCs are not a new feature of the global economy. The East Indian company, for example, had nearly 3000 shareholders by 1773. Many British multinational companies have roots in nineteenth century merchant firms which operated in the British Empire.

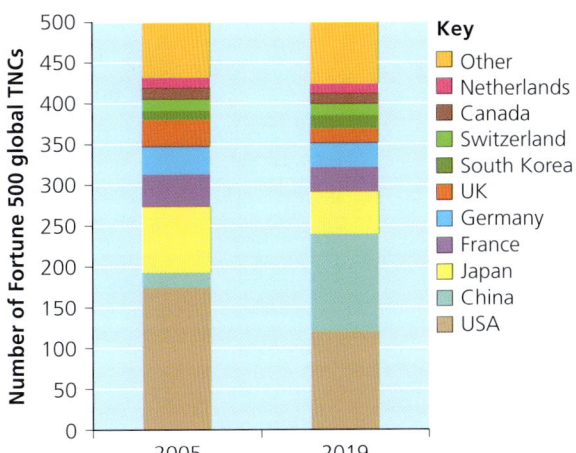

Figure 8.3 Fortune Global 500 TNCs in 2005 and 2019
Data: www.fortune.com/global500

Synoptic themes:

Players

The size, global reach and wealth of TNCs make them key players in maintaining power and wealth. TNCs are economic entities, but their influence on prosperity gives them political as well as economic leverage. Governments compete to attract their foreign direct investment (FDI) by providing them with tax breaks and favourable tax deals. This is why companies such as Apple, Amazon and Google have been investigated by the EU Commission in Europe, as they appear to pay far less tax than might be expected. The favourable tax arrangements they enjoy were of course agreed by individual EU member states.

Table 8.3 The world's largest public and state-owned TNCs

Publicly traded		
Name (country)	Revenue in 2019 (US$ billion)	Employees
Walmart (USA)	514	2,200,000
Royal Dutch Shell (UK/Neth.)	397	82,000
BP (UK)	304	73,000
Exxon Mobil (USA)	290	71,000

Oil & Gas   Construction

State owned		
Name (country)	Revenue in 2019 (US$ billion)	Employees
Sinopec (China)	415	249,000
China National Petroleum (China)	393	1,640,000
State Grid (China)	387	914,000
China State Construction	181	256,000
ICBC (China)	168	453,000

Retail

Data: www.fortune.com/global500

Table 8.4 The top fifteen global brands by brand value, 2019

Brand and 2019 Rank	Sector	Brand value (US$ million)
1. Apple	Technology	234
2. Google	Technology	167
3. Amazon	Technology	125
4. Microsoft	Technology	109
5. Coca-Cola	Beverages	63
6. Samsung	Technology	61
7. Toyota	Automotive	56
8. Mercedes-Benz	Automotive	51
9. McDonald's	Restaurants	45
10. Disney	Media	44
11. BMW	Automotive	41
12. IBM	Business Services	40
13. Intel	Technology	40
14. Facebook	Technology	39
15. Cisco	Business Services	36

Data: Interbrand

> **Key term**
>
> **Brand value, or brand equity:** The value of a brand measured using metrics such as market share, customer opinion of the brand and brand loyalty.

Many would argue that TNCs have been the main beneficiaries of the post-1990 US dominance of the global economic system and the free-market capitalism economy. TNCs are driven to maximise profit by their shareholders, who benefit from these profits. The managers of TNCs receive very large salaries and bonuses for generating profits, but the shareholder institutions (banks, other TNCs, pension funds) receive even more. A key criticism of TNCs as a driver of globalisation is their pursuit of shareholder profit above all else. However, the Chief Executive Officers (CEOs) of TNCs are obliged to look after the interests of the owners of the corporations (the shareholders) and maximise their profits. There has been a dramatic shift in the origin of the world's largest companies since 2005, reflecting the rise of the emerging powers (Figure 8.3). Close to 400 of the world's 500 largest TNCs originate from only seven countries. China's share of these has risen, at the expense of the USA and European countries.

Cultural impact

TNCs bring influence to the countries they originate from. More important than the size of a company is its cultural impact on global consumers. This can be measured using the concept of **brand value**. Table 8.4 shows the 2019 global brand value rankings calculated by the company Interbrand:

- Eleven of the top fifteen companies are from the USA, two from Germany and one each from South Korea and Japan.
- Ten of the top fifteen brands are involved in ICT and communications (mobiles, computers, media), three are car makers and two are food and drink.
- The top Chinese company in 2019 was Huawei telecoms and mobiles at position 74.
- Many of the brands can be recognised instantly from the colour and shape of their logos, which are sometimes free of words, for example, Apple.

The dominance of the USA since 1990, and the economic power of the EU, has led some people to identify the increasing cultural globalisation referred to as 'Westernisation' that involves the arts, food and media. It is difficult to identify exactly what this global culture is, but some characteristics are commonly linked to it:

- a culture of consumerism
- a culture of capitalism and the importance of attaining and accumulating wealth
- a white, Anglo-Saxon culture with English as the dominant language
- a culture that 'cherry picks' and adapts selective parts of other world cultures and absorbs them.

Global culture is most often exemplified by the ubiquity of consumer icons such as Coca-Cola and McDonald's. In the case of McDonald's, 37,800 restaurants worldwide serve about 69 million people every day (750 people buy a McDonald's every second).

Cultural globalisation is not quite as simple as it might first appear, however. In India, McDonald's has had to adapt its menu to suit local tastes and the diets of those of Hindu and Islamic faiths. It does not sell beef or pork, and has more vegetarian options (Table 8.5). Throughout the world this process of local adaptation or hybridisation occurs as Western culture reaches new areas.

American or Western culture is not adopted wholesale around the world, nor is 'cultural traffic' always one way. For instance:

- In the UK the curry, not the American burger, is the most popular takeaway food. There are six times as many curry restaurants in the UK as there are McDonald's.
- Sushi, from Japan, has become an increasingly popular food in European nations and the USA.
- Some cornerstones of American culture, for example, American football and baseball, have had a hard time being exported to the rest of the world.

An important source of influence is the media. Newsfeeds, films, music and TV are dominated by global media brands, most of which are from the USA (Table 8.6). This gives the USA the ability to constantly reinforce its cultural message and values – often in a very subtle, unseen way that fits with Gramsci's concept of hegemony. Only four of the top twenty grossing movies of 2019 were not made by a TNC listed in Table 8.6.

Interestingly, many IT companies and even retailers have branched out to become 'content providers' in recent years. This includes Google, Amazon and Apple. This linkage of cultural content – which is often 'Western' – to IT and communications technology makes it a very powerful delivery system.

Table 8.5 Examples of McDonald's adaptations to local tastes

Country	McDonald's adaptation
India	The Maharaja Mac: a Big Mac made of lamb or chicken
	McAloo Tikki: a vegetarian burger
Japan	Gracoro Burger: korokke (a type of potato croquette), cabbage and katsu sauce
	Ebi-Chili: shrimp nuggets
	Green tea-flavoured milkshake
Israel	Over a quarter of its restaurants are kosher
	Burgers are grilled over charcoal, not fried
	The McKebab, with eastern seasoning, is served in pitta bread

Table 8.6 The top six media TNCs – all originate from the USA

Media TNC	Revenue 2019 (US$ billion)	Brands and businesses
AT&T	181	Warner Bros, CNN, Cartoon Network, HBO
Comcast	108	NBC, Universal Studios, Universal parks and resorts
Disney	69	Disney, Pixar, Marvel Studios, ESPN
Google	66	Google Play, Google Search, YouTube, Android
ViacomCBS	27	CBS news, Paramount Pictures, Comedy Central
Netflix	20	Netflix Studios, Netflix

Innovation and patents

An often overlooked aspect of the role of TNCs is the invention of new technology and the development of new products and brands. TNCs, and governments, invest huge sums in research and development (R&D) to develop new products. Intellectual property law protects these new developments in the form of:

- patents, for new inventions, technologies and systems
- copyright for artistic works, such as music, books and artworks
- trademarks to protect designs, such as logos.

Any person or company wishing to use one of the above has to pay a royalty or licence fee to the inventor or designer. Globally, over 85 per cent of all royalty payments go to the USA, EU and Japan (Figure 8.4).

This domination of global royalties reflects the fact that:

- Existing superpowers and developed countries are paid for inventions and artistic works they created decades ago.
- Developed world TNCs are in the best position to invest in R&D, so patent holders tend to also be new patent developers.

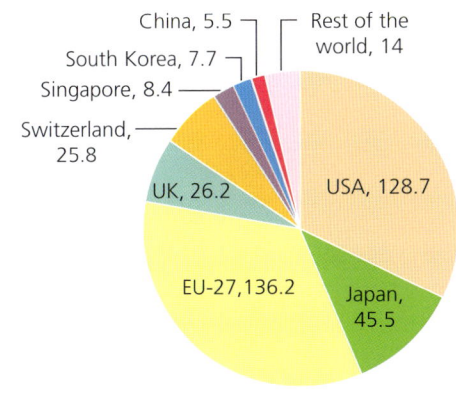

Figure 8.4 Worldwide royalty and licence earnings, 2018 (US$ billion)

- Education levels are higher in already developed countries, as are skill levels.
- Westernisation and cultural globalisation tend to spread US and European music, film and TV (copyright) and brands (trademarks).

Of course, emerging superpowers and developing countries continually pay these royalties, representing a cost to them but a benefit to the USA and EU. In the last 20 years, China has begun to develop many more patents. In 2018 Chinese innovators applied for 1.54 million new patents versus only 597,000 in the USA. However, there are question marks over Chinese patent applications in terms of both quality and the extent to which they can generate royalty revenue in the future for Chinese companies.

8.2 International decision-making

In the event of a global crisis, nations will ask other nations for their help, but which superpowers would they call on? Russia? China? The USA? This is an important question. Its answer reflects a number of aspects of power:

- Which nations are trusted by others to do the 'right thing' and respond when others are in danger?
- Which nations have the global reach to respond to a war or natural disaster in a far-flung place?
- Which nations have the money and technology to be able to help?

The idea that some countries should act as the 'world's police' is an old one. It dates back to the four powers on the winning side at the end of the Second World War (the USA, UK, USSR and China) and led to the setting up of the UN Security Council in 1946.

The Security Council is the primary global mechanism for maintaining international peace and security. It has five permanent member states (marked with asterisks in Figure 8.5) and ten rotating non-permanent members (selected to represent nations from different continents) that change every two years. The Security Council can maintain international law by:

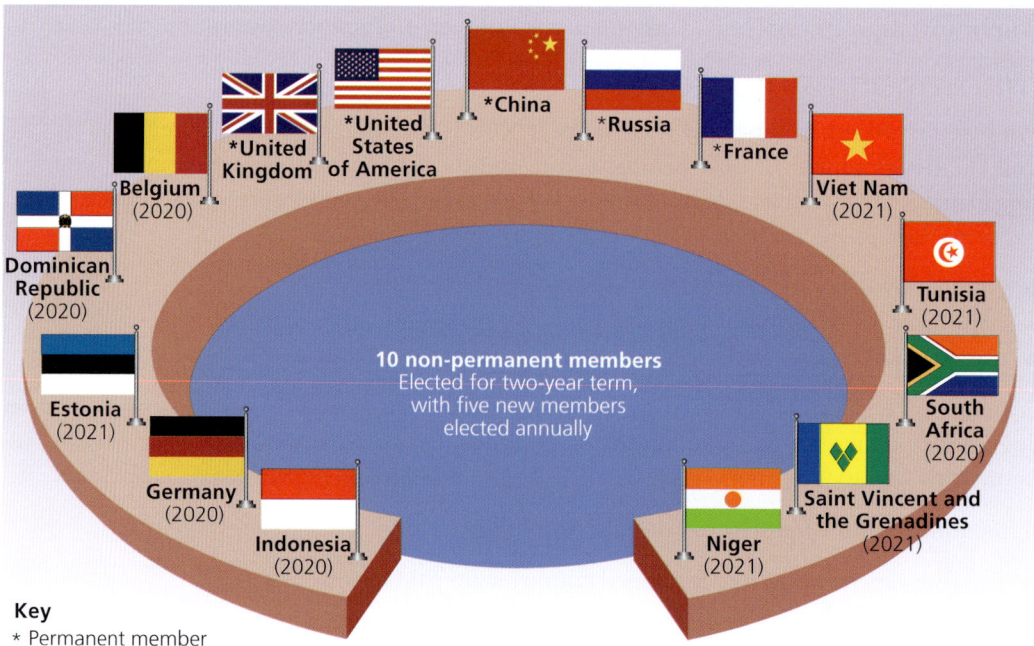

Key
* Permanent member

Figure 8.5 The UN Security Council, 2020

- applying **sanctions** to countries that are deemed to be a security risk, harbouring terrorism, threatening or invading another state or breaching human rights
- authorising the use of military force against a country
- authorising a UN Peacekeeping Force: troops occupy a country or region under the UN flag to keep the peace in a conflict but do not 'take sides'.

In the latter two cases, military forces are pooled from UN member states. The Security Council works in a very imperfect way:

- any one of the five permanent members can veto a decision, preventing it happening
- the USA, UK and France tend to vote 'as one', as do Russia and China, leading to deadlock
- the Security Council has been accused of passing resolutions condemning a country's actions but then failing to take action.

Any actions require a 'yes' vote but also countries willing to act. In the last 40 years it has often been the case that the Western powers (USA, UK and France) have failed to get agreement at the UN and have subsequently taken unilateral action themselves. Arguably this undermines the 'collective security' principle that the Security Council was set up to achieve.

USA: global police?

A striking feature of the last 40 years is the number of times the USA has intervened militarily in foreign countries. It has done this in three ways:

- as part of a UN Security Council action
- together with allied countries as a coalition, but outside a UN remit
- unilaterally, that is, with no support from other countries.

Figure 8.6 summarises these military actions since 1980. In many cases the UK and/or France has also been involved, but very rarely China or Russia. This reflects the fact that the USA and European countries tend to have similar geopolitical concerns and goals that are not shared by Russia and China.

There are many examples of superpowers and emerging powers attempting to solve problems (Table 8.7, page 170), but interventions such as these are also open the criticism of unwanted interference in other countries affairs.

> **Key term**
>
> **Sanctions:** These can be diplomatic, such as ordering staff at a foreign embassy home, or economic, such as banning trade between countries. Military sanctions ban trade in weapons and military co-operation, while sporting sanctions can be used to prevent a country taking part in global sporting events. The aim is to force a country back to the negotiating table without using military force.

> **Synoptic themes:**
>
> **Players**
>
> Some countries, notably the USA, Britain and France, regularly act as 'global police'. They deploy forces overseas to intervene in the affairs of other countries – either unilaterally or as part of a coalition. Russia and China rarely act in such a way, at least beyond countries on their borders. Acting in this way is expensive, politically risky and requires certain assets – particularly a blue water navy and large air force.

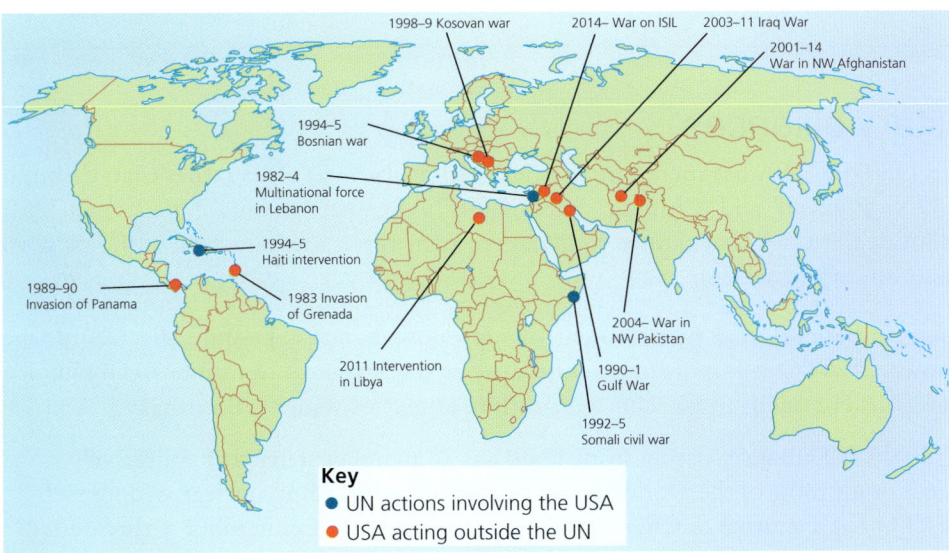

Figure 8.6 Selected US military interventions since 1980

Table 8.7 Global crisis response

Iran's nuclear programme	Rightly or wrongly, Western countries have long been concerned that Iran was developing a nuclear weapons programme. The USA and Israel in particular fear a nuclear-armed Iran would further destabilise the already unstable Middle East. • The USA first applied sanctions to Iran in 1995. • UN sanctions began in 2006 and included an arms embargo and asset freezes. • EU sanctions began in 2007. • Sanctions were placed on Iran's crucial oil exports in 2012. Iran's economic situation slowly deteriorated as it was denied oil export revenue and found importing technical equipment increasingly difficult. Sanctions worsened the humanitarian situation for ordinary Iranians, especially the sick, disabled people and women-led households. Import restrictions caused inflation and many goods, including some medicines, became unavailable. Iran was forced to agree to end its nuclear weapons programme in 2015 and sanctions ended in 2016.
Bosnian War, 1992–5	The Bosnian War resulted from the post-Cold War break-up of Yugoslavia; it was Europe's first ground war since 1945. A toxic mix of ethnic, religious and nationalist divisions led to a war that involved genocide, mass rape and numerous other war crimes. • A UN resolution led to a NATO-enforced no-fly zone, led by US jets. • UN peacekeepers tried to establish safe zones in the conflict area, but this proved difficult to achieve. • A NATO-led, UN-sanctioned naval blockade in the Adriatic was put in place. • Eventually, NATO air-strikes against the Bosnian Serbs forced an end to the conflict. The NATO campaign was deemed crucial in the signing of the 1995 Dayton Accord which ended the war.
Haiti earthquake humanitarian relief effort, 2010	The devastating magnitude 7.0 earthquake on 12 January 2010 in Haiti destroyed 70% of buildings in the capital city, Port-au-Prince, and created a humanitarian disaster of huge proportions. A huge relief operation began, led by the UN and involving numerous countries and NGOs. US military logistical and technical assistance was crucial. Within six days: • The US Air Force restored air traffic control to Port-au-Prince's airport to allow relief flights in. • US Coastguard helicopters began relief flights. • The aircraft carrier *USS Carl Vinson* and helicopter carrier *USS Bataan* arrived to assist with rescue, food and water aid. • 1600 US marines arrived by sea to provide humanitarian aid and technical help.
The Ebola epidemic, 2014–16	The outbreak of Ebola in West Africa in 2013 had become a terrifying epidemic by early 2014. Ebola has no cure or vaccine, and is easily transmitted in insanitary conditions. It has a mortality rate of 50 to 70%. The initial global response was slow and led by NGOs, such as MSF. However, it gathered pace: • The USA, France and UK led the response in Liberia, Guinea and Sierra Leone, respectively. • In Sierra Leone the UK committed £430 million, 1500 troops and 150 NHS personnel to fight the epidemic. • The work of NGOs and the World Health Organization (WHO) was crucial, but so was the support of traditionally powerful countries that can deploy significant assets quickly. By early 2016 the epidemic was over (11,300 people died).

Military alliances

Since 1980 the USA has increasingly acted with its NATO allies rather than through the UN. Russian and Chinese vetoes of UN Security Council proposals have convinced the USA that failure to get agreement there means acting alone or with a 'coalition of the willing'.

Military alliances are a key element of superpower status. Despite the USA's vast firepower, NATO is important to it because of the 'strength in numbers' an alliance brings. NATO needs to be seen within the context of wider US military power:

- The NATO alliance, dating from 1949, has 28 member states that collectively account for most of the world's military capability including nuclear weapons.
- NATO has a mutual defence agreement, meaning if one member is threatened, all others come to its aid.

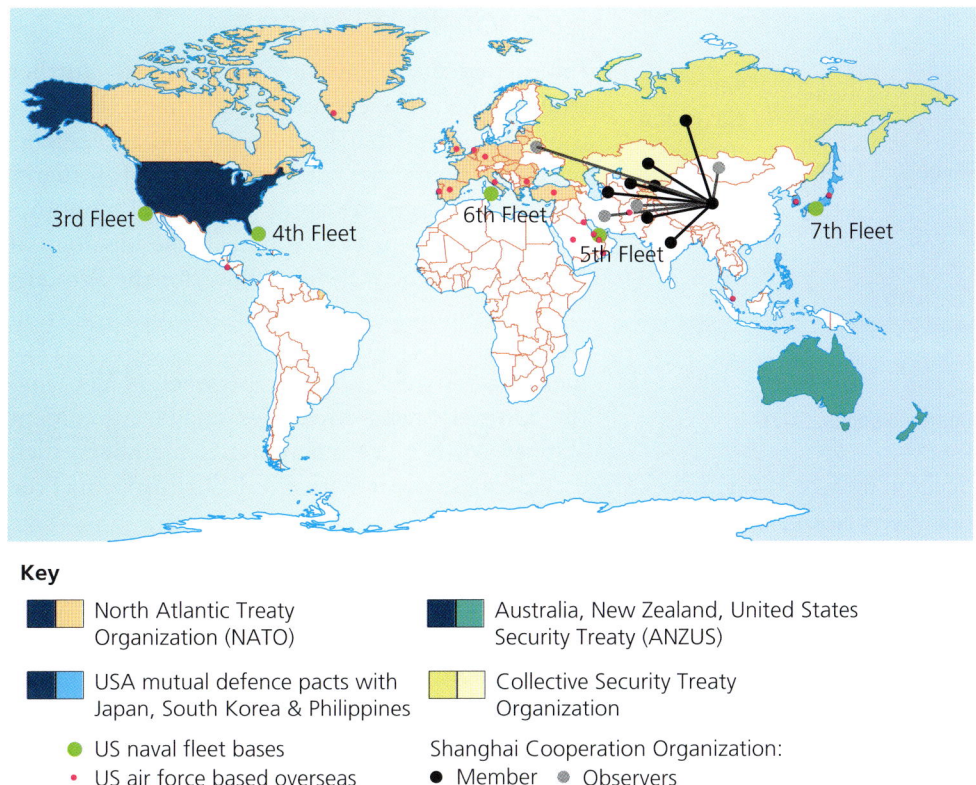

Key

- North Atlantic Treaty Organization (NATO)
- USA mutual defence pacts with Japan, South Korea & Philippines
- Australia, New Zealand, United States Security Treaty (ANZUS)
- Collective Security Treaty Organization
- US naval fleet bases
- US air force based overseas
- Shanghai Cooperation Organization: ● Member ● Observers

Figure 8.7 Key military alliances

- The USA also has alliances across the Pacific, namely the ANZUS Treaty (with Australia and New Zealand), and mutual defence pacts with Japan, South Korea and the Philippines (Figure 8.7).
- US naval and air force bases are spread globally, not just in the USA, giving the USA true global reach.

Russia's military alliance, the Collective Security Treaty Organization, only consists of former USSR republics attached to Russia's borders. China lacks any formal military alliances with other countries that go as far as a mutual defence agreement. However, the Shanghai Cooperation Organization (SCO), set up in 1996 between China, Kazakhstan, Kyrgyzstan, Russia, Tajikistan and Uzbekistan, is becoming a more strategic partnership in Asia. India and Pakistan joined the SCO as full members in 2017, hugely expanding the geography of the organisation. More a security and economic co-operation organisation than a true military alliance, the SCO is broadly against 'western liberal democracy' and so in the future could be geopolitically important as a bloc. The countries involved co-operate on:

- security, including counter-terrorism and cyber-warfare
- military matters, including carrying out joint exercises and 'war games' (which, in 2010, involved 5000 troops from member countries)
- there is also some cultural and economic co-operation between members.

Other countries have 'observer status', including Iran, Belarus, Afghanistan and Mongolia. These are possible future members of the SCO. Further expansion of the organisation geographically and in terms of political and military co-operation could create a powerful rival to NATO.

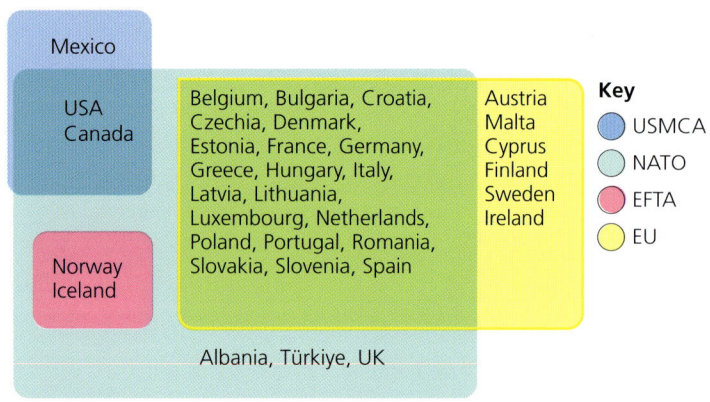

Figure 8.8 Western economic and military alliances, 2020

Economic alliances

Interdependence between nations is further strengthened by economic alliances. Broadly, those countries that have free-trade agreements between them are also members of military alliances. This creates a powerful axis of economic and military security that reflects the ideology of each bloc. Figure 8.8 shows how, for the Western allies, economic and military alliances overlap with many EU countries that are also NATO members, as well as overlap between NATO and the United States–Mexico–Canada Agreement (USMCA). NATO but non-EU members, such as Iceland and Norway, have a free-trade agreement with the EU (European Free Trade Association, or EFTA).

Under President Trump hopes of a free trade agreement between the USA and the EU faded. A future bilateral agreement between the UK and USA may be possible. Trump also questioned the relevance of NATO, and how it is funded – arguing that European countries should contribute more. Trans-Atlantic geopolitical ties have weakened, not strengthened, in recent years.

Economic alliances exist elsewhere. The Association of Southeast Asian Nations (ASEAN) was founded in 1967 and includes Indonesia, Malaysia, the Philippines, Singapore, Thailand, Brunei, Cambodia, Laos, Myanmar and Viet Nam. It has economic, cultural, security and political aims. In 2009 it became a free trade bloc (also including Australia and New Zealand). ASEAN has free-trade agreements with China, South Korea and Japan, and possibly with the EU in the future. The AfCFTA (African Continental Free Trade Area) is a 48-country pan-African free trade area launched in 2018. It has potential to become the world's largest if fully implemented.

The free-trade agreements within trade blocs like the EU, USMCA, AfCFTA and ASEAN encourage economic interdependence because:

- free of import/exports taxes and tariffs, TNCs can operate as truly international entities, moving physical, human and financial resources anywhere within the bloc
- workers find it easier to move between countries, especially when freedom of movement is part of the agreement, as it is within the EU
- the revenues and profits of TNCs, and the smaller businesses that supply them, are highest when the economic health of the whole trade bloc is good.

With this economic interdependence comes demand for political stability and security. It might therefore be seen as 'natural' that security and military co-operation are often found between countries that are economically very interdependent.

Global security

It is interesting to consider the system that is in place to help maintain global security – of both the political and economic kind. Essentially it is a system set up in 1945 – in other words, in a previous century – revolving around the UN. Just as there are 'pillars' of superpower status (see Figure 7.1, page 148), there are pillars of global security, which are run by IGOs (Figure 8.9).

Figure 8.9 The pillars of global security

This post-war system is under strain:

- Its leaders – the USA, UK and France – are not as economically or militarily powerful as they were.
- There is a strong case for emerging powers – India and Brazil especially – to have more of a say in world affairs.
- Currently, neither Africa nor Latin America has a seat at the top table of world security decision-making, despite a combined population of 1.65 billion.
- The global financial crisis of 2007–8, the worst since the Great Depression of the 1930s, strained the IMF and global financial system to the limit and made many economists wonder if there was a 'better way'.
- The ongoing threat of global terrorism from al-Qaeda, Islamic State and the Taliban, among others, might suggest that global security co-operation is not all it could be.

However, the system has promoted stability in some ways. UN agencies, such as the WHO, Food and Agriculture Organization (FAO) and World Food Programme (WFP) have reduced the burden of disease and hunger so that scenes of desperate people are rare today unless caused by war and conflict. The International Criminal Court is charged with bringing war criminals to justice (and acting as a deterrent to would-be despots). This requires complex international law to be obeyed by many countries in order to catch and extradite suspects. It has had some limited success in prosecuting war criminals from the Democratic Republic of the Congo, Uganda, the Central African Republic and Ivory Coast. A related International Criminal Tribunal for the former Yugoslavia in The Hague is prosecuting leaders accused of genocide and other war crimes during the Yugoslav Wars.

> **Synoptic themes:**
>
> **Attitudes and actions**
>
> The actions of IGOs to maintain aspects of global security are largely determined by the willingness of member states to act. In most cases, this requires strong political will from a group of countries with support from a superpower or emerging power. It also requires money, for example, to pay for a UN peacekeeping force or to fight a global pandemic. This money has to come from member states.

8.3 Global environmental concerns

Superpowers have very large resource footprints. Maintaining a large economy, a military machine with global reach and a wealthy population requires energy, mineral, land and water resources. A good example is coal. China accounts for half of the world's coal consumption (Figure 8.10) but only has 19 per cent of the world's population. Coal has fuelled China's dramatic industrialisation since 1990 and is likely to continue to do so.

China is the world's largest producer of iron ore, used to make steel. It is also the world's largest importer of iron ore. The 0.8 billion tonnes mined in China in 2019 only met demand when added to the 1.1 billion tonnes it imported.

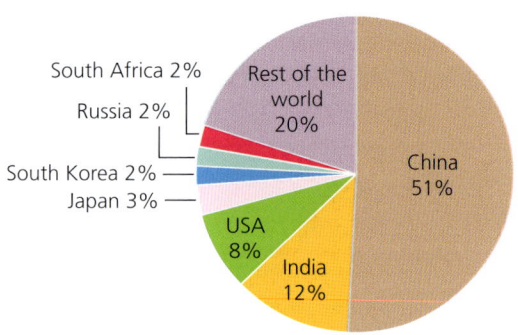

Figure 8.10 Global coal consumption, 2018

The high resource consumption of superpowers and emerging powers generates a range of environmental issues:

- Urban air quality is low in emerging power cities due to coal-burning power stations, the continued use of open stoves (India) and dramatic increases in car use. This has major health implications: air quality in Beijing, Shanghai, Delhi and Mumbai regularly exceeds WHO safe limits.
- Demand for fossil fuel, mineral and food imports, plus manufactured goods exports, accounts for most of the world's CO_2 emissions from shipping (if these were a country, they would be the world's sixth largest).

Skills focus: Proportional symbols

Figure 8.11 shows carbon dioxide emissions estimates for 2014 for selected countries. This is a proportional symbols map, a visually useful way of comparing data. Raw data in millions of tonnes of CO_2 has been made into proportional symbols by finding the square root of the data and multiplying by a constant (any whole number, or fraction, used to produce a convenient measurement for the radius of the circle).

China dominated global CO_2 emissions in 2014, accounting for 29 per cent of global emissions, followed by the USA (15 per cent) and EU (10 per cent). This is a dramatic change from 2002, when China's emissions were only 3780 million tonnes per year compared to more than 10,000 million in 2014. In 2002, the USA accounted for 23 per cent of all global CO_2 emissions and China only 14 per cent.

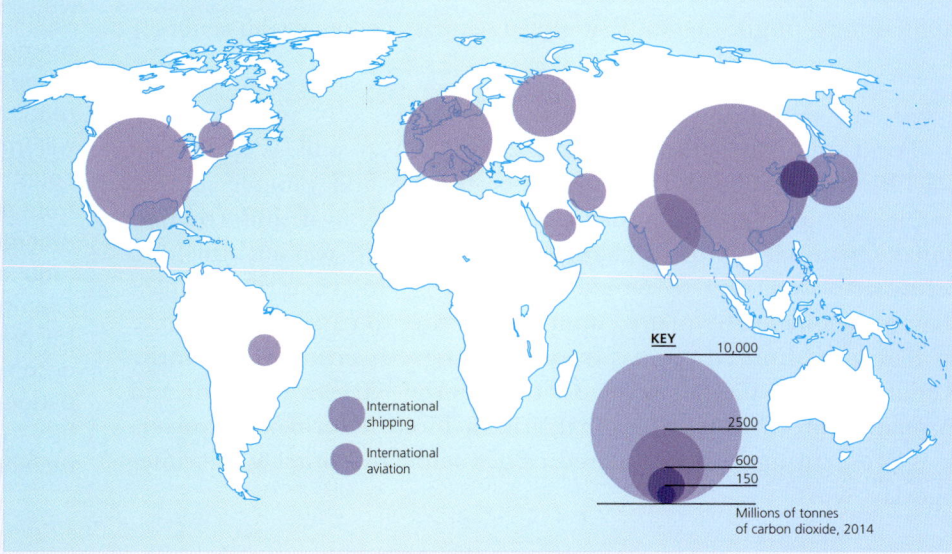

Figure 8.11 China dominated global CO_2 emissions in 2014

- Deforestation and land degradation are issues in some emerging powers as they seek to convert more land into farmland, continue to urbanise, increase demand for water and increase the use of chemicals in farming to increase yields.

Superpower carbon emissions

Data on carbon emissions per person reveal differences in how willing superpowers are to act on reducing emissions. Despite its high-profile decision not to sign the 1997 Kyoto Protocol, per capita emissions in the USA have fallen (Figure 8.12). The USA signed up to the legally binding emissions reduction targets from the COP21 Climate Change Conference in Paris when President Biden took office in 2021.

In contrast, China's per capita emissions are rising. This is hardly surprising given that it is still industrialising and urbanising. China's emissions are likely to rise in the future because today's per capita GDP in China is around US$13,000 whereas in the USA it is US$80,000. China's emissions could grow much higher as affluence increases.

In 1990, China's total emissions were 4.1 billion tonnes of CO_2 equivalent rising to 15.7 billion tonnes by 2022. In contrast, the USA's emissions peaked in 2007 and have since fallen to 6 billion tonnes.

China's current emissions target is that its emissions will peak before 2030 and be 'carbon neutral' before 2060. China might defend rising emissions on the basis that:

- it should be allowed to industrialise and develop in the same way that the EU, North America and Japan did
- emerging countries are not really to blame – at least not yet.

From 1850 to 2007, the USA emitted 339,000 million tonnes of CO_2, or 28.8 per cent of all cumulative historical emissions, compared to China's 105,000 million tonnes or 9 per cent.

Neither China nor the USA appear as committed as some European countries are to emissions reductions, but the evidence suggests that the USA is more willing to act than China as the USA has engaged with the 'net zero' aspirations of the United Nations COP agreements on emissions targets and reductions.

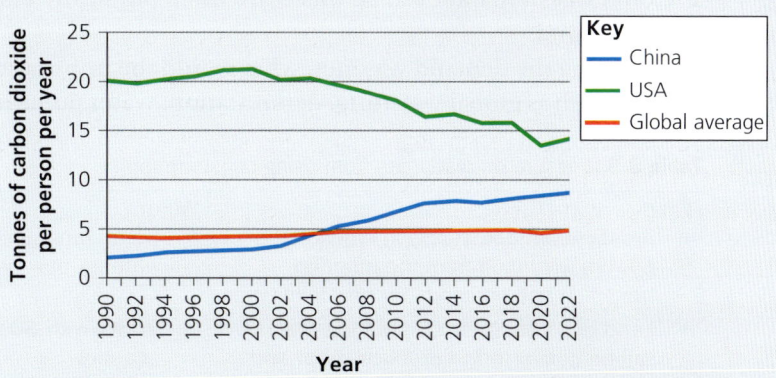

Figure 8.12 Trends in per capita carbon dioxide emissions, 1990–2022

Synoptic themes:

Attitudes and actions

Attitudes and actions towards the threat of global warming are influenced by many factors. The science behind global warming is generally trusted more by Europeans than Americans. Economic growth and personal wealth are prioritised by governments over environmental issues in some countries (notably BRICS). High-income countries can more easily afford renewable and alternative technologies to reduce emissions. Compact, high-density countries may find it easier to develop affordable public transport as an alternative to car use.

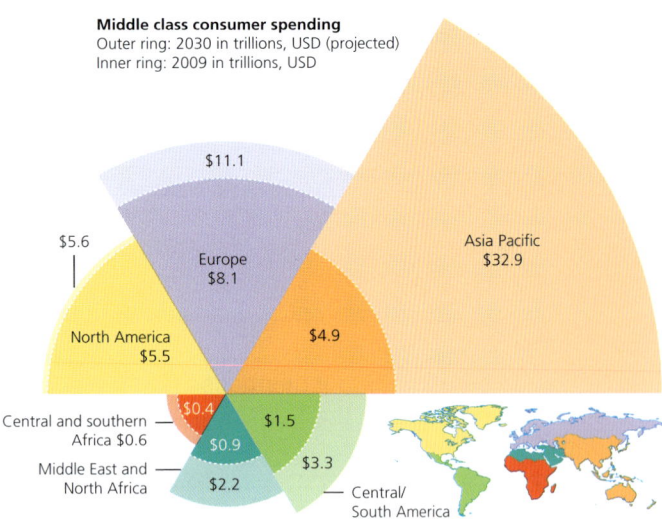

Figure 8.13 The growth of global middle class spending by 2030 (Source: OECD)

> **Key term**
>
> **Middle class:** Globally, the middle class are defined as people with discretionary income. They can spend this on consumer goods and perhaps holidays. The global middle class can be defined as people with an annual income of over US$10,000.

The global middle class

A significant future concern related to the growth of BRICS and other emerging powers is that rising affluence will rapidly increase the numbers of **middle class** consumers. This is obviously positive in terms of development but it may place huge strain on resources. Figure 8.13 shows projected growth in the global middle class by 2030. Numbers of middle-class consumers are expected to increase from about 2 billion today to 5 billion by 2030. In the Asia Pacific region numbers could swell from 600 million in 2015 to 3000 million by 2030.

Middle-class consumption has a wide range of implications (Table 8.8). If, by 2030, the average Chinese consumer has a similar income to a consumer in the EU or USA, it would mean:

- 1 billion cars on China's roads (there were 1.4 billion worldwide in 2019).
- 1350 million tonnes of cereals consumed each year, or two-thirds of global production in 2019.
- 180 million tonnes of meat consumed each year, or 60 per cent of global production in 2019.
- Coal consumption in China would exceed global consumption in 2019.

There are two key questions:

1. Could this demand even be met – are there physically enough resources to supply the demand?
2. Assuming the demand was met, what would the implications be for environmental issues such as global warming, deforestation, water pollution and land degradation?

Table 8.8 Pressure on resources from rising consumption

Food	Water
• Pressure on food supply in emerging powers will result from the nutrition transition and demands for new food types • Land once used for staple food grains will be converted to produce meat and dairy products • Without new land, prices could rise, squeezing the poorest	• Some emerging powers already have water supply problems, notably India • India's situation is likely to be critical by 2030, with 60% of areas facing water scarcity • Water supply in China, Indonesia and Nigeria could be problematic by 2030, especially in urban areas
Energy	**Resources**
• Global oil demand was about 100 million barrels per day in 2019 • By 2030 this is likely to rise, along with coal and gas demand, perhaps by 30% • Meeting this demand may lead to price rises and/or supply shortages • Countries with their own domestic supplies (Russia, Brazil) are likely to be in a stronger position than those relying on imports (India)	• Demand for rare earth minerals – used in LCD screens and numerous other hi-tech gadgets – could increase prices • The demand for lithium-based batteries is very high and could be hard to meet in the future • Even more basic metals, such as copper, tin and platinum, are at risk of supply shortages and dramatic price changes

Superpowers

> **Key terms**
>
> **Nutrition transition:** A change in diet from staple carbohydrates towards protein (meat, fish), dairy products and fat. It often includes eating more processed food. It occurs as people transition from rural poverty to being urban, middle-class workers.
>
> **Staple foods:** Carbohydrates relied on in large quantity and eaten regularly, such as potatoes and wheat for bread in Europe and North America, maize in Latin America, and rice in Asia. Stable supply at affordable prices is important to regional food security.
>
> **Rare earth minerals (or rare earth elements, REE):** A group of metal elements crucial to modern communication, medical and laser technology. Found dispersed in rocks, they are hard to mine, costly and supplies are limited.

> **Review questions**
>
> 1 Define the term 'inter-governmental organisation (IGO)'.
> 2 Outline the key differences between capitalist and socialist planned economies.
> 3 Explain how global IGO and TNCs influence the global trade.
> 4 Outline the ways in which TNCs reinforce the economic and cultural power of the nations they originate from.
> 5 Explain why the USA has acted as the 'global police', even when it is forced to act alone.
> 6 What evidence is there for 'western' and 'eastern' blocs of economic and political alliances in the world today?
> 7 Using Table 8.7 (page 170), evaluate how far global responses to crises have been successful.
> 8 'Any global agreement to reduce greenhouse gas emissions made without China would be pointless.' How far do you agree with this statement?
> 9 Explain why environmental problems are likely to worsen by 2030 as a result of growing middle-class consumption.

> **Further research**
>
> The work of different IGOs can be explored using their websites: www.wto.org, www.weforum.org, www.worldbank.org and www.imf.org.
>
> The work of the UN Security Council, and the resolutions and role of peacekeepers around the world, can be explored at: www.un.org/en/sc/
>
> Military alliances have their own websites, for example, NATO: www.nato.int
>
> The Janes website can be used to explore military technology and national capability: www.janes.com/defence-news

9 Superpower spheres of influence

> **What spheres of influence are contested by superpowers and what are the implications of this?**
> By the end of this chapter you should:
> - understand why some geographical locations and economic spheres are claimed by more than one power, and how this can create conflict
> - recognise that the rise of powerful countries in Asia is causing a fundamental global power shift
> - understand that this power shift affects emerging powers, developed nations and the developing world in both positive and negative ways
> - understand that political and cultural tensions may increase as emerging powers flex their muscles
> - recognise the sources of uncertainty when attempting to predict the future geopolitical balance of power.

Synoptic themes:

Attitudes to resources

Tensions over resources are more likely if countries and/or TNCs have a 'must have' attitude to those resources. If players invest in alternatives (e.g. renewable energy over fossil fuels), or prioritise conservation over resource exploitation, then tensions are likely to reduce.

9.1 Contested geography

Superpowers are fuelled by resources. In the twenty-first century human resources are crucial. Human innovation, skill and entrepreneurship are vital to maintain technological, economic and military hegemony. However, old-fashioned physical resources are still important, especially fossil fuels, ores and minerals. Securing access to these resources is vital for both governments and TNCs.

Some resources are contested. This could be because:

- the land border between two countries is in dispute, such as the border between India-controlled Kashmir and Pakistan-controlled Kashmir
- the ownership of a landmass is in dispute, such as Argentina's claim to the UK-governed Falkland Islands (which may contain offshore oil and gas resources)
- the extent of a nation's offshore **exclusive economic zone** is in dispute or claimed by another nation.

The latter situation is the case in the Arctic, where several nations claim ownership of areas of the Arctic Ocean, which may contain valuable oil and gas reserves.

Intellectual property

A possible source of tension is over intellectual property (IP) rights. A global system of IP has been run since 1967 by the World Intellectual Property Organization (WIPO), which is part of the UN. It ensures TNCs, individuals and government agencies can protect new inventions, trademarks, artistic works and trade secrets from use by others. This is important because:

- Without IP, innovations and ideas can be stolen and used by others.
- This would be a huge disincentive to innovate and invent.
- The costs of developing new medicines or communication technologies could not be recouped through selling products if others could simply copy the idea.

Key term

Exclusive economic zone (EEZ): The area of ocean extending 200 nautical miles beyond the coastline (or to the edge of the continental shelf), over which a nation controls the sea and sub-sea resources. EEZ borders are decided by the UN in the event of a dispute.

Arctic oil and gas

The US Geological Survey (USGS) estimates that 30 per cent of the world's undiscovered gas, and 13 per cent of oil resources, are in the Arctic. This amounts to 90–100 million barrels of oil, worth billions of dollars. Figure 9.1 shows national claims to the Arctic Ocean seabed and areas of dispute. Much of the dispute centres around whether an area of ocean bed known as the Lomonosov Ridge is an extension of Russia's continental shelf (and therefore part of its EEZ) or not.

- Three of the parties in dispute have nuclear weapons (Russia, USA and EU).
- In 2007, the Russians used a submarine to place a Russian flag on the seabed at the North Pole, which inflamed tensions.
- Since then the number of 'scientific' expeditions to the Arctic has increased, as countries seek to have a greater presence in the area.
- Both Russia and Canada have created dedicated 'Arctic Forces' to protect their interests.
- By 2019, Norway, Canada, the USA and the UK had all strengthened their Arctic military capacity and patrols.

Tensions are likely to rise further as global warming makes the Arctic increasingly accessible to shipping for longer periods of the year, and oil and mineral exploration becomes easier. As oil reserves elsewhere run out, Arctic oil will look increasingly tempting. In theory, the UN will decide whose claims stand and whose do not. But will the powerful countries involved respect these decisions?

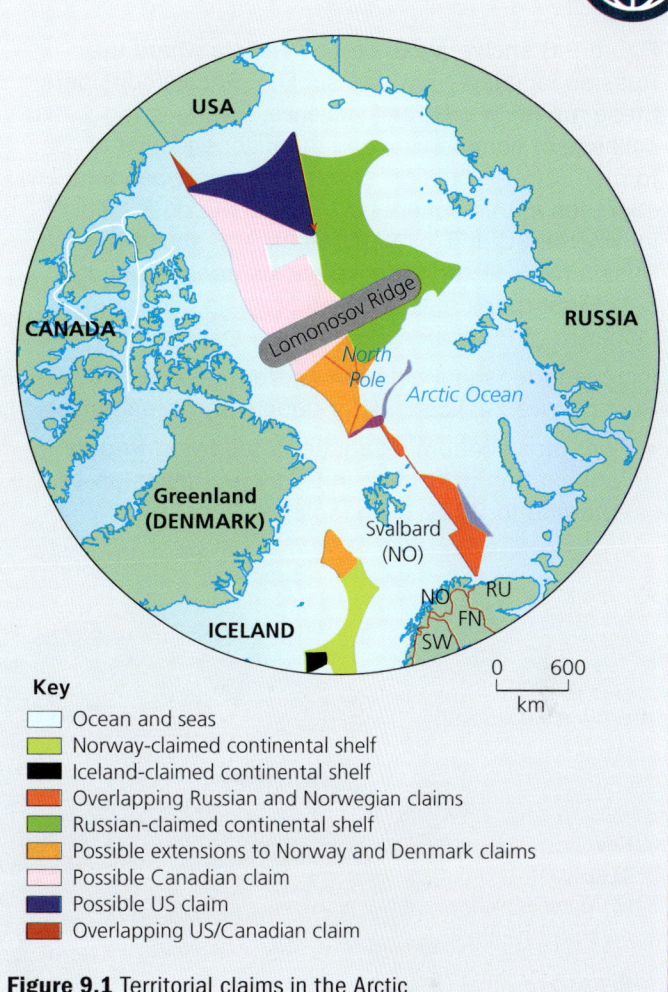

Figure 9.1 Territorial claims in the Arctic

On the other hand, IP has been criticised:

- IP requires users of a product to pay royalties (fees) to the inventor, which is a cost to developing countries.
- IP holders do not have a duty to make a new invention available; in theory, at least, they could prevent a new medicine being made.
- The system can create a monopoly where a patent holder can charge what they like for a new product, denying it to some people on the basis of price.

Chinese companies are well known for infringing IP by producing counterfeit products. Figure 9.2 shows a brazen example of this. Many Western car companies, including BMW, Mercedes Benz and even Rolls-Royce, have seen Chinese companies make copies of their models. Fake Apple products are also common: 22 fake Apple Stores were found in China in 2011. According to the US Chambers of Commerce, 86 per cent of the world's counterfeit goods originate in China – worth close to US$400 billion in 2018.

Figure 9.2 A Range Rover Evoque (top) and the Chinese LandWind X7 (bottom)

9 Superpower spheres of influence

Russia's western border

Figure 9.3 shows areas beyond Russia where the Russian language is widely spoken. Russia considers these areas its sphere of influence. Following the 1991 collapse of communism and the independence of former Soviet republics, several newly independent countries in this sphere have raised the possibility of joining the EU and/or NATO. Figure 9.3 shows how Russia has reacted to these moves, reasserting its belief that these regions should remain under the influence of Russia.

Increased tensions between western Europe and Russia have implications for people and economics:

- Western economic sanctions following Russia's annexation of Crimea in 2014 and invasion of Ukraine in 2022 have isolated Russia economically, but also affected EU exporters who can no longer sell to Russia. Russian consumers have also suffered as imports have dried up.
- The open conflict in Crimea, eastern Ukraine and Georgia led to the forced displacement of tens of thousands of people, as well as hundreds of deaths.
- NATO deployed additional air and ground forces to its eastern border in Estonia, Lithuania, Latvia, Poland, Romania and Bulgaria.

Russia's 2022 invasion of Ukraine has led to dangerously strained geopolitical relations between NATO and EU allies and Russia.

Figure 9.3 Russia's sphere of influence

Key term

Sphere of influence: A geographical area over which a country believes it has economic, military, cultural or political rights. Spheres of influence extend beyond the borders of the country and represent a region where the country believes it has a right to influence the policies of other countries.

Disregard for international IP treaties and counterfeiting sour relations between countries, especially the USA and China:

- TNCs may be reluctant to invest in China, knowing that their profits are likely to be reduced by counterfeiting.
- Lack of action by the Chinese authorities on IP issues might suggest its government is less likely to co-operate on other issues of international law.
- The possibility of trade agreements being made is limited if one side believes the other will not 'play by the rules'.

Spheres of influence

Counterfeiting is unlikely to lead to physical conflict, but there are some situations that might. Several locations in the world have contested spheres of influence. In some cases these are simply disputed borders. Nuclear-armed Pakistan and India have a long-running territorial dispute over the ownership of Kashmir, further complicated by Chinese occupation of a nearby area called Aksai Chin, which is also claimed by India. Japan and Russia both claim ownership of the Kuril Islands, which are currently under Russian control. Such disputes frequently flare up, often because:

- the balance of power changes, such as when Pakistan tested a nuclear weapon in 1998, putting it on a par with India in terms of military capability
- disputed territories are visited by high-level officials, such as when then Russian President Medvedev visited the Kuril Islands in 2010, which incensed Japan
- military 'sabre rattling' occurs, such as flying jets or sailing naval vessels close to a disputed territory: this frequently occurs in the South and East China seas by both China and the USA
- new resources are discovered or suspected, such as the possibility of oil in waters off the Falkland Islands, governed by the UK but claimed by Argentina.

South and East China seas

China has overtly stated its sphere of influence in Southeast Asia in a policy referred to as the Island Chain Strategy. Figure 9.4 shows the line of the First and Second Island Chains. China is actively pursuing a policy of controlling the ocean from its coast to the First Island Chain. It launched its first aircraft carrier, the *Liaoning*, in 2012, a second in 2019, and is likely to launch a third carrier in 2021–23. China may be planning further nuclear-powered carriers for launch in the late 2020s. This policy has problems. Not least the fact that many rocks, islands and areas of continental shelf in the South and East China seas are disputed (see Figure 9.4). These areas are militarily important to China in terms of defending what it sees as ocean areas it should control. The largest naval presence in the area at the moment is the USA, due to its close allies Japan, South Korea and the Philippines. The islands could be economically important as they may harbour oil and gas reserves.

China's recent strategy in the area has been to occupy deserted islands and to artificially build larger, or even new, islands, especially in the Spratly Islands (Figure 9.5):

- In 2014, China began constructing an airport on reclaimed land on Fiery Cross Reef.
- In 2015, China began the construction of a port and possibly an airport on Mischief Reef.
- In 2015, Subi Reef appeared to be being developed into a Chinese military base.

China's actions have been referred to as the 'Great Wall of Sand' – reflecting the fact that its new island bases are built of sand. However, the policy is a long-standing one. It is often called the 'Nine-dash Line' policy, in reference to Chinese maps delimiting the area of claimed control in the South China Sea using nine dashes.

On 18 April 2020 – at the height of the global Covid-19 pandemic – China announced the creation of two new administrative districts in the South China Sea (Xisha in the Paracel Islands and Nansha covering the Spratly Islands, both of which are claimed by Viet Nam). This was a significant raising of the stakes in an already tense region.

Figure 9.4 Tensions in the South and East China seas

Figure 9.5 An artificial island constructed in the South China Sea

9 Superpower spheres of influence

China's Belt and Road Initiative (BRI) 2

China's BRI has a price tag of US$4–8 trillion and a timeline extending to 2049. The BRI consists of huge infrastructure corridors (Figure 9.6) designed to reduce China's reliance on shipping goods by sea through the narrow Strait of Malacca choke point and at the same time expand China's markets and connections to other countries. It consists of:

- **Overland silk road:** rail links from western China to Germany, a corridor of development through Mongolia to Russia, an economic development corridor from China through Central Asia to Türkiye, economic corridors south through Indochina and Pakistan, to the coast.
- **Maritime silk road:** ports, sea routes and infrastructure development around the Indian Ocean.
- **Polar silk road:** The opening of the North Sea Route through the Arctic Ocean to Europe.
- Road, rail and pipeline construction in East Africa.

A successful BRI would create huge wealth, at least for some people, in previously isolated parts of South and Central Asia, but it also has major geopolitical implications such as:

- Strengthening ties between China and Pakistan (through the China-Pakistan Economic Corridor/CPEC) – which makes India nervous.
- Strengthening China's influence in Viet Nam, Thailand and Singapore, Sri Lanka and Türkiye – which troubles the Americans.
- Making China a player in the Arctic, with Russia as a key ally.
- Tying China more closely with Iran, a long-time adversary of the USA.

The BRI could redraw huge swathes of the economic and geopolitical map in Asia, the Middle East and Europe.

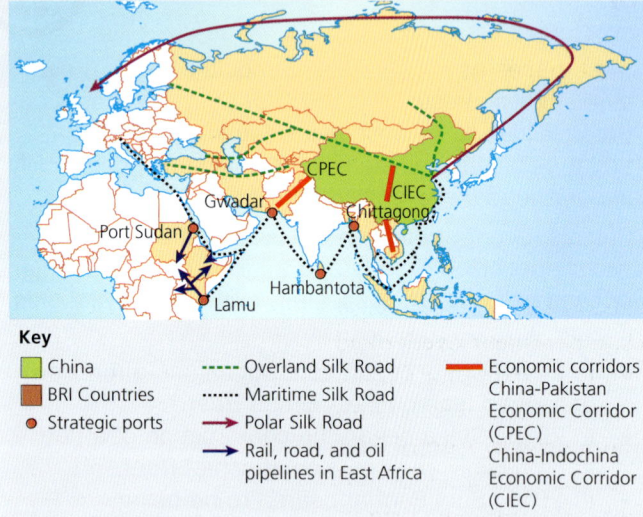

Figure 9.6 China's Belt and Road Initiative

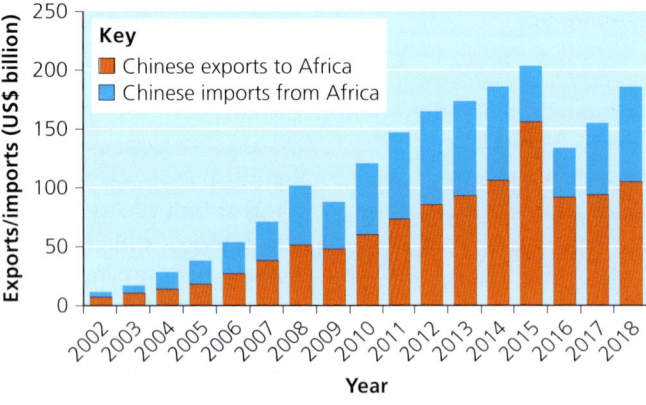

Figure 9.7 Growth in China–Africa trade, 2002–18

Two geographical locations are of particular concern and these are explored in the accompanying place contexts: Russia's western border (page 180) and the South and East China seas (page 181).

China's actions in the South and East China seas have implications for the USA and other countries in the region:

- Arguably, island building is a direct attack on the UN Law of the Sea as construction is happening in areas that are disputed; China is taking de facto control of places claimed by other countries.
- China is challenging US naval and air hegemony in the area, and has begun to question the right of US ships and aircraft to sail and fly in the disputed areas.

The situation is complicated by the unresolved status of Taiwan. It is Chinese, but only ethnically. Politically it exists in a kind of 'limbo', fearful of declaring independence from China but protected by ties with the USA. Any of the island disputes could easily flare up into open conflict due to ill-judged actions by either side.

9.2 Developing countries: opportunities and threats

Existing superpowers, such as the USA and EU, have often been accused of having unfair relationships with developing countries. This means relationships based on:

- neo-colonialism: superpowers pulling the economic and political strings of developing countries, despite not ruling them directly as during the colonial/imperial era
- unfair terms of trade: cheap commodity exports for the developing world (coffee, cocoa, oil, copper) set against expensive manufactured imports from developed countries
- the brain drain of skilled workers from developing countries to boost developed-world economies
- local wealthy elites, who control imports and exports in developing countries, benefiting from the neo-colonial relationship but having no interest in changing it.

An important question is whether the rise of the emerging powers would make any difference to this relationship.

China's African adventure

Since 2000 China has looked beyond its own borders and has become a source of foreign direct investment (FDI), not just a destination for it. Africa, a continent overlooked by the developed world for decades, has become a major trading partner for China. Data from the China–Africa Research Initiative (Figure 9.7) shows that exports from Africa to China and imports from China into Africa have both grown. Annual FDI from China had increased to about US$5 billion by 2018 and the total stock of FDI stood at about US$50 billion. China also provides economic and development aid to Africa in the region of US$2–3 billion each year. By 2015, China had built 2250 km of railways and 3350 km of roads in Africa.

China's involvement in Africa has created greater interdependence:

- China relies on African oil – from Angola, Nigeria and Sudan – as well as minerals such as Zambian copper, and even sugar and biofuels grown in Africa, to fuel its growing economy.
- Africa increasingly imports Chinese-manufactured goods and relies on Chinese investment in infrastructure like roads, rail and ports.

A key question is what sort of relationship China and African developing countries have. Is it:

1. A neo-colonial one, where China exploits Africa for its cheap raw materials but Africa gets little in return?
2. A developmental relationship where African countries, such as Sudan, Angola and Ethiopia, benefit and can progress through trade and deeper connections to the global economy?

As Table 9.1 (page 184) shows there is a continuing debate on this issue. In reality, most Chinese investment focuses on just a few countries, so is not evenly spread across Africa.

> **Synoptic themes:**
>
> **Players**
> The role of emerging powers in Africa could mirror that of the USA in Taiwan, South Korea and Singapore in the 1960s and 1970s, i.e. a political, economic and military ally whose investment aided long-term development towards NIC status. Or, it could be a repeat of the colonial and imperial exploitation of previous centuries.

Table 9.1 China's relationship with Africa

Neo-colonial challenge?	Development opportunity?
Infrastructure investments ensure China can export raw materials as cheaply and efficiently as possible	China has invested heavily in roads, railways and ports to export raw materials – infrastructure that can be used by Africans themselves
Skilled and technical jobs are often filled by Chinese migrant workers, estimated to number 200,000 in 2018	Vital jobs are created, especially by large industrial, transport and energy projects, which also modernise the economy
Cheap Chinese imports (clothes, shoes, etc.) have undercut local producers and forced them out of business	Chinese factories and mines bring modern working practices, and technology, to Africa
Much of the FDI brings only temporary construction jobs; there are few long-term jobs in mechanised mines and oil fields	Chinese finance has funded seventeen major HEP projects since 2000, adding 6780 MW of electricity to the continent by 2013
Aid from China is tied to FDI: allow investment and China provides some aid	Investment deals are often accompanied by aid, so the benefits of Chinese money are more widely spread

There are concerns about the environmental impact of Chinese investment and resource exploitation:

- Chinese imports of tropical timber have been linked to widespread illegal deforestation in Mozambique.
- Oil spills linked to Chinese-funded oil wells have been reported in Chad, Sudan and Angola.
- The extraction of the metallic ore coltan in the Democratic Republic of the Congo has led to widespread forest loss and river pollution, but is vital to Chinese mobile phone and computer manufacturers.

These environmental issues are not new, of course, nor are they wholly Chinese in origin. Shell and other Western oil TNCs have been accused of polluting the Niger Delta in Nigeria for decades.

The centre of gravity shifts

Perhaps the best illustration of the rise of emerging powers is the position of the world's economic 'centre of gravity'. The points on Figure 9.9 can be thought of as the 'most likely' location for economic activity at a particular date.

Figure 9.9 can be interpreted in a number of ways:

- Between 1820 and 1913, the Industrial Revolution in the UK (then other European countries) rapidly pulled the world economy towards Europe.
- Following the Second World War, the dominant US economy pulled it across the Atlantic, but the subsequent rise of the EU edged it back east.
- Since 1990, the centre has shifted dramatically back towards Asia, especially China.
- India and China may see the shift back to Asia as a return to their rightful position held between the years 1 and 1500.

Figure 9.8 Chinese president Xi Jinping shakes hands with South African president Jacob Zuma at the Johannesburg Summit of the Forum on China-Africa Cooperation, 2015

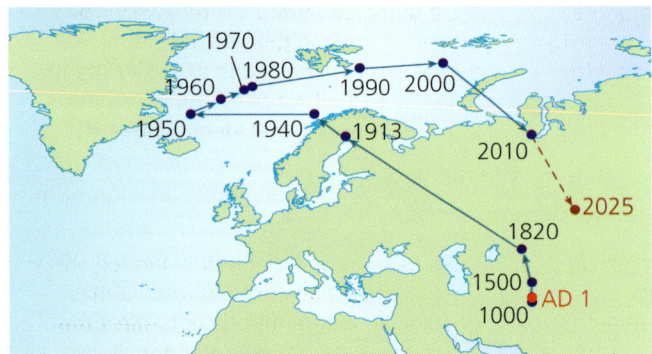

Figure 9.9 The world's economic centre of gravity, 1–2025
(Source: Adapted from McKinsey Global Institute Report)

Superpowers

There is no doubt that Asia is, or very soon will be, the dominant global region. Asian nations include India, China, Indonesia, Japan, South Korea and Malaysia – all powerful countries, some with emerging and even superpower credentials. Figure 9.10 shows the total population and total GDP of Europe, the USA and Asia in 2015, and projected to 2050. While GDP projections are a best guess, it is widely expected that by 2050 Asia will be by far the world's most populous continent and the world's largest by GDP.

Asia could be a very economically and politically crowded continent by 2030. Indonesia, China, India and Japan are all likely to have economies greater than US$5 trillion by that date, giving them all the means to have significant military power. Viet Nam, Japan, the Philippines, Indonesia, Bangladesh, Pakistan, India and China will all have populations of over 100 million people.

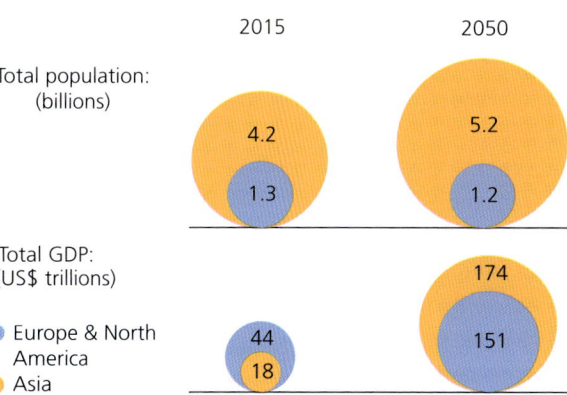

Figure 9.10 Population and GDP, 2015 and 2050

By 2030 there could be a strong case for India, and even Indonesia, to have a permanent seat on the UN Security Council. While China and India may be vying for superpower status by 2030, other Asian economies will be jostling for the third, fourth and fifth places in the regional power rankings.

Superpowers and the Middle East

The Middle East has a number of characteristics that make it a location of frequent tension and conflict (Table 9.2). The 2011 **Arab Spring** uprisings arguably added to instability in the region.

Figure 9.11 (page 186) summarises the complex Middle Eastern situation in 2016. Conflict and tension exist on a number of fronts, including:

- A broad alliance between the USA, Israel and some EU member states and their Middle Eastern allies set against a loose alliance of Iran, Russia and China. The alliances range from overt military support, such as US support for Israel, to suspected supply of arms to Iran by China through illicit channels.
- A refugee crisis caused by **IS** in Syria and Iraq, and the Syrian civil war, which has forced more than 6 million Syrians to flee and put huge strain on neighbouring Jordan, Lebanon and Türkiye, which have accepted these refugees.

> **Key terms**
>
> **Arab Spring:** A series of pro-democracy, pro-human rights civil uprisings in 2011 that affected Syria, Tunisia, Libya, Egypt, Bahrain and Iran. Some governments were overthrown but, in most cases, protracted instability followed the uprisings.
>
> **IS (Islamic State, also known as ISIS, ISIL and Daesh):** A terrorist organisation that rose to prominence in 2013 during the Syrian civil war, occupying parts of the Middle East and carrying out terrorist attacks worldwide.

Table 9.2 Multiple sources of instability in the Middle East

Religion	Oil & Gas	Governance
Differences between Muslim sects (Saudi Arabia is a majority Sunni state whereas Iran is Shia), Christians (large minority populations exist in Lebanon and Syria) and Judaism (a majority religion in Israel) contribute to wider tensions both within and between countries.	65% of the world's crude oil exports originate in the region, as a result western countries maintain economic and military influence in the Middle East.	Most of the countries are relatively new states, at least in their current form; democracy is either weak or non-existent; religious and ethnic allegiances are often stronger than national identity ones.
Resources	**Youth**	**History**
Although rich in fossil fuels, the region is short of water and farmland, meaning territorial conflict over resources is more likely.	Many countries have young populations with high unemployment and relatively low education levels: the potential for young adults to become disaffected is high.	Many international borders in the region are arbitrary; they were drawn on a map by colonial powers and do not reflect the actual geography of religious or cultural groups.

Key term

Israel–Palestine conflict: Conflict between the Jewish state of Israel and Arabs in Palestine (who claim the same territory) is one of the world's longest running and most intractable conflicts. Since 1949, Israel has illegally occupied Palestinian territory and built settlements there, making a long-term agreement harder to achieve. The involvement of Colonial, Cold War and contemporary superpowers (Iran is a long-term supporter of Palestine and the USA has provided Israel with billions of dollars in economic and military aid) has inflated a religious and territorial dispute into a wider geostrategic conflict over regional political and economic influence. Numerous peace initiatives have so far failed to find a sustainable solution to the ongoing conflict.

- Destabilising terrorist groups including al-Qaeda in the Arabian Peninsula, IS in Syria and Iraq, and the Taliban in Afghanistan.
- The ongoing **Israel–Palestine conflict**: the USA has traditionally supported Israel with military and economic aid while some Middle Eastern countries, such as Iran, are openly hostile towards the Jewish state of Israel and actively support Palestinian military groups.

The USA, Russia, China and some EU countries have a long history of supporting different countries, terrorist groups and rebel armies in the Middle East with economic support and arms sales. Over time, alliances and priorities change, but weapons remain. Superpowers and emerging powers fighting against al-Qaeda and IS often face weapons they sold to the Middle East decades ago.

The complexity of Middle Eastern politics, religion, ethnic differences and territorial disputes has led to some intractable and potentially dangerous situations:

- Türkiye has long been fighting a civil war against the Kurds in Türkiye, who want their own Kurdish state. However, the Kurds were a key group fighting IS in Syria and Iraq. Türkiye supported this fight as a member of NATO and the 'Western alliance' against IS.
- The EU and the USA initially supported the 2011 Arab Spring uprising against President Assad in Syria. However, by 2014, they found themselves bombing IS in Syria, effectively acting on the 'same side' as Assad's forces.
- The air-bombing campaign against IS in Syria and Iraq by the USA, the UK, France and others since 2014 meant those countries were acting with Russia (which supported Assad) while geopolitical relations with Russia deteriorated dramatically over the Russia–Ukraine war that began in 2022.
- The 2001–2021 USA-led war in Afghanistan only partly stabilised that country. It remains unclear whether restoration of Taliban power will stabilise Afghanistan in the long-term after US troops withdraw in 2021.

Figure 9.11 Middle East conflict in 2016

Superpowers

- Some of the funding and support for IS, al-Qaeda and other terrorist groups originates from the very countries that are fighting against them, for example, Saudi Arabia, Qatar, Pakistan and Türkiye, straining relations between allies.

The Middle East situation has proved costly both in economic terms (military costs) and human terms (military and terrorist casualties in the region). It is made even more difficult by the fact that the USA, UK, Germany, France and other countries affected by terrorism have large populations with families still in the Middle East and cultural ties to the region.

Conflict continues as of 2025. Civil war in Yemen began in 2014 and has drawn in Saudia Arabia and some western countries. The terrorist attack on 7 October 2023 by Hamas in southern Israel quickly escalated into a wider conflict in Gaza and the West Bank (Palestinian territories) plus against Hezbollah in Lebanon.

In December 2024, the Assad regime was overthrown by the Islamist group Hay'at Tahrir al-Sham (HTS). This prompted Israel to occupy parts of southern Syria along its border. It remains to be seen whether HTS will bring stability to Syria, or whether Israel's actions can stabilise the long-running Israel–Palestine conflict.

9.3 Uncertainty for existing superpowers

Existing superpowers face a number of economic challenges including rising debt, unemployment and high social costs. The EU and USA both have a number of challenges (page 188). The USA is arguably in a stronger position than the EU for two reasons:

1. Although consisting of 50 states, which have their own rights and laws, the differences between states are minor and they are not sovereign, unlike the 27 countries that make up the EU. The former are much more likely to agree on policy.
2. The USA is not ageing as fast as the EU. It has a fertility rate of 1.9 versus 1.6 for the EU, so its population will be more youthful for longer.

Both the EU and USA face the ongoing costs of **economic restructuring**. Traditional manufacturing cities, such as Detroit, Pittsburgh, Sheffield, Newcastle, Lille and Turin, have lost jobs and required major investment in regeneration as well as the social costs of coping with rising unemployment.

The 2007–8 global financial crisis dealt a severe blow to the EU and, to a lesser extent, the USA. The costs of bailing out collapsing banks and then collapsing countries (Greece, Portugal, Ireland, Cyprus) has made the EU more inward looking and focused on its many problems. The long-term economic and social costs of the 2019–21 Covid-19 pandemic could damage future growth prospects for some countries. This perhaps makes it more likely that in the future China will fill the gap that the EU once filled. However, China faces the costs of its ageing population, such as a shrinking workforce and rising healthcare costs.

> **Synoptic themes:**
>
> **Attitudes and actions**
>
> Facets of 'Western ideology' – including capitalism, democracy, individual freedom, gender equality, environmentalism and, to some degree, Christianity – are important in Western countries to a greater or lesser degree. Islamic ideology stresses the primacy of religion and Sharia (Islamic law - see page 197) which has its own interpretation of the rule of law, governance, terms of trade and the roles of men and women. These contrasting attitudes can affect political relationships. Shared economic self-interest and cultural compromise means there are usually good geopolitical relations between Western countries and Jordan, the UAE and Türkiye but the same self-interest (oil supply) and the need for military allies in the Middle East often means turning a blind eye to gender inequality and human rights abuses in countries such as Saudi Arabia.

> **Key term**
>
> **Economic restructuring:** The shift from primary and secondary industry towards tertiary and quaternary industry as a result of deindustrialisation. It has large social and economic costs.

Key concept: Global terrorism threat

The security and international co-operation costs of coping with the threat from terrorism conducted internationally have increased rapidly in the last 20 years since the 9/11 attacks in New York. The threat comes largely from terrorist organisations such as al-Qaeda and IS fighting jihad (struggle) against the non-Islamic West. Their motives are complex and include the ideological belief that war should be fought against all non-Muslims. In addition, many are fighting against what they see as long-term interference by the West in the Middle East. Poverty, unemployment and lack of opportunity also play a role. Terror attacks have affected many countries, including:

- July 2005: the London underground and bus bombings, which killed 52 and injured 700
- November 2008: 166 people were killed in a series of co-ordinated attacks in Mumbai
- December 2014: three killed and four injured during a hostage attack in Sydney
- November 2015: a series of attacks in Paris on cafes and a concert hall that killed 137 people and injured 368
- November 2017: terrorists opened fire on those finishing Friday prayer at the Al-Rawdah mosque in Egypt's northern Sinai region. A bomb then ripped through the mosque as worshipers began to flee. 305 people were killed, including 27 children, and 120 were injured.

These attacks range from large-scale, co-ordinated bombings to lone gunmen, and even bombing airliners. Action to prevent them relies on the intelligence services (MI5, CIA) intercepting plots and in some cases direct military action against the terrorist bases. All of this has a high economic costs as well as the potential to create mistrust and cultural tensions at home.

Challenges for the USA and EU

Challenge	USA	EU
Economic	National debt in the USA in 2020 was US$25.5 trillion, but the US dollar's status as the global currency of choice makes it less vulnerable to economic shocks The USA has many large, innovative global TNCs, for example, Apple, Google, Facebook and Cisco	Debt in the Eurozone amounted to €10.6 trillion in 2020 and £2.3 trillion in the UK, in both cases about 90% of annual GDP; debt is a drag on economic growth EU unemployment was close to 7.4% in 2020, representing a cost to taxpayers and underused economic capacity
Demographic	The USA is ageing less fast than the EU and social costs (pensions, healthcare) tend to be borne by individuals rather than the government The total population will keep growing from 328 million in 2019 to 415 million in 2060	The EU is ageing fast; by 2025 20% of EU citizens will be over 65 The EU's workforce will drop by 14% by 2030, which will place an increasing burden on those in work to fund pensions, healthcare and care homes After 2035, the EU's population is likely to be falling
Political	Race relations in parts of the USA are strained and, at a national level, there is often political deadlock between Democrats and Republicans	The EU's 27 nations do not sing with one voice, despite Eurovision! Tensions between countries wanting deeper union (France, Germany) and those wanting less (UK) have grown (culminating in the UK leaving the EU in 2020) Relationships with Russia are strained and immigration is an increasingly divisive issue
Resources	The USA is increasingly energy secure as a result of oil and gas fracking Water insecurity is an increasing problem in the southwest	Energy security is a key EU issue, as it relies on imported oil and gas, some of which comes from Russia
Social	Health spending swallows 17% of the USA's annual GDP and is a huge cost to families and government 74% of adult Americans are overweight (30% are obese), adding significantly to healthcare costs	Youth unemployment in the EU was 15% in 2020, 36% in Greece and 31% in Spain Long-term youth unemployment risks a 'lost generation' of young people, as well as political disaffection

The costs of being a superpower

Given the economic problems of the EU and USA, it is perhaps not surprising that the costs of being a *global* player are questioned by some. A good example of this is the UK's nuclear arsenal. This consists of:

- 225 nuclear warheads, delivered by submarine-launched Trident II intercontinental ballistic missiles (ICBMs) with a range of 12,000 km, bought from the USA
- four UK-built Vanguard class nuclear-powered submarines, at least one of which is at sea at all times.

The UK is one of only a handful of countries with this capability, the others being the USA, Russia, France and China. The next generation of nuclear armed submarines (Dreadnought class) are currently under construction at a cost of around £30 billion and are expected to enter service in the 2030s. Public and political opinion in the UK does not universally support replacing Trident. The money could be spent on other areas, such as education and skills, infrastructure or health. The UK has also built two new aircraft carriers (HMS *Queen Elizabeth* and HMS *Prince of Wales*), that entered Royal Navy service in 2017 and 2019. The cost was £6–7 billion, a huge sum to pay for a seat at the top table of global geopolitics, and that cost does not include any aircraft.

Figure 9.12 shows the estimated annual expenditure of the USA in 2013–14 on maintaining its global supremacy. This includes all military spending and intelligence services, as well as foreign aid and NASA. The total is over US$900 billion, more than the entire annual GDP of Indonesia or the Netherlands. China spends around US$200 billion by comparison. Emerging powers have lower labour costs and lower salaries and, to some extent, can increase their military power by copying technologies that were initially developed (at huge cost) by the USA.

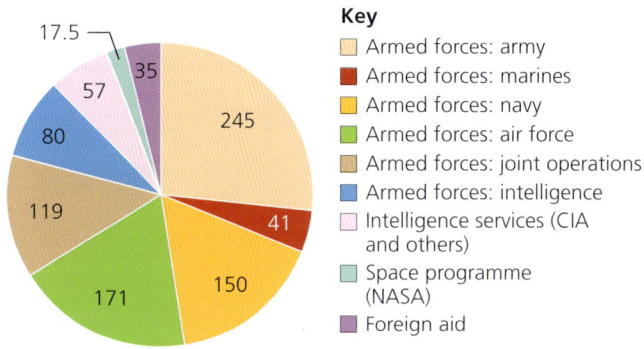

Figure 9.12 US defence, intelligence gathering, space and aid spending, 2013-14 (US$ billion)

The future?

It is, of course, impossible to say what the world will be like in 2050, or even 2030, in terms of the geopolitical situation. Uncertainties over future population size, GDP growth and even the impact of future wars and global warming mean that 'best guesses' are the best that can be done. Table 9.3 (page 190) suggests some possible futures.

Each of the geopolitical structures has different implications for:

- **Stability**: Arguably the 'regional mosaic' structure is inherently unstable as broadly equal powerful countries make complex and competing alliances, with no country acting as the 'global police'. A 'new Cold War' could lead to a period of tense stability.
- **Resources**: The 'Asian century' scenario would see strong economic and population growth in Asia, but continued demand for resources in the old West. An expanding Asian middle class is likely to lead to a 35 per cent increase in demand for food, a 40 per cent increase in demand for water and a 50 per cent increase in demand for energy by 2030 – risking shortages and unrest and/or conflict over remaining resources.

Skills focus: Projecting future GDP

Projecting future GDP is one of the most important determinants of superpower status. It is hard to do, however, because small differences in GDP annual growth rate – for example, 2 per cent versus 3.5 per cent – compound over a number of years, leading to large deviations from projections.

Figure 9.13 shows three future GDP projections all made in 2015 but differing in significant ways.

- The Centre for Economics and Business Research (CEBR) projects much higher growth for India, China and the USA than the US Department of Agriculture (USDA) or PricewaterhouseCoopers (PWC) projections.
- The USDA predicts that the USA will still be the world's largest economy in 2030, unlike CEBR and PWC.
- The difference between the highest and lowest predictions of the Chinese economy in 2030 is US$8 trillion – a vast difference.

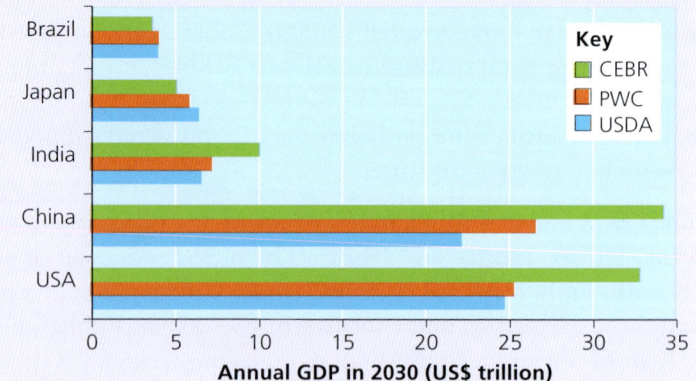

Figure 9.13 Three projections of GDP in 2030

Table 9.3 Alternative superpower futures in 2030

Scenario	Map	Description
US hegemony (unipolar)		US dominance, and economic and military alliances, continue in a unipolar world. China faces an economic crisis, similar to Japan's in the early 1990s, and ceases to grow rapidly.
Regional mosaic (multi-polar)		Emerging powers continue to grow while the EU and USA decline in relative terms, creating a multi-polar world of broadly equal powers with regional but not global influence.
New Cold War (bipolar)		China rises to become equal in power to the USA, and many nations align themselves with one or other ideology, creating a bipolar world similar to the 1945–90 Cold War period.
Asian century (unipolar)		Economic, social and political problems reduce the power of the EU and USA; economic and political power shifts to the emerging powers in Asia, led by China.

- **Military**: A new arms race is a possible outcome of the new Cold War scenario. As China expands its global reach with naval and air power, the USA and its allies may need to react by diverting resources away from economic growth and social programmes and into guns and ships.
- **Economics**: The Asian century scenario would cause a fundamental shift in the world's economy; global economic well-being would depend on the health of NICs in Asia rather than the economies of Europe and North America.

Projecting to 2050 or beyond is almost impossible. A whole range of so-called Black Swan events could occur between now and then, including wars, global pandemics (such as the 2019–21 Covid-19 pandemic), economic crises and even destabilising natural disasters – any of which could change the course of one or more countries.

Synoptic themes:

Futures and certainties

So uncertain is the geopolitical future that the US National Intelligence Council produces a global trends report every five years that attempts to pinpoint emerging global trends and power shifts. All superpower and emerging power governments do this type of 'what if' thinking, as do TNCs and global IGOs. The aim is for key players to have a range of policy options available depending on which 'future' becomes reality.

Review questions

1. Briefly explain what is meant by intellectual property.
2. Explain the political and environmental risks of a race to exploit resources in the Arctic.
3. Which players could be affected in a negative way by counterfeiting manufactured goods?
4. Explain the possible consequences of China's 'island building' strategy in the South China Sea.
5. To what extent should the EU and USA be concerned by China's Belt and Road Initiative (BRI)?
6. What are the reasons behind the EU and USA's military and geopolitical involvement in the Middle East?
7. For a country such as the UK, do the high costs of being a 'global player' actually outweigh the benefits?
8. Evaluate the weaknesses of the USA and EU. Which is the weaker superpower?
9. Look at Table 9.3 on page 190. In your view, which is the most likely power structure in 2030 and 2050? Explain your answer.

Further research

Explore the Arctic, its resources and potential conflicts over them at the Arctic Council website: **www.arctic-council.org/index.php/en/**

View an interactive timeline of events surrounding the crisis in Ukraine at the European Parliament website: **https://www.europarl.europa.eu/topics/en/topic/ukraine**

Explore the US National Intelligence Council's 'Global Trends' report: **https://www.dni.gov/index.php/global-trends-home**

The complex Middle East conflict can be explored using the BBC News website, which archives past stories: **www.bbc.co.uk/news/world/middle_east**

Exam-style questions

1. Explain how 'soft power' can be used to maintain superpower status. [4]
2. Explain the role of superpowers in international crisis response. [4]
3. Assess the extent to which emerging superpowers threaten the economic and political global dominance of the USA. [12]
4. Explain how economic restructuring has affected the economies of existing superpowers. [4]
5. Assess the importance of military and economic alliances in maintaining superpower status. [12]

Topic 8
Option 8A: Health, Human Rights and Intervention

Chapter 10: Human development

Chapter 11: Human rights

Chapter 12: Interventions and human rights

Chapter 13: The outcomes of geopolitical interventions

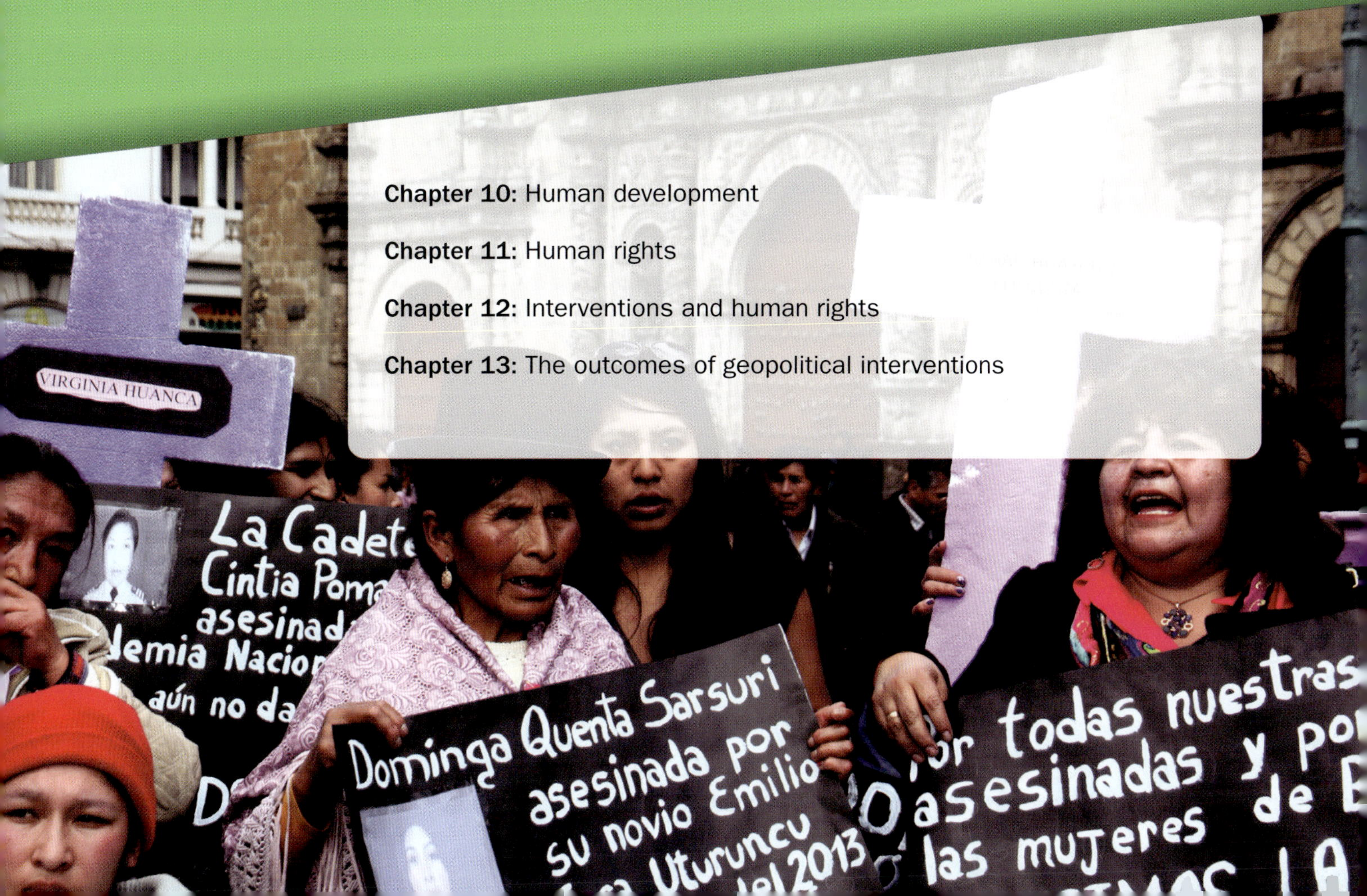

Human development

10

What is human development and why do levels vary from place to place?
By the end of this chapter, you should:
- understand what is involved in human development
- be aware of variations in human health and life expectancy
- be aware of who helps to define development targets and policies.

This Option Topic is about development. Most studies of development highlight its economic dimensions. It is widely acknowledged that economic growth provides much of the power that drives other aspects of the development process.

The course of global development today is guided by the decisions and geopolitical interventions of national governments and international organisations. The interventions take many different forms, from development aid to military campaigns. The motives behind much of this intervention need to be scrutinised. Many interventions are made by powerful and wealthy governments and organisations on the grounds that they should help the weaker and poorer parts of the world. Often there are mixed motives, not all of them commendable.

10.1 Concepts of human development

Measuring human development

Human development is often measured primarily in economic terms, using a country's total gross domestic product (GDP) or per capita GDP (Figure 10.1). This reflects a Western, capitalist perception of development that not all players agree with. It is widely recognised that human development involves much more than economic growth.

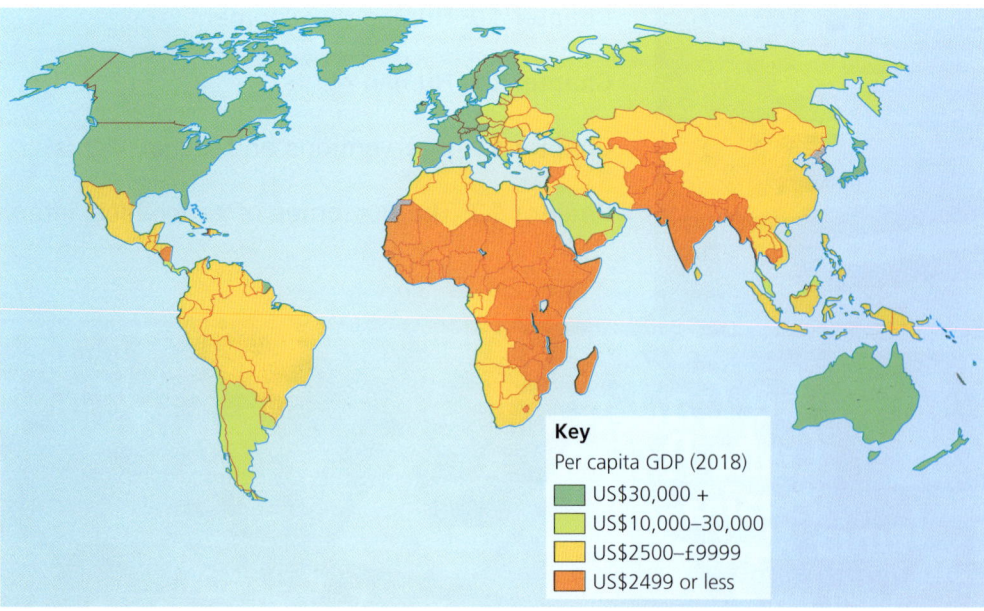

Figure 10.1 Global development levels based on per capita GDP, 2018

Option 8A: Health, Human Rights and Intervention

The concept of development today is altogether wider, more rounded and much more people-aware. As a consequence, the term 'human development' seems much more appropriate. It reminds us that there are other significant dimensions to this process, not just economic (Figure 10.2). These include improving people's well-being, quality of life and contentment. Widely used indicators of human development today include life expectancy, infant and maternal mortality, literacy and healthcare.

Happy Planet Index

A newcomer to the wide range of human development measures is the Happy Planet Index (HPI). This is now claimed to be the leading global measure of sustainable well-being.

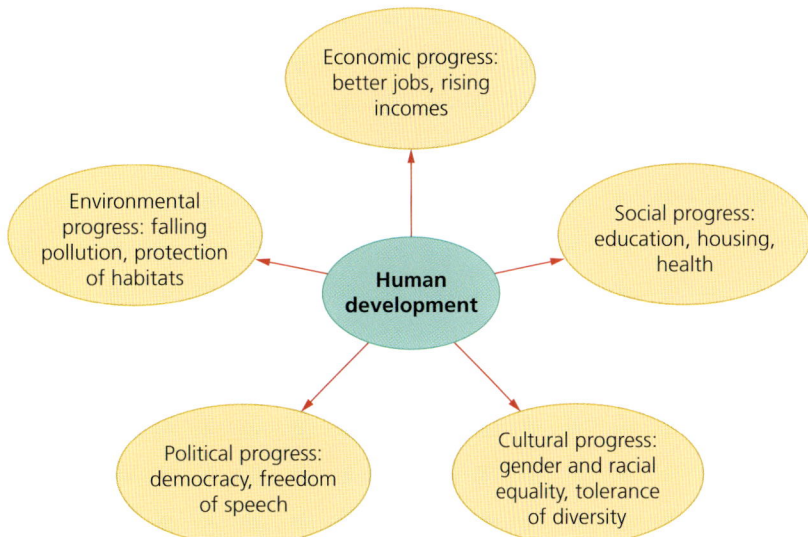

Figure 10.2 Aspects of human development progress

The HPI is calculated using three data sources: life expectancy, self-reported well-being and carbon footprint.
- self-reported well-being – people are asked where they place their present well-being on an imaginary ladder of ten steps, where zero is the worst possible and ten the best, this data is collected in collaboration with the Gallup World Poll
- life expectancy – assumed to be an important indicator of a nation's health, which is collected using data from the United Nations
- carbon footprint – the per capita greenhouse gas emissions caused by consumption and economic activity within a country.

Table 10.1 Happy Planet Index scores, 2024

Rank	Happy Planet Index score	
1	Vanuatu	57.9
2	Sweden	55.9
3	El Salvador	54.7
4	Costa Rica	54.1
5	Nicaragua	53.6
World average		37.3
143	Chad	18.3
144	Afghanistan	16.2
145	Lesotho	15.6
146	Botswana	14.7
147	Central African Republic	13.7

Data: Happy Planet Index

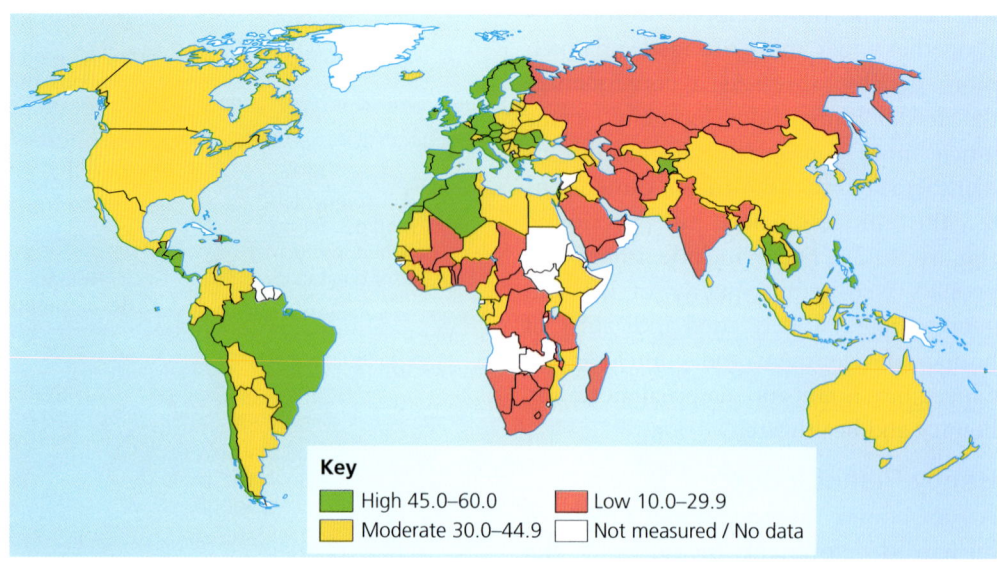

Figure 10.3 Global variation in Happy Planet Index scores

The HPI is an interesting attempt to make the complex link between resource consumption and human well-being. It does have some shortcomings, however. Two of the four measures (well-being and carbon footprint) are based on highly aggregated data. Measuring people's perception of their own well-being is subjective and hard to compare between countries. Only life expectancy data can be considered fairly reliable.

Figure 10.3 shows that wealthy western European countries have high HPI as might be expected. However, so do some middle-income countries in Asia, South and Central America. Here incomes are comfortable but fossil fuel use is relatively low, so HPI is higher than GDP per capita alone might suggest.

Countries with low HPI scores include very low-income countries in Central and southern Africa but also higher income Middle East countries with very high carbon footprints.

Fundamental differences

Differences is history, beliefs and values mean the dominant Western model of 'development' is not accepted everywhere. For example, some very strict implementations of Islamic law or the regime in North Korea are very different than the model of human development in Bolivia under Evo Morales. The place contexts on pages 197 and 198 illustrate this vital point.

Skills focus: Comparisons of ranked data

An interesting exercise is to rank countries according to a range of different measures, such as per capita GDP, HPI and HDI, and then to compare the rank orders. You will be surprised to see how different they often are. Spearman's rank correlation coefficient is a statistical test often used when making such comparisons.

Option 8A: Health, Human Rights and Intervention

Islamic law

Islamic law (*fiqh* in Arabic, often called 'Sharia Law' in English) was developed by religious scholars applying the teachings of the Quran and the practices and sayings of the prophet Muhammed to new situations. Sharia means the path to salvation, encompassing all of the doctrines of Islam, from how to pray and fast to rules for commercial contracts.

Muslim majority countries have different legal systems. Some have laws made by a legislature, for example Türkiye, while others such as Saudi Arabia use Islamic law. In Saudi Arabia some offences can be punishable with extreme measures. Drug trafficking, for example, can be punishable by death. Women's rights in particular have been very limited; for example, Saudi women were not allowed to drive until 2018, or to travel abroad without a male guardian's permission until 2019.

The implementation of Islamic law in countries such as Saudi Arabia does not rest easily with the Universal Declaration of Human Rights (see Chapter 11, page 213). The UK and some other countries are concerned that Muslim citizens may use Sharia amongst themselves as a parallel legal system.

The countries currently using Islamic law includes some of the world's richest (Brunei, Qatar, Saudi Arabia and the United Arab Emirates) and some of the poorest (Afghanistan, Mauritania, Sudan and Yemen).

Figure 10.4 Woman driving a car in Saudi Arabia

Improvement in health, life expectancy and human rights

A prevailing view of development today, but not a unanimous one, is that it should focus on:

- health
- life expectancy
- human rights.

Clearly, the first two are closely linked. Some would add a fourth objective, namely increasing care of the environment. Improvements in environmental quality (such as reducing pollution levels) are seen as being vital to the well-being of both the physical world and its inhabitants.

The focus on economic growth and constantly rising GDP as the main way to deliver improved human development is increasingly questioned, especially by environmentalists and climate scientists. Continued resource exploitation to power GDP growth inevitably leads to environmental degradation (air pollution, climate warming, deforestation) which has a negative impact on human well-being. This risks offsetting the supposed gains from rising GDP.

Access to education

Even those who emphasise the economic dimension of development would agree that access to education is crucial. A literate, numerate, enterprising and skilled workforce is precious human capital. Such capital is vital if a country is to move along the development pathway. Education promises a better job and higher wages, and from this flow material benefits that raise the quality of life.

Bolivia under Morales

Evo Morales, a native Aymara Indian, won an unprecedented third term of office in Bolivia's 2014 presidential elections (Figure 10.5). Not only was he Bolivia's first indigenous president, but he experienced a remarkable rise from humble beginnings growing coca, the source of cocaine.

On the international stage, Morales was widely known for his anti-colonialist and anti-imperialist rhetoric. His popularity was based on the exploitation of Bolivia's natural gas and mineral resources and sharing the derived wealth among the people. As a consequence, half a million Bolivians were lifted out of poverty. But, despite this progress, Bolivia remains one of the poorest countries in Latin America. According to the World Bank, roughly a quarter of all Bolivians still live on only US$2 a day.

As far as development is concerned, the Morales model is a socialist one, but it does not look beyond giving all Bolivians a share (not necessarily an equal one) in the wealth derived from the country's natural resources. The model has a rather limited view of development. It has little to say about the more 'human' aspects of development, such as education, equal opportunities or freedom of speech.

Figure 10.5 Evo Morales, President of Bolivia, 2006–19

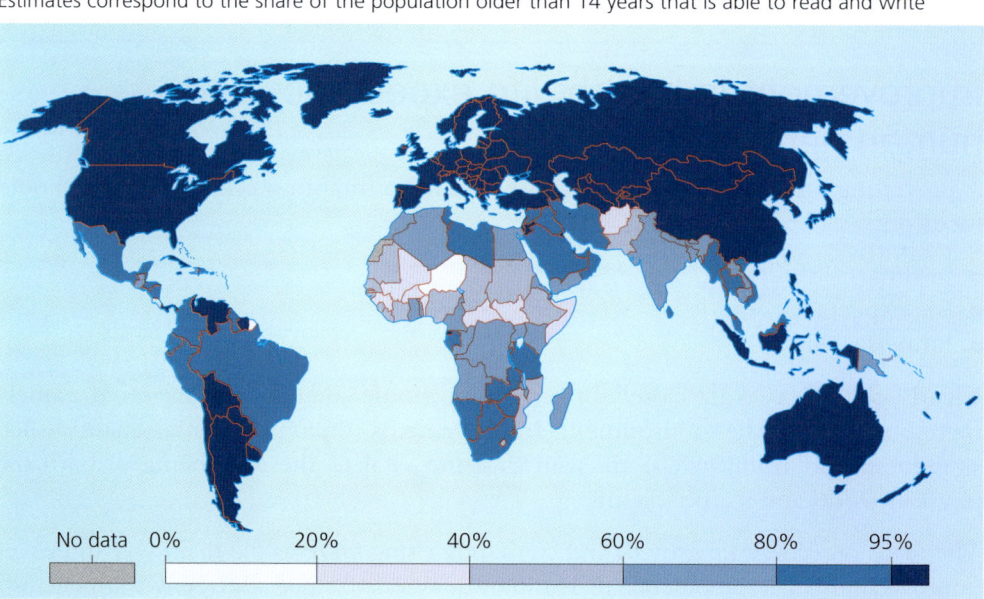

Figure 10.6 Global literacy rates, 2015

For those who adopt a more 'human' view of development, education provides a key to other things that collectively also enhance the quality of life. For example, understanding and asserting your human rights (see Chapter 11, page 213). There is also being informed about personal health, hygiene and diet, and what the individual needs to do under those three headings for a longer life.

Figure 10.6 illustrates how literacy rates vary between countries. In 2015, 86 per cent of the world population older than 14 years were literate. However, there is evidence of inequalities, most notably with Africa and the rest of the world.

There are few countries that do not recognise the human right of access to education, even if it amounts to no more than a few years at primary school. However, there are substantially more countries where there is overt gender discrimination, with females being increasingly barred or deterred from access to other levels of education – secondary and tertiary. Both boys and girls in Pakistan have a low enrollment rate, and education on the whole is underfunded. However, girls are disproportionately affected – 44 per cent of boys do not go to school, where as 56 per cent of girls do not go to school. But Pakistan is not alone in this. Figure 10.7 shows some huge differences between male and female literacy rates. The situation is particularly serious over much of Africa, the Middle East and India.

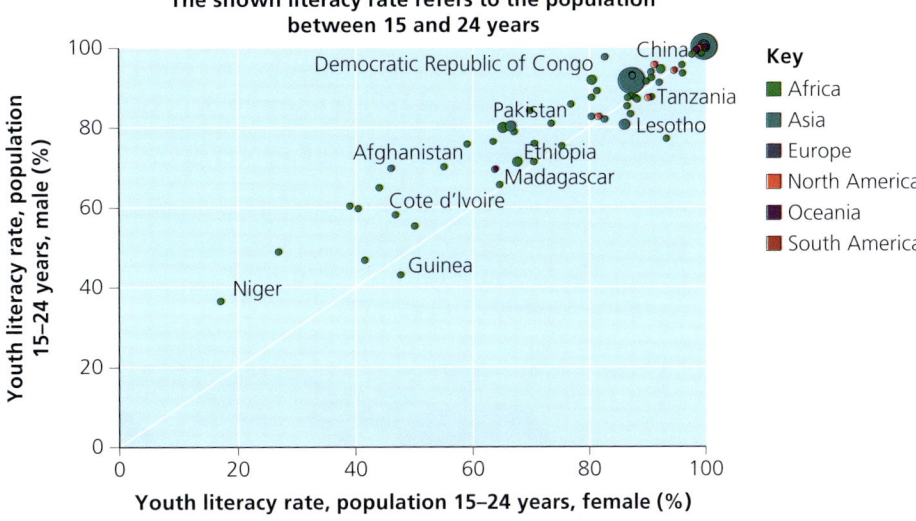

Figure 10.7 Literacy rate of young men and women, 2015

Access to education is impeded by other obstacles, such as:

- ethnicity
- physical and mental disability
- social class
- wealth.

UNESCO

The United Nations Educational, Scientific and Cultural Organization (UNESCO) has done much throughout the world 'to ensure that every child, boy or girl, has access to quality education as a fundamental human right and as a prerequisite for human development' (see Table 10.7 on page 208). It has done much to raise levels of literacy around the world, but Figure 10.7 is a salutary reminder that there is much still to be done. In much of Africa and South Asia the female literacy rate is more than a quarter below that for males.

Human Development Index

The Human Development Index (HDI) provides a good way of rounding off this section on the nature and measurement of human development. It is another widely used measure to show the state of global development. The HDI has breadth, in that it takes into account three important dimensions of the development process:

- life expectancy (an indicator of health and well-being)
- education (years of schooling)
- economic growth (per capita income).

However, there has been some criticisms regarding its accurate depiction of a country's level of development because it fails to consider the role of inequality, environmental quality or human rights.

Figure 10.8 shows the world clearly divided into three according to a scheme of high, medium and low HDI values. It is important that you compare this distribution pattern with that shown by the other two measures: per capita GDP (Figure 10.1, page 194) and HPI (Figure 10.3, page 196).

Two particularly useful aspects of the HDI are:

1 It relies on statistical data that are collected frequently and widely at a national level.
2 Because of this it can be used to monitor development progress over a year or period of years.

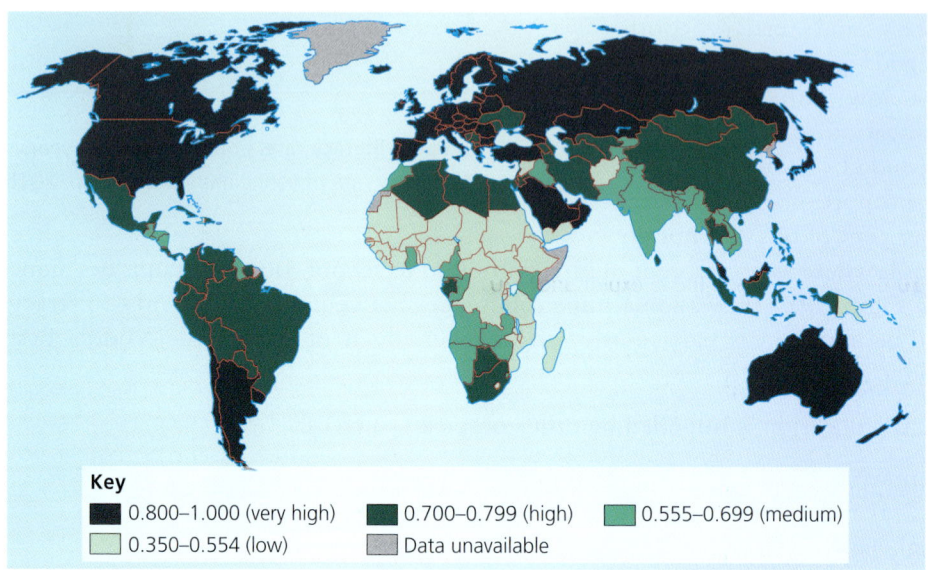

Figure 10.8 Human Development Index (HDI) scores, 2018

10.2 Human health and life expectancy

In this section, the focus is on these two significant aspects of human development. Of course, they are related in that life expectancy depends on health. Data on life expectancy are readily available worldwide and often on an annual basis.

Life expectancy

Figure 10.9 shows that, over much of the world, life expectancy is now over 65 years. The one obvious exception is much of Africa. The traditional subdivision of the world on the basis of economic development into 'developing' and 'developed' (with low life expectancies in the former and high in the latter) is not entirely clear-cut. The picture is more complex with relatively high values in South America, North Africa, the Middle East and throughout most of Asia. These are emerging countries, where life expectancy has recently improved as a result of economic development. Table 10.2 focuses on the countries at either end of the life-expectancy spectrum.

Before moving on, there is an aspect of life expectancy that needs to be stressed, namely that there are gender differences. In nearly all populations, female life expectancy is greater. In many developed countries, the difference can be five

Option 8A: Health, Human Rights and Intervention

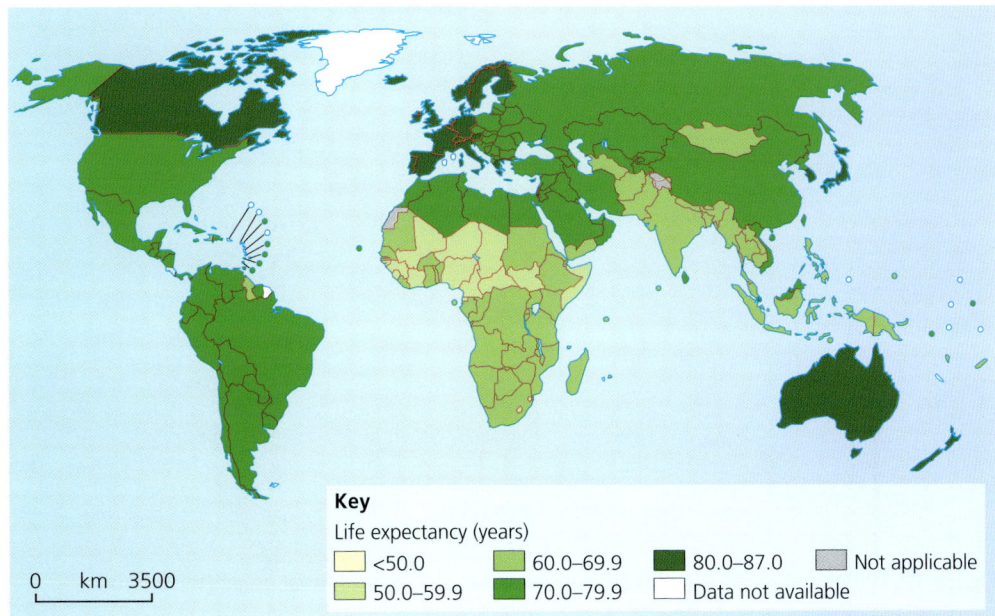

Figure 10.9 Global variation in life expectancy, 2016

years or more. For example, in 2020 life expectancy in the UK for women is 83.3 years and for men 79.8 years. In the least-developed countries, the age differential is less. For example, life expectancy in Bhutan is 70.4 years for men and 70.7 years for women. The key factor here is the high rate of maternal mortality (death during childbirth). The incidence of HIV/AIDS may also be a factor.

Table 10.2 Countries with highest and lowest life expectancies, 2022

Rank	Highest life expectancy	Years	Rank	Lowest life expectancy	Years
1	Japan	84.8	184	South Sudan	55.6
2	Switzerland	84.3	185	Central African Republic	54.5
3 =	Singapore	84.1	186	Nigeria	54.6
3 =	Italy	84.1	187=	Chad	53.0
5	South Korea	84.0	187=	Lesotho	53.0

Data: United Nations

Health

Portraying the global state of health is rather more challenging. There are no easy options in terms of readily available data. Life expectancy is probably as good as any, but there are two more overtly medical measures to be considered:

- The number of doctors per 100,000 people.
- The percentage of the population with regular access to essential drugs (Figure 10.10).

The second measure shows more clearly a threefold global subdivision of developed countries (over 95 per cent), emerging countries (between 50 and 95 per cent) and least-developed countries (less than 50 per cent).

Figures 10.9 and 10.10 illustrate the fundamental point that, at a global scale, levels of both life expectancy and health vary considerably from place to place.

10 Human development

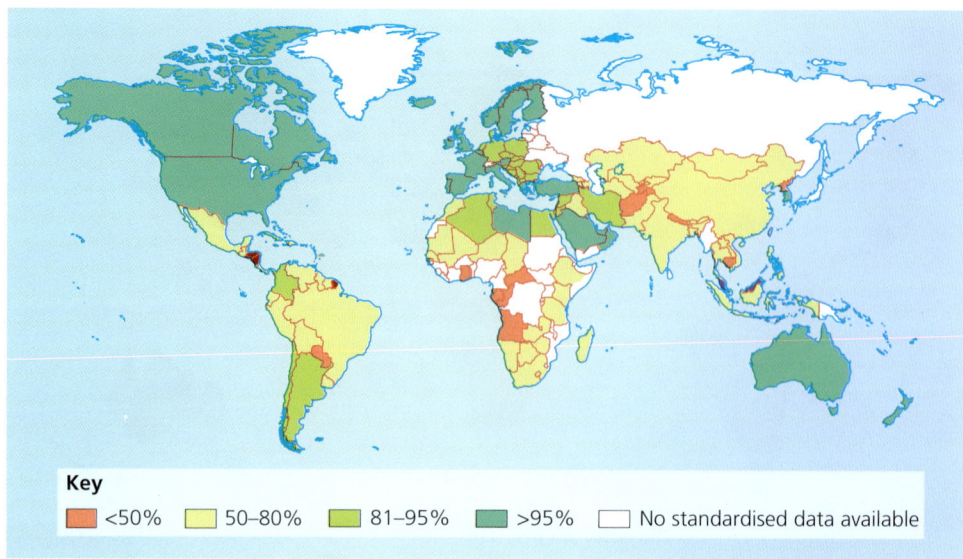

Figure 10.10 Percentage of population with regular access to essential drugs (WHO data)

Spatial variations in the developing world

Figure 10.9 has already illustrated the point that life expectancies are lower in the developing world compared with the developed world. Now look at Africa and note that there are some significant variations in life expectancy in this part of the developing world, particularly between the north and the rest of the continent.

Here, as elsewhere, there is a positive correlation between life expectancy and per capita income. The relationship is explained by the fact that with diminishing income, the following critical necessities of life become less guaranteed:

- food
- safe water
- proper sanitation
- healthcare (Figure 10.10).

A shortfall in any of these necessities immediately increases the risks of disease, ill health and premature death.

Spatial variations in the developed world

Life expectancy and health also vary considerably from place to place within the developed world. Despite higher levels of economic development and income, there are significant national differences. Compare the countries of eastern and western Europe, for example, in Figure 10.9. As in the developing world, the same four access factors come into play. Here the term **deprivation** is used to describe a situation of poor diet, poor housing and poor healthcare. In other words, these symptoms of poverty combine to create health risks that ultimately increase the death rate and lower the life expectancy.

Societies in the developed world are typically polarised, showing extremes of poverty and great wealth. However, it is not only the poor who are confronted by health risks. The lifestyles of the better-off also carry health risks, such as obesity (Figure 10.12), smoking, alcoholism and heart disease. There may be others; we are only just beginning to discover the health costs of modern living.

> **Key term**
>
> **Deprivation:** When an individual's well-being and quality of life fall below a level regarded as a reasonable minimum. Measuring deprivation usually relies on indicators relating to employment, housing, health and education.

Option 8A: Health, Human Rights and Intervention

Skills focus: Scatter graphs and correlation techniques

A scatter graph is a simple and highly visual method used to investigate the relationships between sets of paired data (variables). Is there a relationship, for example, between life expectancy and the incidence of poverty? Basically, the data of one variable are plotted on the x-axis and that of the other on the y-axis. Once the graph has been completed, a pattern may become visible that suggests some sort of a relationship between the two variables. A best-fit line should then be drawn on the graph: this line should pass through the spread of plotted points, minimising the total distance the points are from the line and with roughly equal deviations on either side of the line (Figure 10.11).

A scatter graph is used initially to identify whether or not a relationship exists between two variables. If it does, then Spearman's rank correlation coefficient can be calculated to establish the strength of the relationship and whether it is statistically significant. The calculation of the coefficient is based on the rank differences between the two data sets.

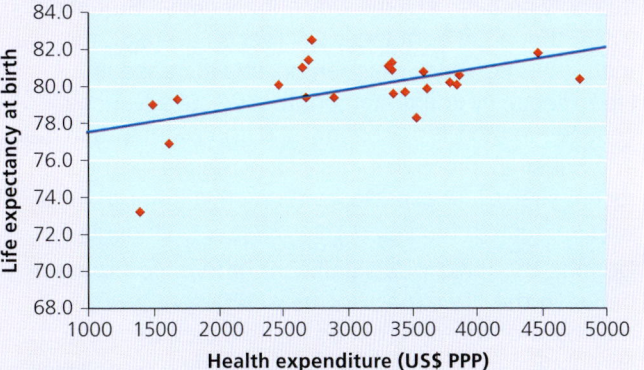

Figure 10.11 A scatter graph showing the relationship between life expectancy and health expenditure

A particularly significant factor is healthcare – its quality and accessibility. A big differential here is between:

- countries with national health services that are 'free', being funded by some form of taxation
- countries where healthcare is largely in the private sector and paid for either through social health insurance or on an 'as and when' basis.

Clearly, the former situation is a better one in terms of the poor accessing healthcare. There are significant variations between countries in terms of spending on health and education as seen in Table 10.6 (page 206).

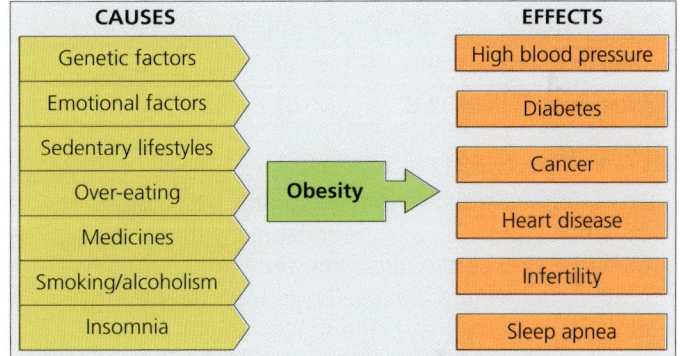

Figure 10.12 The causes and effects of obesity

Democratic, wealthy countries tend to spend more. However, there are exceptions, such as socialist Cuba, that have historically prioritised such spending.

Spatial variations within countries

So far we have looked at spatial variations in life expectancy and health at two spatial scales: globally, between the developing and developed worlds, and internationally, between the countries falling in either of those global subdivisions. But significant spatial variations also occur at a third spatial scale, namely within countries.

These country studies suggest that the following factors play a significant part in causing life expectancy to vary within them:

- ethnicity
- poverty and deprivation
- lifestyle and socio-economic group
- healthcare.

There is, however, another factor – government. Its policies and interventions can have a profound impact on the factors bulleted above.

10 Human development

Life expectancy variations within the UK

A girl born in the UK at the start of the twentieth century had an average life expectancy of less than 50 years. It is now predicted by the Office for National Statistics (ONS) that a girl born today will live, on average, for more than 80 years. This remarkable increase in life expectancy is a testament to medical advances, changes in the UK's economy and improvements in diet and housing, although it has not increased as much in recent years. Life expectancy for men is lower than for women, though the gap is narrowing very gradually.

Despite this progress significant differences remain between the life expectancies of different groups. It may be that the differences in life expectancy are the cumulative outcome of inequalities, such as poverty and deprivation.

By country and county

Table 10.3 Life expectancy estimates in the UK, 2020–2022

	England	Wales	Scotland
Men	78.8	77.9	76.5
Women	82.8	81.8	80.7

Data: ONS

The differences between three of the UK's countries are small (Table 10.3). It is most likely that they reflect differences in lifestyles and the general level of affluence. Figure 10.13 shows that, despite its relatively high values, England remains an unequal country with more than 15 years' difference in the healthy life expectancy of men in Richmond upon Thames (London) compared with those in a part of Manchester. The highest incidences of the UK's big killers – including heart disease and cancer – are found in the most deprived areas.

By socio-economic group

Table 10.4 Life expectancy at birth in the UK by socio-economic group, 2012–2016

Socio-economic group	Men	Women
Higher managerial and professional	85.5	83.6
Lower managerial and professional	84.9	81.9
Skilled non-manual	84.2	81.2
Skilled manual	81.8	79.6
Semi- or unskilled manual	82.3	78.7

Data: ONS

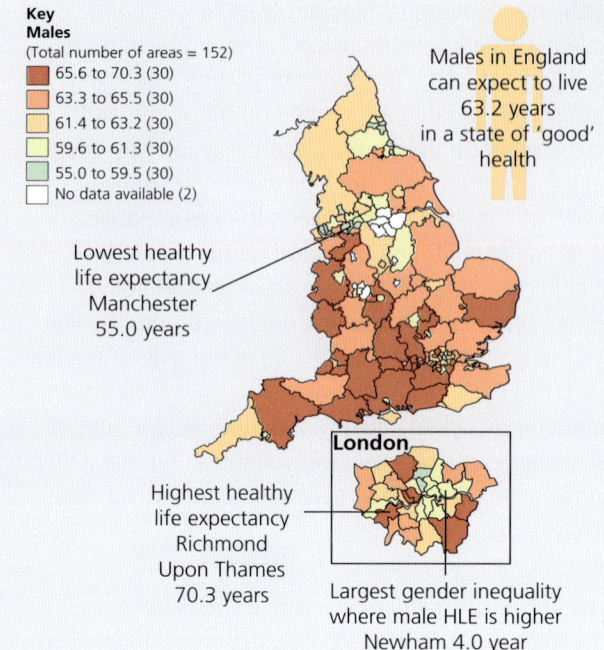

Figure 10.13 England: healthy life expectancy for men, 2009–11

The differences between socio-economic groups are large (Table 10.4). Again, they are partly explained by differences in lifestyle and housing conditions, but attitudes towards health and diet are likely to play a part, as well as different levels of safety at work.

By ethnic group

Table 10.5 Indirect estimates of life expectancy at birth for different ethnic groups, England and Wales, 2011–2014

	Female	Male
Mixed	83.1	79.3
White	83.1	79.7
Black Caribbean	84.6	80.7
Indian	85.4	82.3
Bangladeshi	87.3	81.1
Black African	88.9	83.8

Data: ONS (2021)

When it comes to differences in the life expectancies of different ethnic groups (Table 10.5), it is possible that the explanation may lie in differences in lifestyle (behavioural factors and those relating to diet), or indeed widening inequaliities in access to and quality of healthcare received. Other economic factors, such as low incomes as a result of low educational achievement which influence occupations, are likely to play a part.

Option 8A: Health, Human Rights and Intervention

Life expectancy variations within Brazil

In 2020, life expectancy in Brazil – one of the world's leading emerging countries – stood at 70.8 years for men and 75.6 years for women. Figure 10.14 shows quite considerable variations within Brazil at a state level, albeit some years previously.

The highest life expectancies occur in southeast Brazil, stretching from Minas Gerais to Rio Grande do Sul. Here is the core of the Brazilian economy, and presumably the higher life expectancies can be explained in terms of beneficial spin-offs such as jobs, higher wages and adequate housing. The surprisingly low life expectancy in the small state of Rio de Janeiro reflects the many *favelas* (shanty towns) located in this huge metropolitan area.

The relatively low values in the northern part of the country, particularly in the Amazon lowlands, may reflect its remoteness and relatively undeveloped nature. It may also reflect the fact that this is where many of Brazil's remaining indigenous people live. They occupy great tracts of very sparsely populated forest. They rely on traditional rather than modern medicines for healthcare. It is estimated that about 900,000 Amerindians now live in Brazil, compared with 5 million when the Europeans began to colonise South America.

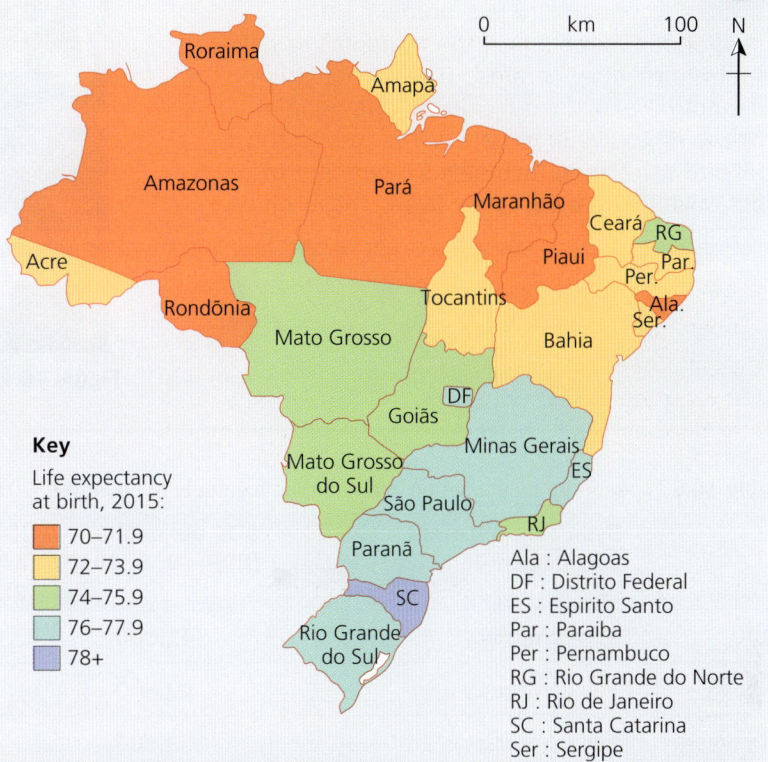

Figure 10.14 Life expectancy in Brazil's states, 2015

In Brazil, as in other countries with significant indigenous populations (Australia, Colombia, Mexico, New Zealand, South Africa and the USA), it is estimated that native peoples have up to 20 years less life expectancy than non-indigenous people.

Life expectancy variations within Australia

Australia enjoys one of the highest life expectancies of any country in the world: 82 years for men and 85.8 years for women. It is currently ranked eighth in the world by the United Nations Development Programme.

Indigenous people make up around 3 per cent of Australia's population of nearly 24 million. Indigenous people's life expectancy for men is currently estimated to be 10.6 years lower than that of non-indigenous men, and 9.5 years lower for women. Over the last five years indigenous people's life expectancy has increased by about one year for both genders.

But why the big difference between the two components of Australia's population? According to the UN, the quality of life for Australia's indigenous people is the second worst in the world. It is widely agreed that contributory factors include:

- poor housing
- low education levels and high unemployment
- weekly household incomes of $830 compared to $1080 for other Australians
- high rates of smoking and alcohol use
- ethnic discrimination and a history of being ignored by politicians.

While there is more spending per capita on the indigenous than the non-indigenous populations, access to healthcare is a problem. Many indigenous people lack the transport to get them to medical centres. This particular problem cannot be explained by indigenous people choosing to live in remote areas (the Outback) (Figure 10.15); in fact, only 25 per cent of them do – over 30 per cent now live in major cities.

Figure 10.15 Indigenous men in Queensland, Australia

Table 10.6 Expenditure on education and health in a sample of countries, 2017–19

	Type of government (Economist Democracy Index)	GDP per capita (PPP) (US$, 2019)	Health expenditure as a percentage of GDP, 2017	Education expenditure as a percentage of GDP, 2018 or most recent
UK	Full democracy	41,000	9.6	5.5
Japan	Full democracy	40,800	10.9	3.2
Denmark	Full democracy	59,800	10.1	7.6
South Africa	Flawed democracy	6,100	8.1	6.2
India	Flawed democracy	2,200	3.5	3.8
Türkiye	Hybrid regime	9,000	4.2	4.4
Bangladesh	Hybrid regime	1,900	2.3	1.9
China	Authoritarian	10,100	5.2	4.1
Cuba	Authoritarian	8,000	11.7	12.8
Ethiopia	Authoritarian	950	3.5	4.7

Data: Economist Intelligence Unit, World Bank

10.3 Defining development targets and policies

In most countries economic development provides the means (capital and human resources) that drive and sustain human development. The link between these two types of development is critical. The link is, in effect, in the hands of government. It is government that determines how much of a country's wealth should be spent on providing those vital, enriching components of human development, such as education, health and other social services. This, in turn, hinges very much on governmental attitudes towards social progress.

Different attitudes to social progress

There is no universally agreed classification of different types of government. However, it is possible to think in terms of two continua, one running from left wing (socialist) to right wing (capitalist), and the other from democratic (a regularly elected governmental body) to authoritarian and totalitarian. As a generalisation, it may be said that where a government is located along these two continua will be reflected in its attitudes towards human development or **social progress**.

Table 10.6 looks at expenditure on education and health in a sample of countries selected to represent different types of government.

In only four countries is total expenditure on education greater than 5 per cent of GDP. It is interesting that one of the poorest countries in the world, Ethiopia, spend just under 5 per cent. This might suggest that despite a lack of democracy that country prioritises education as a crucial aspect of development. Cuba's figure is also impressive and reflects a socialist commitment that all children should have equal access to schooling and that all people have access to healthcare. At the other extreme, one can understand that authoritarian and **totalitarian regimes** might be reluctant to see society becoming too well informed. The relatively low figure for Japan is a surprise, particularly when compared to that for health. The difference between the two expenditures reflects Japan's rapidly ageing population and the need to spend large sums on health and social care for the over 65s.

Countries that spend similar amounts on education compared to healthcare tend to be those, like Bangladesh, Ethiopia and India, that have youthful populations so there are large numbers of children to be educated.

Broadly, greater democracy leads to higher social spending. Authoritarian regimes may spend a higher percentage of GDP on military, police and security leaving a smaller slice for education and healthcare. Equally overall wealth plays a role. Developing and emerging countries may be spending proportionally more on infrastructure (roads, electricity, water supply) as part of attempts to stimulate economic development. Figure 10.16 (page 209) gives data for the leading emerging countries – Russia, Brazil, Mexico, India and China. Their welfare spending is hardly pitched at the same level. Brazil is clearly leading the way, followed by Mexico.

> **Key term**
>
> **Totalitarian regime:** A system of government that is centralised and dictatorial; it requires complete subservience to the state with control being in the hands of elites. These may be the military or powerful families or tribes. For some, 'totalitarian' and 'authoritarian' are taken to mean more or less the same thing.

> **Key concept: Social progress**
>
> Social progress is the idea that societies can and do improve their economic, political and social structures. It is about meeting basic human needs, raising well-being and creating opportunities for people to improve their lot. It is a normal part of socio-cultural evolution, but its pace can be very slow. It can be accelerated by deliberate inputs of:
>
> - government intervention – for example, creating a national health service, providing subsidised housing for the poor, ensuring free education for all children
> - social enterprise – businesses that trade for a social or environmental purpose
> - social activism – intentional actions aimed at bringing about social change, for example, the empowerment of women.

IGOs' views of development

Among the major players involved in the promotion of global development are the World Bank, WTO and IMF (Table 10.7). Their efforts are very much focused on economic development, seeing it as the springboard for advances on the broader front of human development. They believe that economic growth will come from a free market that will eventually mean the wealth trickles down to the poorest in society. This, in turn, relies on such things as free trade, privatisation and the deregulation of financial markets. However, in more recent times there has been a greater attempt to focus on promoting programmes that contribute towards improving the life chances of citizens through addressing issues around healthcare, education, human rights and the environment.

Skills focus: Proportional circles and pie charts

Proportional circles are a good way of comparing locations (countries, regions, cities, etc.) in terms of a single variable (population, per capita GDP, HDI, etc.). The square root of each piece of data is calculated and the radius of each circle is drawn proportional to square root value. The use of square roots has the effect of visually diminishing the differences between circles. Proportional circles may be located on a map to give a visual impression of distribution, or they can simply be set side by side in a diagram.

Pie charts are drawn to show the make-up of a particular variable (for example, the relative importance of the economic sectors in a country's economy, the commodities involved in exports or imports, the ethnic composition of a population). Each component of the data is calculated as a proportion of 360 (because there are 360 degrees in a circle) and shown as a slice of so many degrees (Figure 10.16).

Table 10.7 The missions of major IGOs

Organisation	Founded	Member countries	Mission
World Bank	1944	189	Originated as a facilitator of post-war reconstruction and development. Now committed to the alleviation of poverty. It is a vital source of financial and technical assistance to developing countries around the world. It is not a bank in the ordinary sense but a unique partnership to reduce poverty and support development.
World Trade Organization (WTO)	1995	164	Succeeded the General Agreement on Tariffs and Trade (GATT), set up in 1948, and is the only global organisation dealing with the rules of trade between nations. At its heart are the WTO agreements, negotiated and signed by the bulk of the world's trading nations and ratified in their parliaments. The goal is to help producers of goods and services, exporters and importers, to conduct their business.
International Monetary Fund (IMF)	1948	189	Aims to foster monetary co-operation, secure financial stability, facilitate international trade, promote high employment and sustainable economic growth, and reduce poverty around the world. It seeks to improve the economies of member countries through data collection and analysis, monitoring economic performance and, where necessary, recommending self-correcting policies.
United Nations Educational, Scientific and Cultural Organization (UNESCO)	1945	193	Its purpose is to contribute to peace and security by promoting international collaboration through education, science and culture to further universal respect for justice, the rule of law and human rights, along with fundamental freedom as proclaimed in the UN Charter (1945). It also promotes cultural diversity and aims to secure the world's cultural and natural heritage.
Organisation for Economic Co-operation and Development (OECD)	1960	37	Promotes policies that will improve the economic and social well-being of people around the world. It provides a forum in which governments can work together to share experiences and seek solutions to common problems. It works with governments to understand what drives economic, social and environmental change. It recommends policies designed to improve the quality of people's lives.

Option 8A: Health, Human Rights and Intervention

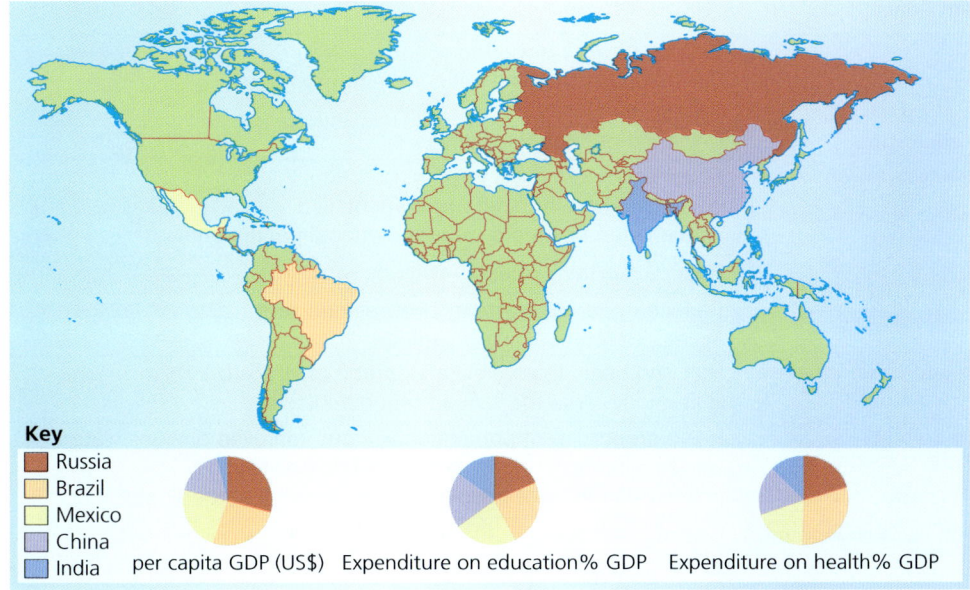

Figure 10.16 The leading emerging economies by per capita GDP and expenditure on education and health, 2014

Table 10.7 serves as a reminder that not all IGOs are so focused on economic development. The OECD and UNESCO have agendas that are more to do with the human condition, quality of life (including health and education) and human rights.

Development targets and goals

Since 2000 the United Nations has set ambitious targets that have attempted to reduce the **development gap** between the poorest and richest countries. There have been two sets of goals:

- United Nations Millennium Development Goals (MDGs) agreed in 2000 covering the period 2000 to 2015
- United Nations Sustainable Development Goals (SDGS) running from 2015 to 2030.

Eight MDGs were agreed in 2000 (Table 10.8). These focused on issues such as poverty, health, education and food supply. It was hoped that progress towards achieving these goals would narrow the development gap. By 2015 the MDGs had achieved some success:

- The proportion of people living in extreme poverty in the developing world fell from 47 per cent to 23.5 per cent.
- The proportion of girls and boys in education was almost the same by 2015.
- The percentage of the world's population without access to safe drinking water fell from 24 per cent to 12 per cent.

However, progress was geographically patchy. Asia, China and some other emerging countries made great progress in terms of poverty reduction. There was less progress in Africa and Oceania. Some goals, such as improving access to reproductive health services and reducing maternal mortality, were not achieved but there was progress in many countries.

> **Key term**
>
> **Development gap:** The widening income and prosperity gap between the global 'haves' of the developed world and the 'have-nots' of the developing world, especially the least developed countries.

Table 10.8 The Millennium Development Goals

Goal 1: Eradicate extreme poverty	Reduce poverty by half Create productive and decent employment Reduce hunger by half
Goal 2: Achieve universal primary education	Universal primary schooling
Goal 3: Promote gender equality and empower women	Equal girls' enrolment in primary school Women's share of paid employment Women's equal employment in national parliaments
Goal 4: Reduce child mortality	Reduce mortality of under-fives by two-thirds
Goal 5: Improve maternal health	Reduce maternal mortality by three-quarters Access to reproductive healthcare
Goal 6: Combat HIV/AIDS, malaria and other diseases	Halt and begin to reverse the spread of HIV/AIDS Halt and reverse the spread of tuberculosis
Goal 7: Ensure environmental sustainability	Halve proportion of population without improved drinking water Halve proportion of population without sanitation Improve the lives of people living in informal settlements
Goal 8: Develop a global partnership for development	Use of internet

Sustainable Development Goals

A high-level meeting in 2010 set in motion the definition of a post-2015 development agenda to build on MDGs. This involved consultations with major groups and stakeholders in 70 countries. The outcome, the 2030 Agenda for Sustainable Development, was agreed by world leaders at a summit meeting on 25 September 2015.

The agenda set out 17 Sustainable Development Goals (SDGs) to end poverty, fight inequality and injustice, and tackle climate change by 2030 (Figure 10.17). The new SDGs and the broader sustainability agenda go much further than the MDGs. They address the root causes of poverty and the universal need for a style of development that works for all people. The SDGs are also connected to the three strategic focus areas of the UN Development Programme (UNDP):

- sustainable development
- democratic governance and peace building
- climate and disaster resilience.

The MDGs could be seen as focusing on the basic needs and rights of people in developing and emerging countries. The SDGs recognise that sustainable use of the environment and action to tackle climate change and environmental degradation are fundamental if humans are to have high quality of life. The SDGs can be viewed as a more holistic agenda for sustainable human development. While the focus remains on the developing world, little is said about the contribution that might be made by the developed world, particularly in the broad context of aid.

Table 10.9 outlines the progress that has been made so far towards some of the SDGs. It is important to note that major events such as wars and natural disasters can hinder national progress towards the SDGs, making them much harder to achieve. In its 2023 report on progress towards the SDGs, the United Nations described progress as being 'in peril' with only around 15 per cent of goals and targets set to be achieved by 2030.

 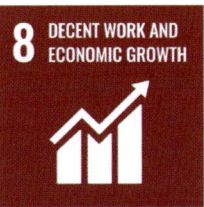  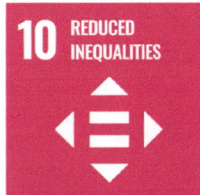 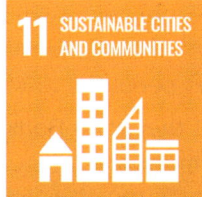 

Figure 10.17 Sustainable Development Goals for 2030

Table 10.9 Progress towards the SDGs, 2023

SDG Target	Progress	Details
1. No poverty	Severely off track	On current trends over 575 million people will still live with extreme poverty by 2030
2. Zero hunger	Severely off track	Numbers living with hunger increased 2019–2022, and 600 million are likely to be living with hunger in 2030
4. Quality education	Off track	70–100 million children are likely to not be in school by 2030
7. Affordable and clean energy	Off track	By 2030 an estimated 660 million people will not have access to electricity and 2 billion will use polluting fuels
15. Life on land	Off track	Deforestation rates are still increasing in many countries and degraded and eroded farmland area is increasing

Review questions

1. Explain the importance of economic growth in human development progress.
2. What are the limitations of the Happy Planet Index?
3. Compare the global views provided by Figures 10.1 (page 194), 10.3 (page 196) and 10.8 (page 200).
4. Suggest why the type of government might affect policies on education and healthcare.
5. Give reasons for the lower life expectancies of indigenous peoples.
6. Identify what the World Bank, WTO and IMF have in common.
7. Describe how the Sustainable Development Goals (SDGs) differ from the Millennium Development Goals (MDGs).
8. Outline the reasons for different attitudes to social progress.
9. Summarise the roles of the World Bank, WTO and IMF.
10. Assess progress towards achieving the UN Sustainable Development Goals (SDGs).

Further research

Explore the factors that explain differences in life expectancy within the developed world: www.who.int/data/gho/data/themes/mortality-and-global-health-estimates/ghe-life-expectancy-and-healthy-life-expectancy

Find out more about the five organisations that make up the World Bank Group: www.worldbank.org

Human rights

Why do human rights vary from place to place?
By the end of this chapter, you should:
- be aware of the importance of human rights
- understand that countries differ in their definition and protection of human rights
- be aware that there are significant variations in human rights within countries.

11.1 The importance of human rights

Key concept: Human rights

Human rights are moral principles that underlie standards of human behaviour. They are commonly understood as inalienable and fundamental rights 'to which a person is inherently entitled simply because she or he is a human being' and which are 'inherent in all human beings' regardless of their nation, location, language, religion, ethnic origin or any other status. They are universal in the sense of being applicable everywhere, and they are egalitarian in the sense of being the same for everyone.

There are relatively few countries today that deny the importance of **human rights**. However, there are significantly more that give economic development precedence over human development. The great concern here is what history tells us, namely that a disregard for human rights has led to 'barbarous acts which have outraged the conscience of mankind' (UDHR). Unfortunately, there have been many examples. In the 2010s the world has witnessed the brutal actions of Islamic State (IS). But it was the Nazi attempt to exterminate Jewish people during the Second World War that was the immediate catalyst for the Universal Declaration of Human Rights (UDHR) proclaimed by the UN General Assembly in 1948 (Figure 11.1).

Figure 11.1 Human rights poster

11 Human rights

Universal Declaration of Human Rights

The UDHR is a statement of intent and a framework for foreign policy statements to explain economic or military intervention. It sets out 30 universal human rights which are wide ranging, from freedom of speech and movement to education and justice (Table 11.1). They are vital strands in what is widely recognised as constituting human development.

The UDHR was adopted by the UN General Assembly by a vote of 48 in favour and eight abstentions (from the former Soviet Union and four of its satellites, the former Yugoslavia, South Africa and Saudi Arabia for various reasons). Honduras and Yemen, both members of UN at that time, failed to vote or abstain. In 1945 the UN had 51 members, in 1948 it had 58 members; today it has 193.

It is important to understand that the UDHR is not a treaty – it is not legally binding; there are no signatories. Some regard this as a fundamental weakness as the articles are unenforceable – however, the declaration does define the meanings of two key terms: 'fundamental freedoms' and 'human rights'. These terms are embedded in the UN Charter and, by implication, all 193 members of the UN are bound to recognise and respect all the articles of the declaration.

Table 11.1 A sample of the UDHR's 30 articles

Article	Summary statement
1	All human beings are born free and equal in dignity and rights.
2	Everyone is entitled to all the rights and freedoms set forth in this declaration, without distinction of any kind, such as race, colour, sex, language, religion, political or other opinion, national or social origin, property, birth or other status.
3	Everyone has the right to life, liberty and security of person.
4	No one shall be held in slavery or servitude; slavery and the slave trade shall be prohibited in all their forms.
5	No one shall be subjected to torture or to cruel, inhuman or degrading treatment or punishment.
7	All are equal before the law and are entitled without any discrimination to equal protection of the law.
9	No one shall be subjected to arbitrary arrest, detention or exile.
13	Everyone has the right of freedom of movement and residence within the borders of each state. Everyone has the right to leave any country, including their own, and to return to their country.
16	Men and women of full age, without any limitation due to race, nationality or religion, have the right to marry and to found a family.
19	Everyone has the right to freedom of opinion and expression.
26	Everyone has the right to education. Education shall be free, at least in the elementary and fundamental stages. Elementary education shall be compulsory.
27	Everyone has the right freely to participate in the cultural life of the community, to enjoy the arts and to share in scientific advancement and its benefits.

It will become apparent in Chapters 12 and 13 that, since 1948, violations of the UDHR have been used to justify a number of military interventions. Equally, the promise of aid, particularly of an economic kind, has been used as a lever to persuade other countries to improve their human rights record. In short, the UDHR has been a significant factor influencing foreign policies and international relationships.

Other international agreements

There are two particularly important conventions, namely the European Convention on Human Rights (ECHR) and the Geneva Conventions. The Human Rights Act 1998 is a version of the ECHR that has been adopted by the UK (Figure 11.2 and Table 11.2). The point needs to be made that the UK recognises the importance of the human rights set out in the ECHR; however, there is a body of opinion in the UK that is less convinced about the need for the European Court to interfere in our legal and parliamentary proceedings.

Table 11.2 Two conventions and the UK's Human Rights Act (1998)

Geneva Conventions	European Convention on Human Rights (ECHR)	UK Human Rights Act
A series of four treaties applied at times of armed conflict to protect people not taking part in the conflict (including prisoners of war). The first treaty was signed by 16 European countries and some American states in 1864. This was followed by treaties in 1906, 1929 and 1949. The 1949 Geneva Conventions has been ratified by 196 countries, but not all have agreed to the three subsequent protocols.	This convention, like the UDHR, comprises a number of articles, each setting out a specific human right. Coming into force in 1953, it has played an important part in developing an awareness of human rights in Europe. It was a response to: 1. the serious violations of human rights that occurred in Europe during the Second World War 2. the post-war spread of communism in Central and Eastern Europe and the threat of communist subversion. Violations of the convention come before the European Court of Human Rights.	This act, passed in 1998, incorporated into UK law the rights contained in the ECHR (Figure 11.2). It means that any breach of the convention's rights can be heard in UK courts and need not go to the European Court of Human Rights. However, appeals related to the verdicts of UK courts in such cases can be sent to, and possibly overturned by, the European Court. This has led some to believe that the UK has lost some of its sovereignty.

Article 2: Right to life

Article 3: Right not to be tortured or treated in an inhuman or degrading way

Article 4: Right to be free from slavery or forced labour

Article 5: Right to liberty

Article 6: Right to a fair trial

Article 7: Right not to be punished for something which wasn't against the law

Article 8: Right to respect for private and family life, home and correspondence

Article 9: Right to freedom of thought, conscience and religion

Article 10: Right to freedom of expression

Article 11: Right to freedom of assembly and association

Article 12: Right to marry and found a family

Article 14: Right not be discriminated against in relation to any of the rights contained in the European Convention

Article 1, Protocol 1: Right to peaceful enjoyment of possessions

Article 1, Protocol 13: Abolition of the death penalty

Article 2, Protocol 1: Right to education

Article 3, Protocol 1: Right to free elections

Figure 11.2 The rights contained in the UK's Human Rights Act 1998

11.2 Differences between countries

Human rights versus economic development

There are few, if any, countries in today's world that give human rights real precedence over economic growth. The inescapable reality is that material prosperity and global influence come more from economic development. This is not to say that human rights do not matter. When it comes to human rights, countries may be located on a continuum running from 'no regard' at one end of the scale to 'healthy respect' at the other end.

> **Key term**
>
> **Democracy:** Countries with a system of government in which power is either held by regularly elected representatives or directly by the people.

Most **democracies** are committed to the principles of human rights, but almost inevitably there are occasions when the interpretation of one of those principles by a particular government does not fall in line with that made by the international community at large. A recent example involving the UK is the government being chastised by the European Court of Human Rights for denying most prisoners the right to vote in elections. The UK abides by the principle that anyone convicted of a serious crime loses certain civil rights (including their physical freedom) while completing their sentence. Who is to say which interpretation of human rights is right? So, there will often be tension between the autonomy of the individual government and the rules and interpretations made at an international level.

While there are plenty of statistics relating to the economic wealth of countries, there are few measures relating to human rights. Figure 11.3 is a map of global freedom produced by an independent organisation known as Freedom House. The map rates the level of political rights and civil liberties in 210 countries. Based on these ratings, each country is broadly classified as:

- free – there is broad scope for open political competition and a climate of respect for civil liberties
- partly free – there are some clear restrictions on political rights and civil liberties
- not free – basic political rights and civil liberties are absent or systematically violated.

Unfortunately, Figure 11.3 shows a lack of freedom over large areas of the world, for example, much of Asia, Africa and the Middle East. The map certainly gives the impression of a polarised world, with few countries falling into the 'partly free' category.

Countries with the worst 'Freedom Rating' of 7/100 or below in 2019 were largely those where political unrest either prevails or is firmly suppressed. The lowest recorded country with a score of 0/100 was Syria. There were seven in all – three in Africa, two in Asia and two in the Middle East. In terms of economic status, all but one of these countries has very low per capita GDP. The notable exception is oil-rich Saudi Arabia, which ranks in the top ten richest countries.

At the other end of the scale there were 43 countries with the best 'Freedom Rating' of 70/100 or more. Most are what are widely recognised as 'developed' or 'advanced' countries. But there are some surprises: the list contains countries that are not necessarily renowned for their economic wealth.

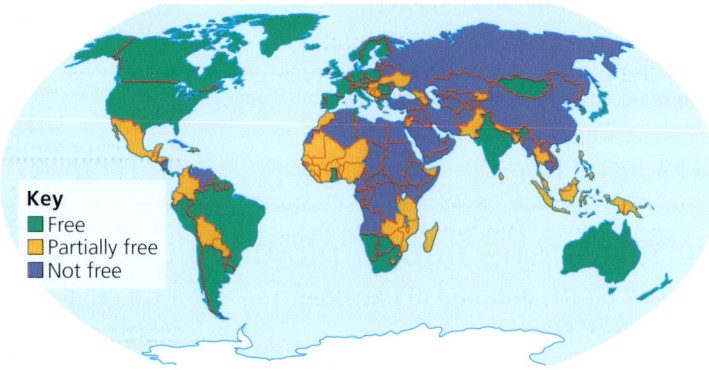

Figure 11.3 Levels of freedom in the world, 2019 (Freedom House)

The two Koreas

The Korean Peninsula extends some 1100 km south from mainland Asia into the Pacific Ocean (Figure 11.4). Just after the end of the Second World War, what had been a Japanese colony was divided into two states, the boundary running along the 38th parallel of latitude. To the north lies the Democratic People's Republic of Korea (usually referred to as North Korea); to the south lies the Republic of Korea (South Korea). Since their creation in 1948, the relationship between the two states has ranged from a persistent propaganda war, through provocative incidents, to outright war (the Korean War of 1950–3). Since then the states have followed two diametrically opposed ideologies.

Figure 11.5 A well-orchestrated and rehearsed military parade in front of Kim Jong-un

Figure 11.4 The Korean Peninsula

North Korea

North Korea describes itself as a self-reliant socialist state. It holds elections, but they only involve one political party. North Korea is in effect a totalitarian state with an elaborate personality cult based on the dictatorship of the Kim Jong family. This family has now supplied three generations of leader (Figure 11.5). It is without doubt a highly authoritarian regime.

Today, North Korea lives in self-imposed isolation from the rest of the global community. Its people are forbidden to use the internet. It is widely recognised as a 'rogue state' through its possession of nuclear weapons and its vast military forces, its frequent threats against South Korea, as well as its frequent violations of human rights. The last include the summary trials and executions of dissidents and the arrest of foreigners on the grounds of espionage. There is no freedom of speech.

Because of its isolation, there are few statistics about conditions in North Korea. We do know that it spends a huge amount of its GDP on its military forces and armaments. We also know that North Koreans suffer food shortages, malnutrition and occasional, but severe, famines. The last of these in the 1990s killed an estimated 2.5 million people, approaching 10 per cent of the population. This is hardly surprising when the regime makes feeding its people a much lower priority than the 'defence' of the country and its strategic industries.

South Korea

The story of South Korea is very different. It has embraced capitalism and has transformed itself from a war-torn country into a high-income advanced economy. The key to this economic success lies in firm government; increasingly powerful high-tech *chaebols* (large family-owned businesses) such as Samsung, Hyundai and LG; a committed labour force and a rich human resource of enterprise and technological innovation. But, interestingly, the first free elections were not held until 1987.

Figure 11.6 Student protests in South Korea meet a strong-armed response

Today, the Democracy Index ranks South Korea number two in Asia. Large-scale public protests and confrontations with heavy-handed police are not unknown (Figure 11.6), but it also ranks among the highest in the world in terms of education, healthcare and ease of doing business. It does, however, live in the shadow of a belligerent and menacing neighbour.

A comparison of the two countries today in terms of per capita GDP sees South Korea ranked 32nd in the world and North Korea 179th. While development is much more than economic performance, taking account of human rights and health, the post-war history of the Korean Peninsula speaks volumes for democracy and capitalism and very little for totalitarianism. The fact is that North Korea has hardly made a move towards real democracy. The World Democracy Audit in 2014 ranked North Korea as the most corrupt country in the world and as the least democratic.

The transition to democracy

The ten economic superpowers shown in Table 11.3 fall into two groups: those that have 'arrived' (the USA, Japan, Germany, France, the UK and Italy), and those that are 'emerging' (China, Brazil, India and Russia). If we focus on the latter group, it is interesting to compare authoritarian China, which has yet to begin the transition to democracy, with India, which has made the transition.

Table 11.3 The world's largest economies in 2019

	GDP (US$ billion)	GDP per capita (US$)
USA	21.4	65,100
China	14.1	10,100
Japan	5.2	40,800
Germany	3.9	46,600
India	2.9	2,200
UK	2.7	41,000
France	2.7	41,700
Italy	2.0	32,900
Brazil	1.8	8,800
Canada	1.7	46,200
Russia	1.6	11,200

Data: World Bank

Political corruption

When the term 'political corruption' is mentioned, most of us think in terms of election rigging, but it can assume other forms:

- Allowing private interests to dictate government policy.
- Taking decisions that benefit those who are funding the politicians.
- Diverting foreign aid and scarce resources into the private pockets of politicians.

All these and other malpractices result in corrupt politicians who can all too easily steer a country away from good government. Such a movement is often accompanied by a serious threat to human rights.

Unfortunately, there are rather too many countries suffering from political corruption. We have already taken a look at North Korea. Myanmar and Zimbabwe have the same CPI score and have been ranked equally by Transparency International as the eighteenth most corrupt countries in the world. But their case histories are very different.

Two emerging superpowers: China and India

These two giant countries, with their vast extents and huge populations, have made remarkable economic progress over the last 25 years. Together they account for 8.5 per cent of the world's land area and 36 per cent of its population. China is roughly three times larger than India in terms of area, but India's population in 2018 was only 50 million short of China's 1.4 billion.

China

Modern China was founded as a communist country with a one-party government following the Second World War. However, economic reforms introduced in 1979 have seen China become increasingly involved in the global economy, so much so that today China is described as a 'socialist market economy'. An increasing willingness to trade with the capitalist world has led to it rivalling the USA as the world's largest economy. But this economic success has been driven by a form of government that has a scant regard for human rights (Figure 11.7).

Figure 11.7 Tiananmen Square, Beijing, April 1989 – a pro-democracy demonstration ruthlessly put down by the military

China's human rights record has been widely criticised. Listed below are just some of the human rights abuses that make China a target for international protest:

- Re-education through labour – this is frequently handed out to critics of the government and followers of banned beliefs (e.g. the treatment of Uighur Muslims).
- Suppression of the internet and media freedom – hundreds of websites are blocked or banned in China. China has been described as 'the world's leading jailer of journalists'.
- Unfair trials – the Chinese judicial system falls a long way short of international standards.
- Torture – this, and the ill-treatment of detainees, is widespread. It is particularly directed at human rights activists and people detained because of their political or religious beliefs.
- Workers' rights – trade unions are illegal. Workers are not allowed to protest about low wages, poor working conditions, mass lay-offs or corrupt management.
- Death penalty – it has been estimated that China accounts for nearly three-quarters of all the world's executions each year; 46 offences are eligible for the death penalty.

The Chinese government is a highly authoritarian one-party state. Human Rights Watch (HRW) claims that it 'places arbitrary curbs on expression, association, assembly and religion'. It also prohibits trade unions and human rights organisations, and controls the judicial system. The government obstructs domestic and international scrutiny of its human rights record, insisting that any such scrutiny is an attempt to destabilise the country.

The key questions here are:

- Would China's economic development have been more or less had there been a greater respect for human rights?
- How vital to its economic success has the availability of a huge labour force of suppressed, exploited and forcibly compliant workers been?

India

India is a democratic republic (Figure 11.8) with a parliamentary system of government. It involves a union of 29 states and seven territories. In 2019, the general election led to the Bharatiya Janata Party taking control. The election was fought mainly on five issues: the stalled economy, rising prices, corruption, national security and infrastructure.

Figure 11.8 India: a social democracy going to the ballot box

11 Human rights

While it is shown in Figure 11.3 (page 216) as a 'free' country, it still has some human rights issues. These include:

- incidents of violence against religious minorities, especially Muslims
- caste-based discrimination and neglect of tribal communities
- sexual abuse and other violence against women and children.

Perhaps most worrying is the fact that members of India's security forces continue to enjoy impunity for serious human rights violations. The new government (2015) has expressed a commitment to freedom of speech, however it has not yet ended state censorship.

The Indian economy has not performed as spectacularly as the Chinese economy (Table 11.3, page 218). Whether that difference can be put down to the fact that economic growth in China has had the backing of an authoritarian government is not entirely clear. Equally, has the slightly less sparkling performance of the Indian economy anything to do with the fact that India is a democracy and broadly respects human rights?

Skills focus: Indices of corruption

Transparency International is a global organisation with one vision, namely to rid the world of corruption. It has a presence in more than 100 countries. Working with governments, businesses and citizens, its mission is to stop the abuse of power, bribery and secret deals. It was set up in the 1990s and, since 1995, has produced an annual Corruption Perceptions Index (CPI), which scores each country on how corrupt its public sector is seen to be. The methodology used to arrive at this score is not too transparent however, other than that it is determined by expert assessments and opinion surveys. Figure 11.9 shows the results of the 2019 survey.

It is hardly surprising that there are so few reliable measures of corruption. By its very nature, corruption prefers to remain hidden from public scrutiny and mostly does so successfully.

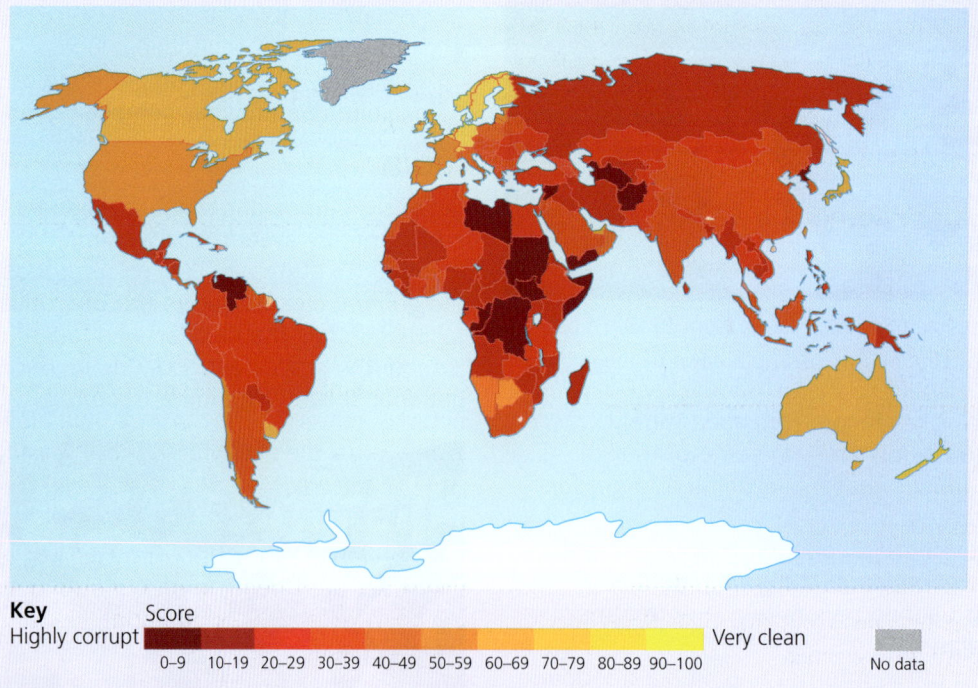

Figure 11.9 A global view of corruption based on the Corruption Perceptions Index, 2019 (Source: www.transparency.org)

Two cases of corruption: Myanmar and Zimbabwe

Myanmar

For more than 40 years Myanmar has been ruled by an unelected military junta. Political violence and systematic repression of democratic opposition have been rife. The country's political and economic environment has continued to deteriorate. Myanmar faces major challenges of endemic corruption. Little is known of the specific forms and patterns of corruption in the country, but the scale of the informal and illicit economy suggests strong links between the ruling elite (the military junta) and organised crime activities, such as drugs, human trafficking and illegal logging.

In 2011 a nominally civilian government was introduced in response to international pressure and growing civilian demand for democratic government. In 2015, the first reasonably fair general election for over 50 years was held. The results offer the country some hope. Aung San Suu Kyi's National League for Democracy won a landslide victory and now has control of parliament. However, the military-drafted constitution (2011) guarantees that unelected military representatives take up 25 per cent of the seats in parliament. In 2016, Aung San Suu Kyi took on the role of State Councilor (similar to a Prime Minister). However, since then her image has been tarnished due to inaction over the genocide of the Rohingya people in northern Myanmar. She even defended Myanmar's military at the International Court of Justice in 2019, to the consternation of many of her former supporters. In February 2021, the military seized control, following a landslide victory by Suu Kyi's NLD party.

Figure 11.10 Aung San Suu Kyi – winner of the Nobel Peace Prize and Myanmar's 2015 election

Zimbabwe

Myanmar has never enjoyed much economic prosperity. Zimbabwe, during its time as the British colony of Southern Rhodesia (1880–1980), became one of the most prosperous parts of Africa. This was thanks to productive agriculture, a profitable mining industry and a strong manufacturing sector, including iron and steel. The region's wealth benefited the UK and white colonists almost exclusively with indigenous Zimbabweans at best working as labourers and domestic servants.

Figure 11.11 Robert Mugabe – a president for life but the 'death' of a once prosperous country

Robert Mugabe (Figure 11.11) came to office in 1980, first as prime minister and then as president of newly independent Zimbabwe. His control of the country rested on questionable elections, corruption and the denial of human rights.

Land reforms undertaken in the name of redistributing land to black Zimbabweans from former colonial white settlers could have had positive outcomes, but instead was often a corrupt process with the best land going to Mugabe's supporters. Mismanagement, lack of training and investment for new farmers actually contributed to an economic collapse in Zimbabwe's once profitable farming sector. By 2019 per capita GDP was US$1400, barely above the US$1000 at independence in 1980.

In September 2019, Robert Mugabe died aged 95, two years after being forced to step down by the army over concerns he was preparing his wife Grace to be his successor.

Key term

Land reform: Most often this involves the redistribution of property and agricultural land as result of government-initiated or government-backed actions.

11.3 Differences within countries

The spotlight now sharpens from the global to the national. Are there significant variations in human rights within countries?

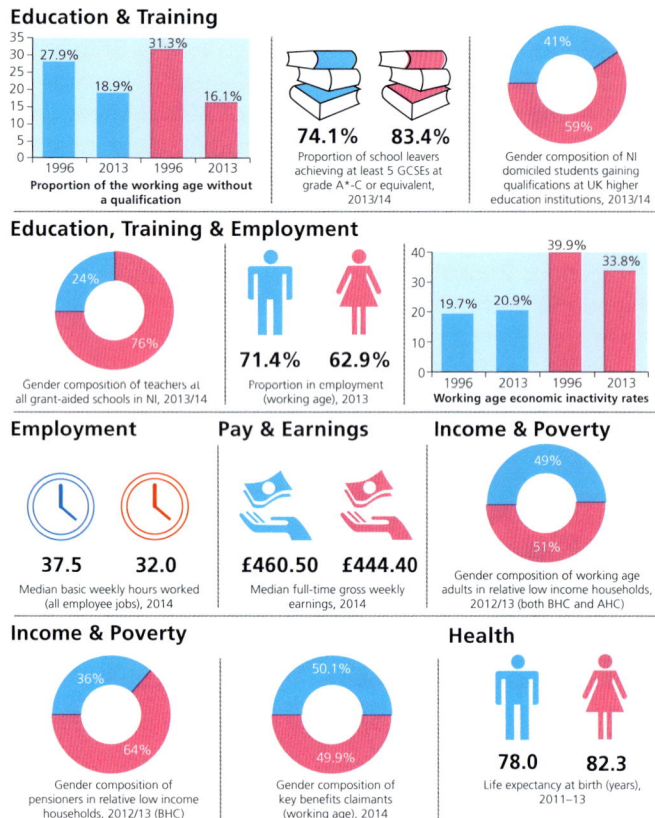

Figure 11.12 Gender equality in the UK, 2015 (Source: Northern Ireland Statistics and Research Agency)

Discrimination based on gender and ethnicity

Gender

Globally there are differences in gender equality measured using metrics such as the Gender Inequality Index (GII). Figure 11.13 shows one such measure, the SDG Gender Index. Although gender equality is often enshrined in law and government policy, it can vary within a country because of cultural and attitudinal differences between groups, e.g. minority ethnic, immigrant groups or age strata within a society.

Figure 11.12 suggests that elements of gender discrimination persist in the UK, although the country has clearly been moving in the right direction. Despite legislation, more needs to be done in terms of ensuring equal pay. But does gender discrimination vary within the UK, say on a regional basis? Sadly, there are no official statistics to shed light on this issue.

Ethnicity

Large parts of the world were under militarised colonial control in the first half of the twentieth century. Most colonial powers were European, with the UK assembling the most extensive colonial empire of all.

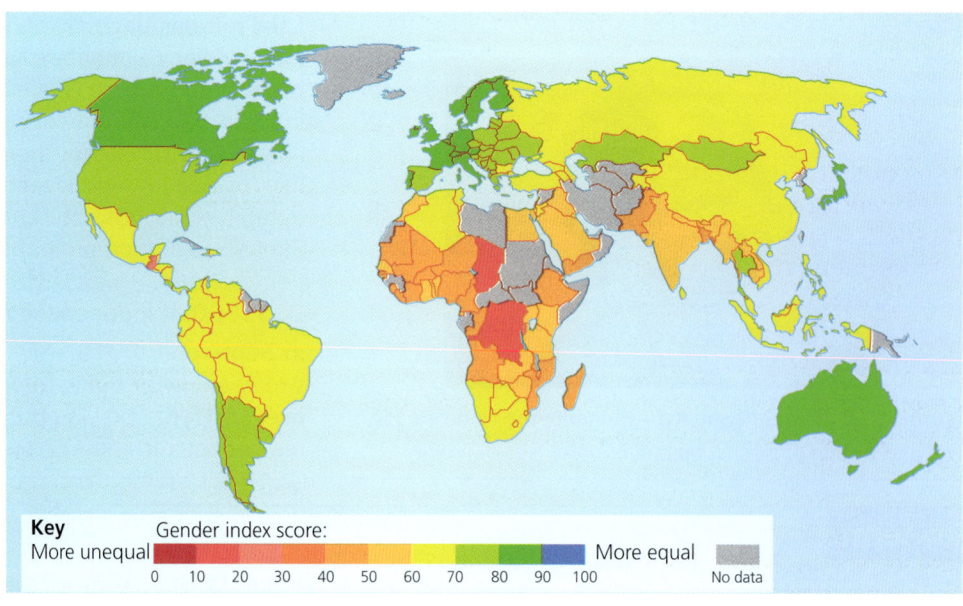

Figure 11.13 SDG Gender Index scores, 2019

Option 8A: Health, Human Rights and Intervention

Africa was probably the continent most fragmented by militarised colonial control, having been divided up by at least seven European powers.

After the Second World War came the era of de-colonisation and independence. Sadly, in many cases, independence brought internal conflict and division rather than what had been hoped for – freedom and prosperity. There were three main reasons:

- Although various forms of government had been set up in the colonies, indigenous people had been largely excluded from their administration. The result was that, when independence came, most countries inherited their political institutions directly from the colonial state, which they had little experience of, and so independence often led to ineffective governance.
- This lack of governance meant that opportunist insurgent groups were able to vie for political control. Much violence ensued.
- Perhaps most important of all, the colonial borders did not recognise or realise the importance of traditional ethnic and religious borders. The colonial boundaries often cut across these deeply engrained lines. One supposes that the colonial powers were arrogant in thinking that colonial rule would soon neutralise any differences between traditional groupings. They could not have been more wrong.

Rwanda

Possibly one of the worst examples of the costs of ignoring deep-rooted tribal divisions comes from the small African country of Rwanda. Rwanda became a German colony in 1884 as part of German East Africa. Following Germany's defeat in the First World War (1914–18) it became a mandated territory under Belgian control.

All the indigenous people of Rwanda were believed to be drawn from just one cultural and linguistic group, the Banyarwanda, so the Germans thought there was nothing wrong with its boundaries when they colonised it. But what was not understood at the time was the existence of three distinct sub-groups within the Banyarwanda: Hutu (84 per cent of the population), Tutsi (15 per cent) and Taw (1 per cent). During pre-European times, Rwanda had been a unified state largely controlled by the minority Tutsi. The Hutu were treated as second-class citizens by the dominant Tutsi.

Before and since independence in 1962, the Hutu have struggled for supremacy. They were keen to convert their superiority in numbers into power. After 30 years of tension, matters came to a head in April 1994 when the Hutu decided to literally eliminate the Tutsi. Many Tutsi fled in terror to neighbouring countries but, within 100 days, around 800,000 of them were massacred. The Rwandan Patriotic Front (RPF) – a rebel group composed mainly of Tutsi refugees based in neighbouring Uganda – invaded the country and regained control by July, whereupon 2 million Hutu fled the country.

Today some stability has returned to Rwanda and the economy has recovered. But it would be wrong to say that relations between the Hutu and Tutsi have healed. There is some irony in the present situation, however: while ethnic discrimination continues between Hutu and Tutsi, the country has one of the best gender equality records in the world.

Figure 11.14 Hutu refugees cross the Rwanda–Tanzania border in 1996, on their way home from exile in Tanzania

Human rights, health and education

There is plenty of evidence to suggest a broad correlation between human rights on the one hand and access to health and education on the other. Indeed, the UDHR has decreed such access to be one of the most basic human rights. The relationship probably also works in the other direction, namely that education at least should lead to a greater respect for other human rights. But does everyone want access to education and healthcare?

The growing demand for equality

In many countries the struggles of women and minority ethnic groups for equality have been long and persistent. The pace of progress varies from country to country: in some the goal has been achieved, but in others success is still a long way off.

Afghanistan exemplifies discrimination on the basis of gender. In Bolivia the need is for more equality for both indigenous groups and women.

In conclusion, it should have become evident that there are variations in human rights within countries, and that the degree of variation may reflect different levels of social development. All people may be equal, but they are not so when it comes to wealth, freedom and opportunity. This is because there are other factors at work, such as the type of government, the distribution of political power and deeply rooted cultural traditions.

First Nations in Canada

There are an estimated 370 million indigenous people living in more than 70 countries. They account for about 5 per cent of the world's population. In the USA, the figure is just under 2 per cent of the population, while in Canada it is just over 4 per cent.

First Nations in Canada describes indigenous people who lived in Canada prior to colonisation by the British and French. First Nations lived below the tree-line, so the Inuit that inhabit Canada's tundra are not part of First Nations. In 2021 there were around 1.8 million Canadians who identified as being from an indigenous group and 1.1 million identified as First Nation. Some 350,000 are Cree or have Cree heritage.

European colonists considered First Nations a threat:

- Indigenous people were driven off their ancestral lands in favour of European settlers.
- The 1876 'Indian Act' banned a range of First Nation cultural practices and made them non-citizens that could not vote until 1960.
- Even today in Canada, First Nations are not governed by the same regulations as other Canadians and parts of the controversial Indian Act are still in force.

Figure 11.15 National Chief of the Assembly of First Nations Cindy Woodhouse (centre)

Discrimination against First Nations has, inevitably, contributed to different human development outcomes compared to the wider Canadian population. In 2020, median income for First Nations people was in the range CA$32,000–42,400 compared to $50,400 for non-indigenous Canadians. Around 25 per cent of First Nations live in poverty (below an annual income of $26,000) but only 10 per cent of non-indigenous Canadians.

Option 8A: Health, Human Rights and Intervention

First Nation health

- First Nation life expectancy is 3.5 years less than for non-indigenous Canadians.
- The prevalence of some diseases is much higher: 10 per cent higher for Type 2 diabetes.
- Rate of smoking and drug misuse are higher among First Nations.

This partly reflects differences in income as well as the fact that First Nation people living in remote locations are isolated from high-quality healthcare provision.

First Nation education

- Ninety-one percent of non-indigenous Canadians have completed high school but for First Nations this is 60–83 per cent, depending on where First Nations live. Those living on designated reserves have the lowest completion rate.
- Only 10–15 per cent of First Nation people have a university degree compared to 37 per cent for non-indigenous Canadians.
- Differences in educational attainment lead to reduced opportunities and lower incomes for First Nations people.

In the last 20 years there have been improvements in First Nation health, education and income but there is still a significant gap, reflecting centuries of discrimination and marginalisation. The Indian Act has been widely criticised by the UN and Amnesty International as an ongoing human rights abuse because it governs First Nations in a different way to non-indigenous Canadians.

The place of women in Afghanistan

Before the 1979 invasion by the USSR, women's rights in Afghanistan were progressing similarly to elsewhere in the world. Afghan women gained the vote in 1919 (only a year after women in the UK), and in the 1950s, *purdah* (gender separation) was abolished. In the following decade, a new constitution gave women equality in several important aspects of life, including being able to stand in elections.

USSR power collapsed in 1992 and a period of civil war ensued in Afghanistan, ending with the creation of the Taliban-led Islamic Emirate of Afghanistan (1996-2001). Under Taliban rule, the rights of women in Afghanistan were severely restricted as the Taliban enforced their version of Islamic law (see page 197). Women and girls were banned from:

- going to school beyond primary, studying and working
- leaving their home without a male chaperone
- showing bare skin in public (so a full body veil, or burqa, had to be worn)
- accessing healthcare delivered by men (as women could not work, healthcare became inaccessible and female health deteriorated)
- involvement in politics or speaking publicly.

The punishments for breaking these rules including flogging and execution.

In 2001, a military coalition led by the USA invaded in direct response to the 9/11 attacks and defeated the Taliban government. The position of women was used as one of the justifications for the military intervention and women's rights were restored to an extent. Despite ongoing conflict, female school enrolment increased from 100,000 girls in 2000 over 3.5 million by 2019 and female literacy doubled from 2011 to 2018.

In 2021, the US-led military forces withdrew and the Taliban returned to power. Subsequently women's rights have been removed, returning them to their situation prior to 2001. Most of the 1,880 girls' secondary schools in Afghanistan had been closed by 2024. Afghan women led public street protests against their oppression, but these have become increasingly rare due to the brutality of the Taliban response. Women and girls try to pursue education at home, in secret, and some women have set up secret businesses as the only way to earn some income.

Figure 11.16 Afghan women protesting about the Taliban taking away their human rights

A mixed picture in Bolivia

With a population of about 9 million and 35 different ethnic groups, Bolivia is one of the most ethnically diverse countries in Latin America. Over 60 per cent of the population is classified as indigenous while 30 per cent are *mestizo* (of mixed European and Amerindian descent). Since the early 1990s there has been a strong move to recognise indigenous identity and culture, as well as the rights of these people. Attempts have been made to involve indigenous people in national policy making.

While progress has been made culturally, socially and politically, the indigenous people remain marginalised in economic terms. Most continue to live in extreme poverty. President Morales (2006–19), who is indigenous himself, was recognised as a champion of indigenous rights and the environment (see Chapter 10). However, he is increasingly criticised for failing to deliver on the reduction of poverty among indigenous people, as distinct from the *mestizo*. The challenge here is that, despite the sale of primary products, Bolivia is still the poorest country in South America. Per capita GDP stands at only US$3550.

The record with respect to gender equality is worse. For example, gender violence causes more death and disability among women aged 15 to 44 than do cancer, malaria, traffic accidents or war. Around 50 per cent of Bolivian women from a variety of backgrounds have admitted to having been subject to physical abuse in their lifetime (Figure 11.17). Bolivia also has the highest rate of **maternal mortality** in South America. Women have little access to services such as cancer screening and sexual health education.

In short, women still play a subordinate role in Bolivian society. This, combined with their lack of education, means that they do not have a voice to call for their human rights. Some ask, why does President Morales not take up this cause?

Figure 11.17 Women's rights activists in La Paz carrying the names of recent victims of violence

Key term

Maternal mortality: The death of a woman while pregnant or within 42 days from the end of the pregnancy.

Progress in Australia

It is remarkable to note that, up to the Second World War, Australia pursued a so-called whites only immigration policy. Its main aim was to prevent Chinese miners and Pacific Island labourers from entering the country. The policy was slightly misnamed, however; while it favoured people from English-speaking countries, it discriminated against those from Southern and Eastern Europe.

The policy was gradually dropped after the Second World War. While immigration is still carefully controlled, today Australia has become one of the world's most multicultural countries. Entry to Australia has been largely governed by its need for specific skills in its labour force. Its citizens come from literally around the world. More than 40 per

Figure 11.18 An asylum seekers' boat arrives at Christmas Island, but is subsequently escorted back to Indonesian waters

Option 8A: Health, Human Rights and Intervention

cent of Australians now have national origins other than British or Irish, while just over 2 per cent are indigenous peoples.

Australia is widely recognised as having one of the best records relating to human rights. There is little discrimination, perhaps because of its rich mix of ethnic groups. However, there are currently three areas of concern:

- The treatment of asylum seekers and refugees, particularly preventing them from entering the country and processing their claims offshore (Figure 11.18).
- The need to do more to protect the rights of disabled people.
- The long-running issue of the treatment of the indigenous Australians.

With regard to the last of these, there is the feeling that Australian society should be more inclusive towards indigenous people. As we saw in Chapter 10 (page 205), it seems that the plight of the indigenous people is not responding to the relatively high levels of expenditure earmarked for their education and healthcare. Is it possible that the indigenous people feel, like their counterparts in the Americas, that they do not wish to be included? Would they rather be left alone and allowed to follow their traditional lifestyles? If this is their wish, can Australia deliver it?

Review questions

1 What do we mean by human rights?
2 Compare China and India in terms of:
   a economic growth
   b human rights.
3 Suggest reasons why there has been so much conflict in post-colonial Africa.
4 Compare the distributions shown in Figures 11.3 (page 216) and 11.9 (page 220). Are you able to identify any possible links between the two distributions?
5 Briefly explain why inequality is worse in some countries than others.
6 Bolivia is a mixed success in terms of development progress. To what extent do you agree?
7 Outline the impacts of Australia's immigration policy.

Further research

Examine the arguments made for scrapping the UK's Human Rights Act: https://www.telegraph.co.uk/news/general-election-2015/politics-blog/11598319/Michael-Goves-attempt-to-repeal-the-Human-Rights-Act-faces-almost-insurmountable-odds.html

Research the factors giving rise to political corruption: https://www.transparency.org/en/

Investigate the link between basic education and gender equality: www.unicef.org

Interventions and human rights

How are human rights used as arguments for political and military intervention?
By the end of this chapter, you should:
- be aware of different forms of geopolitical intervention
- understand the different forms and motives of development aid
- understand the motives and outcomes of military intervention.

12.1 The nature of geopolitical intervention

The focus in geopolitics is on political power in relation to geographic space. Geopolitical interventions are the exercise of a country's power in order to influence the course of events outside its borders. For most countries, geopolitical interventions are very much the nuts and bolts of foreign policy. It should also be made clear from the start that the sort of political power we will be talking about is rooted in economic strength. So, the broad scenario to be investigated is one where the most powerful countries (the superpowers) seek to assist, mould or control less-powerful countries.

Interventions and motives

The motives behind geopolitical interventions are many; they vary from country to country and between the different international organisations. Possible motives include:

1. Offering development aid to the poorest and least-developed countries.
2. Protecting human rights.
3. Encouraging education and healthcare.
4. Strengthening security and stability.
5. Promoting international trade and protecting trade routes.
6. Accessing resources.
7. Encouraging inward investment.
8. Providing military support.
9. Increasing global or regional influence.

The list of possible motives given above has been deliberately ordered and crudely runs from humanitarian (points 1, 2 and 3), through mutual benefit (4, 5, 6 and 7) to self-seeking (8 and 9). These and other possible motives are delivered through three different mechanisms:

- development aid
- economic support
- military power.

In this topic we are only really concerned with geopolitical interventions to do with human development and human rights. It should be remembered that significant interventions are also made in these areas by IGOs and NGOs (see below).

Option 8A: Health, Human Rights and Intervention

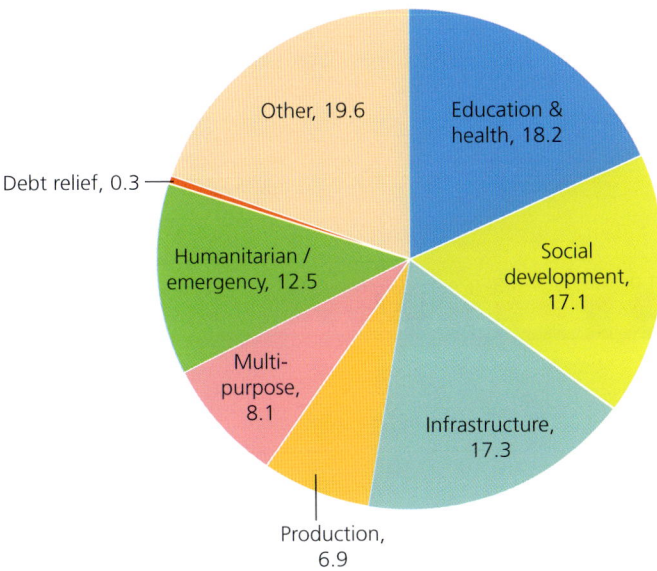

Figure 12.1 OECD development assistance spending by type (%), 2019

Development aid

Development aid has two main delivery routes: **bilateral** and **multilateral**. Bilateral aid may allow a donor country to pursue its own agenda and to target aid at preferred countries and objectives. With multilateral aid, in theory, the donor country has less direct control. Powerful countries *do* control the UN and World Bank, and so influence their policies and development agenda.

Much of the financial aid given by developed countries today is recorded by the OECD under the heading of **Official Development Assistance (ODA)**. Most of this money is bilateral aid, so the donor country has some say as to how the money is spent by the receiver country. ODA reached a total of US$152.8 billion in 2019. The USA was the largest donor country contributing US$34.6 billion. ODA increased from US$55 billion in 2000 to US$180 billion by 2020 because of the need to make progress on the MDGs and SDGs. Figure 12.1 shows that ODA is mainly spent of social development programmes and economic development. A significant amount is spent on humanitarian and emergency aid to deal with disease outbreaks and natural disasters. Only a tiny fraction is spent on debt relief.

Figure 12.2 illustrates how some developed countries are much more generous than others. Only a few have ever met or exceeded a target of 0.7 per cent of GDP given as ODA which was agreed as long ago as 1970. The UK is generous compared to many countries, but some politicians and some voters see this generosity as misplaced. Why should 0.7 per cent of the UK's GDP be spent abroad when it could be spent at home on schools and the NHS? Table 12.1 perhaps provides some of the justification. Much ODA is given to countries experiencing conflict, with the aim of reducing human suffering but also in an attempt to contain conflict and prevent escalation – which could ultimately be more costly to developed countries in terms of economic disruption (lost export markets) and humanitarian crises such as refugee flows.

> **Key terms**
>
> **Bilateral aid:** Aid that is delivered on a one-to-one basis between a donor and a recipient country.
>
> **Multilateral aid:** Aid (usually financial, sometimes technical) given by donor countries to international aid organisations such as the World Bank or Oxfam. These organisations distribute the aid to what they deem to be deserving causes.
>
> **Official Development Assistance (ODA):** A term used by the OECD to measure aid. It is widely used as an indicator of flows of international aid. Flows are transfers of resources, either in cash or in the form of commodities or services.

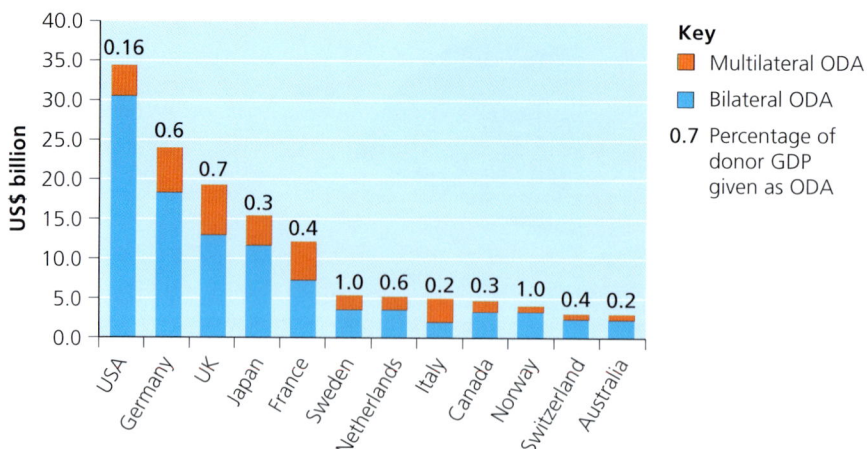

Figure 12.2 The largest ODA donors, 2019

Table 12.1 Top 10 recipients of ODA in 2019 (darker blue cells indicate states involved in conflict)

USA		UK	
Afghanistan	1070	Pakistan	481
Jordan	1022	Ethiopia	413
Ethiopia	927	Nigeria	411
Kenya	848	Syria	365
Nigeria	788	Afghanistan	313
South Sudan	785	Somalia	311
Syria	664	DRC	243
Tanzania	631	Yemen	243
Uganda	629	Bangladesh	241
South Africa	588	South Sudan	210

Data: OECD

There are different forms of aid but much is in the form of loans. Unfortunately for the borrowing country, loans attract interest and, ultimately, loans have to be repaid. It is all too easy for receiver countries to enter a downward spiral of increasing debt. More acceptable to receiver countries is what is known as technical assistance. This involves the transfer of expertise, technology and education. It is thought that this can contribute more to human development than capital loans. Certainly, technical assistance is more effective in supporting 'bottom-up' approaches to development.

Economic support

Two types of intervention are particularly significant drivers of economic development: trade and investment.

Trade

It is widely agreed that increased trade can give less-developed countries a leg up. That is, provided the terms of trade are favourable and that the strategy is to encourage exports rather than imports. There are many different forms of trade intervention, some of which have the potential to help the least-development countries (Table 12.2). However, some – notably embargoes and sanctions – can be used to force 'bad' regimes to change. This was the case with South Africa during the Apartheid years (1948–94) and more recently with Iran because of its violation of the Treaty on the Non-proliferation of Nuclear Weapons.

Table 12.2 Main types of trade intervention

Tariffs	Taxes levied on imports
Quotas	Restrictions on imports
Exchange rates	Deliberate lowering to increase the competitiveness of either imports or exports
Trade blocs	Free trade between member countries
Embargoes	Bans on trade in specified commodities
Sanctions	Restrictions on trade imposed by countries against others for political reasons

Some particularly successful trade interventions include:

- the setting up and workings of the Association of Southeast Asian Nations (ASEAN), a regional agreement relating to free trade and economic co-operation
- the Fairtrade Foundation, which seeks to obtain a fair price for a wide range of goods exported by developing countries
- the Doha Development Agenda, aimed at lowering trade barriers, for example, by allowing agricultural products from developing countries to enter the EU and the USA in return for opening their doors to manufactured goods and services.

Skills focus: Flow-line maps

These maps show the pattern of movement between places, for example, of people, goods, traffic or even aid. Direction of movement is indicated by an arrow, while the width of the arrow is drawn proportional to the volume of movement. They can be very effective in showing the broad pattern of movement at a global scale (Figure 12.3).

Investment

As with trade, investment is largely undertaken for economic motives, such as:

- securing primary resources
- facilitating private investment
- providing technical know-how.

However, there may well be beneficial spin-offs from the resulting economic development. These would include improved living standards and the provision of better education and healthcare rather than improved recognition of human rights.

Military intervention

Interventions of a military kind can take a number of different forms and be undertaken for a number of motives. They can range from training and equipping a developing country's armed forces, through sending troops to help deal with insurgents and terrorists, to all-out military occupation. These different motives will be scrutinised in Section 12.3.

All three of these mechanisms – development aid, economic support and military action – can be, and are, used to make interventions on behalf of human rights and human development. However, matters are not always transparent when it comes to looking at the motives and reasons.

International intervention players

In addition to those of individual governments, geopolitical interventions are also made by:

- IGOs, such as the UN, EU, World Bank and WTO
- NGOs, such as Amnesty International and Human Rights Watch.

IGOs

Five IGOs with an interest in human development were introduced in Table 10.7 (page 208): the World Bank, WTO, IMF, UNESCO and OECD. One more might be added to the list: the United Nations Conference on Trade and Development (UNCTAD).

Given that there are so many IGOs operating in the development arena, there is occasionally a sense of competition between some of them, which is no surprise when consensus may be lacking.

NGOs

For the most part, these are charities. They are free to act and are not subject to government intervention. They fall broadly into two groups:

- Those concerned primarily with human rights (for example, Amnesty International, Human Rights Watch).
- Those more focused on human development and aid, including emergency aid in response to natural disasters (for example, Oxfam, Médicins sans frontières).

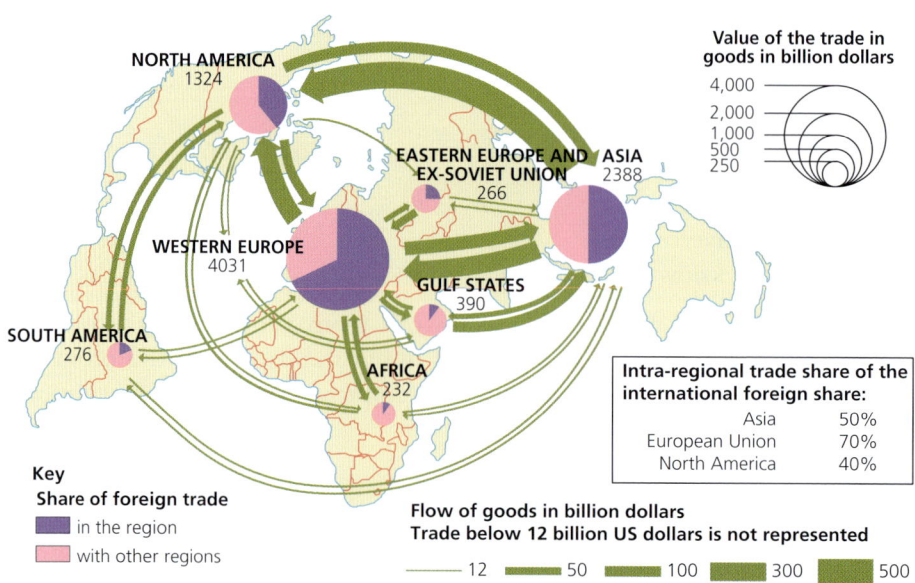

Figure 12.3 The global trade in goods

Table 12.3 Some major NGOs

Organisation	Founded	Mission
Amnesty International	1961	Founded in the UK and focused on the investigation and exposure of human rights abuses around the world. Takes on both governments and powerful bodies, such as major companies. Today it combines its considerable international reputation with the voices of grassroots activists on the spot to ensure that the UDHR is fully implemented. It also provides education and training so that people are made aware of their rights.
Human Rights Watch	1978	Founded under the name of Helsinki Watch to monitor the former Soviet Union's compliance with the Helsinki Accord (aimed at reducing Cold War tensions). Like Amnesty International it is constantly on the lookout for violations of the UDHR. It is not frightened to name and shame non-compliant governments through media coverage and direct exchanges with policymakers.
Oxfam	1942	Founded in the UK to help deal with the hunger and starvation that prevailed during the Second World War. Today it has three main targets: development work aimed at lifting people out of poverty and improving health (safe water and sanitation); assisting those affected by conflicts and natural disasters; and campaigning on a range of issues, from women's rights to the resolution of conflicts.
Médicins sans frontières (Doctors Without Borders, MSF)	1971	Founded in France with the belief that all people have the right to medical care regardless of race, religion or political persuasion. Today it provides healthcare and medical training in about 70 countries and has a reputation for providing emergency aid in conflict zones. It remains independent of any economic, political or religious influences.

Representatives of each group are examined in Table 12.3.

All are major players with truly global networks and widely respected reputations.

12.2 Development aid

The range of development aid

The nature of development aid was examined in Section 12.1. All that is necessary here is to stress that it is a broad term covering a range of interventions. They vary:

- in scale, from installing a village well to constructing a vast irrigation project
- financially, from a small charitable gift to a global appeal raising millions
- in timescale, from short term (for example, emergency aid) to long term (for example, disease-eradication programmes)
- in the mix of aid providers, from local charities to major IGO and NGO players.

Most development aid is aimed at human development. Safeguarding human rights and improving human welfare are more specific but recurrent targets. In many instances, development aid has an economic dimension, in that the creation of regular employment is thought to be important to a better standard of living.

A look at Haiti provides an opportunity to see the major aid players (IGOs and NGOs) in action.

Haiti: the challenges of aid

Haiti in the Caribbean is one of the poorest countries in the world. Indeed, it is the poorest country in the western hemisphere. Around four in every five Haitians live on less than US$2 a day; nearly one-third of adults are illiterate. During the last 200 years, Haiti has suffered from:

- exploitation of its resources and people by foreign companies and business interests
- violations of civil rights by a succession of dictatorships
- widespread corruption
- a highly polarised society, with 1 per cent of the population controlling nearly half of the country's wealth
- large-scale emigration
- poor healthcare and lethal outbreaks of contagious diseases
- a high level of aid dependency.

Between 1990 and 2009, Haiti received aid amounting to well over US$5 billion, most of it coming from the USA, Canada and the EU. Little seems to have resulted from this. The parlous state of Haiti continues to make it particularly vulnerable to any economic downturn or hazard. The most recent hazard to hit Haiti was a 7.0-magnitude earthquake on 12 January 2010. Its epicentre was located near the capital Port-au-Prince (Figure 12.4).

- The disputed death toll was between 100,000 and 316,000: poor record keeping and the need to bury bodies immediately may be partly to blame for the imprecision.
- More than 300,000 people were injured.
- Nearly 200,000 dwellings were badly damaged and 100,000 were completely destroyed (Figure 12.5).
- Around 1.5 million people were displaced from their homes, many taking refuge in emergency camps that were at risk from storms, flooding and contagious diseases, most notably cholera (Figure 12.6).

Figure 12.5 A valley of homes destroyed by the 2010 Haiti earthquake

Global appeals for help were soon being answered. Within months Haiti's plight had generated aid pledges valued at over £12 billion.

As of 2015, five years after the earthquake, only half the promised aid had been received. Over 500,000 victims were still living in temporary shelters without electricity, plumbing or sewerage. A prolonged outbreak of cholera was caused by the failure to provide proper sanitation. By then Haiti's aid programme should have moved on from emergency relief and reconstruction to longer-term objectives, such as dealing with serious human rights abuses, corruption, poor governance and poverty.

It is sobering to realise how little the people of Haiti have benefited from the huge outpouring of goodwill, donations and offers of help that immediately followed the earthquake. So, what lies behind this? What lessons are to be learnt?

- Too many unqualified and small-scale NGO relief organisations and charities were involved in the relief effort. Many had no language skills or interpreters, or any previous experience of working in a developing country.

Figure 12.4 The location of Haiti and the 2010 earthquake

Figure 12.6 Women and children in emergency tents after the Haiti earthquake

- Many aid pledges were never fulfilled. It may be that a longstanding track record of corruption discouraged some donors from delivering what they had promised in the immediate aftermath of the earthquake.
- Aid was unequally distributed, being too focused on the emergency camps and on the 'safe' parts of Port-au-Prince. Rural areas were largely ignored.
- The weak Haitian government was rather left outside the loop by the major aid players, such as the UN agencies, MSF, Oxfam, Christian Aid and the Red Cross.
- During 2016–18, Haiti received around US$1 billion in foreign aid, this is about 10 per cent of the country's total GDP. In other words, Haiti is highly dependent on foreign aid. Has the aid made a difference? To some extent it has as income per person increased from about US$500 in 2005 to US$850 by 2019. However, HDI barely increased at all over the same period.

Positive impacts of development aid

The case of Haiti is certainly a disappointing one but, happily, development aid *has* had its successes, both on a global scale as well as in specific parts of the world.

Progress in the fight against disease

Much development aid has been targeted at healthcare. One highly contagious disease – smallpox – has been eliminated as a result of global vaccination campaigns, and another – polio – nearly so. Success in the human battle against infectious diseases has not been universal, however. A number of diseases that once prevailed throughout the world have largely been eradicated in developed countries but still persist in developing countries. Examples of such diseases are cholera and typhoid.

Malaria

Despite advances in the fight against the disease, malaria remains the world's number one killer; in 2018, there were an estimated 405,000 deaths recorded globally. Up to 2 million people still die each year from malaria and its complications. The number of cases remain high each year with an estimated 28 million in 2018. Approximately 85 per cent of these cases were in Africa and India. The battle against malaria has been fought on three fronts:

- Draining the swampy areas where the *Anopheles* mosquito (the carrier of the disease) breeds, or spraying those areas with DDT or similar chemicals.
- Encouraging those at risk to take preventive medicine; these drugs do not cure the disease – they simply reduce the risks of contracting it.
- Distributing mosquito nets for people to sleep under; people are most likely to be bitten when they are asleep.

Development aid has done much over the years to make both anti-malarial drugs and mosquito nets freely available. Certainly, the mortality rate due to malaria has come down. In 2015 there was the exciting news that a malaria vaccine was undergoing trials and was soon expected to be licensed.

An important player in improving health and healthcare is education. Teaching basics such as personal hygiene and the critical need for safe water and proper

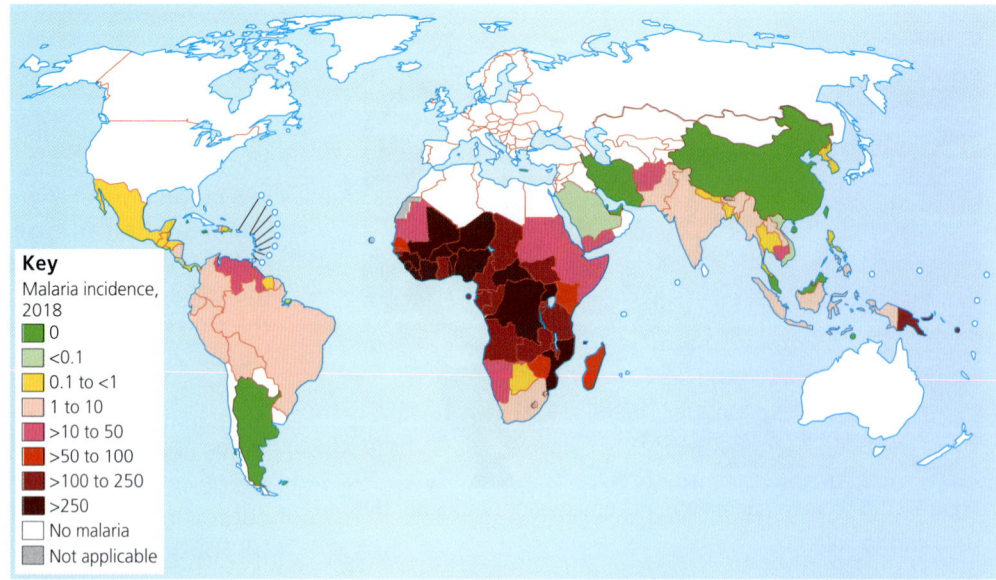

Figure 12.7 Malaria risk areas (new cases per 1000 people), 2018

sanitation can do so much to contain killer contagious diseases such as cholera and typhoid.

Success with poverty and human rights

Progress against the MDGs and SDGs, plus economic development and globalisation, has reduced the number of people living in extreme poverty (living on less than $1.90/day) from 1.9 billion in 1990 to an estimated 690 million in 2024. Development aid helps reduce extreme poverty, as does economic development. However, rising average income often masks increasing inequality. The percentage of undernourished people worldwide was 9.1 per cent in 2023, higher than the 8.7 per cent in 2010.

Progress on human rights has been slow. Development aid has had some success – for example, in terms of improving gender equality and access to primary education – but the slow rate of progress sometimes reflects the difficulty of changing attitudes that have become deeply engrained over the centuries.

It is important to recognise that war and conflict often reduce human rights and human development progress. According to the Geneva Academy there were 110 armed conflicts ongoing in 2024. The rights of women, the right to a fair trial and the right to food and water are usually dramatically eroded by conflict. For example, the rise and spread of IS threatened to reverse 50 years of development in parts of the Middle East (see Section 12.3, page 239 and Chapter 13, page 246).

Growing concerns about development aid

Development aid is coming in for increasing criticism on a number of different counts.

- Aid in the form of capital grants and loans is seen to be inappropriate by some. They argue that it is better to donate technical assistance and skills training.
- In some countries there is concern about the size of the aid budget. The UK is no exception here. The aid budget in 2019 was £15.2 billion. This figure is criticised from two sides, as being either too much or too little.
- There is criticism about the actual distribution of aid. For example, up until 2015 a large amount of the UK's aid went to India, but India is a much-lauded emerging economy. Are there not other countries more in need of aid?

Option 8A: Health, Human Rights and Intervention

- Another concern is that too much aid money is being spent on the military and siphoned off to fill the pockets of corrupt officials in receiver countries (Figure 12.8). As a consequence, even less is being directed towards the poor, minority groups and human rights.
- Development aid is thought to encourage aid dependency rather than economic progress. Inflows of aid make governments economically lazy and encourage corruption.

Negative impacts of economic intervention

The record of interventions of a more economic nature certainly shows some positives in the form of jobs and taxes. However, foreign direct investment from the developed world and the presence of TNCs are guilty of generating some serious negatives, not least of which is that profits leak back to company headquarters in the developed world. More specific negatives relate to the environment, minority groups and human rights. All three are illustrated by events in the Niger Delta of Nigeria (West Africa).

Reference has already been made to corruption in the context of humanitarian aid. It also flourishes in economic development. Bribes are offered for a variety of different reasons and are also given in different forms: usually cash but sometimes land. Land grabbing is a symptom of bribery and corruption.

Figure 12.8 Much aid money is lost to corruption

Skills focus: Source materials – newspaper articles and marketing materials

There are two particular challenges presented by most sources that might be used to evaluate the impacts of development aid or, indeed, any other form of intervention.

- The need for even coverage or treatment of the situation before and after the intervention. In many cases, coverage of before will be less comprehensive than that of after. The reasons are understandable: given that, in most cases, the focus will be more on what has been done and achieved rather than giving a detailed picture of what the situation was like. In trying to come to a verdict, they will naturally focus on what has been done. Once action has been taken, detail of the preceding situation will be lost.
- The need to be aware of bias. Political bias is particularly omnipresent in newspaper articles. Only those articles that toe a particular newspaper's line are likely to be published. The answer here is to look at more than one newspaper and, if possible, choose newspapers that are known to hold opposing views on issues. Be aware also that the material published by IGOs and NGOs is not without bias. They will wish to present a positive image showing how good and effective their interventions have been.

Oil and human rights

The Anglo-Dutch and British oil companies Shell and BP were granted an oil exploration licence covering the entire country in 1938 when Nigeria was still a British colony. This preferential treatment by colonial authorities had given Shell and BP a monopoly over oil exploration in the country. Oil was discovered in the Niger Delta some 40 years ago (Figure 12.9). Royal Dutch Shell is the largest oil company working the oilfields and obtains about 10 per cent of its crude oil supply from them. The Nigerian government earns some £10 billion a year from oil revenues. However, negligence from these oil companies has led to a strong feeling among the local people who live on the delta that they have been cheated both environmentally and economically.

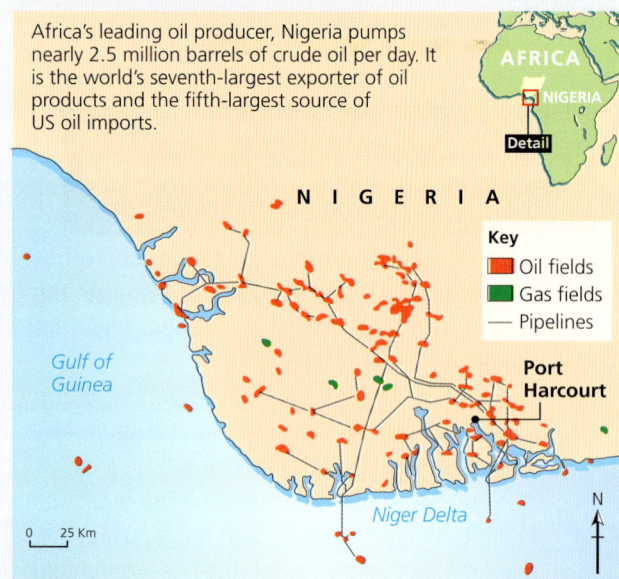

Figure 12.9 The Niger Delta

Damage to the environment and human health

The production of oil is having a devastating effect on Nigeria's largest wetland region, its wildlife and its inhabitants. Human health is threatened by the pollution of air, water and farmland by many unattended oil spills, as well as by gas flares and frequent fires.

Minorities

Local people have benefited little if at all from the oil industry. Rather, they have had to suffer its negative impact on their traditional means of livelihood, fishing and farming. Roads remain poor; schools and medical services are grossly underfunded. The root of the problem is that the population of the delta is made up of several minority ethnic groups, while the oil revenues derived from the delta are largely in the hands of the major ethnic group, the Yoruba.

Human rights

It is small wonder that a number of militant groups have formed to pressure both the government and the oil companies to compensate them for the loss of farmland and the health risks, and to share the delta's wealth. Putting a stop to the appalling environmental pollution is no less a need. Tensions are rising and outbreaks of violent protest are increasing, so too are the kidnappings of foreign oil company workers and incidents of 'oil bunkering' (illegally taking oil from pipelines and selling it abroad).

Pipeline construction

Even in the developed world the economic power of oil can mean the human rights of minority groups are disregarded. The Dakota Access (2017) and Keystone XL (shelved in 2021) oil pipelines in the USA were planned to pass over or close to Sioux Nation tribal lands in North and South Dakota. Fears of oil spills, water pollution and destruction of sacred sites were frequently rejected by the US government and courts in favour of pipeline construction. Keystone XL's permit was revoked by President Biden in 2021.

Land acquisitions in Ethiopia

In 2013 the Dutch TNC Heineken built a brewery in the village of Kilinto near Addis Adaba, Ethiopia's capital city. The brewery employs 1000 people so there are economic benefits, however 200 villagers were evicted from their farms in Kilinto. Compensation was paid and villagers were given new land, but the new parcels of land were tiny in comparison to those lost. This is an example of **land grabbing** where foreign investors and TNCs, aided by government officials, buy or rent land in developing countries (Figure 12.10). It is estimated that about 7 million hectares of land in Ethiopia has been leased to investors usually for rent of US$2 per hectare/year.

In African developing countries, land is often leased by foreign TNCs for industrial development, like Heineken in Ethiopia, but more usually to grow cash-crops for export, as:

- land in developing countries is cheap
- labour costs are low, keeping production costs down
- governments and local officials are only too keen to gain foreign exchange money through land grab deals.

However, there are significant down-sides:

- Indigenous people often have no written title to their traditional lands, so cannot prove ownership; this makes eviction very easy to achieve.
- Corruption means that promised compensation is either not paid or is partial.

In 2022, in Ethiopia over 20 million people were food insecure and needed help from the UN World Food Programme to survive, with 7.4 million women and children malnourished. Land grabs are likely to worsen this food insecurity.

Figure 12.10 Land grabbing in developing countries

12.3 Military interventions

Finally in this chapter, we turn to the third and perhaps most contentious of the three types of intervention. It can certainly be the most brutal, and the motives range from benevolent to sinister. The aim now is to look at three different intervention scenarios: defending human rights, providing military aid and the '**war on terror**'.

Defending human rights

Defending human rights has been a persuasive motive behind many military interventions.

Intervention in defence of human rights certainly puts the interventionist on the moral high ground. However, there are instances where such a defence has been a pretence and provided cover for other less-laudable motives. One of the more recent instances is the action of Russia in Ukraine.

> **Key terms**
>
> **Land grabbing:** A contentious issue involving the acquisition of large areas of land in developing countries by domestic and transnational companies, governments and individuals. In some instances, land is simply taken over and not paid for.
>
> **War on terror:** The ongoing campaign by the USA and its allies to counter international terrorism, initiated by al-Qaeda's attacks on the World Trade Center in New York and the Pentagon on 11 September 2001.

12 Interventions and human rights

Providing military aid

This scenario is a familiar one to the superpowers. Basically, it involves providing military aid to less-powerful countries to keep them on the same side. There are various motives for wishing to do this, including:

- Because the country's location has a strategic value in a wider power struggle, for example, US aid to Pakistan to help in dealing with its troubled neighbour, Afghanistan and the Taliban (Figure 12.11).
- To deal with incursions that threaten a country's stability and allegiance, for example, UK aid to Kenya to help protect it against attacks from Somalia.
- To ensure access to valuable resources, for example, UK aid to oil-rich Saudi Arabia.

Figure 12.11 US aid to Pakistan: food aid may be tied to the use of Pakistani territory by the US for military operations

Libya

A recent example of military intervention was the overthrow of President Gaddafi in Libya in 2011. He and his immediate supporters were thought to be complicit in a number of terrorist acts, including the downing of Pan Am Flight 103 over Lockerbie, Scotland, in 1988. They were also guilty of seriously abusing the human rights of many Libyan civilians in the course of maintaining their political grip on the country.

A multi-state coalition began military intervention in the form of an arms embargo and the imposition of a 'no-fly zone' over the whole of Libya. The latter meant that Gaddafi could not conduct air strikes against those who were trying to dislodge him from power. The intervention did not involve sending in troops, but air strikes were undertaken by British and French air forces against Libyan army tanks and vehicles, and they secured the country's air space as well as its inshore waters so that there was no external support for Gaddafi's forces.

Gaddafi was deposed in 2011 but, since then – as in many of the countries involved in the Arab Spring (see Chapter 13, page 246) – the removal of one regime has so destabilised the situation that rebel factions or militias are now fighting one another to gain the political upper hand.

Figure 12.12 Anti-interventionist protest in Benghazi, Libya

Option 8A: Health, Human Rights and Intervention

Skills focus: Source materials – images

Anyone using images to investigate the impacts of development, whether it is on the environment, minority groups or whatever, needs to be aware of the issues that were identified in the previous Skills focus box (see page 237):
- Giving an equal and fair portrayal of before and after situations.
- Being aware of photographer bias. Few professional photographers just go out and shoot at random and simply hope that they will take some saleable images: most are commissioned. They will be paid to take images that support the case the person commissioning the images wishes to make.

An added problem is that an image is a frozen moment in time at a particular point in space. How sure can you be that this one image fairly reflects the situation surrounding it?

Russia in Ukraine

Russia's invasion of Crimea, a sovereign part of Ukraine, in 2014 was justified by Russia on the basis that a majority of people in Crimea are Russian speaking and ethnically Russian (Figure 12.13). Russia's invasion of Ukraine in 2022 came with similar justifications – protecting ethnic Russians' human rights in eastern Ukraine from being eroded or attacked by the Ukrainian government.

In reality this concern for human rights was really a cover story for aggressive territorial expansion by Russia and an attempt to occupy parts of Ukraine and prevent it from joining the EU and/or NATO. Figure 12.13 shows the Russian-speaking territory Russia is unlikely to give up easily (Crimea, Luhansk, Donetsk) now that it is annexed or occupied, despite Western countries' financial, political and military support for Ukraine. Russia views these regions as strategically important for its access to the Black Sea.

Figure 12.13 Ukraine's ethno-linguistic zones

UK military aid to Saudi Arabia

The UK and Saudi Arabia have been allies since 1915, when Saudi Arabia became a British protectorate. In 1927 Saudi Arabia became an independent state.

In 2005, the UK and Saudi Arabia concluded a military agreement whereby the UK would equip Saudi Arabia with fighter planes (Figure 12.14). Since then the UK has sold Saudi Arabia nearly £10 billion worth of defence equipment, and Saudi Arabia has invested over £60 billion in the UK, mainly in joint ventures and real estate. Over 30,000 UK nationals live and work in Saudi Arabia and it is the UK's largest trading partner in the Middle East.

In recent years, relations between the two countries have become strained because:

Figure 12.14 Typhoon jets like those sold to Saudi Arabia

- Saudis are mainly Sunni Muslims; supplying arms to them is seen by Iran as the UK taking sides.
- Saudi Arabia has a poor record with respect to human rights, most notably free speech, women's rights and capital punishment.
- The murder of Saudi dissident and journalist Jamal Khashoggi in Istanbul in 2018 has been widely linked back to Saudi crown prince Mohammed bin Salman.
- Rumours state that Saudi princes receive millions of pounds in 'commissions' as a result of awarding arms contracts to British firms.

Reluctance on the part of the UK government to apply too much pressure on human rights issues for fear of losing lucrative military contracts was overturned by the Court of Appeal in 2019 which ruled that the sales were illegal on human rights grounds. However, in 2020, the UK government quietly restarted sales at the height of the Covid-19 pandemic.

Key term

Rendition: The practice of sending a foreign criminal or terrorist suspect covertly to be interrogated in a country where there is less concern about the humane treatment of prisoners.

Waging 'war on terror' and torture

A few years ago, the Taliban and al-Qaeda were reckoned to be the world's most loathed terrorist organisations. Today it is IS (also known as ISIS, ISIL and Daesh). It is causing much trouble in the Middle East and has mounted many terrorist attacks in other parts of the world. As a consequence, the Western superpowers find themselves increasingly embroiled in a war on terror.

It is clear that the international military campaign against IS is motivated by three main concerns:

- The political stability of the Middle East
- Safeguarding access to the region's great oil reserves
- The serious abuse of human rights.

Given the subversive nature of IS, there can be little doubt that surveillance of suspects and intelligence gathering are going to play an important role in the fight against it. Indeed, this is likely to play as critical a part as overt military action. On this more murky battlefield, it may be tempting to resort to one of the activities that figured prominently in the UDHR in 1948: torture and **rendition**.

The issue of torture really does raise a minefield of moral issues. Whose rights are more important: the rights of suspects not to be tortured, or the right to life of those who could become the victims of a suicide bombing? To say that all humans and all human rights are equal is morally correct, but it does not always help to resolve issues like the one just posed. The same applies to military intervention. Who is to say what is right and what is a contravention of territorial integrity and human rights?

IS in the Middle East

IS is an opportunist terrorist organisation with no respect for human rights. Its roots lie in al-Qaeda's operations in Iraq. It took advantage of the power vacuum in Iraq created by the withdrawal of Allied troops from that country and the civil war in Syria. Quite by surprise in 2014–15 it took control of large parts of Iraq and Syria and proclaimed an 'Islamic caliphate' (Figure 12.15). This became the base for IS to wage its jihad, or defence of Islam, against all other religions. This included numerous terrorist attacks in other countries, as well as a strategy of annihilating minority communities in the Middle East such as Christian Assyrians, Kurds, Shabaks, Turkmens and Yazidis. In Syria, the victims were Ismailis and Alawis. IS carried out numerous summary executions, forced conversions and rapes. Such activities are war crimes and genocide.

Figure 12.15 IS controlled territory, 2015

Beginning in 2014, an international military campaign was launched against the IS caliphate, which drove IS out of Iraq by 2017 and left it controlling less than 5 per cent of Syria. In March 2019 IS was finally defected in Syria. Anti-IS forces included the Islamic Military Counter Terrorism Coalition (41 Muslim countries, led by Saudi Arabia) and many 'western' countries including the USA, the

Figure 12.16 US air strike on Tikrit, Syria, March 2015

UK, France, Canada and Denmark. Air strikes (Figure 12.16) of troops on the ground and the crucial role of Kurdish Peshmerga fighters all contributed to the defeat of IS. IS proved a very tough enemy. It showed itself skilled at using modern communications to brainwash, groom and recruit young Muslims to the 'jihad'. It is also very good at creating militant cells in distant major cities and activating them to kill large numbers of innocent civilians.

Today, the situation in the region is still highly fluid and complex. IS has gone underground rather than gone away. Both Iraq and Syria remain very dangerous places and there are millions of vulnerable, displaced people and refugees that could be breeding grounds for future militant groups. Powerful countries – Russia, the USA, Iran and even China – continually intervene in the region's politics usually at the expense of the human rights of ordinary people.

12 Interventions and human rights

Torture and extraordinary rendition

International terrorism has increased the need to identify terrorist groups and discover their plans and this is a high priority for Western governments. Thanks to modern communication and intelligence-gathering technologies, much can be done to identify terrorist plots and prevent them. But there is still a need to apprehend terrorist suspects and to gain as much information from them as possible. The key word here is 'possible' because most governments have signed up to the UN Convention against Torture (1987). This prohibits physical or mental duress being used to extract a confession or important information from individuals.

It is suspected that many signatories to the convention still use torture and acts of cruel, inhumane and degrading (CID) treatment in their questioning of terrorist suspects. The USA did so after the terrorist attack on the twin towers of the World Trade Center in New York on 11 September 2001, where 2753 people were killed. In the immediate aftermath of the attack, the US government was under considerable pressure to track down those responsible. The imprisonment of suspects without trial at US Naval Station Guantanamo Bay, Cuba, was one outcome (Figure 12.17).

A number of prisoners and detainees have claimed to have been subjected to torture and other forms of ill-treatment by the US government and its agencies at Guantanamo Bay detention camp and in Afghanistan, Egypt, Iraq, Morocco and Pakistan. This includes extraordinary rendition – the act of sending terrorist suspects covertly (and illegally) to be interrogated in a country where the humane treatment of prisoners is less of a concern. Extraordinary rendition is a way of subverting the UN Convention against Torture (Figure 12.18). Both the UK and Poland have been accused of helping the USA carry out extraordinary rendition.

It is difficult to justify the actions of countries that have set aside their obligations under international treaties. In the case of Guantanamo Bay detention camp, 90% of detainees were released without charge, calling into question the purpose and usefulness of the camp. The brutal conditions for detainees were widely condemned by organisations such as Amnesty International.

Figure 12.17 Inside Guantanamo Bay detention camp

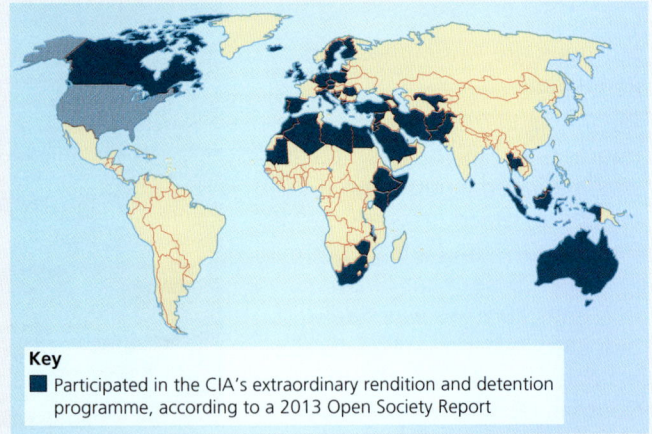

Key
■ Participated in the CIA's extraordinary rendition and detention programme, according to a 2013 Open Society Report

Figure 12.18 Countries that reportedly participated in the US CIA's extraordinary rendition programme

Option 8A: Health, Human Rights and Intervention

Review questions

1. Define the term 'geopolitical intervention'.
2. Identify the different reasons for geopolitical interventions.
3. Give some examples of governments using human rights as a reason for making interventions.
4. Outline the role of NGOs in development aid.
5. Identify some of the successes and failures of development aid.
6. Evaluate the success of strategies designed to reduce the impacts of malaria.
7. To what extent do you agree that Haiti needs development aid?
8. Examine the success of military interventions in protecting human rights.
9. Whose human rights do you think matter more, those of the terrorists or those of their civilian victims?

Further research

Research the advantages that NGOs have over IGOs when it comes to making geopolitical interventions: www.eolss.net/sample-chapters/c14/e1-44-03-00.pdf

Research the abuse of human rights in the Peruvian Amazon: http://news.mongabay.com/2014/05/new-report-reveals-human-rights-abuses-by-corporations-governments-in-the-amazon/

Evaluate the case for the UK continuing to give aid to India: www.bbc.co.uk/news/world-asia-india-34398449

13 The outcomes of geopolitical interventions

What are the outcomes of geopolitical interventions in terms of human development and human rights?

By the end of this chapter, you should:
- be aware that there are different ways of evaluating geopolitical interventions
- understand that development aid has mixed outcomes
- understand that military interventions have mixed outcomes.

13.1 Evaluating geopolitical interventions

Three broad types of geopolitical intervention were recognised in Chapter 12 in the context of human rights and human development. Having made an intervention, there inevitably comes a time to evaluate its outcome. Did it achieve the desired result? If it didn't, then perhaps some attempt might be made to find out what went wrong.

A range of measures

Given the diversity of interventions and hoped-for outcomes, it is hardly surprising that there are many possible yardsticks for assessing whether or not the outcome has been successful. Let's keep the focus on human development and human rights (Table 13.1).

Table 13.1 Possible measures for evaluating intervention outcomes

Intervention target	Possible measure
Human development	Life expectancy Provision of healthcare (doctors per 100,000) Literacy rate (% of population) Quality of physical infrastructure (% with access to safe water and sanitation) Per capita GDP or GNI
Human rights	Freedom of speech Gender equality (gender index) Democratic elections Respect for minorities Recognition of refugee status

There is no one measure that stands out from the rest. Indeed, there is much to be said for using more than one measure. However, the measures that are most frequently used are those for which statistical data are readily available. In this respect, progress in human

Skills focus: Data errors

Evaluations of actions and outcomes require accurate and reliable data. Often such data simply do not exist. This can be for a variety of reasons. Take, for example, the large numbers of refugees from the Middle East and North Africa into Europe during the present decade. Here accurate counts are handicapped by:
- the failure of border guards to keep proper records in the prevailing chaos
- the 'porous' national borders, which allow people to cross unnoticed
- the activities of people traffickers and the smuggling of migrants across the Aegean and the Mediterranean.

The only reasonably reliable figures will be those collected by the countries that happen to be the final destinations. But even here there is a problem, in that the host countries need to distinguish between genuine refugees escaping persecution and economic migrants who are exploiting the chaos to make their way illegally into what they perceive to be 'good' destinations.

So, when it comes to investigating migration (numbers, origins and destinations) triggered by the unrest in the Middle East, data errors are inevitable. But they are not deliberate ones; rather, they reflect the physical impossibility of comprehensive and relevant data gathering. In such circumstances any data that is circulated can only be regarded as best estimates, and many of those estimates may well be not particularly good. So, how on earth can we assess the true enormity of this human exodus?

development is much easier to measure than progress in human rights. Human rights indicators are rather 'slippery'; they are qualitative rather than quantifiable, but are no less significant because of that.

The importance of democracy

Today's world is crudely divided between those countries in which democratic government is deeply rooted and those in which there are much more authoritarian forms of control. Broad respect for human rights is more likely to flourish in a democracy than in a one-party state. Indeed, the essence of one-party government is that it safeguards against opposition.

Following the Bolshevik Revolution in 1917 and the emergence of the Soviet Union and its satellites in Eastern Europe, much of the Eurasian continent came under communist and authoritarian rule. There was little room for democracy at that time, but the seismic events of the 1980s were to profoundly change that situation. The 'free' world looked on and kept its fingers crossed that democracy would prevail.

The collapse of the USSR

For nearly 45 years after the end of the Second World War, the Soviet Union was a superpower locked in a so-called Cold War with the Western superpowers. It was a vast country stretching some 10,000 km from Eastern Europe to the east coast of Asia (Figure 13.1). The Soviet Union was a single-party state governed by the Communist Party. It was, in fact, a union of 15 sub-national republics. It also controlled 'satellite countries' in Eastern Europe, such as East Germany, Poland, Czechoslovakia, Hungary and Bulgaria. All of these satellites (known as the Eastern Bloc) had communist governments.

Figure 13.1 The former Soviet Union

In the late 1980s, the satellite countries began to shake off Soviet control. Change began in Poland and spread to Hungary, East Germany, Bulgaria, Czechoslovakia (since split into the Czech Republic and Slovakia) and Romania.

Perhaps the most publicised act was the pulling down of the Berlin Wall in 1989, which had separated communist East Berlin from capitalist West Berlin. The peace treaty that had ended the Second World War saw the whole of Germany divided into two. The length of the border between East and West Germany was marked by a huge and impenetrable fence to stop East Germans defecting to the West. In Berlin, it took the form of a wall. The fall of the Berlin Wall paved the way for Germany's reunification.

In December 1991, the world watched in amazement as the Soviet Union disintegrated into 15 separate countries. The collapse was hailed by the West as a victory for freedom and democracy, and as proof of the superiority of capitalism over communism. Three of the republics – Estonia, Latvia and Lithuania – quickly aligned themselves with the West.

A truly remarkable feature of this great political rupture was that it was achieved without any significant military intervention or bloodshed. Its outcome was that some ten states in Eastern Europe achieved independence and became democracies and market economies. All that is left of the Soviet Union is the Russian Federation. It has moved towards capitalism but retains a one-party government. Political power rests largely with one man, Vladimir Putin, who has held power since 2000, alternately serving as president and prime minister.

China: economic growth rules, OK?

Ranked 124th in the world according to the size of its GDP in 1976, China has since made gigantic economic strides. A figure of US$13.6 trillion in 2019 makes it the second largest economy in the world and puts it well ahead of its nearest rivals, Japan and Germany. Remember, however, that China is not only a vast country – it also has a huge population. Its population has expanded by just over half since 1976, from 0.9 to 1.4 billion. For this reason, per capita GDP growth has been much more modest. Indeed, China has only climbed up the global rankings from 160th to 66th place. Today's figure is US$10,200 per capita.

Although China remains a communist country, there is no doubt that a major factor in its economic progress has been its gradual involvement in the capitalist global economy. This in itself inevitably creates a political tension, as the population becomes increasingly aware of a very different (much freer) world outside their boundaries. Within China, economic success has come with a price: extensive environmental pollution, largely thanks to its expanding manufacturing industries, and the persistent abuse of human rights (see Chapter 11, page 213).

In 2015, the Chinese **economic miracle** showed the first signs of slowing down. By this time many Chinese people had come to enjoy consumerism. Might this economic downturn lead to widespread protests and, once again, sharpen the focus on human rights?

Another critical question needs to be posed: how important has authoritarian government been in explaining China's economic success? Would the economic progress have been so swift under a rather different political regime with more respect for human rights?

Figure 13.2 Shanghai's waterfront: a symbol of China's economic rise

> **Key term**
>
> **Economic miracle:** An informal term commonly used to refer to a period of dramatic and fast economic development that is unexpectedly strong.

The importance of economic growth

Economic growth promises power and prosperity, but not necessarily respect for human rights. A serious tension can exist between economic growth and human rights, particularly if a country is keen to fast-track that growth. But even less ambitious governments are tempted to give economic growth precedence over human rights. One country in the economic fast lane is China.

13.2 Evaluating development aid

Huge amounts of development aid have been extended by developed countries over the last 50 years. The obvious question to ask is: have receiver countries fully benefited from this aid? Unfortunately, there is little data available on which to base a reliable answer to this question. In addition, as indicated in Section 13.1, there is no universal acid test to tell us whether or not the aid has literally paid dividends.

The following are five highly generalised verdicts on development aid.

- The outcomes of development aid have not matched the inputs. Three factors might go some way to explaining this discrepancy:
  - the inappropriateness of some forms of aid
  - the siphoning of funds by corruption
  - a lack of sound governance and, related to this, the civil and political unrest that has characterised the recent histories of too many developing countries.

Option 8A: Health, Human Rights and Intervention

The Ebola outbreak

The Ebola outbreak in West Africa was first reported in March 2014 and rapidly became the deadliest occurrence of the disease since its discovery in 1976. The epidemic that swept across the region in 2014–15 killed five times more than all other known Ebola outbreaks combined (Figure 13.3). Eighteen months on from the first confirmed case, recorded on 23 March 2014, nearly 12,000 people had been reported as having died from the disease in six countries: Liberia, Sierra Leone, Guinea, Nigeria, Mali and the USA (one aid worker). The bulk of the deaths were in the first three countries.

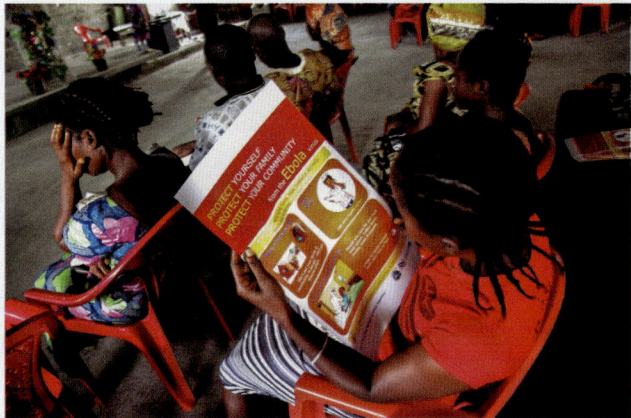

Figure 13.4 Getting the message across: a vital part of the fight against Ebola

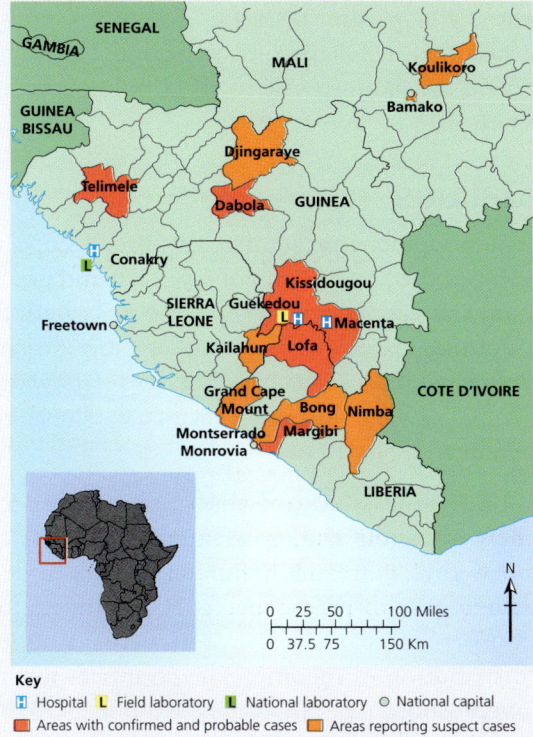

Figure 13.3 Ebola outbreak in West Africa, 2014–15

Six months after the first confirmed case, and with the spread of the disease apparently out of control, the UN Mission for Ebola Emergency Response (UNMEER) was set up. Its purpose was to lead and co-ordinate the international responses of NGOs such as MSF. Other tasks assigned to UNMEER included:

- informing everyone about the disease (Figure 13.4)
- locating and monitoring everyone who had come into contact with an Ebola patient
- overseeing the safe burial of Ebola victims
- establishing and equipping treatment centres for the safe treatment of the sick
- organising the transport of medical supplies and personnel
- accessing adequate food supplies in a situation of decreasing food production, rising food prices and closed national borders.

Thanks to the combined efforts of UNMEER, the governments of the Ebola-stricken countries and a number of NGOs, there was a sharp fall in the number of new cases of Ebola during the first half of 2015. It was beginning to look as if the epidemic was under control. However, it soon became clear that the decline in new cases had stalled, particularly in Sierra Leone. The following factors were allowing the virus to continue to spread:

- Fear of reporting to the authorities that a family member or friend might have contracted or died from Ebola.
- Fatigue with the 24/7 task of carefully following the necessary stringent precautions.
- Denial by a hardcore of the population who were strongly resistant to the idea that they needed to change their behaviour.

The region was finally declared Ebola-free in January 2016. UNMEER is the first ever mission deployed by the UN to tackle a huge health security challenge. It has since been admitted that its creation came rather late in the day, however, and that action needs to be taken much sooner in any similar future emergencies.

It is clear that such outbreaks will continue until ways are found to totally eradicate the disease. It will involve the commitment and co-operation of governments as well as the efforts of medical research scientists to come up with a vaccine. It bodes well if this sort of co-operation continues and possibly expands into a much broader concern about other aspects of human development.

13 The outcomes of geopolitical interventions

Key terms

Emergency aid: Rapid assistance given by organisations or governments to people in immediate distress following natural or man-made disasters. The aim is to relieve suffering and the aid includes such things as food and water, temporary housing and medical help.

Ebola: A highly contagious and fatal disease spread through contact with body fluids infected by a filovirus. Its symptoms are fever and severe internal bleeding. The host species of the virus has not been confirmed but fruit bats and primates have been implicated.

- In the earlier years of development aid there was an emphasis on economic development and on prestige projects, which meant that there was little trickle-down of benefits to the most needy people.
- In more recent times, development aid has been directed more at a grass-roots level and focused on education, skills training and healthcare. The verdict on such aid is altogether more positive.
- There is a school of thought that argues that no matter what form development aid takes, it encourages developing countries to become dependent on donor countries.
- Over the years there has been much debate about the relative merits of bilateral and multilateral aid. The jury is still out on this one.

Contrasting outcomes

Whether **emergency aid** should be considered as an integral part of development aid is debatable. Certainly, even in the opening decades of the twenty-first century, the world has been confronted by some major natural disasters. The aid response to the Haiti earthquake of 2010 was examined in Chapter 12 (see page 234). The evaluation was not altogether positive. Did the aid response to the **Ebola** outbreak in West Africa four years later do any better?

Trends in economic inequalities

One of the widespread concerns about development aid has been its broad impact. Has it widened or narrowed the gap between the receiving country's rich and poor? It has been suggested that top-down aid has tended to increase the polarisation, while bottom-up aid has done rather more for the poor in terms of access to basic services (safe water and proper sewage disposal), primary education and healthcare.

Table 13.2 shows some of the small amount of data there are about the changing distributions of income within countries. It uses the Gini index or coefficient: a decrease in the Gini index for any country indicates a move towards a more even distribution of income. The countries in the table were chosen because they happened to be developing and emerging countries where suitable data were available. The table shows that the most unequal countries in the 1990s have tended to become a little more equal, but countries that were more equal have become more unequal. To put the values in this table into perspective, in 2016 the UK's Gini index was 34.8.

Time sets of data about life expectancy and health are more readily available than that required for calculating Gini indices. Because of this they are frequently used in monitoring the impacts of development aid, particularly among the poor. In the case of Botswana, however, such indicators need to be treated with caution.

Table 13.2 Changing income inequality in developing and emerging countries

Country	Gini coefficient in 1990–3	Gini coefficient in 2015–18	Income inequality change
Brazil	60.5	53.9	↑
Panama	58.2	49.2	↑
Kenya	57.5	40.8	↑
Mexico	53.7	45.4	↑
Bolivia	49.1	42.2	↑
Russia	48.4	37.5	↑
Costa Rica	45.3	48	↓
Thailand	45.3	36.4	↑
Uganda	41.1	42.8	↓
Cote d'Ivoire	39.4	41.5	↓
Tanzania	35.3	40.5	↓
Sri Lanka	32.4	39.8	↓
China	32.2	38.5	↓
Indonesia	31.2	39	↓
Bangladesh	27.6	32.4	↓
↑ Income inequality improving		↓ Income inequality worsening	

Data: World Bank

Option 8A: Health, Human Rights and Intervention

The unusual case of Botswana

Botswana is a beacon of hope in southern Africa. It is:

- relatively prosperous
- politically stable
- fairly free of corruption
- reasonably respectful of human rights.

But there is a very uneven distribution of income – a wide gap between rich and poor.

Aid and development

Botswana is a sparsely populated, arid and landlocked country with large areas of wilderness. At independence in 1966, it was one of the poorest countries in the world, with a per capita income of just US$70 a year. In the first few years of independence about 60 per cent of government expenditure came from international aid. Agriculture (mostly cattle farming for beef production) accounted for 40 per cent of GDP.

Since 1966, however, Botswana has maintained one of the world's highest economic growth rates. Through fiscal discipline and sound government, Botswana has transformed itself into a middle-income country with a per capita income of around US$8250 in 2019. The exploitation of one resource – diamonds – has underlain most of this remarkable economic development.

Botswana is the world's largest producer of diamonds (Figure 13.5). Diamonds account for a third of GDP and about three-quarters of its exports (by value). Upmarket tourism, financial services, subsistence farming and cattle rearing are other significant sectors of the economy. An expected levelling off in diamond production within the next 20 years overshadows the country's long-term prospects.

Figure 13.5 Workers polish Botswana's best friend – diamonds

Figure 13.6 shows how Botswana's HDI has improved since 1990, actually overtaking that of South Africa. Despite its economic success, it might seem strange that Botswana still receives overseas aid. In financial terms this accounts for only 3 per cent of GDP, and much of it is to do with Botswana's number one health problem: HIV/AIDS.

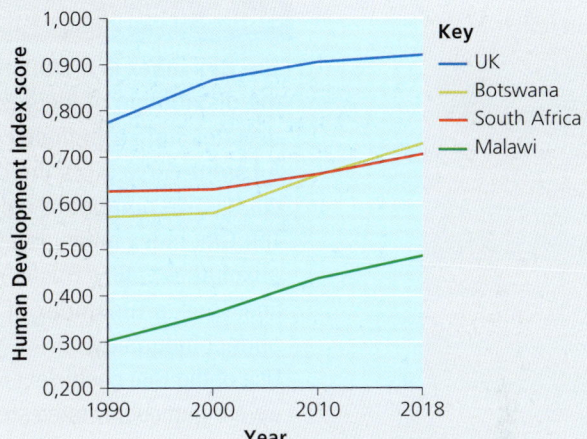

Figure 13.6 Trends in Botswana's HDI, 1990–2018

Health

The prevalence of HIV/AIDS in Botswana is the second highest in the world and threatens the country's impressive economic gains. The UN estimates more than one in three adults in Botswana is currently infected with HIV or has developed AIDS. The disease has orphaned many thousands of children and has dramatically cut life expectancy. However, with overseas aid, Botswana now has in place one of the most advanced treatment programmes. Thanks to the economic growth, the government can afford to make anti-retroviral drugs readily available.

Human rights

Botswana has the reputation of speaking out against human rights abuses in Africa. Often it is the only regional voice to do so. However, not everything in its own backyard is rosy. For example:

- The same political party has been in power for over 40 years. State control of the media makes it difficult for opposition parties to campaign in elections on an equal footing.
- Botswana retains the death penalty, which it exercises in cases of murder, treason and assassination.

13 The outcomes of geopolitical interventions

- LGBT groups face widespread discrimination, despite homosexuality being decriminalised in 2019.
- The government is accused of dispossessing the indigenous San people from their lands and traditional water supplies in the Kalahari Desert, mainly because of the vast diamond fields, but also in order to create large game reserves for the benefit of tourists.
- Expressions of cultural diversity are not encouraged.

As with so many countries, Botswana's record on human rights has its blemishes. But at least the country does have a stable, fairly democratic government that has invested some of the diamond profits in public goods and infrastructure.

> ### Skills focus: Gini coefficient or index
> The Gini index is a statistical measure that can be used to assess the extent to which the distribution of income among individuals or households within an economy deviates from a perfectly equal distribution.
> A Lorenz curve plots the cumulative percentages of total income received against the cumulative number of recipients, starting with the poorest individual or household. The Gini index measures the area between the Lorenz curve and a hypothetical line of absolute equality. This is expressed as a percentage of the maximum area under the line. Thus, a Gini index of 0 represents perfect equality while an index of 100 implies perfect inequality.
> Use of the Gini coefficient is limited by the lack of income data for those countries in which we are particularly interested, i.e. the least developed countries receiving development aid.

Superpower objectives

Superpowers are defined by great economic wealth, military strength, reliable access to resources and a dominant ideology (see Topic 7, page 148). But no superpower can afford to rest on its laurels. It needs to be constantly scheming to reinforce or enhance its global status. More specifically, it needs to be constantly securing:

- strategic locations
- future supplies of resources (food, energy, minerals and water)
- alliances (economic, political and military) with other countries
- technological advances
- a global sphere of influence.

Aid can be used to pave the way to achieving most of these objectives. Aid can open doors and can create a sort of halo effect that donor countries are able to exploit for their own ends. Aid is rarely offered without strings attached. In short, any superpower that neglects to extend aid to less-developed countries does so at its peril.

The USA has done much to create the image of a magnanimous and benevolent power – a sort of philanthropic fatherly figure providing support to all those in need. But most can see through that image and spot its real motives, namely to stay a superpower and outmanoeuvre its rivals. But what about one of those emerging superpower rivals, China?

The interesting thing is that, unlike the USA, China does not seem to have felt the need to set up any sort of larger military alliance. We can only speculate about the reasons. Is it simply because China is such a giant of a country?

> **Skills focus: The misuse of data**
>
> In any conflict it is inevitable that there are going to be at least two different sides or viewpoints. It is human nature to exaggerate the positives of your side and to amplify the negatives of the opposition.
>
> In a situation of military intervention, such as Afghanistan, victory (if indeed it yet exists) will be claimed in terms such as gaining territory, winning minds, driving out terrorist insurgents, and restoring political stability and respect for human rights. In other words, conflicts commonly create situations in which propaganda and misinformation prevail. Facts will either be deliberately distorted or ignored to suit one particular side in the conflict.
>
> In such a situation, the only data that we can begin to trust is that collected and collated by independent bodies, such as the various agencies of the UN and NGOs, for example, Médecins sans frontières and the Red Cross.

13.3 Evaluating military interventions

Costs of military interventions

Table 13.3 summarises details of five recent military interventions. It makes for sober reading, especially when the political outcomes are considered along with the immense human costs. The outcomes have mainly been negative. It needs to

Table 13.3 Summaries of some recent military interventions

Location	Date	Intervention	Outcomes	Estimated total deaths	Estimated refugees
Afghanistan	2001–21	US-led invasion following the 9/11 attacks believed to have been planned in Afghanistan. Need to counteract Taliban seizure of country as well as address its appalling record of human rights.	Civilian government restored in 2004, but much unrest – assassinations and suicide bombings. The USA and UK withdrew all armed forces in 2021 and the Taliban returned to power.	175,000–200,000	2,670,000
Syria	2011–2024	Democratic uprising against the Assad regime started in 2011. Use of chemical weapons led to US-led coalition launching air strikes against Syria and IS. Joined by UK in 2015. Russia also started air strikes, but in support of Assad.	Huge civilian casualties; Assad used chemical weapons. Migration of refugees into Europe, Türkiye, Jordan, Lebanon and Iraq, of which Türkiye received the most. The oppositional forces finally regained territory in December 2024 with Assad fleeing the country to Russia.	580,000–620,000	13,000,000+
Libya	2011 and 2014	Libyan civil war and civilian casualties led to UN intervention involving 19 states, a naval blockade, no-fly zone and air strikes against Gaddafi's forces.	Gaddafi killed and his regime ousted. Breakdown of country into opposing military forces – much violence and disagreement. Possible IS infiltration.	120,000	644,000
Ukraine	2014–	By Russia on grounds of support for minority ethnic Russians in Ukraine. By 2022 it was clear the whole of Ukraine was the target.	Crimea annexed by Russia in 2014, followed by an attempted invasion of the entire country in 2022.	300,000–500,000 by 2024	14,000,000+

Syrian suffering

Since 2011, Syria's civil war has led to almost unimaginable human suffering. UNHCR estimates that 6.7 million Syrians have been internally displaced and a further 6.6 million have fled abroad as refugees - more than 50 per cent of the population of 21 million in 2010. Civilian deaths exceed 200,000 and total deaths around 600,000 as of 2024.

Protests against the brutal regime of President Bashar al-Assad in 2011 developed into a full-scale civil war by 2012. Assad fought to cling onto power against the Free Syrian Army and Syrian Democratic Forces. Deaths, internal displacements and refugee numbers quickly escalated as civilians were caught up in an increasingly complex conflict. Adding to the complexity of the diverse rebel groups struggling to overthrow President Assad's regime, the conflict became even more complicated when:

- IS invaded and occupied part of Syria in 2012
- the USA and its allies mounted air strikes against IS in 2014 and later against Assad
- Russia began to support Assad militarily in 2015.

In 2024, unexpectedly, a military offensive led by the Islamist political and military group Hay'at Tahrir al-Sham (HTS) swept across Syria, captured Damascus and Assad fled. HTS formed the Syrian transitional government in December 2024.

It is too early to say what this will mean for the Syrian people. Fighting continues as of 2025, at a much-reduced level. Many refugees hope to return home. But large areas of cities have been reduced to rubble and education and health systems are severely degraded. A key question is whether HTS, a Sunni Islamist group, can unite and govern a country which is 10% Kurdish, and has large Christian, Shia, Alawite and Druze religious minorities.

Figure 13.7 Syrian refugees push and pull their boat through the Mediterranean Sea as they approach the shore of the island of Lesvos, Greece

Figure 13.8 Aleppo: a city pulverised by Assad

be stressed that accurate data about civilian deaths and displacement are simply not available. But there are best estimates made by neutral IGOs and NGOs. If anything, these estimates are thought to err on the side of understatement.

The sad fact is that if we look at the military interventions of the last 25 years, the successes have been few and far between. Some would argue that the positive effects of mitigating potential mass-killings and genocide do not outweigh the long-term negative effects of destabilising the Middle East and North Africa. Perhaps this is because the military interventions have not been long enough to achieve reconciliation between warring groups and the reconstruction of countries that have been badly damaged, both physically and politically.

The record of non-military interventions

Brought into focus here are the interventions by the UN under the heading of peacekeeping. According to the UN: 'Peacekeeping has proven to be one

The Arab Winter

The Arab Spring of 2011 was a time of great euphoria as seemingly spontaneous public uprisings in Tunisia, Libya and Egypt overthrew their corrupt leaders and undemocratic governments. Western governments were not directly involved and simply applauded from the sidelines. Sadly, in the wake of the Spring has come the Arab Winter. Parts of North Africa are now beginning to be plagued by terrorists, particularly those related to IS (Figure 13.9). The sad truth is that the removal of autocratic leaders and regimes with little respect for human rights did lead to some short-term gains, but it also created political destabilisation, civil wars and the appearance of regimes with even worse human rights records.

Underlying some of this unrest is the deep-rooted animosity between Shia and Sunni Muslims. How to neutralise this animosity is the challenge. Is it a challenge just for the Arabs or for the whole global community? Some might ask where the UN is in all of this? Should it be taking more of a lead?

Figure 13.9 The distribution of Shia Muslims in the Middle East. The distribution of the Sunni Muslims is the mirror image of this map. The lower the percentage values, the higher the incidence of Sunnis.

of the most effective tools available to the UN to assist host countries navigate the difficult path from conflict to peace.'

UN peacekeeping is guided by three basic principles:

- consent of all parties in the conflict
- impartiality
- non-use of force except in self-defence and defence of the mandate.

The UN is able to draw on troops and police from around the world to provide its peacekeeping forces. Civilians are also integrated into those forces. So the costs of peacekeeping are shared by UN member states.

UN peacekeeping began in 1948 when military observers were deployed to monitor the armistice between the newly created Israel and its Arab neighbours. Since then, 69 peacekeeping operations have been deployed, 56 of them since 1988.

Since the end of the Cold War at the end of the 1980s, the nature of UN peacekeeping has changed from inter-state conflicts to intra-state conflicts and civil wars. UN peacekeepers are now asked to undertake a wide variety of complex tasks, from disarmament and reintegration of former combatants, to helping to build sound governance and monitoring human rights. Countries that have been the scene of UN peacekeeping operations include Angola, Mozambique and Namibia in Africa; Cambodia in Asia; El Salvador in Central America; and Croatia and Bosnia-Herzegovina in Europe.

As of 2020, there were 14 different peacekeeping operations (Figure 13.10). Four of those operations are in the Middle East; seven of them in Africa. A contemporary, and completed, mission was in Ivory Coast between 2004 and 2017.

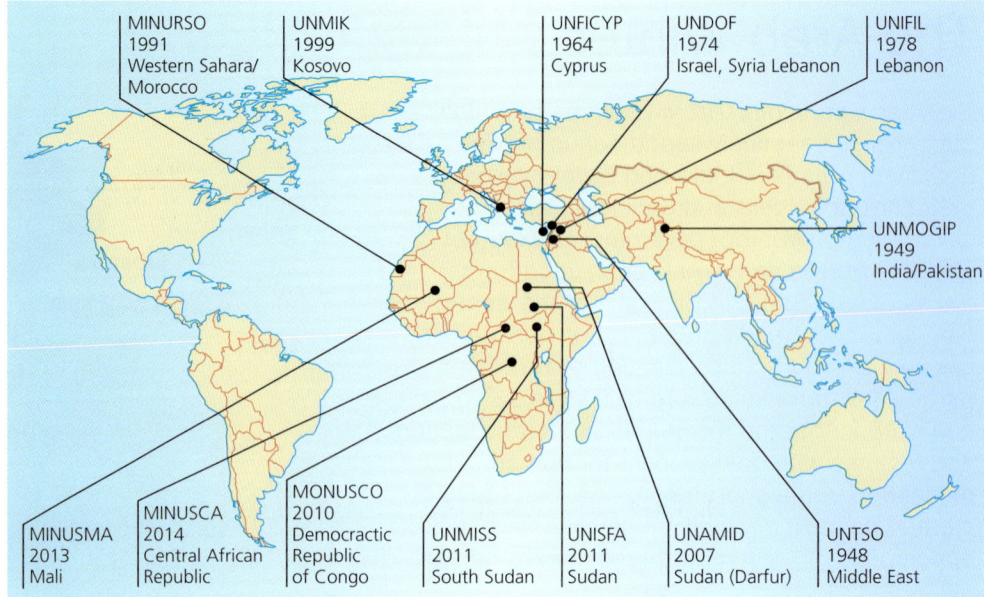

Figure 13.10 The locations of UN peacekeeping operations, 2020

Tribal rivalries have been a common source of conflict over much of Africa. Such conflict has not been helped by the failure of European powers to recognise tribal borders when drawing up their colonial boundaries. But conflict has other roots, such as political and religious differences, as well as the recurring desire of some groups to hijack sources of economic prosperity for themselves.

UN peacekeeping has certainly struggled in its 69 missions to resolve conflicts and improve human rights. The brutal reality is that peacekeepers cannot be everywhere all the time. Furthermore, remember that the UN can only take action where and when all parties to a dispute say 'yes'. Improving development has not been in the UN peacekeeping mandate. Rather, its task has been to put in place the right sort of governance needed for successful development.

Costs of non-intervention

Faced with threats to human well-being and human rights, the global community has three options:

- Turn a blind eye and do nothing.
- Make a limited military intervention to deal with the short-term threat and encourage local people to take matters into their own hands at the earliest possible opportunity.
- Make an extended military intervention that includes the longer-term tasks of reconciliation and reconstruction.

The first option does not find favour today. There is now global acknowledgement of the huge risks associated with turning a blind eye. The genocide in Rwanda in 1994 is a salutary reminder of the dangers of doing nothing.

Non-military intervention: UNICEF

UNICEF, the United Nations Children's Fund, is a specialised agency of the UN set up in 1946. Its aim is to promote the health and well-being of children and it has a budget of about US$8 billion per year.

Most of UNICEF's work is in the field, frequently in areas of crisis and conflict. In 2023, the main areas of UNICEF spending were:

- US$3.6 billion on child food poverty across 158 countries
- US$1.7 billion on education across 144 countries
- US$1.25 billion on WASH (water, sanitation and hygiene) services
- US$1 billion on child protection across 150 countries.

About 70 per cent of UNICEF's annual budget is spent in a humanitarian context, such as providing an immediate response to a war, natural disaster or civil disturbance where children's health and lives are under threat.

Protecting children's human rights is an increasingly challenging task because:

- many wars include child-soldiers, with catastrophic impacts on children's physical and mental health
- children are put to work before they finish their education, reducing opportunities later in life
- our digital, internet driven age exposes children to inappropriate content online
- climate change threatens water supply, food production and the prevalence of some diseases and, as dependents, children are especially vulnerable to these shocks.

Figure 13.11 A vaccination clinic

UNICEF works with other UN agencies, NGOs and national governments to try and improve children's lives within the wider context of the United Nations Sustainable Development Goals.

In recent years, UNICEF has worked with partners to deliver malaria vaccines to children. Malaria kills roughly 500,000 children under five in Africa each year. Since the RTS,S/AS01 malaria vaccine has been approved for use by children, UNICEF has helped to vaccinate over 1.7 million children in Ghana, Kenya and Malawi and plans to expand this to 18 million children across 12 more African countries by 2025.

Clearly, this particular situation calls for the third option. History tells us that too many military interventions have belonged to the second option. Hopefully we have learned the lesson that, while such interventions might bring short-term relief, more often than not they create longer-term costs.

There are, however, major issues facing us today that also require intervention, but not of a military kind. The issues include eradicating poverty, advancing social development and, indeed, progressing many of those rights contained in the UDHR. The appropriate interventions here are either economic or aid-related. One major issue not covered by such interventions is the need for greater care of the environment. Lack of global action here threatens the very survival of the human race and possibly heralds a world where human rights will no longer matter or be relevant.

Review questions

1. Explain what is meant by 'geopolitical interventions'.
2. Give reasons why it is difficult to measure human rights.
3. Should economic growth be given precedence over human rights? Give your reasons.
4. Explain why Botswana is a 'most unusual case'.
5. Examine the possible motives of superpowers in providing development aid.
6. Identify what you think are the costs of development aid.
7. Explain why there are not more interventions by UN peacekeepers.
8. Examine the possible tensions between military intervention and human rights.
9. Assess the relative merits of limited and extended military interventions.
10. 'Development aid has contributed towards narrowing the gap between the rich and poor.' Discuss.
11. Study Table 13.2 (page 250). To what extent does it show a narrowing of the gap between rich and poor?
12. Examine the costs of non-military intervention.
13. 'Military intervention has more costs than benefits.' Discuss.

Further research

Can military intervention be humanitarian? Start your search for an answer by visiting: https://merip.org/1994/03/can-military-intervention-be-humanitarian/

Visit the Foreign Policy Journal and research the motives and outcomes of military intervention in Iraq: https://www.foreignpolicyjournal.com/

Exam-style questions

1. Study Table 13.4 below, which shows the five highest- and five lowest-ranking countries according to the Happy Planet Index (HPI). Human Development Index (HDI) rankings are also given for the ten countries.
   a. Name the country with the smallest difference between the two rankings. [1]
   b. In which part of the world are most of the highest-ranking countries located? [2]
   c. Name the statistical test widely used when comparing rankings. [1]
   d. Suggest reasons for the apparent differences between the HPI and HDI rankings. [6]

Table 13.4

	HPI rank	HDI rank		HPI rank	HDI rank
Costa Rica	1	97	Bahrain	147	44
Viet Nam	2	116	Mali	148	179
Colombia	3	97	Central African Republic	149	187
Belize	4	101	Chad	150	185
El Salvador	5	136	Botswana	151	126

2. Explain why the level of spending on healthcare and education varies from country to country. [8]
3. Evaluate the view that some forms of geopolitical intervention are more successful than others. [20]

Topic 8

Option 8B: Migration, Identity and Sovereignty

Chapter 14: The impacts of globalisation on international migration

Chapter 15: Nation states in a globalised world

Chapter 16: Global organisations and their impacts

Chapter 17: Threats to state sovereignty

14 The impacts of globalisation on international migration

What are the impacts of globalisation on international migration?
By the end of this chapter you should be able to:
- explain how globalisation has led to an increase in migration, both within countries and between them
- understand why the causes of migration are varied, complex and subject to change
- evaluate the varied consequences of international migration, and the differing perspectives that feature in migration debates.

14.1 Economic systems and labour flows

Globalisation has led to a rise in migration, both within countries (internal migration) and between them (international migration). A record number of people migrated internationally in 2019: more than 280 million people now live in a country they were not born in.

- Modern transport networks enable truly global labour flows to operate. Every year many young Australians and South Africans travel to the UK to work.
- Migration should not be seen as an inevitable consequence of globalisation, however. The vast rise in trade between some countries, such as the UK and India, has actually coincided with a decline in migration between them.
- Also, much international migration is relatively regionalised. In general, the largest labour flows connect neighbouring countries like the USA and Mexico, or Poland and Germany.

National and international migration patterns

Changes in the pattern of demand for labour at the national scale are often linked with globalisation. National **core–periphery systems** develop and strengthen over time because of positive feedback effects. Uneven economic growth may be linked originally to a natural advantage that one 'core' region enjoys over others (such as raw material availability or the presence of a coastline). Over time, any initial imbalance becomes exaggerated due to the perpetual outflow of migrants, resources and investment from peripheral regions towards the core (Table 14.1, page 265). Collectively, these flows are called **backwash** effects.

Global systems encourage rural–urban migration within countries in various ways, including the introduction of mechanised agriculture and land grabs by states and agribusinesses. Complementing this are the employment pull factors found in urban areas in developing and emerging economies, often linked with global supply chain growth in export processing zones (EPZs).

The process of core-periphery polarisation is sometimes repeated at larger spatial scales (see EU Schengen place context on page 263).

> **Key terms**
>
> **Core–periphery system:** The uneven spatial distribution of national population and wealth between two or more regions of a country, resulting from flows of migrants, trade and investment from periphery to core.
>
> **Backwash:** Flows of people, investment and resources directed from peripheral to core regions. This process is responsible for the polarisation of regional prosperity between regions within the same country.

Rural–urban migration within China

China is an example of large-scale internal migration. The 2020 census showed 375 million lived away from their place of birth. This is the largest migration in history. Figure 14.1 shows that the net losers are China's central provinces, whereas coastal areas to the east are major gainers. Most of the migrants have moved from rural provinces to urban areas on the coast. In 1978, on the eve of economic reforms, 20 per cent of China's population lived in cities. By 2020, this had risen to 60 per cent. The relocation of millions of rural people gave many Chinese cities a 'site factor' certain to attract foreign direct investment: a large, cheap labour force. The Chinese government's authorisation of free movement can be viewed with hindsight as a rational economic decision allowing China to benefit from globalisation. The 'migrant miracle' that followed underpinned 30 years of rapid economic growth.

Key
- + 2 million or more
- + 1 to 2 million
- Net change of less than 1 million
- – 1 million or more

Figure 14.1 Net internal migration in China, 2010–15

Rural–urban migration within Spain

Core–periphery growth in Spain has accelerated to such an extent in recent years that the Celtiberian Highlands, a rural region east of Madrid, has been all but abandoned. Following decades of depopulation, just eight people per square kilometre now remain in one of Europe's least-populated areas. Population density in the Celtiberian Highlands is now as low as Lapland in the cold far north of Finland. Provinces such as Terual and Soria have 600 villages with fewer than 100 people and an average age of 57. High rural unemployment has meant that young people continue to seek out new opportunities in Madrid and Barcelona (both cities are important hubs in the global economy). Spain's low birth rate means rural recovery is unlikely. When local schools shut, a threshold or tipping point is often reached in Celtiberia's villages, from which there is no return.

Figure 14.2 An abandoned village in Spain

Option 8B: Migration, Identity and Sovereignty

International migration inside the EU Schengen area

The process of core–periphery polarisation is sometimes repeated at larger spatial scales. Within the EU, free movement of labour and capital has fuelled a core–periphery pattern. The EU core of northern France, the Netherlands, Belgium and western Germany includes the world cities of Paris, Brussels and Berlin. **Economic migrants** from eastern and southern Europe flow towards this core (Figure 14.3).

Internal national border controls within most of the EU were removed in 1995 when the Schengen Agreement was implemented. Schengen enables free movement of people and goods within the EU, meaning that passports and goods declarations are not required at borders. Eastern European nations implemented the agreement in 2007–8. Schengen brings benefits, as EU labour can move to where there is most demand, but also costs – once someone is in one EU country, they can move to others. In recent years, most EU states have witnessed the growth of nationalist movements opposed to free movement such as National Rally in France and the AfD in Germany. Racist attacks on and hostility towards immigrants have become more common in Italy, Germany and other EU countries.

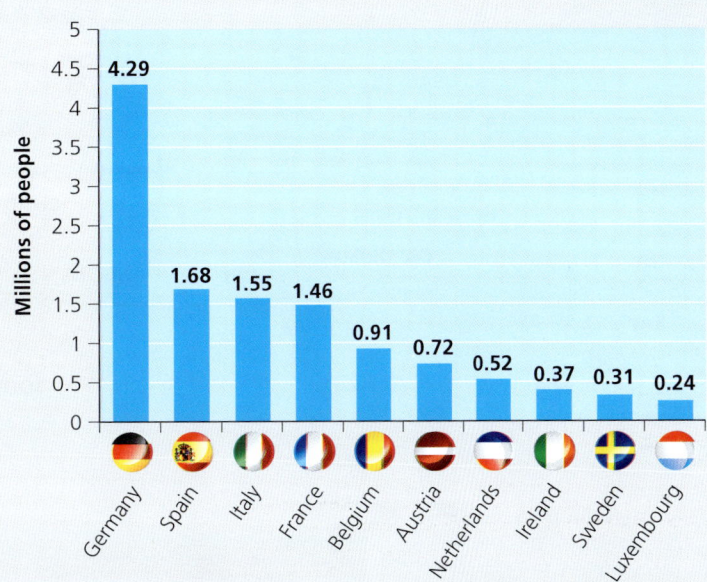

Figure 14.3 EU states with the biggest populations of people born in another EU state, 2019

Variations in migrant population sizes

Between 3 and 4 per cent of the world's population lives outside the country of their birth. This percentage has not changed greatly over time despite the fact that the number of people migrating internationally has risen. This is because the total size of the world's population has grown too. Between 1950 and 2020, world population grew from 2.5 billion to 7.8 billion.

Important changes have taken place in the *pattern* of international migration in recent years, however. As recently as the 1990s, international migration was directed mainly towards developed world destinations such as New York and Paris. Since then, world cities in developing world countries, such as Mumbai (India) and Lagos (Nigeria), have also begun to function as major global hubs for immigration.

Individual nations vary enormously in terms of the number or proportion of their population that is comprised of migrants (Figure 14.4). Differences in the level of political engagement with the global economy is one major reason for this. In order for a state to become deeply integrated into global systems, its government may need to adopt liberal immigration rules (Table 14.1, page 265). Inward investment from TNCs may depend in part on the ease with which a company can transfer senior staff into a particular nation.

> **Key concept: 2019–21 Covid-19 pandemic**
>
> The Covid-19 pandemic dramatically curtailed international migration flows during 2020 as countries closed borders to prevent the spread of the virus. Global air travel fell by 60 per cent during 2020. Even internal migration, such as rural–urban migration, was prevented in some countries due to travel restrictions. The longer-term impact of the pandemic on all forms of migration is as yet unknown.

14 The impacts of globalisation on international migration

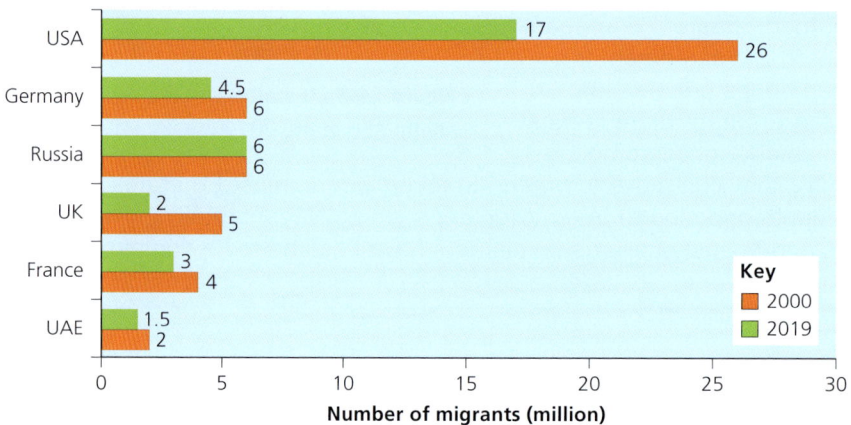

Figure 14.4 Major world migration flows and the number of international migrants living in selected countries, 2019

Many of London's leading law firms have regional offices spanning the globe, from Singapore to Moscow. In order to maintain their global networks, these companies depend on foreign states granting UK lawyers permission to relocate to their overseas offices.

Key term

Economic migrant: Most migrants move for economic reasons such as a better job or hope of higher pay. Many economic migrants move voluntarily (they choose to move) but not all. In some cases migrants are forced to move by people traffickers or even family members and put to work in a new location.

Changes in the global pattern of migration

Figure 14.6 (page 266) is a representation of the global pattern of international migration between 2010 and 2015. Important features that are clearly visible include:

- Large volumes of intra-regional migration (most international migration originating in Africa is directed towards other African nations).
- Significant inter-regional flows linking North America with other regions including South Asia and Central America.

Skills focus: Use of flow lines

Apply your geographical knowledge and skills to the unusual representation of global migration flows shown in Figure 14.6. Make an estimate of the number of people (in millions) who migrated to the USA from each Asian region between 2010 and 2015.

Migration policies in Singapore, Japan and Australia

Table 14.1 Policies for international migration in Singapore, Japan and Australia

Singapore (liberal migration rules)	Singapore, an economically developed city-state of 5.7 million people, has an ethnically diverse population as a result of its British colonial past and well positioned as a trading hub in Asia. It has had a large ethnically Chinese population for centuries. It has an open policy towards economic migrants but also a two-tier system recognising 'foreign workers' (low-skill labour often from India, the Philippines, Thailand) and 'foreign talents' (high-skill from developed countries). Singapore's treatment of 'foreign workers' has been questioned as many live in crowded dormitories, are paid poorly and have few rights.
Japan (stricter migration rules)	Less than 2 per cent of the Japanese population is foreign or foreign-born. Despite the growing status of Japan as a major global hub from the 1960s onwards, migration rules have made it tough for newcomers to settle permanently. Nationality law makes the acquisition of Japanese citizenship by resident foreigners an elusive goal (the long-term pass-or-go-home test has a success rate of less than 1 per cent). Japan faces the challenge of an ageing population, however. There will be three workers per two retirees by 2060. Many people think that Japan's government will need to loosen its grip on immigration.
Australia (stricter migration rules)	While Singapore has a high percentage of foreign workers, the proportion found in Australia is lower due to a recent history of restrictive migration policies. The country currently operates a points system for economic migrants called the Migration Programme. In 2018–19, only 160,000 economic migrants were granted access to Australia (this figure included the dependants of skilled foreign workers already living there). Almost 50% of these immigrants came from just three countries: China, the UK and India. Until 1973, Australia's government selected migrants largely on a racial and ethnic basis. This was sometimes called the 'White Australia' policy.

Figure 14.5 Singapore

The main factor explaining the pattern shown is the uneven distribution of economic opportunity within global systems. As well as being triggered by economic inequality, migration also reproduces it. This is because the 'brain drain' of talent away from source countries represents an economic loss that may only be partially offset by the receipt of remittances.

In the future, the patterns of migration that globalisation helps drive are likely to change on account of other factors growing in importance (Table 14.2).

Table 14.2 Additional factors that give rise to complex global patterns of migration

	Environmental change	**Economic events**	**Political events**
Influence of factor	Climate change is already causing refugees to leave regions where agriculture is threatened. Syria's refugee crisis has in part been attributed to desertification by the US Pentagon's security analysts.	The global financial crisis (GFC) of 2007–8 had an unprecedented effect on migration. For the first time since 1945, world GDP shrank. Net migration from Poland fell to its lowest level since the 1950s.	New conflicts can unexpectedly trigger or diversify global migration flows. On many occasions since the Second World War, political regime changes have prompted ethnic groups to flee states.
Evaluation of importance	Climate change acts to intensify rural poverty in some countries. Movers who might previously have been classed as economic migrants become refugees due to an increasingly hostile environment.	The realisation that globalisation has a 'reverse gear' means future projections for global migration and urbanisation should be treated with caution. Economic systems can become unstable.	In parts of North Africa, Central Africa and the Middle East, political factors are now a more important influence on migration than globalisation; 14 million people have been displaced by the conflict in Syria since 2011.

Key term

Net migration: Net migration is the difference between the number of immigrants (arrivals) and emigrants (departures) usually over a period of a year. If immigration is higher than emigration then there is positive net migration. The term can be applied to a country, or a region within a country.

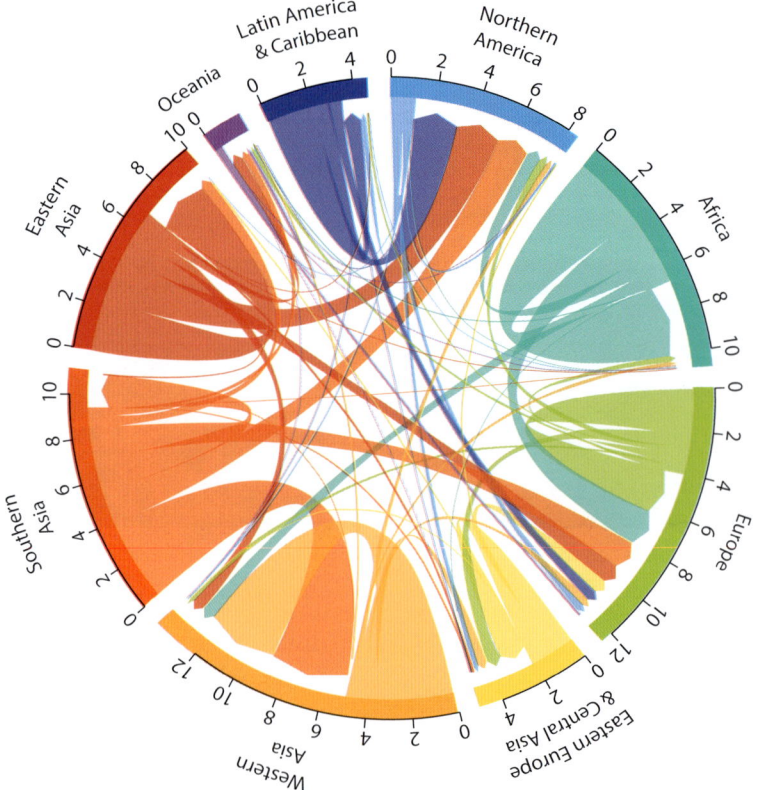

Figure 14.6 A chord diagram of global migration patterns

Visualising migration flows

Movements of people are often shown as lines or arrows on a world map, but circular chord diagrams can also be an effective way of visualising global migration flows. This diagram shows migration between 196 countries between 2010 and 2015, broken down into nine world regions. The colour shows which region each flow came from, the width of a flow shows its size, and numbers indicate the total migration in and out of a region, in millions.

Option 8B: Migration, Identity and Sovereignty

14.2 The complex causes of migration

Most international migrants move for work-related reasons. In Book 1 we learned about the factors driving these international flows of people. They include poverty in source regions, low primary commodity prices and poor access to markets within global systems. In addition, in 2020 there are about 26 million refugees that have been forced to leave their home country. It is worth noting that when people flee extreme poverty, the dividing line between forced and voluntary migration may not always be clear.

Economic, social and political reasons for movement

In general, migrants are not the poorest citizens of the states they leave behind. This is because money is needed to make an international journey. Migrants moving from Africa to Europe have been known to pay well over £1000 to traffickers. The world's poorest people simply do not have access to this capital.

In addition to labour flows, other important reasons for migration include:

1 **Family:** spouses and children may follow workers overseas. In time, extended family members may move too as part of a process called **diaspora** growth. **Post-colonial migrant** flows to the UK involved the movements of large numbers of people from former colonies of the British Empire.
2 **Conflict:** war, conflict and persecution are responsible for the international displacement of millions of people. Close to 1 million Rohingya Muslims have fled persecution in the Buddhist majority Rakhine state in Myanmar and migrated to Bangladesh since 2012. Over 10 million Syrians have been displaced within their own country since the conflict began there in 2011 (internally displaced people or IDPs).

> **Key terms**
>
> **Diaspora:** A dispersed group of people with a shared cultural background who have spread internationally from their original homeland.
>
> **Post-colonial migrants:** People who moved to the UK from former colonies of the British Empire during the 1950s, 1960s and 1970s.

European migrant crisis, 2014–19

Between 2014 and 2019, 2 million migrants and refugees entered Europe by crossing the Mediterranean Sea or overland between Türkiye and Greece. Some 20,000 died or went missing during their perilous journey on rafts and unsafe boats, many organised by ruthless people traffickers. Over 1 million arrived in 2015 alone. Mediterranean migrants are a diverse group of people. They include both economic migrants and refugees of varying faiths and ethnicities (Figure 14.7). Many travel large distances to reach the shores of the Mediterranean. Sending regions include the Middle East and parts of Africa. The issues arising from this movement of people include:

- **Migrant status:** many migrants arrive with no papers or ID, so their origins and travel history are unclear. Most are genuine refugees, but some may be economic migrants. This presents a difficult challenge to receiving countries in terms of who to provide safe haven to.
- **The political reaction:** Most Mediterranean migrants arrive in Greece or Italy where, under EU rules, any asylum claims must be processed. Neither the Greek nor Italian government wants large numbers of migrants to settle there permanently, however. Both countries want to see the burden of resettlement shared with other EU members. Consequently, plans were established to distribute genuine refugees across Europe. Germany has taken in many hundreds of thousands.

- **The ethical debate:** The number of accepted asylum seekers in 2015 amounted to less than 0.1 per cent of the EU's population. However, it remains an emotive issue that divides people's views. A suicide bomber in the Paris attacks of December 2015 was later revealed to be a Syrian refugee who had travelled to France via Greece. Since then the political debate has intensified over whether border passport checks should be reinstated within the EU as part of an emergency situation.

To what extent is globalisation responsible for Mediterranean migration? It is certainly not the only cause. However, global economic disparities are an important factor; so too are 'shrinking world' technologies. Increasingly, international migrants communicate with one another using smartphone apps. In 2015, a Facebook group called 'stations of the forced wanderers' helped over 100,000 migrants exchange advice on how to avoid authorities and find routes across European borders using GPS information.

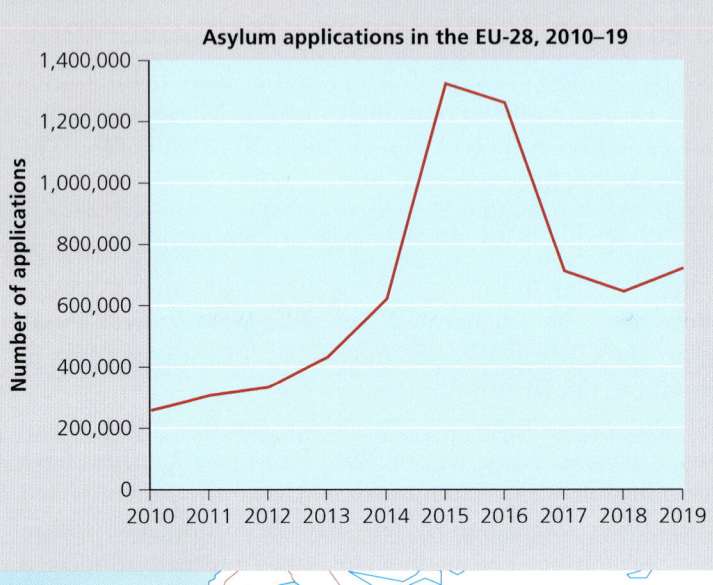

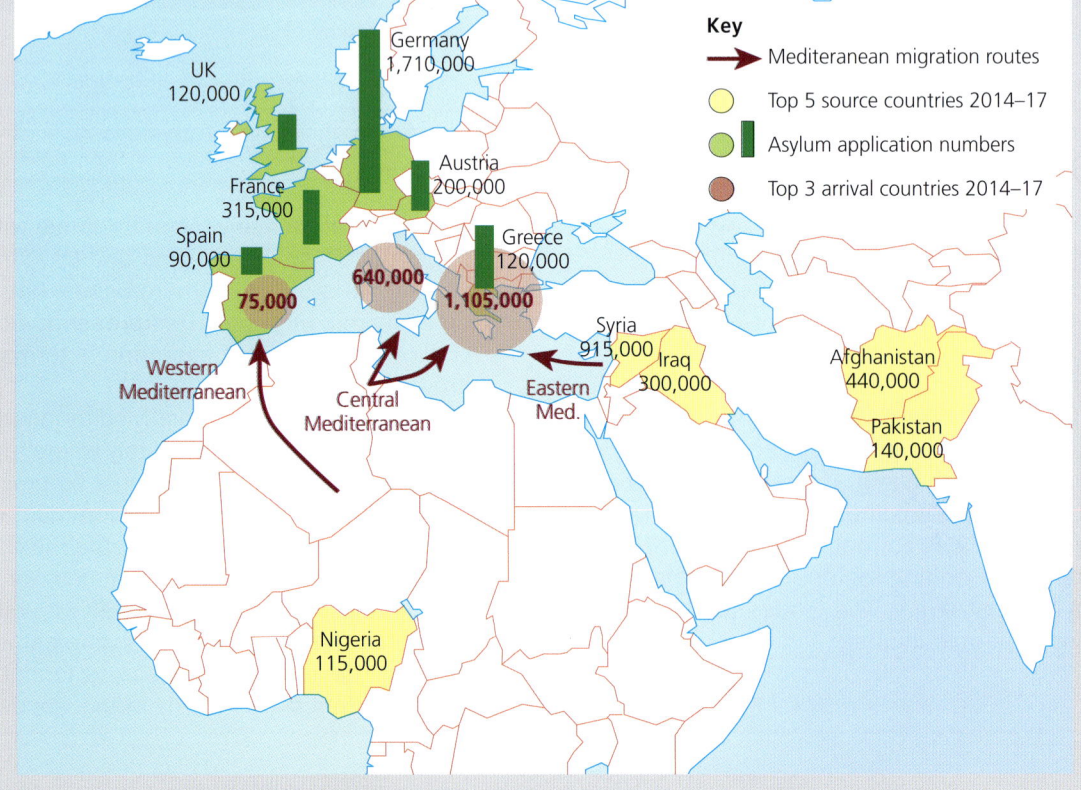

Figure 14.7 Mediterranean migration source countries, routes and destinations, 2014–17

The economic rationale for permitting free movement of people

The core–periphery model of economic development depends on a process called backwash, as we have already seen. Migrant labour flows become focused on core regions at varying geographical scales. Within the EU, this movement takes place at both the national and international level. The latter has been encouraged by the **Schengen Agreement** since 1985 (Figure 14.8). The majority of the EU's 27 members are now Schengen countries. The logic of the agreement is rooted in an economic theory that views human beings as an economic resource that businesses need to make use of. People should therefore be allowed to move to where work is available. Migration is viewed as an efficient way of making sure that the economic output of a territory is optimised.

In order to aid this process, most states allow workers to migrate freely from peripheral to core regions. While it may appear that this is causing spatial disparities to increase at first, some economists argue that the exact opposite is true and that the core–periphery system is in everyone's interest. The supporting argument used by core–periphery theorist John Friedmann is that backwash effects are balanced out by spread effects, also known as the **trickle-down** of wealth (Figure 14.9).

Critics of this model refute the proposition that spread 'gains' ultimately exceed backwash 'losses' for peripheral regions. In reality, it is hard to either accept or reject the hypothesis due to the sheer complexity of the economic and demographic processes taking place. Also, even if the economic rationale for Friedmann's core–periphery model is sound, he did not account for the **negative externalities** that backwash and spread movements sometimes generate.

> **Key terms**
>
> **Schengen Agreement:** An international agreement that aims to make it easier for people to move freely within the EU. Passports do not usually have to be shown by citizens at the borders of the 26 EU and non-EU countries that have agreed to this.
>
> **Trickle-down:** The positive impacts on the peripheral region of wealth creation in core regions. These may include investment (in the form of back offices and branch plants), regional aid and grants, and the diffusion of innovations, technology and infrastructure from the core to the periphery.
>
> **Negative externalities:** Costs suffered by people and places as a result of changing economic activity. These may be unintended social or environmental impacts, such as unemployment or pollution.

Figure 14.8 EU citizens enjoy free movement at airports within the EU

14 The impacts of globalisation on international migration

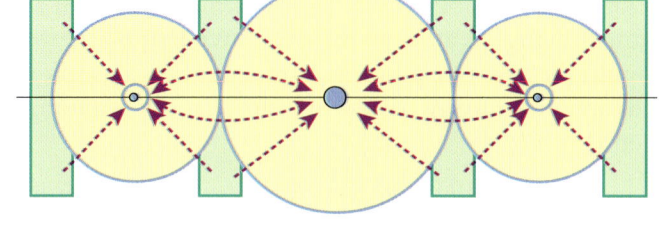

Figure 14.9 Backwash processes in the Friedmann core–periphery model

A strong economic core develops fuelled by the in-migration of people (workers and investors) from the peripheral regions of a state.

In economic theory, additional core regions form as part of the development process over time. The growth of these cores is fuelled by flows of raw materials and workers from neighbouring areas.

Evaluating the benefits of free movement at varying scales

Is it really true that unrestricted economic labour flows, at varying geographical scales, result in an efficient allocation of human resources that is in society's best interest? Evidence from China is sometimes used to support this view, as we saw at the start of the chapter. The place contexts on internal migration within the UK and migration from Poland to other EU states evaluate further evidence from the EU at two different geographical scales.

Free movement at the national scale: internal migration within the UK

The UK's 'North–South' population drift accelerated during the 1980s. The deindustrialisation of northern cities such as Liverpool and Sheffield triggered the exodus of many young people towards the UK's economic core of London and the Southeast. Since then, this trend has continued (Figure 14.10). London's population reached a record high of 9 million in 2019. Its house prices have tripled in value since 1995 as a result of high demand from incomers and investors.

Since 1945, the gap between house prices in northern and southern England has grown and lessened several times. Rising costs of doing business in the capital have sometimes triggered out-migration of people and businesses. Regenerated post-industrial cities including Cardiff and Bristol offer an attractive alternative to London; the BBC relocated to Manchester in 2011. However, London's global hub status means the UK's core–periphery imbalance is likely to persist.

Figure 14.10 Net migration flows between London and other settlements, 2009-12

Option 8B: Migration, Identity and Sovereignty

Free movement at the international scale: migration from Poland to other EU states

Does EU migration benefit all member states? Poland's government has encouraged its population to work overseas and make the most of EU membership. However, Poland has lost population every year since the 1960s, with the trend accelerating since 2004. A low birth rate and ageing population mean that Poland's population is forecast to age and shrink. Few people migrate into Poland: it has the lowest foreign-born population percentage of any EU state (Figure 14.11).

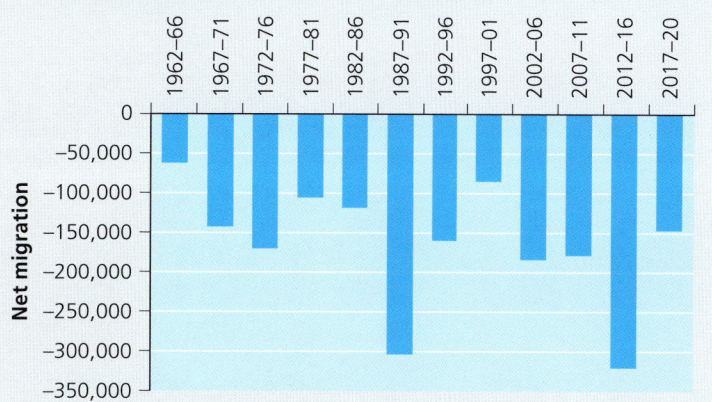

Figure 14.11 Net international migration for Poland, 1962–2020

While remittances help compensate for the labour loss and brain drain in the short term, there is no guarantee remittances will continue to be sent in the long term. Children of Polish migrants born in the UK could feel less connected to Poland and may send less money in the future. By 2050, there may be only 1.3 working people for each child or elderly dependent in Poland, compared with 1.75 today. Continued emigration may become economically and socially unsustainable.

14.3 Differing perspectives on the consequences of migration

Migration has become one of the defining issues of the twenty-first century. Accelerated by economic development, shrinking world technologies and political interconnectedness, migration is a major management challenge for most governments, especially those in developed countries towards which large flows are directed.

National cultures and the processes of migration and integration

Migration changes the cultural and ethnic composition of states. These are not the same thing: French or Irish migrants in London are sometimes viewed as groups who bring cultural *but not ethnic* diversity. Definitions of **ethnicity** may vary, but most people view ethnic differences as relating to variations in religion or race, in addition to the language that people speak. For this reason, some migrant flows are popularly believed to introduce a greater degree of cultural 'otherness' to places. Figure 14.12 shows that in Greater London some places have a high degree of clustering by ethnicity (or segregation) shown by high concentrations of Black/Black British or Asian/Asian British populations. However, other places are much more mixed suggesting a process of **integration** has taken place through inter-marriage, social mixing and different cultures living and working with each other. Highly segregated places may be problematic because some ethnicities suffer above average rates of poverty, unemployment, health problems and deprivation as a result of inequality and discrimination.

> **Key terms**
>
> **Ethnicity:** The shared identity of an ethnic group which may be based on common ancestral roots or cultural characteristics such as language, religion, diet or clothing.
>
> **Integration:** The process by which migrants (individuals and groups) become accepted into society. It is a two-way process of adaptation by migrants and host societies involving understanding and respecting cultural values. It involves migrants entering the labour market, finding places to live and accessing services.

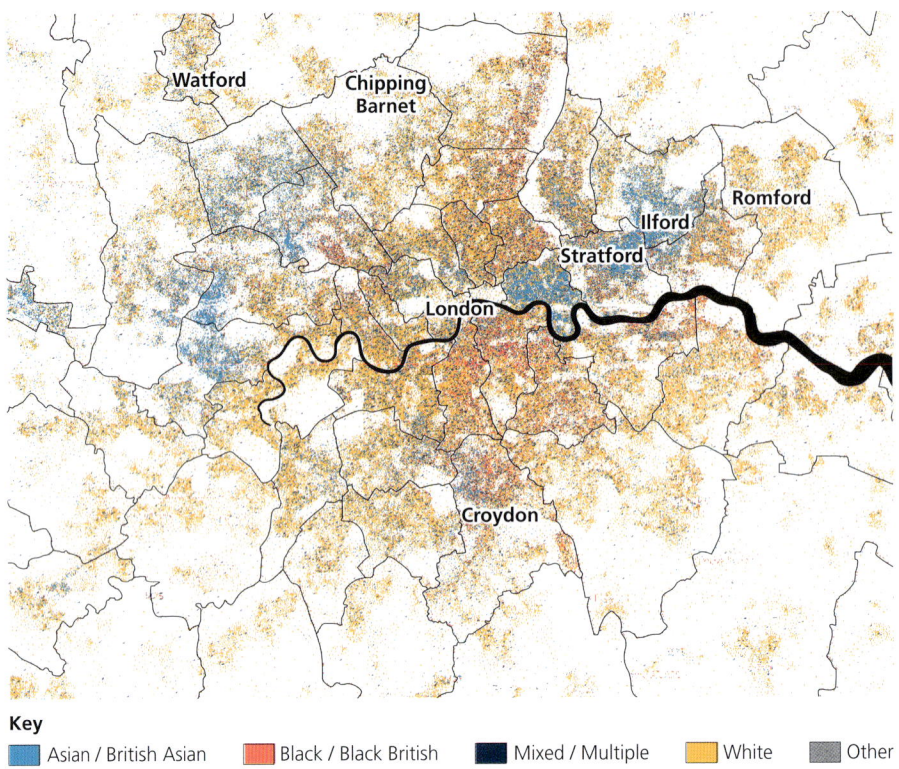

Key
■ Asian / British Asian   ■ Black / Black British   ■ Mixed / Multiple   ■ White   ■ Other

Figure 14.12 Spatial variation in ethnicity in Greater London (ONS, 2021)

In contrast, earlier Jewish migrants and more recently arrived Muslim and Sikh communities have sometimes chosen to hold onto their own culture and traditions. Places of worship continue to 'anchor' some diaspora groups to particular places (Figure 14.13). The persistence of dietary differences is important too: in north London Jewish communities cluster around kosher meat providers. Similarly, British Asians in Tooting and Woodbridge may want to live in close proximity to halal butchers.

Figure 14.13 A view of Sheffield city skyline featuring the blue dome of Madina Masjid (mosque)

History shows that levels of diversity often lessen over very long time periods, however. Almost 80 per cent of people in England today identify themselves as being 'white English'. Yet this community was originally far from culturally homogenous. In the past, varied ethnic communities of Viking, Anglo-Saxon, Celtic, Roman and Norman descent lived in different parts of England. Black people have lived in Britain since Roman times, and during the reign of Henry VII, there were hundreds of black migrants living in England. Over time, these diverse migrant groups combined in a cultural 'melting pot' that gave rise to the English language as it is spoken today. Aspects of the same process are being repeated today.

There was a rise in the growth of 'dual identity' or 'mixed heritage' citizens in the 2021 census (Table 14.3).

Table 14.3 Change in population in England and Wales, 2001 and 2021

Minority ethnic groups in England and Wales	Percentage of total	
	2001 census	2021 census
Black or Black British	2.2	4.0
Asian or Asian British	4.8	9.3
Mixed or multiple	1.4	2.9

Differing perspectives and viewpoints in the migration debate

Migration sometimes creates political tensions due to differing perceptions of, and viewpoints on, the cultural changes it brings. The USA is an interesting and important example, given the number of migrants and their descendants who live there. Between 1900 and 1920 alone, 24 million new arrivals were registered thanks to the 'open door' attitudes and policies of that era. Migration restrictions have subsequently been introduced and the coveted US Green Card has become harder to gain. Nevertheless, around 40 million people live in the USA who were not born there.

In the UK, public opinion about migration is similarly divided. Figure 14.14 shows some newspaper headlines and the facts used to either support or criticise migration.

Newspaper headlines	Where do the newspapers find their facts?
How immigrants have cost Britain £140 billion since 1995 *Daily Mail*, 12 March 2014	Citing an analysis by University College London (UCL) covering the period 1995 to 2011.
EU migrants pay £20 billion more in tax than they receive *Financial Times*, 5 November 2014	Citing another study by UCL which showed a positive net financial contribution to the UK of £20 billion over the decade 2001 to 2011 made by EU migrants.
Immigration from outside Europe cost £120 billion *Daily Telegraph*, 5 November 2014	Citing a different fact from the same UCL report covered by the *Financial Times*, which showed that immigrants who came to Britain from *outside* Europe cost the government £120 billion in NHS, education and welfare costs between 1995 and 2011.

Figure 14.14 Three competing views about migration in UK newspaper headlines which make use of data from University College London in varying ways

Skills focus: Interpreting a range of opinions

Figure 14.14 shows qualitative data in the form of newspaper headlines. What does the language and tone of these headlines suggest about the differing views or possible bias of different publications? How does the focus of the reporting vary in terms of the specific migrant groups reported on, or the timescale?

Migration across the Mexico–US border

For the USA today, the issue of illegal migration across the Mexican border is a major policy issue that divides the public and politicians alike. In 2014 President Obama called for work permits to be issued to many of the estimated 8 million illegal workers in the USA. In contrast, in 2018 President Trump began the construction of a border wall along the USA–Mexican border in order to prevent illegal immigration from Mexico and Central America.

The spatial distribution of unauthorised workers in the USA is highly uneven (Figure 14.15). One reason why US citizens often have different views on migration is clearly because some experience its effects more than others. The issues that commonly divide US public opinion include:

- **Economic impacts:** One view is that migrants are a vital part of the US economy's growth engine. From New York restaurant kitchens to California's vineyards, migrants work long hours for low pay. However, high unemployment in some cities has led to calls for American jobs to be given to American citizens instead.
- **National security:** The terrorist attacks on the USA in 2001 ushered in an era of heightened security concerns. Support grew for the anti-immigration 'Tea Party' movement. In 2016, the Republican presidential contender Donald Trump suggested that Muslims should be banned from entering the USA on the basis that the global terror group IS (also known as ISIS, ISIL and Daesh) pledge allegiance to Islam. Many people found this deeply offensive. Trump insisted he was simply thinking of ways to safeguard national security.
- **Demographic impacts:** In the USA and other developed countries, youthful migration helps offset the costs of an ageing population. However, the higher birth rates of some immigrant communities is changing the ethnic population composition of the USA. In 1950, 3 million US citizens were Hispanic; today, the figure has reached 60 million. This is around one-fifth of the population.

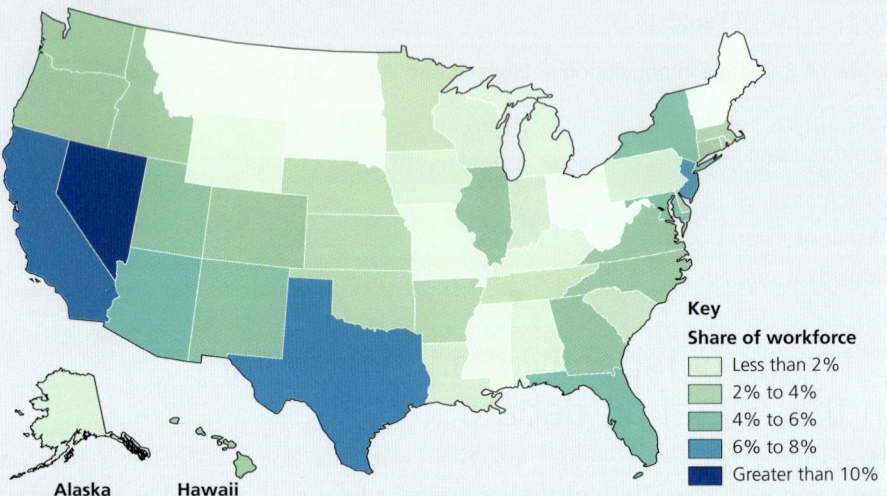

Figure 14.15 Unauthorised migrants in the USA as a share of the workforce in US states, 2017

- **Cultural change:** Migrants change places when they influence food, music and language. Hispanic population growth is affecting the content of US media as programmers and advertisers seek out a larger share of the audience by offering Spanish-language soap operas on services such as Netflix.

In 2016, the US Department of Homeland Security strengthened its national operation to deport unauthorised Central American families. The number of families crossing the border rose steeply in 2015 as a result of drought and gang-related violence in El Salvador and Honduras. There were about 11.3 million unauthorised immigrants in the USA in 2017, down from a peak of 12.2 million in 2007. Over half are from Mexico, with another 15 per cent from the rest of Latin America.

Complex patterns of global migration

Migration is a political and personal process. It is political because government rules dictate whether or not movement can take place freely. It is personal because each person's decision to migrate is influenced by many considerations and is often difficult to make.

Migration as a political process

Both in this book and in Book 1, we have encountered many examples of how states attempt to control flows of migration and the extent to which they succeed. The paradox of globalisation is that, while it promotes a global way of living, it simultaneously causes concern that local identities may be eroded and lost. Globalisation commentators increasingly talk of 'de-globalisation' or the new process of 'tribalisation' wherein people seek to defend local cultural identity from change. Barriers to migration are rising in many places, not falling (Figure 14.16).

Migration as a personal process

The world's poorest people are some of the least likely to become economic migrants. Isolated African subsistence farmers and rainforest tribes are largely 'switched off' to globalisation. Many live in hard-to-reach areas with poor transport links and no mobile phone access. The personal decision to migrate depends on a person's perception of better opportunities elsewhere as much as it does on the actual hardships of everyday life. This may be why some of the world's poorest African countries gain little income from remittances (Table 14.4, page 276): relatively few people have emigrated from these countries yet, compared with emerging economies such as Mexico and other countries in Latin America and the Caribbean with easier access to developed nations close by.

The personal decision to migrate is also influenced by the perception of how difficult the experience of making a move will be. Psychologists tell us that moving house is one of the most stressful events that a person experiences in his or her lifetime. For this reason, strong economic push and pull factors do not always trigger migration. For personal reasons, the vast majority of people alive today have chosen *not* to migrate away from home, despite the economic opportunities available in other places.

Figure 14.16 New barriers to migration introduced during 2015

Table 14.4 Remittance data for selected nations, 2018

Migrant remittance inflows as a share of GDP in 2018 (%)			
Latin America & Caribbean		Africa	
Haiti	32.5	Rwanda	2.7
El Salvador	20.7	Malawi	2.6
Honduras	19.9	Djibouti	2
Jamaica	15.9	Mozambique	2
Guatemala	12.1	Burundi	1.6
Nicaragua	11.5	Sierra Leone	1.5
Dominica	8.9	Benin	1.4
Dominican Republic	7.9	Sudan	1.1
Belize	4.9	Algeria	1
Mexico	2.9	Cameroon	0.8

Data: World Bank

Review questions

1 Outline the role of globalisation in increasing people migration both internally and across international borders.

2 Using Figure 14.6 (page 266), outline the major global migration flows and try to explain the reasons for the migrations.

3 Explain what the Schengen Agreement is and how it has both costs and benefits to countries that are part of the Schengen area.

4 Using examples, describe strategies and polices that different countries have adopted to help manage inflows of international migration. To what extent have these strategies succeeded?

5 To what extent has Poland benefited from its high net emigration rate?

6 Using examples, explain why immigration has become an increasingly political issue in the last two decades.

Further research

Find out more about the ways in which climate change could affect migration flows: http://time.com/4024210/climate-change-migrants

Explore how the EU's Schengen Agreement works: www.bbc.co.uk/news/world-europe-13194723

Take a look at the website of the US Customs and Border Protection Agency, part of the Department of Homeland Security: https://www.cbp.gov/

Read this summary of Europe's migration crisis in 2015: https://www.unhcr.org/uk/news/stories/2015-year-europes-refugee-crisis

15 Nation states in a globalised world

How are nation states defined and how have they evolved in a globalising world?
By the end of this chapter you should be able to:
- explain how demographic and political processes have shaped the populations and borders of states
- assess the role of nineteenth-century nationalism and colonialism in the development of the modern world
- understand why some states have adopted particular economic strategies in response to globalisation.

15.1 States and the processes that shape them

A state is a territory over which no other country holds power or **sovereignty**. There were up to 206 sovereign states in 2020, depending on which exact definition is used.

- The term 'nation' refers to a territorialised group of people who may or may not have sovereignty. This includes the Scottish and Welsh nations that are part of the UK, which is a sovereign state. Although the Scottish Parliament and Welsh Assembly have autonomy over some areas of governance, including education and welfare spending, they lack full control (Scotland cannot declare war on another state, for instance).
- The situation is complicated further by the existence of dependent territories, or dependencies, including Greenland (which belongs to Denmark), Hong Kong (which belongs to China) and Jersey (a Crown dependency of the UK). These also have varying degrees of autonomy for many aspects of governance but lack full sovereignty.

> **Key term**
>
> **Sovereignty:** The ability of a place and its people to self-govern without any outside interference.

Nation states and cultures

The majority of readers of this book will be citizens of 'The United Kingdom of Great Britain and Northern Ireland'. UK citizens do not share a single cultural identity, however. Scottish, Welsh and Northern Irish people have cultural identities that differ clearly from the English. The cultural traits that distinguish and define these different identities encompass language or dialect and the support of national sports teams, music and literature. More complex yet, the UK is a multi-faith society following centuries of migration. Asian British and Black British identity adds further complexity to the cultural mix.

The situation in the USA is even more complicated. Prior to the arrival of Europeans, the continent of North America was home to a heterogeneous mix of indigenous peoples, including the Sioux and Navajo tribes. Today, 328 million people occupy the same territory; the majority are the descendants of a global mix that includes Italians, Greeks, Scandinavians, Scots, Irish, Mexicans, Cubans, Africans, Indians,

Pakistanis, Vietnamese, Puerto Ricans, Koreans and many more besides. Yet from this mix an 'American culture' has developed over time due to a 'melting pot' effect (Figure 15.1). American culture is both inclusive (allowing new arrivals to participate) and dynamic (it becomes modified in turn by each wave of new arrivals).

Figure 15.1 US school children share an identity but come from diverse cultural backgrounds

Some of the world's more recently established post-colonial states are also home to a broad mix of cultures. However, not enough time has passed yet for them to develop a shared cultural identity as a national people.

Other states have maintained a relatively **homogenous culture** over time. Their citizens mostly belong to a single ethnic group and share the same cultural traits (Figure 15.2). Reasons for this include:

- physical isolation (e.g. Iceland's North Atlantic location was a barrier to migration for many centuries)
- political isolation (e.g. North Korea's government limits its citizens' interactions with the outside world).

> **Key term**
>
> **Homogeneous culture:**
> A society where there is very little cultural or ethnic diversity and most people share cultural traits with one another, including language, religion, dress and diet.

Establishing national borders

Prior to today's 'shrinking world' era, mountain ranges and rivers sometimes formed natural barriers to population movements and provided the basic geometry for nations to develop in particular places. Over time, long-settled ethnic groups formed a strong association with their land. In Europe, for instance, borders formed organically over centuries or millennia. Today's European geopolitical map corresponds broadly with its cultural and linguistic map.

In Africa, the situation is very different. Political boundaries correspond poorly with the distribution of different cultural and ethnic groups. This is a legacy of the partition of Africa by European nations in the eighteenth and nineteenth centuries. Space was divided between competing powers. For instance, the boundary between Egypt and Sudan is a straight line drawn by Great Britain in 1899. It is part of the 22nd parallel north circle of latitude.

> 'Spain's borders generally enclose the Spanish-speakers of Europe; Slovenia and Croatia roughly encompass ethnic Slovenes and Croats. Thailand is exactly what its name suggests. Africa is different, its countries largely defined not by its peoples' heritage but by the follies of European colonialism.'
>
> *Max Fisher, The Atlantic*

Cultural unity inside Iceland compared to Singapore

Until the mid-1900s, the physical isolation of Iceland ensured that its population experienced a strong sense of common identity and cultural homogeneity. Anthropologists point to the tough life early Icelanders experienced in a volcanic and glacial island sited atop the mid-Atlantic Ridge in the North Atlantic Ocean. Many cultural traditions born of hardship in earlier times have survived to the present day. Persisting rituals and traditions help to foster community cohesion. For instance, Icelandic people like to share 'rotten shark' at parties. The meat of the Greenland shark is naturally poisonous because the shark contains fluids that allow it to live in cold water without freezing. To make the meat safe, centuries ago Icelanders used to bury chunks of the meat for months at a time before eating. During this time, the fluids drain from the shark making it safe. Although modern Icelanders no longer need to rely on it for survival, rotten shark remains a popular festival food and valued tradition.

In 2008, McDonald's closed its restaurants in Iceland: its high prices had deterred customers. The empty premises were taken over by a locally owned company selling exclusively Icelandic food and ingredients instead. This could be seen as a sign of global culture in retreat. However, there are signs that Iceland's culture is changing. Many young Icelanders are avid consumers of global culture. Ninety-nine per cent of Icelanders are connected to the internet, among the highest rates in the world. Also, the island of 365,000 people is swelled by an average of 2 million tourists each year, who flock to Iceland to see its geography. The ratio of tourists to inhabitants is among the highest in the world and may lead to more visitors wanting to settle there permanently, leading to greater cultural diversity.

Figure 15.2 Selfoss is a large residential area in South Iceland which lacks cultural diversity

In complete contrast, Singapore is a cultural melting pot of Malay, Chinese, Indian and European influences, all of which have intermingled (producing, for instance, a variant form of English called 'Singlish'). What explains the coexistence of Asian and Europeans with so many different origins – including Buddhists, Taoists, Christians, Muslims, Hindus and Sikhs – in Singapore? From 1926 to 1946 it played a major strategic role as a military and trading hub under British administration. The political decision to make it a free port (where no taxes were collected) encouraged migration from China, the Indian sub-continent, Indonesia, the Malay Peninsula and the Middle East.

After independence, Singapore became a fast-growing 'Asian Tiger' economy and today has a per capita income GDP of US$65,000 (2019). As a result of this prosperity, it remains a magnet for new waves of migrants who bring their customs, religions and festivals with them (see also Table 14.2, page 266).

The colonial borders drawn by European colonists took no account of the people actually living there (Figure 15.3). The colonial powers were more concerned with dividing up Africa's raw materials and water resources among themselves. It is hard to imagine a worse approach to state building.

- By 1900, many African ethnic groups found themselves living in newly formed nations that in no way represented their own heritage.
- Some long-established ethnic regions were split into two or more parts, with each becoming part of a different newly established state territory.

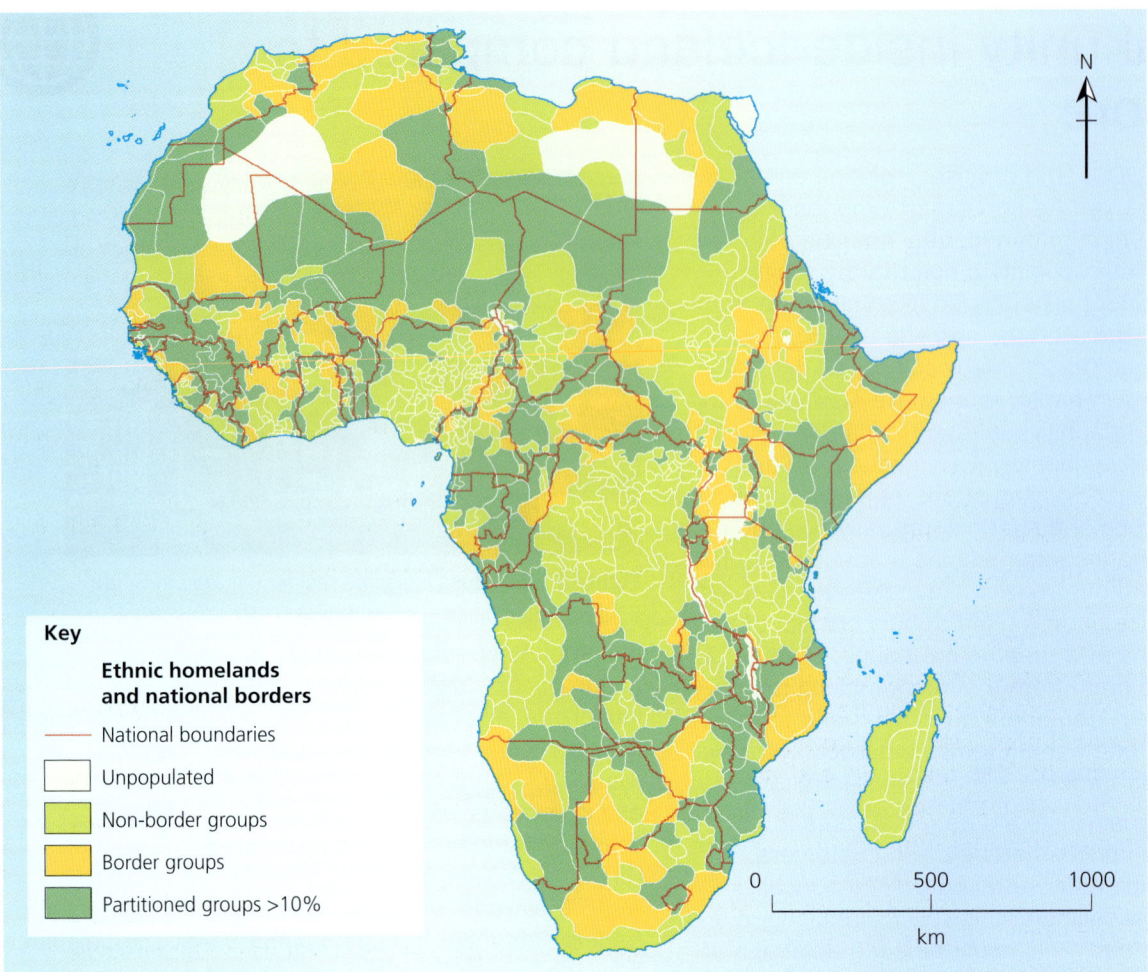

Figure 15.3 An ethno-linguistic map of Africa with modern nation states superimposed

Contested borders and states

It is not only in Africa where contested borders have brought conflict or where geopolitical strife is perpetuated by a boundary that dissects an ethnic group's homeland.

Past wars in Europe were waged over resource-rich lands. Possession of the mineral-rich regions of Alsace and Lorraine passed from France to Germany and back again between 1871 and 1919. Other examples are shown in the place contexts on Ukraine and Russia (page 281), and Iraq and Syria (see page 282).

The issue of non-recognition

The UN recognises over 190 sovereign states, but there are some states which are not regarded as sovereign entities. This includes Kosovo and Taiwan. Kosovo broke away from Serbia in 2008 and although it is recognised as a sovereign state by most EU countries, Serbia and Russia do not accept this. Neither do most ethnic Serbs inside Kosovo.

Ethnic conflict and contested borders in central Africa

European powers began to colonise Africa in the 1700s. They aimed to create a system of raw material extraction for export and made little productive investment in African countries. The legacy of colonialism is a host of unstable states that often lack cultural coherence. The enormous territory of Democratic Republic of the Congo (DRC) is home to 240 ethnic groups who jointly came under Belgian rule in the late 1800s and finally gained independence in 1960. This cultural diversity has posed a huge challenge to post-colonial unity and has been a major factor contributing to conflict in DRC.

Enduring problems also arise from the way that boundaries between DRC, Uganda and Rwanda were established by Belgium, Great Britain and Germany. The geographical regions traditionally occupied by ethnic Tutsi and Hutu people became fragmented. The resulting 'transnational' identity of both groups is a cause of ongoing political instability and violent territorial skirmishes in Central Africa.

- Conflict in Rwanda between Tutsi and Hutu people in the 1990s (see page 223) spread quickly into neighbouring Uganda and DRC. Between 1998 and 2008, over 5 million people died in the so-called Africa World War that ensued.
- During this time, armies and militia groups from DRC's nine neighbour states repeatedly entered DRC on the grounds that the ethnic groups with whom they claimed kinship required support (Figure 15.4).

Figure 15.4 DRC shares a border with nine neighbour states, many of which have invaded

Contested borders in Ukraine and Russia

The boundaries of Russia have changed several times, most recently when Russia annexed part of Ukraine in 2014 (Figure 15.5). The rationale provided by Russia's President Putin was that many ethnic Russians live in Ukraine. From the late 1700s onwards, Crimea belonged to Russia (or the Soviet Union). Its eventual inclusion as part of the independent state of Ukraine in 1991 was controversial due to the large number of ethnic Russians still living there. In 2014, a brief period of civil conflict in Crimea ended with the territory being annexed by Russia. The international community condemned this but no actual steps were taken to prevent it. Emboldened by this lack of international response, Vladimir Putin's Russia launched a full-scale invasion of Ukraine in 2022.

Figure 15.5 Crimea, Ukraine and Russia

15 Nation states in a globalised world 281

Ethnic conflict and contested borders in Iraq and Syria

The Middle East provides another instance of an unhappy fit between state borders and a region's ethnic map. One root cause of many ongoing conflicts in Iraq and Syria is the Sykes–Picot line. This boundary was drawn by Great Britain and France in 1916 in order to define their own spheres of influence in the Middle East. As a result, a large Kurdish population of 25 million is now distributed between four states: Türkiye, Iran, Iraq and Syria. Large Sunni and Shia Muslim populations were also divided by the Sykes–Picot Agreement (Figure 15.6).

For much of the twenty-first century, the entire region has been mired in conflict linked with the inherent instability of its states. The terrorist organisations Daesh (IS) and al-Qaeda have shown little respect for either territorial borders or human rights. Daesh took advantage of the power vacuum in Iraq created by the withdrawal of Allied troops from that country, and the civil war in Syria. Quite by surprise, and almost overnight, it grabbed some corridors of territory in both countries and put together what it declared as a caliphate in 2014. From there, Daesh waged its so-called jihad against all other religions. Its soldiers have pursued a strategy of annihilating minority communities including Christian Assyrians, Kurds, Shabaks, Turkmens and Yazidis. In Syria, the

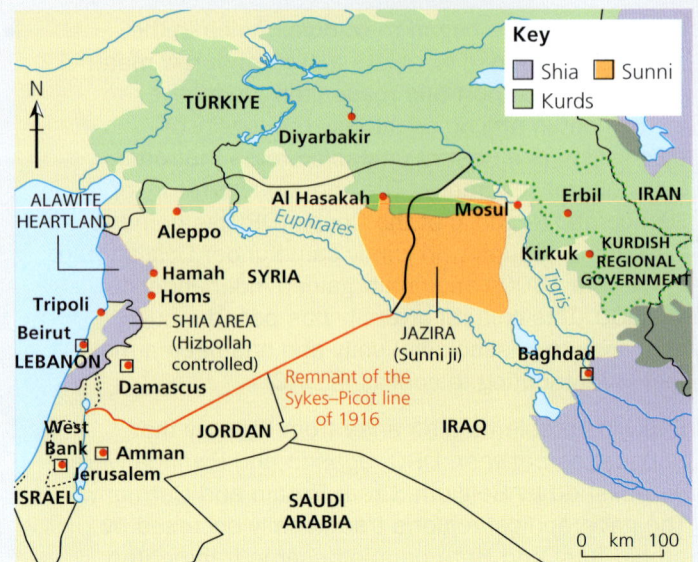

Figure 15.6 The modern-day Middle East and the historical Sykes–Picot line

victims have been Ismailis and Alawis. Its activities are tantamount to war crimes and genocide.

Since 2014, an international coalition including the USA, UK, Saudi Arabia, Türkiye and many Kurdish fighters have gradually pushed back against the Islamic State caliphate. By 2019, it was reduced to a few pockets of ISIS fighters in desert areas, but at a cost of at least 200,000 deaths on all sides.

Non-recognition of Taiwan

China has claimed sovereignty over Taiwan since 1949. The issue of Taiwanese sovereignty has strained relations between China and other global superpowers, including the USA and the EU. The dispute dates back to 1949 when, following the Communist victory in mainland China, two million Chinese Nationalists fled to Taiwan and established a government of their own on the island. Mainland China maintains that Taiwan is part of China. However, Taiwan claims independence as a distinct state and wants to become a member of the United Nations.

When Taiwan first broke away, the UN decided that it could not recognise both Taiwan and the People's Republic of China. Since 1971, it has recognised only the latter. The Taiwanese government hopes in time to be recognised too. Relations between China and Taiwan remain complex. For instance, Taiwanese firms like Foxconn are major investors in China – thereby making both countries interdependent – despite the Chinese government's insistence that Taiwan should not remain independent.

Figure 15.7 The contested state of Taiwan

Option 8B: Migration, Identity and Sovereignty

15.2 Nationalism, colonialism and the modern world

Many people alive today view themselves as being internationally minded, or global, in their outlook. In contrast, others continue to view **nationalism** as a better philosophy. In the past, the nationalistic outlook of European powers was demonstrated by the colonisation and subjugation of the peoples of Latin America, Asia and Africa. Today, powerful states still often pursue nationalistic goals. In contrast with the direct colonial actions of the past, geopolitical influence is now exerted in more indirect ways. This is done by adopting **neo-colonial** strategies when dealing with less powerful states.

Nineteenth-century nationalism and empire building

Between approximately 1500 and 1900, leading European colonial powers built global empires. The invasion and colonisation of South America by Portugal and Spain was well underway by the 1600s. In the 1700s and 1800s, the pattern was repeated in Asia and Africa by Great Britain, France, Belgium and Holland. Traditional territorial names such as 'Kongo' and 'Akan' were overwritten on the map of Africa, with names such as 'Belgian Congo' and 'Gold Coast' appearing in their place. In a remarkable act of hubris, modern-day Zimbabwe gained the name Rhodesia in the late 1800s in honour of a British colonist called Cecil Rhodes.

The end of empire and new nation states

Latin American countries including Brazil and Mexico gained their independence from Portugal and Spain in the 1800s. Independence for Asian and African countries

> **Key terms**
>
> **Nationalism:** The belief held by people belonging to a particular nation that their own interests are much more important than those of people belonging to other nations.
>
> **Neo-colonial:** The indirect actions by which developed countries exercise a degree of control over the development of their former colonies. This can be achieved through varied means including conditions attached to aid and loans, cultural influence and military or economic support (either overt or covert) for particular political groups or movements within a developing country.

The British Empire

As more and more territories came under their direct rule or influence, rival European states each built a global empire. Greatest in extent was the British Empire. By 1880, it controlled over one-third of the world's land surface and one-quarter of its people. The Empire was a vehicle for the imposition of the English language and British laws, customs, arts and sports on a global scale (Figure 15.8). Between 1853 and 1920, some 9.7 million people migrated from the British Isles.

The British Empire was maintained by communications technology such as the telegraph and military power. The Royal Navy allowed Britain to dominate world trade routes from 1700 until the 1930s and to rapidly reinforce military power in its colonies. Most colonies were allowed no political power or local autonomy. In the 21st century ex-colonies have complex relationships with the UK. Many became republics and rejected post-colonial ties with the UK (Egypt, Sudan). Over 50 are members of the Commonwealth of Nations (a voluntary association) and have shared cultural (sport, language, legal and political systems) and population ties. However, these ties have weakened over time. For more on the British Empire see Chapter 7, page 151.

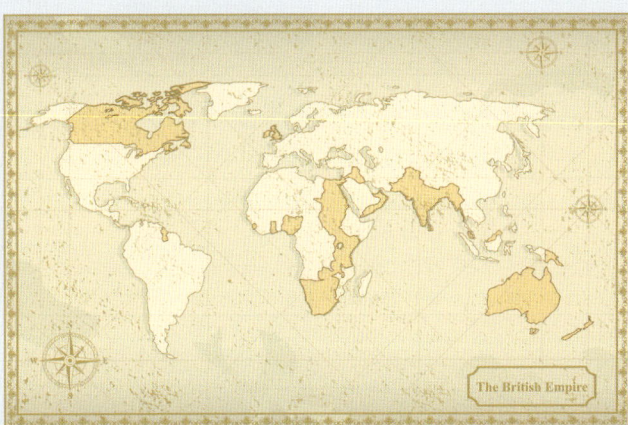

Figure 15.8 In 1880, the British Empire controlled a third of the world's land surface

arrived more recently. Between 1945 and 1970, most colonised nations finally gained their freedom and became independent sovereign states (Figure 15.9).

Why did the era of European colonial empires end? Reasons included:
- the high cost of waging two world wars (the UK was left effectively bankrupt)
- growing resistance to foreign rule (Gandhi campaigned for, and won, independence for India in 1947; the Kenyan Mau Mau rebellion signalled the start of what would be called a 'wind of change' sweeping across Africa by the 1960s)
- growing concern for the injustice of militarised colonial control among young European citizens (protests and demonstrations against colonial rule were growing at home too)
- national liberation wars to end colonisation occurred in many countries, including Indonesia (Netherlands), Algeria (France) and Mozambique (Portugal)
- Europe's shift towards post-industrial economic activity (European countries were becoming less dependent on raw materials from their colonies).

The data in Figure 15.9 suggest that the speed with which European empires were dismantled greatly exceeded the time spent building them. The rapidity of de-colonisation often left a 'power vacuum' and a troubled transition towards independence that did not bring economic stability and development. In many countries, power was seized by the army (DRC, Uganda, Nigeria, Indonesia) or

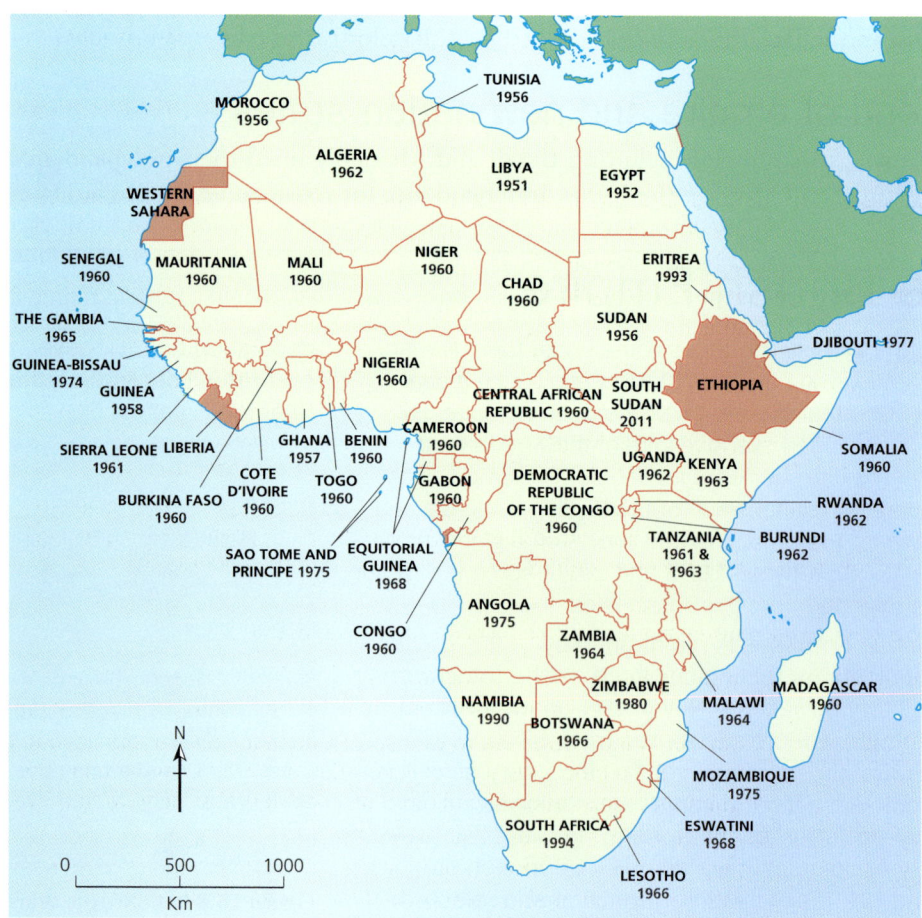

Figure 15.9 Dates of independence of former colonies in Africa

Option 8B: Migration, Identity and Sovereignty

The 'wind of change' for Democratic Republic of the Congo (DRC)

Table 15.1 shows a timeline of changes affecting DRC before and after its independence in 1960 (Figure 15.10). Colonial Belgian Congo was particularly brutal. King Leopold II was known as 'the butcher of the Congo' and his personal rule over the Congo Free State, 1885–1908, is estimated to have led to 1.5–13 million deaths. DRC has suffered persistently from a 'resource curse'. Its enormous raw material wealth attracts numerous outsiders who eventually find local collaborators to help them loot the country's natural resources. Table 15.2 shows the economic, environmental and human costs of conflict for DRC during the post-colonial period.

Figure 15.10 Independence in Congo, 1 July 1960

Table 15.1 Factors affecting the evolution of DRC as a state

Geopolitical changes	Factors influencing change
Colonised by Belgium, 1870–1960 Millions of Congolese were killed or worked to death by King Leopold II of Belgium.	**Raw materials** drew Europeans to the region. Timber, rubber and gold made the country an important prize.
Independence as Zaire, 1960–90s After a power struggle, Joseph Mobutu took power and renamed the country. He created a difficult regime for TNCs to operate in, while amassing a US$4 billion fortune for himself. Zaire eventually defaulted on loans, resulting in a cancellation of development programmes.	**Geopolitical strategies** followed by Belgium and the USA during the Cold War helped put Mobutu in charge (he was favoured because of his generally pro-Western stance). However, aid and loans designed to help alleviate poverty in DRC were siphoned off by Mobutu's family while leaving ordinary people in great debt.
Regime change and conflict, 1990s–2005 Mobutu was removed from power when neighbouring states Uganda and Rwanda assisted a rebel leader, Laurent Kabila, to become president in 1997 (he renamed Zaire as DRC). Later, they turned on him only to discover that other neighbours (Zimbabwe, Angola and Namibia) had sent troops to 'help' Kabila. The six-nation war claimed 5 million lives.	**Cross-border ethnic ties** between some Rwandans and some Congolese is a legacy of colonialism and an important reason why the conflict came to involve more than one nation. Raw materials once again proved to be a source of prolonged conflict. The UN believes many occupying forces were motivated in part by a desire to grab DRC's resources.
Attempted conflict resolution, 2005–present UN peacekeepers are trying to bring stability. The World Bank has approved US$8 billion in debt relief.	**Displaced refugees** and traumatised civilians have yet to be rehabilitated. Armed militia groups still operate in the east of the country.

Table 15.2 Economic, environmental and human costs of conflict for DRC

Economic costs	In 2019, DRC's GDP per capita was just US$500 per year despite its rich resources. Political mismanagement and conflict squandered the country's early development opportunities.
Environmental costs	Conflict in the 1990s led to widespread abandonment of farmland and the re-growth of secondary forest. Simultaneously, loss of vegetation occurred in and around the extensive refugee camps where poor sanitation has allowed diseases such as cholera to thrive.
Human costs	DRC has one of the world's lowest HDI scores (179th place). Life expectancy is just 60. The majority of deaths among younger people are due to infectious diseases and malnutrition linked with the loss of health services and food security due to conflict.

15 Nation states in a globalised world

by ethnic groups who had worked alongside the colonial powers, such as the Tutsi people in Rwanda. This sometimes bred resentment among non-represented social groups, sowing the seeds of future conflict.

Lack of expertise also hindered development. One estimate suggests there were only 16 university graduates left in DRC after the Belgians left in 1960, due to a lack of provision of secondary and higher education for indigenous people under colonial rule. This created a damaging skills shortage. Interference by the USA and USSR sometimes exacerbated post-colonial instability further. In the 1950s and 1960s, proxy wars between the two great superpowers brought enormous loss of life in Korea, Viet Nam and Indonesia.

Post-colonial Viet Nam

Viet Nam is the easternmost country of the Indo-China peninsula in Southeast Asia. In the north, Viet Nam is bordered by China (Figure 15.11). It is this particular aspect of its geopolitical situation that accounts for most of what has happened to Viet Nam in its post-colonial period.

During the two decades up to 1887, Viet Nam was gradually annexed by France as one of its colonies. It became known as French Indochina and for the next 50 or so years it flourished. During the Second World War, however, the colony was invaded and taken over by Japanese forces. With the defeat of Japan in 1945, a communist and nationalist liberation movement, led by Ho Chi Minh, took over the northern city of Hanoi and proclaimed a provisional government independent of French rule. The French started a campaign to reclaim the lost territory, but were defeated in 1954. The UN intervened and French Indochina was dissolved. Viet Nam was split at the 17th parallel north into communist North Viet Nam and capitalist South Viet Nam (now relieved from French rule). However, this was just the beginning. The communists, with Soviet and later Chinese military backing, started a guerrilla campaign (involving the Viet Kong) to infiltrate and capture South Viet Nam. In 1964, and with South Viet Nam on the brink of being overrun, the USA was persuaded to intervene.

So the Viet Nam War continued, but now clearly against the backdrop of the Cold War between the two superpowers of the USA and the Soviet Union.

After nearly twenty years of fighting and eight years of US intervention, growing opposition to the war in the USA eventually led to US troops being withdrawn in 1973. Two years later, communist forces seized control of Saigon, the former colonial capital city in the south and renamed it Ho Chi Minh City. As

Figure 15.11 Map of Viet Nam

a consequence, the country became re-unified in 1976, but this time as the Socialist Republic of Viet Nam.

The damage done by this prolonged struggle was immense. More than 3 million people, including 58,000 US military personnel, were killed and 150,000 wounded. More than half of these deaths were innocent Vietnamese civilians. The agrarian economy was so disrupted that starvation became widespread.

Strategic bombing in the north destroyed Viet Nam's modest industrial infrastructure.

Not only were the human costs of the war immense, so too was the damage done to the environment. Heavy artillery and bombing left 26 million craters. Agent Orange, a herbicide and defoliant, was sprayed over much of South Viet Nam in an attempt to remove the cover being used by the Viet Kong (Figure 15.12). It was applied at such great strengths that it resulted in an estimated 400,000 deaths and disabilities; 500,000 children were born with defects. Napalm, a flammable liquid, was widely used to burn off vegetation and to flush out the Viet Kong from their bunkers and tunnels.

The economic cost of the war to the USA has been put at US$950 billion (at today's prices). The USA also paid a high political cost. Its failure to push back the infiltration of the south by the Viet Kong brought into question its superiority as the world defender of freedom and justice. The USA appeared unable to push back the spread of communism.

Viet Nam recovered slowly and painfully from war, but since 2005 its economy has boomed; incomes have risen to about US$8000 per person. It has good international relations with the USA, and its government, while still a one-party socialist state, has embraced capitalist economics and globalisation. Many TNCs have invested in Viet Nam as a lower-cost alternative to China and international tourist arrival increased from 2 million in 2000 to 18 million by 2019.

Viet Nam survives as a sort of geopolitical cushion between the USA and China, which has now taken over from the Soviet Union as the superpower in

Figure 15.12 From 1962 until 1971 US military sprayed defoliant over Viet Nam's forests to deny communist forces the use of jungle cover and food sources.

this global region. From a Chinese perspective, the border with Viet Nam is a vulnerable one in that it does not coincide with any sort of physical obstacle. This is in stark contrast to its borders with Laos and Myanmar, both of which involve mountainous jungle terrain. But Beijing is unlikely to see Viet Nam as a threat, if only because, having become unstuck once, the USA is unlikely to be drawn into another confrontation here. So a sort of superpower standoff prevails.

Post-colonial conflict in Sudan and South Sudan

Ever since becoming an independent state in 1955, Sudan has lacked internal cohesion, partly due to its large size. An ethnically and culturally diverse nation, Sudan suffered a strong sense of internal separation between its northern and southern regions from the outset. This was due to colonial intervention. Sudan was divided into two based on ethnic characteristics, an example of the 'divide and rule' policy of the British.

By 2015, almost 2 million lives had been lost to conflict (Table 15.3), some of the worst violence happening in the semi-arid Darfur region. On the fringe of the Sahara desert, Darfur is home to many different Black African and Arab groups, creating a potential for ethnic conflict. Following Sudan's independence in the 1950s, members of ethnic Arab tribes came to dominate the national government based in Khartoum. As a result, the Black African groups living in Darfur (settled farmers of the river valleys, such as the Zaghawa people) began to feel politically and economically marginalised. Some formed the Sudan Liberation Army (SLA) and began attacks on government targets. The SLA claimed that the Khartoum government was oppressing Black African farmers in Darfur while simultaneously providing support for the local Arab cattle-herding people.

Shortly afterwards, Arab militia groups known as Janjaweed began to attack Black villages in Darfur. The Janjaweed were widely believed to be supported by the Khartoum government. Huge displacements of people resulted, with hardly a village in Darfur left intact. Following intervention

Table 15.3 Sudan's post-independence timeline

1956	Sudan revolts and gains independence from Britain and Egypt. Its government is based in Khartoum
1962	Civil war breaks out in the Christian south of the country
1978	Oil is discovered in the south of Sudan
1983	President Nimeri introduces Islamic law across Sudan. Civil war begins between the government and the southern-based rebels
2003	Rebels in the northwestern region of Darfur rise up, claiming that they have been neglected by the government
2004	The UN reports that the Janjaweed are systematically committing genocide in Darfur
2008	The UN estimates 300,000 deaths in Darfur since 2003
2011	People in the south of Sudan vote for independence
2013–20	Civil War, ended by a peace deal in 2020 but at the cost of 380,000 lives and 2.5 million internally and externally displaced people

by the African Union and the United Nations, a first peace agreement was reached in 2005. However, conflict in Darfur continued sporadically, resulting in the displacement of 2 million people from their homes.

In 2011, Sudan finally abandoned the struggle for unity. It divided in half when the new state of South Sudan broke away. To the north, the remainder of Sudan, shrunken in size, retained its name and capital city of Khartoum. However, a civil war subsequently broke out in South Sudan in 2013. Today, South Sudan has some of the worst health and development indicators in the world.

Post-colonial patterns of migration

Since the 1950s, patterns of migration between former colonies and the imperial core country have played an important role in changing the ethnic composition of the latter. Between the 1950s and 1970s, the UK received migrants from former colonies of the British Empire including Jamaica, India, Pakistan, Bangladesh and Uganda. Smaller numbers came from Nigeria, Kenya and other ex-British African territories (Figure 15.13). The movement of Australians to the UK is a type of post-colonial migration too. These movements increased British **cultural heterogeneity**, especially in London and other urban areas including Liverpool and Bristol.

The migrants came to fill specific gaps in the labour force that had opened up after the Second World War. Sometimes migrants were recruited directly (London Underground held interviews for bus drivers in Kingston, Jamaica). At the time

> **Key term**
>
> **Cultural heterogeneity:** A society where there is a high level of cultural and or ethnic diversity among its citizens, often resulting in a multi-lingual and multi-faith community.

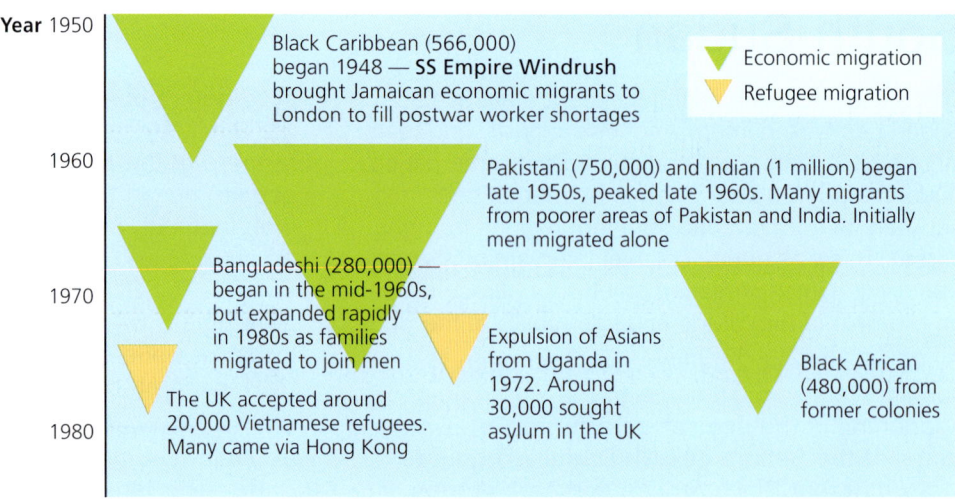

Figure 15.13 UK post-colonial migration, 1950–80 (Source: Data from UK census, 2001)

there was still a strong demand for workers in manufacturing industries, such as the textile mills of Lancashire and Yorkshire. The populations of ex-colonies spoke fluent English and had some knowledge of the British way of life. Many young Indians and Jamaicans were excited to move to the UK, following an education in schools where British history and culture were taught.

There were gaps in the skilled labour market too, notably within the ranks of the newly established National Health Service. UK doctors had not been trained in sufficient numbers during the 1930s and 1940s to fully staff hospitals and surgeries. Many doctors travelled to the UK from India, Pakistan and parts of Africa. Under colonial rule, medical schools in India used the same textbooks as British teaching hospitals. This meant that Indian doctors were sure to fit in well with the way medicine was practised in the UK.

Despite the invitation to come to the UK, newly arrived migrants experienced high levels of discrimination, prejudice and racism – as many second and third generation migrants still do today.

The same pattern was repeated in other European countries. France received migrants from ex-French territories such as Algeria and Morocco, for instance.

15.3 State strategies in response to economic globalisation

The world's capitalist states are constantly jostling with one another as they try to optimise their own position within global systems. Some states use low business tax rates as a strategy. By attracting large TNCs to relocate or do more business within their borders, the governments of low-tax states hope to prosper financially. Low government taxes nonetheless generate a considerable amount of revenue when levied on a very high overall volume of capital flows.

Low-tax states and TNCs

The developed countries where many TNCs are domiciled are major beneficiaries of globalisation on account of the corporation taxes they can raise. Apple, which is headquartered in California (Figure 15.14), paid US$15.8 billion to the US government in 2017, but despite these large tax bills many TNCs are frequently accused of using complex corporate structures to avoid paying tax.

TNCs may consider leaving their traditional homes if tax regimes in other states look more attractive. In recent years, some European-based TNCs have relocated their headquarters to Ireland, Switzerland, Luxemburg and the Netherlands where corporate taxes are low (in 2020, the UK rate was 19 per cent, whereas in Ireland it was 12.5 per cent. TNCs often move parts of their business abroad in an effort to reduce their tax bill in their home country.

Figure 15.14 Apple Headquarters, Apple Park, Cupertino, California

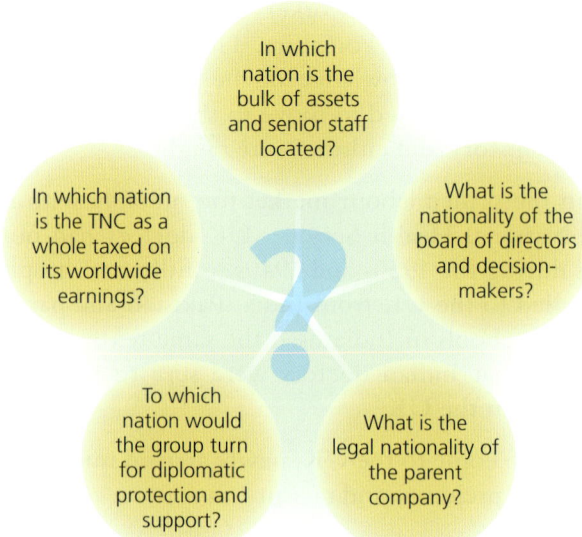

Figure 15.15 Investigating where a TNC is domiciled and the reasons for this choice

Key terms

Transfer pricing: A financial flow occurring when one division of a TNC based in one country charges a division of the same firm based in another country for the supply of a product or a service. It can lead to less corporation tax being paid.

Parent company: The original business that a global TNC has developed around and whose directors still make decisions that affect the organisation as a whole. Both Starbucks and Google are parent companies to global networks of subsidiary businesses, including Ritea Ltd (Starbucks Coffee Company, Ireland) and Google Ireland Ltd.

Tax haven: A country or territory with a nil or low rate of corporation tax, such as Bermuda.

Expatriate: Someone who has migrated to live in another state but remains a citizen of the state where they were born.

However, there are practical reasons why most TNCs choose *not* to relocate. These relate to brand authenticity, corporate responsibility, public perception and their current ability to declare profits in low-tax states (Figure 15.15). Security is another concern. TNCs sometimes look to the national government where they are headquartered for support during a financial crisis, or when their overseas assets become threatened by conflict or nationalisation.

- The oil company Repsol sought support from its country of origin, Spain, when Argentina's government seized control of its Argentinean investments.
- During the global financial crisis of 2008–9, General Motors and Chrysler looked to the US government for support, while the Royal Bank of Scotland was 'bailed out' by the UK Treasury (Figure 15.16).

Instead of moving home, many TNCs use the strategy of **transfer pricing** to reduce their tax burden. This involves routing profits through subsidiary (secondary) companies owned by the **parent company**. These subsidiaries will be based in a low-tax state such as Ireland or possibly an offshore **tax haven**. Around 40 so-called tax havens offer nil or nominal taxes. Some are sovereign states, such as Monaco. Another, the Cayman Islands, is an overseas territory of the UK that has its own tax-setting powers. It is not just companies that route money in this way. Some wealthy **expatriates** try to limit their personal tax liability by migrating to a tax haven.

The global governance of tax havens

A remarkably large number of tax havens have links to the UK. Many are what might be called 'not-quite sovereign states'. These include Crown Dependencies (Isle of Man, Jersey and Guernsey) and British Overseas Territories (Bermuda, Caymen Islands) as well as former colonies (Malta, Hong Kong). Often small islands with limited resources, many have used tax haven status as a way of earning income. Frequently their only alternative income is seasonal tourism. This situation raises governance questions for those territories, as well as for the UK government which is the sovereign entity in some cases.

Business taxes play a vital role in providing governments with much-needed money for health, education, welfare and defence spending. As a result, states suffer economically when TNCs try to minimise their taxes through transfer pricing. The world's richest nations have much to lose from corporate tax avoidance and, as a result, they have begun to collaborate on taking preventive measures. Several IGOs, including the OECD and G20 groups, want to see stricter regulations. Other players, including citizens and NGOs, have an important role in raising awareness of the social costs of corporate tax avoidance (Table 15.4).

Alternative economic models for nation-state building

The global economic system is far from perfect. Leaders of several South American, African and Asian nations have criticised repeatedly the ways in which some states gain disproportionately from global trade and use their power to reproduce this system to their own advantage. Bolivia's Evo Morales, Zimbabwe's Robert Mugabe and North Korea's Kim Jong-un have all spoken out against what they see as global capitalism's rampant inequality. There is certainly plenty of evidence to support their challenges:

- In 2017, Oxfam reported the eight richest billionaires (all men) had wealth equivalent to that of the poorest 50 per cent of the world's population.

Figure 15.16 The Royal Bank of Scotland Group (RBS) was 'bailed out' by the UK government during the global financial crisis of 2008–9. As of 2020 it was still 60 per cent owned by the government.

Table 15.4 Evaluating the views and actions of different players in relation to tax avoidance by TNCs

	Views and actions	Evaluation
Intergovernmental organisations	Since 2009, the OECD has maintained a 'blacklist' of countries operating as tax havens. In 2019 the EU published its own list of 15 states that lack tax transparency. Action has been taken by the G20 countries to 'name and shame' tax havens.	While the Cayman Islands and Bermuda remain legitimate places to legally register a business, some TNCs have become uncomfortable operating there because of scrutiny of their operations. Firms have become more aware of the brand risk associated with being seen to be avoiding taxes (customers may boycott their products).
Non-governmental organisations and citizens	NGOs such as WarOnWant, Oxfam and UK Uncut have campaigned against tax havens and the TNCs that us them. NGOs organised protests, consumer boycotts and demonstrations in order to pressure governments to clamp down on tax avoidance by TNCs (Figure 15.17).	High street protests are a public relations challenge for TNCs, in addition to the profit loss this action brings. Both Starbucks and Google volunteered to pay additional taxes to the UK government following protests, although campaigners said the amount paid was still too little.

- The global financial crisis of 2008 and its after-effects raise questions about the competency of Western powers (including the IMF and World Bank) to shape global markets.
- The all-too-evident lack of environmental sustainability of global economic growth – and the ongoing failure of leading developed nations to shoulder more of the costs of fighting global climate change – have tarnished the reputation of 'business as usual' capitalism.

Life expectancy in North Korea, which has effectively 'opted out' of globalisation, is more than ten years lower than in neighbouring South Korea. Zimbabwe has only recently recovered from an episode of hyperinflation. Economic mismanagement there led to interest rates reaching 79 billion per cent.

Venezuela has managed to fund a non-capitalist development path thanks to its large oil reserves and revenues. However, falling oil prices in 2015–16 threatened the sustainability of Venezuela's development model.

Figure 15.17 Protests outside Starbucks

Alternative development in Latin America

Latin America has a long history of governments that like to 'do things differently'. This can be broadly called Latin American socialist populism. Populism is a political approach that appeals directly to ordinary people, arguing that their concerns have been ignored by the political elite. Populism tends to be associated with a charismatic leader, and even a 'cult of personality'.

Political leaders in Cuba, Venezuela, Bolivia and to a lesser extent Ecuador and Argentina have at various times attempted to reject the 'Western' capitalism followed by Europe and North America and instead follow left-wing socialist principles:

- The 1959 Cuban revolution brought Fidel Castro to power, allied with Russia over their much closer neighbour the USA.
- Hugo Chavez, president of Venezuela 1999–2013, promised to redistribute Venezuela's considerable oil wealth in favour of the poor.
- Evo Morales, president of Bolivia 2006–19, an indigenous Bolivian, nationalised many parts of the Brazilian economy including the gas industry.
- Ecuador's president Rafael Correa (2008–17) effectively declared Ecuador's national debt illegal and refused to pay it back as part of his 'Citizens Revolution' movement.

These leaders have often rejected the help of international organisations such as the World Bank, IMF and WTO and also attempted to reduce the role of Western TNCs in their economy. Tax rises for corporations and the rich were used to improve health and education for the poor, often alongside government subsidised fuel, food, electricity and water supply. Latin American populism has a very mixed record of success:

- Widespread street protests and riots occurred in 2019 in Ecuador in response to President Moreno reversing some of the Correa reforms (especially fuel subsidies, and higher wages).
- After 2014 Venezuela's economy descended into chaos: hyperinflation, food shortages, unemployment and widespread corruption were the legacy of the Chavez reforms and those of his successor Nicolás Maduro. Over 2 million people have fled Venezuela since 2014.
- Morales was forced to flee Bolivia in 2019 due to protests after an election widely viewed as 'rigged', although Bolivia's economy has proved fairly robust with rising incomes and an improving poverty situation.

Overall, attempts to diverge from the neo-liberal, Western capitalist model have proved unsuccessful. Some would argue this is because the USA, IGOs and TNCs are simply too powerful to resist – the USA imposed damaging economic sanctions on socialist Venezuela – but others would argue that populist leaders have simply lacked the skills of good governance.

Figure 15.18 Nicolás Maduro

Review questions

1. Explain what is meant by the terms 'sovereignty' and 'nation'.
2. Briefly outline why Iceland might be considered culturally homogenous.
3. Use a table with two columns, headed 'states' and 'countries', to outline the differences between these two categories.
4. Suggest reasons why the state boundaries established in Africa by European colonial powers have often become a cause of migration and conflict.
5. Using examples, explain why European empires were dismantled during the second half of the twentieth century. To what extent have colonial strategies of control been replaced by neo-colonial strategies?
6. Assess the costs and benefits of the use of tax havens by large TNCs. As part of your answer, consider the costs and benefits for different players, including states, TNCs and citizens.
7. Explain what populist leaders in Latin America were trying to achieve by taking a different development pathway.

Further research

Learn about nation states and how they are changing: https://www.theguardian.com/news/2018/apr/05/demise-of-the-nation-state-rana-dasgupta

Explore a timeline of the British Empire: https://www.historic-uk.com/HistoryUK/HistoryofBritain/Timeline-Of-The-British-Empire/

Explore the DRC in more depth from the pre-colonial era to post-colonial conflict: https://www.warchild.org.uk/history-democratic-republic-congo-drc

Find out more about the geography of tax havens: www.bbc.co.uk/news/business-23371564 and https://www.imf.org/external/pubs/ft/fandd/2019/09/tackling-global-tax-havens-shaxon.htm

16 Global organisations and their impacts

> **What are the impacts of global organisations on managing global issues and conflicts?**
> By the end of this chapter you should be able to:
> - explain how the UN and other global organisations have evolved since 1945
> - assess the role of global organisations in establishing rules for world trade and financial flows
> - understand why some neighbour states have chosen to join regional groups such as the EU.

16.1 Global organisations in the post-1945 world

Global agreements and organisations existed in the past. The League of Nations was established after the First World War in 1919, for instance. The post-war period since the end of the Second World War in 1945 has seen an acceleration towards even greater **global governance**. At the forefront of this is the UN, an umbrella organisation for many global agencies, agreements and treaties.

The UN and global governance

The UN was the first post-war IGO to be established (Figure 16.1). Over time, its remit has grown to embrace a whole range of areas of governance spanning human rights, the environment, health and economics. The timeline in Figure 16.2 (page 296) shows when selected agencies were established (although the World Bank and IMF were set up in 1944, they have been recognised as independent specialised agencies of the UN since 1945).

Also included in Figure 16.2 are selected peacekeeping missions that UN troops have been involved with.

> **Key term**
>
> **Global governance:** The term 'governance' suggests broader notions of steering or piloting rather than the direct form of control associated with 'government'. 'Global governance' therefore describes the steering rules, norms, codes and regulations used to regulate human activity at an international level. At this scale, regulation and laws can be tough to enforce, however.

Figure 16.1 The UN headquarters in New York

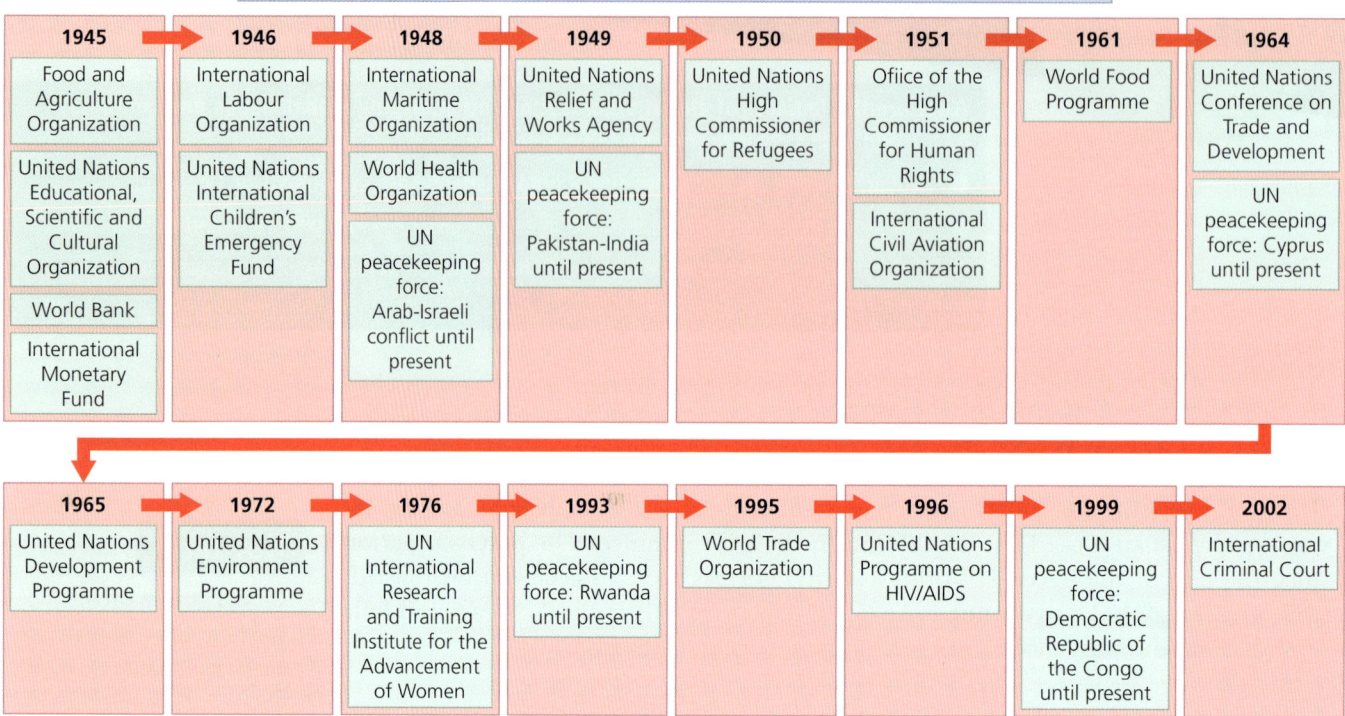

Figure 16.2 A timeline showing the evolution of the UN, including selected peacekeeping actions

The UN has also been responsible for the establishment of important global conventions:

- The Universal Declaration of Human Rights (UDHR), which defines global human rights in considerable detail (Figure 16.3, page 297). A Human Rights Council has been established to press for improvements in states and contexts where human rights are denied to people.
- The UN Convention on the Rights of the Child (CRC) was set up in 1990.
- The 1992 Conference on Environment and Development (the 'Earth Summit') established a plan of action for sustainable development and laid the groundwork for the Kyoto Agreement in 1997 and subsequent climate change conferences, accords and agreements.
- The Millennium Development Goals (MDGs) were introduced in 2000 and their successor, the Sustainable Development Goals (SDGs), followed in 2015. The MDGs were created at the Millennium Summit in New York and aimed to 'free all men, women and children from the abject and dehumanising conditions of extreme poverty'.

The UN's system is based on five main organisational bodies. The General Assembly is made up of voting representatives for all 193 member states. At any time, a smaller number of states are represented – some permanently, others in rotation – on the Security Council, the Economic and Social Council, the Secretariat, and the International Court of Justice.

Option 8B: Migration, Identity and Sovereignty

No one has the right to hold you in slavery.

No one has the right to torture you.

You have the right to recognition everywhere as a person before the law.

We are all equal before the law and are entitled to equal protection under the law.

You have the right to seek legal help if your rights are violated.

No one has the right to wrongly imprison you or force you to leave your country.

You have a right to a fair and public trial.

Everyone is innocent until proven guilty.

You have the right to privacy. No one can interfere with your reputation, family, home or correspondence.

You can travel wherever you want.

You have the right to seek asylum in another country if you are being persecuted in your own country.

Everyone has the right to a nationality.

All adults have the right to marriage and to raise a family.

You have the right to own property.

Everyone has the right to belong to a religion.

Freedom of expression: You have the right to free thought and to voice your opinions to others.

Everyone has the right to gather as a peaceful assembly.

You have the right to help choose and to take part in governing your country, directly or through chosen representatives.

You have the right to social security and are entitled to economic, social and cultural help from your government.

Worker's rights: Every adult has the right to a job, a fair wage and to join a trade union.

You have the right to leisure and rest from work.

Everyone has the right to an adequate standard of living for themselves and their family.

Everyone has the right to education.

Your intellectual property as artist or scientist should be protected.

We are all entitled to social order so we can enjoy these rights.

Figure 16.3 The main ingredients of the Universal Declaration of Human Rights

UN support for human rights

The UN uses a variety of methods to protect human rights globally. International humanitarian law comes into force as soon as an armed conflict begins. The UN Genocide Convention has been in place since 1948, and the updated 1949 Geneva Convention lays out clearly the rights of prisoners of war and civilians in combat zones. For international law to work, however, means of enforcement is required. What actions can the UN take when serious human rights breaches clearly contravene the UDHR, Genocide Convention or Geneva Convention? Table 16.1 assesses several options.

Table 16.1 Assessing the effectiveness of UN actions in support of human rights in conflict zones

Action	Assessment
Sanctions	The UN has, on occasion, approved the use of economic sanctions. Member states have agreed to restrict trade or cultural exchanges (including sports sanctions) with a particular country in the hope that it will bring about a change in that government's domestic or foreign policy. Under UN sanctions pressure, the Libyan government renounced its weapons of mass destruction (WMD) programme.
UN troops	UN troops are drawn from the armed forces of many different member states, including the UK, Germany, India and China. Their roles vary from conflict prevention to peace enforcement in situations where actual fighting has been taking place.
War crimes trials	A permanent war court was established in The Hague in 1988. In 2008, proceedings began against Bosnian Serb leader Radovan Karadžić, who was accused of genocide during the siege of Sarajevo. He was finally convicted in March 2016 and sentenced to 40 years. The slow progress reflects the difficulty of enforcing aspects of international law. More worryingly, many war criminals continue to evade capture.
Healthcare and shelter for refugees	The UN High Commissioner for Refugees (UNHCR) and the WHO provide support to people who have been affected by conflict, including the establishment of camps for internally displaced people (IDPs). Although this does not tackle the cause of conflict, the UN's presence helps to protect vulnerable people from further human rights abuses.

UN forces in Democratic Republic of the Congo

Page 285 looked at the causes of conflict in DRC and central Africa. The UN has been deeply involved with attempts to resolve this conflict. Since 1999, around US$10 billion has been spent on the deployment of UN peacekeeping troops (Figure 16.4). Globally, it is the largest allocation of UN funds for conflict management there has ever been, by a considerable margin.

Under this operation, named the UN Organization Stabilization Mission in the Democratic Republic of the Congo (MONUSCO), 18,300 (in 2017) peacekeepers have been stationed in DRC under Article 7 rules. This means that they can fire weapons if it becomes necessary to do so. On the one hand, MONUSCO has provided Congolese people with protection from rebel and militia groups. On the other hand, it has left the country dependent on external support to guarantee political stability. Additional actions taken by the UN include:

- The 2005 International Court of Justice ruling that Uganda must compensate DRC for the plundering of natural resources during the conflict that claimed over 5 million lives.
- The war crimes trials at The Hague of people involved in the conflict. In 2012, Thomas Lubanga, leader of the Union of Congolese Patriots, was convicted of conscripting children aged under 15 to kill ethnic Lendus during the conflict in DRC.

Figure 16.4 UN peacekeepers in DRC

Figure 16.5 The war on terror has cost the lives of more than 2000 US troops in Afghanistan since 2001

Key term

Unilateral intervention: Military intervention undertaken by a state (or a group of states) outside the umbrella of the UN.

Unilateral intervention by UN member states

Since the formation of the UN, its member states have continued to conduct their own military operations independently. Sometimes a **unilateral intervention** takes place. This may be connected to a particular territorial dispute or a human rights issue. In such situations, the UN Security Council will take a view on events but may stop short of voting to involve peacekeepers.

- In 1982, the UK was briefly at war with Argentina due to rival claims over the Falkland Islands. The UN Security Council passed Resolution 502 condemning Argentina's invasion of the Falkland Islands and called for a diplomatic solution to be sought by the two warring states.
- US and UK armed forces invaded Afghanistan in 2001 and Iraq in 2003. In the case of Iraq, the US and UK governments justified their action around allegations that illegal weapons of mass destruction (WMDs) were present. The UN wanted to send non-combative weapon inspectors to Iraq. The USA and UK were unwilling to wait and took their own military action instead. The move was opposed by UN secretary-general Kofi Annan and no WMDs were found.
- The USA – sometimes assisted by the UK and France – has taken military action in several **fragile states** including Yemen, Somalia and Syria. Increasingly, unmanned drones are used, which raises a challenge for international law makers by arguably breaking Article 51 of the UN Charter. US drone strikes have also killed thousands of people in northern Pakistan. These actions are all part of the so-called 'war on terror' and have had profound impacts on global geopolitical stability (Figure 16.5).

Option 8B: Migration, Identity and Sovereignty

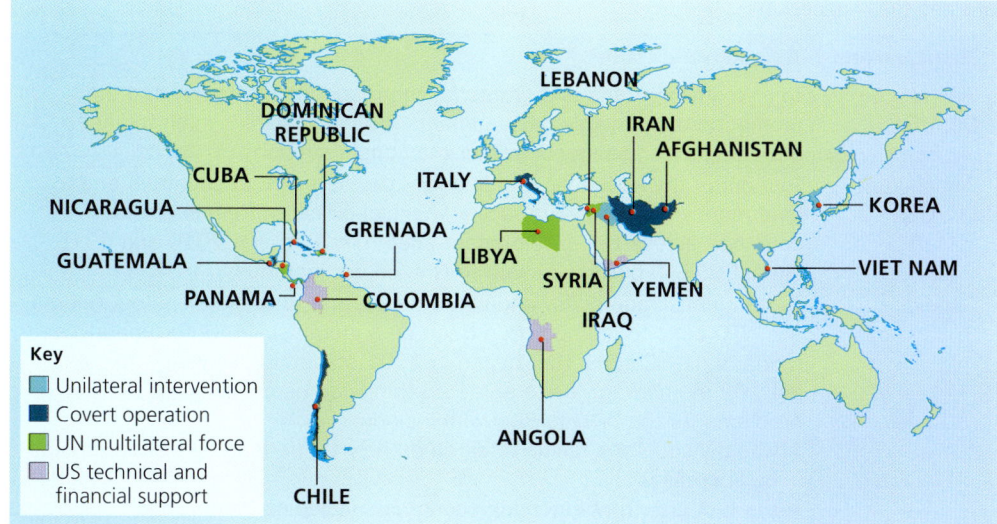

Figure 16.6 Some major US interventions abroad since 1945

The USA has sometimes been described by its critics as a 'rogue superpower' on account of its foreign policy decisions (Figure 16.6). William Blum, a former US State Department official, estimates that his country has tried to overthrow more than 40 foreign governments since 1945 using a variety of overt and covert methods.

16.2 Intergovernmental organisations and world trade

Three IGOs established in the period 1944–7 have had an enormous influence on world trade and economic development in the post-war period. They are the World Bank, the IMF and the WTO. Known as the **Bretton Woods institutions**, they featured previously in Book 1 (page 195).

Western power and influence over the global economic system

The IMF, World Bank and WTO were established at the end of the Second World War by the victorious allied nations and other leading industrialised nations (Table 16.2). Over time, these institutions have been important in maintaining the dominance of 'Western' capitalism. This has been achieved through a combination of global economic management and free-trade policies.

The chief aim of the Bretton Woods conference was to avoid a return to the protectionism of the 1930s. This was viewed as having been extremely harmful to world trade and a major contributing factor to the Great Depression of the 1930s. That decade was marked by mass unemployment and hardship for working people in the USA and Europe. Economic and social instability had also helped fascism to flourish in Germany and Italy, leading ultimately to the outbreak of war in Europe in 1939.

> **Key terms**
>
> **Fragile state:** A country with weak governance, limited rule of law, humanitarian crises and social tensions (often the legacy of armed conflict or civil war). The basic social and economic needs of most people are not being met.
>
> **Bretton Woods institutions:** The IMF and the World Bank were founded at the Bretton Woods conference in the USA at the end of the Second World War to help rebuild and guide the world economy. The General Agreement on Tariffs and Trade (GATT) was set up soon afterwards and later became the WTO.

Table 16.2 Bretton Woods institutions and their successors

Organisation	Founding date	Headquarters	Role in world trade
World Bank	1944	Washington DC, USA	To give advice, loans and grants for the reduction of poverty and the promotion of economic development. The World Bank's main role is to offer long-term assistance rather than crisis support.
IMF	1944	Washington DC, USA	To monitor the economic and financial development of countries and to lend money when they are facing financial difficulty. Help is provided to countries across the development spectrum. Between 2010 and 2015, almost US$40 billion was lent to Greece, for instance.
WTO (previously GATT)	1995 (previously 1947)	Geneva, Switzerland	To formulate trade policy and agreements, and to settle disputes. Overall, the WTO aims to promote free trade on a global scale. Unfortunately, a round of negotiations that began in 2001 stalled for 14 years. Trying to get 162 member states to agree anything can be challenging. Difficult problems for the WTO to deal with include: • wealthy countries failing to agree on how far trade in agriculture should be liberalised • the fast growth of emerging economies including China (which makes it harder to agree on fair policies for so-called developing countries).

> **Key term**
>
> **Hegemonic power:** The ability of a powerful state or player to influence outcomes without reverting to 'hard power' tactics such as military force. Instead, control is exercised using a range of 'soft' strategies including diplomacy, aid, and the work of the media and educational institutions.

To ensure there was no return to protectionism, industrialised nations agreed to implement several key principles to help regulate and foster the growth of the global economic system and trade flows. These were:

- *The establishment of a fixed rate exchange system based on gold and the dollar.* The aim was to make trade and investments easier and to help global financial flows grow over time.
- *Use of the IMF and World Bank to stabilise global systems of finance and trade.* Assistance from either lending organisation would help states experiencing financial difficulty to correct economic imbalances. Over time, the remit of the IMF and World Bank has broadened to include offering long-term development assistance to low-income nations.
- *The establishment of GATT by 23 leading trading nations.* The aim was to encourage the removal of barriers to flows of trade and investment around the world. This goal has been pursued since then, with mixed results, by GATT and its successor, the WTO, through a series of 'rounds' or meetings.

The influence of the USA

The USA had disproportionate influence at Bretton Woods over the design of the new global economic system and the principles that informed it. This was because it was the only Second World War power with substantial financial resources remaining (in contrast, the UK had been left effectively bankrupt by the 1939–45 conflict).

American power over the proceedings can be inferred from the fact the headquarters of both the newly founded IMF and the World Bank were placed in Washington (Figure 16.7). To date, all World Bank presidents have been American citizens. Even though the IMF has a European president, the USA holds a larger share of policy-making votes than any other state (Figure 16.8).

Over time, the free-market principles that the Bretton Woods institutions promote have become known as the 'Washington consensus'. This phrase suggests the USA has used its considerable **hegemonic power** to gain the support of other states for the economic policies it prefers to promote. Under US influence,

Figure 16.7 The World Bank is sited in Washington DC, around 5 km from the Pentagon

alternative economic and development models for global systems (such as the Latin American approaches mentioned on page 293) have been marginalised.

Controversial borrowing rules

World Bank and IMF lending and support have helped many states to develop economically. Figure 16.9 shows that major recipients of World Bank loans are all rapidly growing emerging economies. Of the US$592 billion in World Bank loans in 2020, 45 per cent was lent to just these six countries. However, since the 1970s progressively tougher rules and conditions have been attached to large-scale lending. As a result, **structural adjustment programmes (SAPs)** and **HIPC policies** are now used when states experience severe financial difficulties. Borrowing countries must agree to make concessions in return for new lending. This might involve privatising poorly run government services, or withdrawing costly state support for inefficient industries. For some countries, SAPs were devastating with devaluing of currency, reduction of the role of the government, reduction in government spending and a shift to privatisation.

> **Key terms**
>
> **Structural adjustment programmes (SAPs):** Policies promoted by the World Bank and IMF to help developing countries overcome their debt problems. These are now superseded by poverty reduction strategy papers (PRSPs) as for many countries SAPs resulted in unacceptable hardship and little progress with solutions to debts.
>
> **HIPC policies:** The Heavily Indebted Poor Countries (HIPC) Initiative was launched in 1996 by the IMF and World Bank, with the aim of ensuring that no poor country faces a debt burden it cannot manage. Countries must meet certain criteria, commit to poverty reduction through policy changes, and demonstrate a good track record over time.

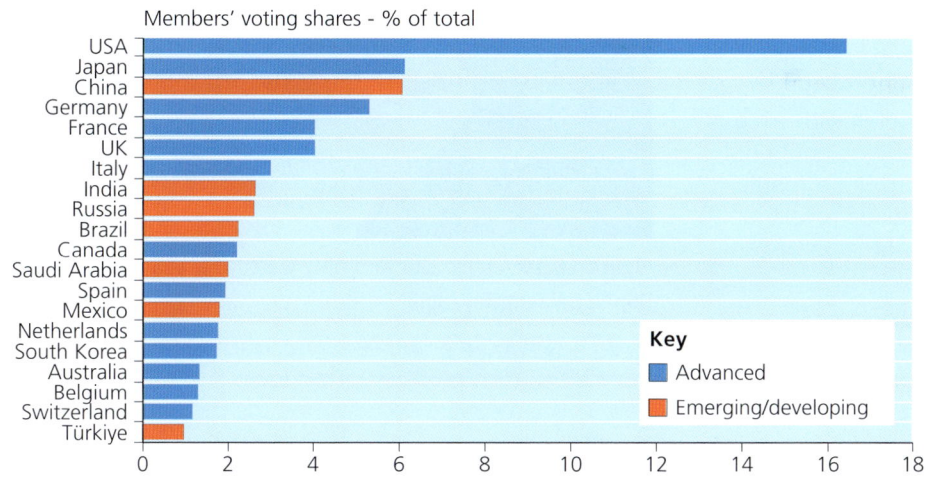

Figure 16.8 The 20 states with the greatest IMF voting share, 2019

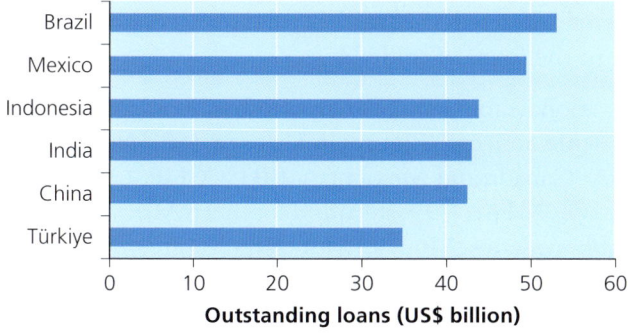

Figure 16.9 The top six borrowers from the World Bank, 2020

Critics of these concessions say they sometimes exacerbate poverty instead of solving it and further undermine the economic sovereignty of borrowing states. In critical theory, SAPs are regarded as a neo-colonial strategy used by developed countries to maintain influence over how the global periphery develops (see page 302).

SAPs and water supplies in Tanzania

Government-run water services in Tanzania had fallen into disrepair in the 1990s but still managed to deliver safe water to some of the poorest households in capital city Dar es Salaam's informal settlements. Clean water is essential if social development goals are to be met, such as improved school attendance for both boys and girls. Unsafe water results in illness and school absences. In the past, before the water system existed, girls frequently missed school because they spent their days carrying buckets of water from wherever it could be found to their family home.

Tanzania approached the World Bank for help. As part of the deal, the World Bank insisted that Tanzania privatise its water services in return for a new US$143 million loan. Consequently, Dar es Salaam's water services were sold to a British-led consortium called City Water, which took over the day-to-day running of the city's water supplies. For the first time, water bills were issued to all households with access to drinking water. When some households could not pay their bills, they were disconnected. As a result, the poorest and most vulnerable families reverted to the use of unsafe water sources and girls began missing school again (Figure 16.10).

In 2005, the Tanzanian government successfully cancelled the contract with City Water. Today, Dar es Salaam's water services are run locally once again but with support from several external players, including the African Development Bank. In 2012, the Indian government provided a US$178 million loan for water projects in Dar es Salaam. This is symptomatic of a wider shift among poor countries towards seeking support from emerging superpowers like China and India alongside, or in place of, the Bretton Woods institutions designed by the USA and its allies.

Figure 16.10 A child sits with dirty water at a well outside Dar es Salaam, Tanzania

Key term

Economy of scale:
If the production of a commodity is expanded then the unit cost price may fall. This is because certain fixed costs (such as the cost of lighting and heating a factory) are spread over more units of output (the product). As a result, the products can be sold more cheaply, which increases revenues further.

Regional groupings of nations as trade blocs

Membership of global IGOs and financial IGOs is almost universal, such is the dominance of these organisations. However, the relative failure of the WTO to deliver on its promise of free trade has led the world's states to create regional groupings. These have emerged in the form of trade blocs.

Book 1 (page 197) explained the rationale for trade bloc growth based on comparative advantage and the **economy of scale** concept (Figure 16.11, page 303). Many trade blocs – such as USMCA (the USA, Canada and Mexico), MERCOSUR (Argentina, Brazil, Paraguay and Uruguay) and COMESA (19 eastern and southern African nations) – are composed of groups of neighbour states. More ambitious trade agreements have, in recent years, tended to fail or be only partly successful:

- The USA withdrew from the proposed Trans-Pacific Partnership (TPP) in 2017, leaving 11 Pacific Rim countries to sign the smaller Comprehensive and Progressive Agreement for Trans-Pacific Partnership (CPTPP) in 2018. CPTPP does not include China.
- The Transatlantic Trade and Investment Partnership (TTIP) would have created a free trade bloc between the EU and North America, but these talks were halted by President Trump in 2018 and all but dead by 2019.

In the 2020s, protectionism seems to be about as popular as free-trade liberalism.

Pre-trade bloc

All four countries produce their own goods and services across all four sectors (vehicles, textiles, financial services and wine); trade between one another at first is non-existent. As a result, output is limited in all cases and product costs are high for consumers. In some cases the goods and services may well be of poor quality, possibly due to inefficiencies resulting from a lack of high-calibre physical or human resources.

Trade bloc (with customs union)

A trade agreement is now in place. National borders are rendered permeable for trade flows through removal of tariffs. In theory, each country discovers it has a **comparative advantage** in one of the four sectors; these firms emerge as leaders in an enlarged market. Scales of production increase and costs of goods and services for consumers fall. A common external tariff may protect these 'winners' from overseas imports. In reality, things are rarely as simple as this model suggests, however.

Common external tariff wall

Figure 16.11 Comparative advantage: the development of trade specialisation within a trade bloc

There are now over 30 major trade blocs and agreements in existence, all exhibiting varying degrees of market liberalisation and customs harmonisation (Figure 16.12). Any decision by states to participate openly in free trade is taken knowing that the promise of easier international sales for firms sits alongside an increased risk of foreign goods flooding home markets. However, the overarching logic of the agreement dictates that all member state companies and citizens should, on balance, find themselves net beneficiaries of the new economic order. Thus, a degree of economic sovereignty is willingly ceded by all governments. Not all of their citizens may support this, however.

In fact, protectionism (restricting imports using import taxes and tariffs) has become increasingly popular in the last decade. In his first term of office President Trump, and others, argued that free-trade agreements are unfair and disadvantageous to some. Trump has argued that a better policy is the protection of US jobs by having bilateral, not multi-lateral, trade agreements with individual countries. Trump's populist rhetoric on trade proved to be an election winner in 2016.

At the simplest level, USMCA (2020) is a trade bloc that encourages **free trade** between the USA, Canada and Mexico by **removing internal tariffs**

A further step involves adopting a **common external tariff**; the MERCOSUR pact (1995) is an example of this type of **customs union**

The EU is highly integrated, moving beyond a **common market** with freedom of movement towards **full economic union** with the introduction of a **common currency**, and sharing some **political legislation**

INTEGRATION

Figure 16.12 The different types of trade bloc and their degree of integration

16 Global organisations and their impacts

16.3 Global environmental governance

The UN has attempted to manage many of the world's pressing environmental problems with varying degrees of success. Its role has expanded over time with the establishment of numerous agencies and agreements to deal with particular issues as they have arisen. In addition, other IGOs including the EU have broadened their remit to include environmental governance (Figure 16.13).

Atmosphere and biosphere governance

The atmosphere and biosphere are important shared resources. All states and their citizens depend on the continued functioning of Earth's atmosphere and on the services that ecosystems provide (see Table 6.2, page 131). Table 16.3 (page 307) shows global actions and agreements relating to the quality of the atmosphere and biosphere.

Figure 16.13 Global action is required to maintain biodiversity and protect endangered species like the giant panda

Global agreements and actions on the atmosphere

Montreal Protocol on Substances that Deplete the Ozone Layer

In the late 1960s, the UN Environment Programme first called for an international response to the issue of ozone depletion caused by worldwide use of a group of chemicals called chlorofluorocarbons (CFCs) in fridges and aerosol sprays. The 1977 'World Plan of Action on the Ozone Layer' gave the UNEP responsibility for promoting and co-ordinating international research and data-gathering activities.

The Montreal Protocol was signed in 1987. It was a remarkable agreement on account of the number of individual governments that were prepared to back an important global goal ahead of narrower economic self-interest. Within a decade, irrefutable proof that CFCs were to blame led to all UN states ratifying the treaty. CFC use was phased out rapidly as a result of this exceptional international co-operation.

Climate change agreement

Climate change was first raised as an urgent issue in 1992 at the UN Earth Summit conference. Many of the meetings that followed were plagued by uncertainty over the evidence and also wrangling over which nations should be held responsible for the majority of the anthropogenic carbon stock that has been added to Earth's atmosphere.

On the whole, international co-operation on climate change has taken place very slowly, which many people deem to be a failure of international governance. Although a new international agreement on action was reached in Paris in 2015, critics say that the pledges that were made to reduce carbon emissions do not go far enough (Figure 16.14, page 305). These pledges cannot be enforced either.

Global agreements and actions on the biosphere

Convention on International Trade in Endangered Species of Wild Fauna and Flora (CITES)

CITES entered into force in 1975. It banned trade in threatened species and their products. Now adopted by 181 countries, it has effectively saved some species but not others. Success stories include the recovery of the Hawaiian neˉneˉ bird and also the Arabian oryx.

Rising wealth in China, Indonesia and South Korea has actually increased illegal trade in some prohibited substances such as ivory and rhino horn. The problem can be summed up as: 'new money, old values'. Without a cultural shift away from the use of these products, CITES will not be able to protect some species.

Millennium Ecosystem Assessment

This international collaboration helped popularise the 'ecosystem services' approach to biodiversity management. A financial value is calculated for threatened biomes and species, thereby strengthening the rationale for their preservation.

'Ecosystem services' is a philosophy that fits well with the capitalist values of the global economic system and is a pragmatic approach for the UN to have adopted and helped promote globally.

Paris was a failure

Two certainties existed entering the Paris climate talks. The first was that the world's heads of state were not prepared to act as is necessary. The second is that it was never going to be up to them anyway.

The richest governments – politically captured by a fossil fuel-wedded corporate class – were hobbled from the outset. It was the movement being built by citizen activists around the globe that shaped the best of the Paris agreement. These civil movements' growing power at home forced a tremendous promise abroad: an aspiration to keep the world to 1.5 degrees of warming, above which spells disaster for indigenous peoples, low-lying island nations, and the African continent. But the limits of their power meant everything shoddy about this agreement: that the US and others could choose their own emissions targets; wouldn't be legally bound or on a strict timeline; and wouldn't pay the poorest countries, who have done practically nothing to create the problem but now suffer practically everything, what is needed to adapt to climate change. And as the world hurtles toward 3 degrees of warming, they have been stripped even of the right to legally demand compensation.

The Guardian, 15 December 2015

Figure 16.14 How one UK newspaper reported the new 2015 global climate change agreement

Skills focus: Source materials – newspaper articles

Figure 16.14 shows source material from a UK newspaper article. What can we determine from this about the impact of the Paris agreement on the management of global climate change issues? This article contains a mixture of fact and opinion. Which is which?

Key terms

Global commons: Global resources so large in scale that they lie outside of the political reach of any one state. International law identifies four global commons: the oceans, the atmosphere, Antarctica and outer space.

Transboundary water: A water resource, including rivers, lakes and aquifers, that occupies a territory shared by more than one state.

Global governance of the Earth's oceans and rivers

IGOs have been involved in developing laws for managing oceans and international rivers.

Earth's oceans, like the atmosphere, are a **global commons**. This means it is in the best long-term interest of individual states to collaborate on making sure that sustainable use of the oceans, and oceanic resources, is achieved over time (see place context on page 306).

IGOs have also developed laws for monitoring the state of the environment.

Laws for managing international rivers

Various agreements exist in relation to rivers, including:
- the UN Convention on the Protection and Use of Transboundary Watercourses and International Lakes (Water Convention), which aims to protect and ensure the quantity, quality and sustainable use of **transboundary water** resources by facilitating co-operation
- the EU's Water Framework Directive, which was established in 2000 with the goal of improving water quality across Europe.

16 Global organisations and their impacts

Managing Antarctica

IGO management also includes responsibility for Antarctica (Table 16.4, page 308) – the Earth's fifth-largest continent – as 'a continent of peace and science'.

> ### Antarctic Treaty System
>
> Antarctica has special status following the 1959 Antarctic Treaty, to which 54 nations are now signatories. Nobody owns Antarctica, although around 20 nations have permanent scientific bases there (seven of whom previously made a formal territorial claim). Antarctica lacks indigenous people.

However, as non-renewable resources become exhausted, might global pressure eventually begin to mount to exploit the coal, oil, copper, silver, gold and titanium that are known to lie beneath the Antarctic wilderness? Mining was banned under the 1991 Madrid Protocol but this is due for review in 2041. So far, the protocol remains popular with the general public in the countries that agreed to it, but that could change in the future should existing stores of oil or other resources become depleted.

Other economic activities are also largely restricted. Whaling in Antarctic waters is one form of activity that survives, but only at a very restricted level following the banning of commercial whaling in Antarctic waters in 1987. All Antarctic seals are protected under international agreements.

Regulating the use of oceans and marine ecosystems as a global commons

The oceans cover more than two-thirds of the surface of the Earth (Figure 16.15). The UN Convention on the Law of the Sea (UNCLOS) is a vast global treaty covering navigational rights, territorial sea limits, economic jurisdiction, legal status of seabed resources beyond the limits of national jurisdiction, conservation and management of marine ecosystems, protection of the marine environment, and a binding procedure for settlement of disputes between states. 'Possibly the most significant legal instrument of this century' is how the UN secretary-general described the treaty after its signing in 1982. Table 16.3 (page 307) highlights elements of UNCLOS that are particularly relevant for geographical study.

Figure 16.15 The oceans are a 'global commons' shared by all states

Option 8B: Migration, Identity and Sovereignty

Table 16.3 Selected UNCLOS provisions that are pertinent to the study of geography

Protection of marine biodiversity	UNCLOS requires that the 167 nations that have signed it must follow International Whaling Commission (IWC) guidelines, which have been in place since 1946. Many species of whale were hunted almost to extinction during the twentieth century but have since been helped to recover (Figure 16.16). In 1982 the IWC issued an indefinite ban on commercial whale hunting. However, some states have not agreed to this. Japan has a long history of defying international whaling laws and continues to hunt whales. Norway has objected on the grounds that whale hunting is part of its culture. Whales are also protected globally under CITES, as are many other marine species.
Regulation of global shipping flows	Ninety per cent of all global trade flows between countries involve sea travel. Container shipping has been described as the 'lifeblood' of the global system. However, substandard ships and illegal activities can cause marine pollution, especially along busy shipping lanes. Vessels are expected to follow international rules set out by UNCLOS. One success has been the progressive retirement of single-hulled oil tankers after the *Prestige* went down off the coast of Spain in 2002, causing great damage to ocean species. A global phase-out of these old tankers is now underway. It is also illegal for ships that have recently delivered oil to use seawater to wash out their tanks. Flushing tanks clean causes significant oil pollution. It also assists the movement of invasive species, which swim inside the ships and become stowaways.
Exclusive economic zone (EEZ)	The EEZ is the area of water extending 200 miles from a state's shoreline. The coastal state has the right to exploit, develop, manage and conserve all resources found in its adjacent waters, including both biotic (fish) and abiotic (oil, gas or minerals) resources found in the water or on the ocean floor of the EEZ.
Rights of landlocked states	Forty-two states have no sea coast (fifteen in Africa, thirteen in Europe, twelve in Asia and two in Latin America). Under UNCLOS, landlocked states have a right of access to and from the ocean for the purpose of enjoying 'the freedom of the high seas'. An interesting case that the UN is arbitrating on involves landlocked Bolivia's claim that Chile should return to it an area of coastline lost during a war 150 years ago. The matter will be adjudicated at the International Court of Justice at The Hague.

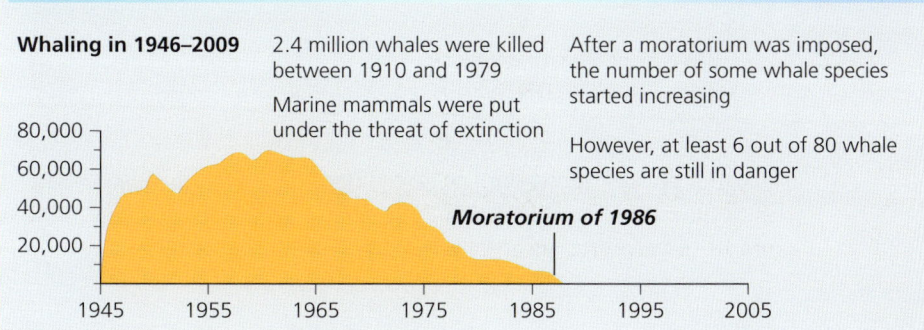

Figure 16.16 Global governance of whaling and trends in commercial whaling

Figure 16.17 Antarctica

Antarctica is, however, under increasing pressure as a result of the 'shrinking world' effect (see Book 1, page 191). Modern transport now gives easy access to previously inaccessible areas, while rising affluence in emerging economies means more of the world's people have the financial means to visit 'bucket list' destinations (which for many includes Antarctica).

Between 1993 and 2003, visitor numbers rose from 6500 to an estimated 13,500. By 2018–19, 56,000 tourists visited Antarctica, most arriving by cruise ship. More than 250 flights land at the South Pole every summer and McMurdo Station, the largest settlement on the continent, has a summertime population of over 1200. It even has its own newspaper (now a website), the *Antarctic Sun*. This is all a far cry from 1899 when the very first Norwegian explorers set foot on the continent, having discovered its existence just 80 years earlier.

Table 16.4 Pressures facing Antarctic governance

Political governance	The number of countries with decision-making powers under the ATS grew from 13 in 1980 to 29 by 2019, so decisions are harder to reach.
	No new, binding, protocol has been signed in more than 20 years.
	The ATS is a Cold War era treaty, but the world is economically and geopolitically very different today.
Resource demands	While currently banned, there is pressure to exploit minerals and fishing grounds.
	Rising populations, the promise of profits and possible trade offs (e.g. profit sharing to fund scientific research) could all increase pressure to exploit.
Tourism	Tourists each spend US$10,000–100,000 visiting Antarctica, but despite this numbers are rising.
	There is pressure for more landing sites, and even hotels.
Emerging powers	China now has five Antarctic scientific bases, built since joining the ATS in 1983.
	Newer parties to the ATS may view Antarctica and its resources differently to the original 1959 signatories.

Review questions

1 Briefly explain the role of the United Nations in global governance.
2 Use an example to illustrate the role of UN Peacekeepers.
3 Explain what is meant by 'unilateral intervention'.
4 Outline the advantages and disadvantages of trade blocs.
5 Explain how the USA and other powerful states have been able to influence global economic rules and policies in ways that work to their own advantage.
6 Assess how far the issue of global warming is being successfully managed at a global decision-making level.
7 Suggest why international lending can be both good news and bad news for the world's poorer states.
8 To what extent do you think the unique governance system of Antarctica is under threat from trends in geopolitics and demand to use the region's resources?

Further research

Explore the specialised agencies and programmes that make up the United Nations: https://www.un.org/en/sections/about-un/funds-programmes-specialized-agencies-and-others/

Investigate some examples of UN Peacekeeping in action: https://peacekeeping.un.org/en/where-we-operate

Find out more about the IMF's HIPC initiative for debt relief works: https://www.imf.org/en/About/Factsheets/Sheets/2016/08/01/16/11/Debt-Relief-Under-the-Heavily-Indebted-Poor-Countries-Initiative

Investigate the CITES treaty on trade in endangered species: https://www.worldwildlife.org/pages/cites

Explore the development of the Antarctic Treaty system: https://www.ats.aq/e/antarctictreaty.html

17 Threats to national sovereignty

What are the threats to national sovereignty in a more globalised world?
By the end of this chapter you should be able to:
- understand why national identity is an elusive and contested concept
- explain what the main challenges to national identity are in a more globalised world
- understand what consequences may follow from disunity within nations.

17.1 The elusive concept of 'national identity'

The theory of 'hyperglobalisation' proposes that the relevance and power of countries will reduce over time. Global flows of commodities and ideas may result ultimately in a shrinking and borderless world. Hyperglobalisers envisage a 'global village' where individual group attachments to ethnic and religious identity will be replaced by a shared identity based on the principles of **global citizenship**. In theory, this should increase global and local prospects for peaceful coexistence by reducing possible opportunities for prejudice to arise.

In contrast, sceptics argue that barriers to globalisation have been rising, not falling, in recent years. Think of the **nationalist** calls for the UK to leave the EU, or for the USA to build a wall along its border with Mexico, for instance. In many states and nations there are ongoing attempts to reassert national identity.

Factors that reinforce contemporary nationalism

Nationalism remains a powerful force in the world. It is reinforced through education, sport and by political parties stressing loyalty to both the institutions and ideals of nation states (Table 17.1). In many places, nationalism is on the rise. There is a clear correlation between the growing interconnectedness of states and the desire of many citizens to sever ties with other places.

> **Key terms**
>
> **Global citizenship:** A way of living wherein a person identifies strongly with global-scale issues, values and culture rather than (or in addition to) narrower place-based identity.
>
> **Nationalist:** A political movement focused on national independence or the abandonment of policies that are viewed by some people as a threat to national sovereignty or national culture.

Table 17.1 Factors that help to reinforce nationalism

Education	Fundamental British Values, defined by the UK Government in 2014 as democracy, the rule of law, respect and tolerance and individual liberty, are taught via the National Curriculum in UK schools. These values are contested, in terms of which national stories and symbols to focus on.
Sport and culture	Sport can serve as a national unifying force, for example through the London 2012 Olympics or Football and Rugby World cups. Union flags, painted faces and traditional songs (Figure 17.1, page 311) are all part of sporting rituals. However, this nationalism is often not that of the UK but specifically Welsh, Scottish and English. This fragmentation of identity does not occur in countries such as France or Germany.
Political parties	UK political parties often use national symbols such as the rose, lion, Union Flag or oak trees as part of their visual message and logos. These symbols appeal to a particular type of national identity which is of course not supported by all people, and is actively off-putting to some voters.
History	Nationalism may focus on the past, and the real or imagined 'greatness' of past military victories (the Second World War) or even the colonial past (British Empire), although this is highly contested.

Option 8B: Migration, Identity and Sovereignty

The rationale for retreating from globalisation is rooted, for some people, in the valid concern that actual sovereignty has been surrendered. It is certainly true that IGOs such as the EU and UN agencies have more power than in the past over law making in areas ranging from human rights to environmental rules. The renewal of nationalism is also linked with a much broader of interpretation of 'loss of sovereignty', which is synonymous with 'loss of control'. The shrinking world, so the argument goes, has brought chaos to national life and culture due to new and sometimes unchecked flows of people, information and ideas.

Figure 17.1 An English football fan

Finding a focus for identity, loyalty and citizenship

Aspects of national identity are sometimes tied to distinctive legal systems and methods of governance. For instance:

- Many US citizens find a common focus for identity in the rights and freedoms granted to them by the US Constitution Bill of Rights, which was established after the American Revolution. The First Amendment's guarantee of freedom of speech is often seen as especially important for US citizenship. The Second Amendment right to self-defence (and hence to own a firearm) is more divisive.
- Cultural cohesion in France is sometimes said to rely on a shared belief in the importance of *liberté* (freedom) more than anything else. This is a legacy of the French Revolution of 1789–99. When a controversial ban on wearing the burqa (a full body veil) was introduced in France in 2009, the rationale offered by President Sarkozy was that the garment was incompatible with French culture, not because it is a symbol of religion but because it is a symbol of oppression (of women).
- The UK's Magna Carta (Great Charter) dates from 1215 and is widely viewed as the foundation of British laws, liberties and principles.

British values

The values I'm talking about – a belief in freedom, tolerance of others, accepting personal and social responsibility, respecting and upholding the rule of law – are the things we should try to live by every day. To me they're as British as the Union Flag, as football, as fish and chips. Of course, people will say that these values are vital to other people in other countries. And, of course, they're right. But what sets Britain apart are the traditions and history that anchor them and allow them to continue to flourish and develop. Our freedom doesn't come from thin air. It is rooted in our parliamentary democracy and free press. Our sense of responsibility and the rule of law is attached to our courts and independent judiciary. Our belief in tolerance was won through struggle and is linked to the various churches and faith groups that have come to call Britain home.

These are the institutions that help to enforce our values, keep them in check and make sure they apply to everyone equally. And taken together, I believe this combination – our values and our respect for the history that helped deliver them and the institutions that uphold them – forms the bedrock of Britishness. (Source: David Cameron, *Mail on Sunday* 15 June 2014)

Figure 17.2 UK Prime Minister David Cameron's 2014 article about British values (page 311) was delivered to celebrate the 799th anniversary of Magna Carta

Table 17.2 2016 EU Referendum results by political party of voters

Leave	Usual party of voter	Remain
58%	Conservative	42%
37%	Labour	53%
30%	Liberal Democrat	70%
25%	Green Party	75%
36%	Scottish National Party	64%
96%	UKIP	4%

Data: YouGov

Appeals to nationalism can lead to difficult political issues. Having opened the Pandora's Box of nationalism, David Cameron then held the referendum on EU membership in 2016. Cameron campaigned for 'remain' but the final result was 52 per cent/48 per cent in favour of 'leave'. As Table 17.2 shows, the referendum split traditional political party allegiances. This suggests that on issues of sovereignty and national independence – perhaps even national identity – there are complex patterns which cut across traditional voting patterns.

Table 17.3 How English national identity has changed over time

	Early twentieth century	Twenty-first century
Religious beliefs	Generally widespread, with high levels of Anglican or Catholic church attendance	Largely secular and non-religious, although some minority faiths are prospering
Food	Locally sourced seasonal food; native herbs preferred to foreign spices	Global, varied tastes in food; strong spices are widely used in cooking
Identity	People had a strong sense of local belonging (either to a town or county); regional dialects were stronger than today; most were also extremely patriotic and would fight for their country	Many would be less willing to fight for their country, although they are often strong supporters of national football teams; younger people may see themselves as 'global citizens'
Roots of vocabulary	Celtic, Saxon, Scandinavian (Norse), Roman, Greek, French	Additional Indian, Jamaican and American influences (due to migration and TV)

Cultural cohesion in the era of globalisation

Many states have a multicultural population, as Chapter 16 explained. This may be an accident of history or due to more recent migration flows. Where many contrasting ethnic groups make up the citizenry of a state, questions of national identity and loyalty become complex, especially in the era of globalisation. Debates arise over **cultural cohesion** and the means of achieving it (Table 17.4).

> **Key term**
>
> **Cultural cohesion:** The capacity of different national and ethnic groups to make a mutual commitment to live together as citizens of the same state.

Table 17.4 Investigating cultural cohesion in the USA, UK and New Zealand

USA	National identity is extremely complex given the varied histories of the people who live there (see page 277). White European, Hispanic, Asian and Black Americans have developed distinctive cultures from one another to the extent that it becomes hard to generalise about a single 'American culture'. In addition, there are hundreds of distinct Native American tribes and communities whose settlement predates the arrival of the first Europeans and who retain their own indigenous culture.
UK	National identity in the UK is complex because the UK is a culturally diverse country. Figure 17.4 shows the results of the 'national identity' question from the 2021 England and Wales Census. A large number of people identify as 'British', but millions do not. Some consider themselves 'English' only, whereas others don't identify with any UK identity or identify with a 'mix' of identities. This is not unexpected because many people believe in 'British values' but also identify with values and beliefs derived from their own family heritage – be that from other parts of Europe, Asia, the Caribbean or elsewhere. Therefore, it should not be a surprise that many people in the UK identify with and have loyalty to more than one country.
New Zealand	Like Australia, New Zealand is a country that was once part of the British Empire and where white settlers usurped established indigenous populations. New Zealand has been independent since 1947. It is home to many minority ethnic groups including Chinese, Korean and indigenous Maori people. To aid cultural cohesion, New Zealand recently agreed to change its flag from a colonial design featuring the Union Jack to a new design that incorporates elements of contemporary New Zealand and Maori culture.

The English countryside

Representations of rural life and landscapes play an important role in the reproduction of English culture. Rural life is portrayed in a range of different media in ways that draw a connection between the countryside and English identity:

- Some important national myths have a strong association with a rural sense of place, such as the Arthurian legends.
- Iconic English classical music, including the works of Elgar and the hymn 'Jerusalem', is strongly linked with 'pastoral' images of the countryside. This music is sung or heard regularly at national sporting events or festivals.
- The rural landscape paintings of Constable and Turner portray beautiful, idyllic places. Paintings of the English landscape were used by the government in the Second World War to foster patriotic feelings and behaviour (Figure 17.3).
- However, this idyllic English countryside was (and is) for many people a myth that fails to mention endemic rural poverty in the past and contemporary social deprivation.

At a time when national cultures are changing rapidly due to globalisation, the countryside can signify stability. Rural landscapes provide a comforting sense of the past in people's imagination. The age of rural buildings, a perceived stability in the appearance of the rural landscape and the proliferation of important historical landmarks and antiquities found in the countryside: all help to evoke nostalgia.

Figure 17.3 Wartime propaganda often used images of the English countryside to evoke patriotism

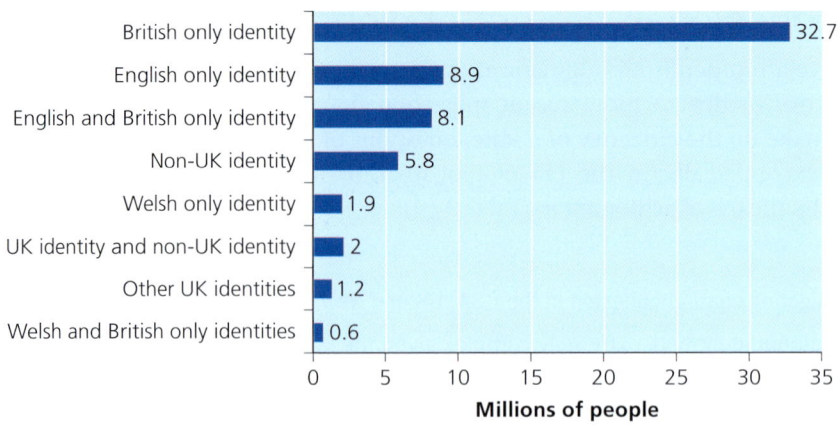

Figure 17.4 National identity results, England & Wales Census 2021

> **Skills focus: Critical analysis of source material**
>
> Figure 17.4 provides source material for critical analysis. Given how personal and complex questions of national identity are, how reliable are these results? Could they change over time?

17.2 Challenges to national identity

In many countries, state boundaries are increasingly permeable to global flows of investment (from companies and rich individuals) and ideas (from global media corporations). Many prized UK assets, from power stations to football clubs, have recently passed into foreign ownership. Anti-globalisation movements and nationalist political parties view such changes as yet another threat to national identity.

Western power and influence over the global economic system

Many UK-based companies are now foreign-owned.

- In some cases, the owner is an overseas TNC. US-based Kraft acquired Cadbury in 2010, for instance. The UK car industry is now almost entirely foreign-owned (Table 17.5, page 315).
- In other cases, a foreign government is the buyer. Both China and Qatar have enormous **sovereign wealth funds (SWFs)** that their governments draw on to purchase assets in other states. The Chinese government owns a large stake in Heathrow Airport (see Book 1, Figure 12.15, page 197).

The UK is highly liberal in this respect; the shareholder acceptance threshold level (for a foreign bid) is set at just 50 per cent. It is set higher in some other countries, making it harder for companies to be bought by overseas investors.

> **Key term**
>
> **Sovereign wealth fund (SWF):** Government-owned investment vehicles that invest the wealth of nations in companies and projects around the world. They are typically associated with China and countries that have large revenues from oil, such as Qatar. Many SWFs are worth hundreds of billions of dollars. They are controversial because they allow foreign governments to effectively own assets in another country.

The UK car industry

The UK used to be the second-largest global manufacturer of cars, after the USA. In the 1950s, exports by car companies such as Leyland, Triumph, Rover and Jaguar were the world's largest. However, rising production costs and strong German and Japanese competition led to dwindling mass car production under UK domestic ownership during the 1970s and 1980s. More recently, UK car manufacturing has rebounded, albeit largely under foreign ownership. While the following may still be widely perceived as iconic 'Made in Britain' brands, their national identity is actually more complicated.

- The Jaguar Land Rover (JLR) Company combines two of the UK's most well-known motoring brands under the foreign ownership of India's Tata Motors. Buying these names cost Tata about £1.6 billion in 2008.
- The rebirth of the Mini has been a success story for German owners BMW. The UK remains the base for Mini assembly: 'If you interfere with the authenticity of the brand, you lose the brand', according to BMW's global head of sales. This 'authenticity' is clearly contestable, however. The company is foreign-owned and the car is assembled using parts made in other EU countries (see Book 1, page 204).
- Rolls-Royce has also been run by Germany's BMW since 2003 (Figure 17.5). Between 1000 and 3000 luxury Rolls-Royce cars are sold worldwide each year at prices starting at £235,000. Manufacturing remains based at the Goodwood plant in West Sussex.
- MG Motor UK Ltd is headquartered in Birmingham but owned by China's Shanghai Automotive Industry Corporation (SAIC). Since 2017, all MG cars have been made in China and imported into the UK leaving only the MG badge and some car design work as 'British'.
- Lotus cars are still made in Norfolk but owned by Geely, a Chinese carmaker.
- The Bentley brand is produced by another German TNC, Volkswagen, in Crewe.

Figure 17.5 The Rolls-Royce 'Spirit of Ecstasy': an iconic ingredient of British identity that is now German-owned

Study of the car industry also demonstrates the complexity of so-called industrial decline in the UK. While employment has declined, output has increased but profits go offshore.

Table 17.5 The global pattern of ownership of car brands manufactured in the UK: 'Made in Britain' is an increasingly complex idea

Brand	Location in the UK	Products	Ownership
Aston Martin	Gaydon	Cars	UK
Bentley	Crewe	Cars and engines	Germany
Caterham Cars	Dartford	Cars	UK
Ford	Dagenham and Southampton	Engines and coaches	USA
Honda	Swindon	Cars and engines	Japan
Jaguar Land Rover	Castle Bromwich and Halewood	Cars	India
London EV Company	Coventry	Taxis	China
Lotus	Norwich	Cars	China
McLaren	Woking	Cars	UK
MG Motor	Longbridge	Cars	China
Mini	Oxford and Birmingham	Cars and engines	Germany
Morgan	Malvern	Cars	UK
Nissan	Sunderland	Cars and engines	Japan
Rolls-Royce	Goodwood	Cars	Germany
Toyota	Burnaston and Deeside	Cars and engines	Japan
Vauxhall	Ellesmere Port and Luton	Buses and coaches	USA

Westernisation and Americanisation

A further challenge to national identity comes from the soft power (see page 150) of large global media and retail corporations. These TNCs often promote European and North American cultural values as part of a process called 'Westernisation'. When it is solely US cultural values that are being promoted, it is called 'Americanisation'.

It is important to acknowledge that 'Western' TNCs do not spread their country of origin's culture as part of some politically motivated global power play. The guiding principle of these firms is simply to build market share on a worldwide scale; in doing so, however, they often bring cultural change to places. The continued success of major Western music and media providers, or fast-food franchises such as Pizza Hut, depends on cultural acceptance and approval of their products.

> **Skills focus: Proportional circles and pie charts**
>
> The data in Table 17.5 can be used to create a proportional circle showing the level of foreign ownership. Some countries have only one or two brands. It is important to decide whether they should be combined in a segment called 'other nations' or should each be given their own segment.

The largest Anglo-American TNCs are experts in designing and advertising aspirational products. The most successful companies, including Apple and Disney, enrol people from a spectrum of different local cultures as their consumers (Table 17.6).

Table 17.6 Large retailing and entertainment companies that contribute to 'Westernisation'

McDonald's	Serving 70 million customers a day in 37,000 restaurants across 100 countries, McDonald's is ideally placed to spread the idea of American fast food. McDonald's advertising is often attached to major sporting events as well as movie franchises such as Minions. These linkages reinforce American and Western culture.
Disney	Disney's numerous films, TV channels and resorts (in France, China, Hong Kong, Japan and the USA) have frequently been accused of portraying a sanitised and unreal vision of the world rooted in an unachievable 'American Dream' that erodes and replaces traditional cultures by its appeal. This is not a deliberate policy, more an unintended consequence.
Apple	Apple, and other major tech companies such as Microsoft and Google, tend to make people 'connect' in very similar ways regardless of where they are from. Indirectly, these connections spread Western news, music and other media which may contribute to Westernisation.

It is not inevitable that national identities are changed by the global actions of retail and media corporations, however. Actual outcomes – as experienced at the local level – can be more complex. Anglo-American TNCs increasingly gain new ideas from the world's myriad local cultures, rather than attempting to supplant them. Japanese, Indian, Korean and Nigerian influences, among others, increasingly drive innovation in global creative industries. Film, music and food industries have all thrived by mixing together Asian, South American and African influences with European and American ideas.

Western cultural influences are resisted by some players in certain places. On 14 February 2016, the local government in Kohat, Pakistan, told police officers to stop shops from selling Valentine's Day cards and items. Kohat is run by a religious political party that worries about the growing popularity of a Western tradition honouring a Christian saint. The move was very unpopular with local citizens, who have come to enjoy the ritual of present giving. The Western cultural traits that travel furthest are arguably those which generate profits for businesses and help the dominant capitalist global system to prosper.

Property markets and place identity

Growing foreign ownership of property and land is one further perceived threat to national identity.

This challenge is experienced unevenly at the local level. In the UK, for instance, non-national ownership of property affects London disproportionately. Even within London, foreign buyers are attracted to particular neighbourhoods and postcodes. Often the buyer has no intention of living in the property that they have bought. Rather, the purchase is seen as a good investment that may gain in value over time. Foreign nationals whose home state is suffering from an economic crisis or conflict may also look to the UK property market as a safe haven for their capital. The value of the Russian rouble plummeted in 2015; Russians who had invested their money in London property were insulated from this financial shock.

Figure 17.6 Walt Disney theme parks have introduced Western culture to many new places

UK citizens are responsible in turn for changing place identity in parts of the Mediterranean. Coastal areas in France, Italy and Spain have experienced in-migration from British retirees, younger entrepreneurs and sun-seekers. Collectively they have contributed to the growth of British-owned residential enclaves in cities such as Nice and smaller villages like Villefranche-sur-Mer. Sometimes, visible landscape changes occur which show that the French, Spanish or Italian **cultural landscape** has been replaced with a new British **ethnoscape** (Figure 17.7).

Figure 17.7 A British ethnoscape on the Mediterranean coastline

Key terms

Cultural landscape: The distinctive character of a geographical place or region that has been shaped over time by a combination of physical and human processes.

Ethnoscape: A cultural landscape constructed by a minority ethnic group, such as a migrant population. Their culture is clearly reflected in the way they have remade the place where they live.

Foreign ownership of London property

As a global city it is no surprise that foreign money flows into London's property market. Figure 17.8 shows that the majority of foreign buyers of new-build property are from Asia. However, overall foreign buyers represent 5–10 per cent of all buyers in London. In prime locations like Kensington and Chelsea, Westminster and the City this rises to about 30 per cent. There is very little evidence that these foreign-owned homes are left empty, so they do not deprive Londoners of homes. However, there are impacts:

- Foreign investment may serve to drive up house prices overall, because it increases demand especially in central London where housing supply shortages are most acute.
- The average house price in Kensington and Chelsea in 2020 was £1.85 million or about 50 times the average London salary.
- Some locations, such as Chelsea, Mayfair and Belgravia (Russians), Kensington (French) and St John's Wood (Americans) are seeing their character change as services such as shops and restaurants increasingly cater for foreign property buyers.

However, this is really no different to the way in which inner city and suburban areas of many UK cities have changed in response to successive waves of post-colonial and EU migration since the 1950s. The key issue is whether foreign property investment makes it harder for local residents to buy property, plus the related issues that some of the money being invested may be linked to criminality and be 'hidden' in London's property market.

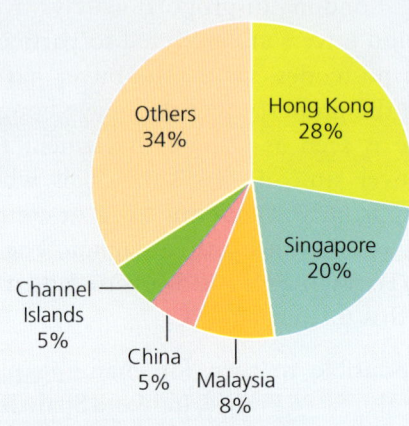

Figure 17.8 Nationality of foreign buyers of new-build property in London, 2014–16

17.3 The consequences of disunity within nations and states

> **Key term**
>
> **Secession:** The act of separation for part of a state to create a new and fully independent country.

As Chapter 15 explained, there is an important distinction between a nation and a state. A little confusingly, the term nationalism is used widely to describe both the promotion of sovereign states (such as the UK, Spain or Nigeria) and those smaller nations which lack sovereignty but form part of those states (Scotland, Catalonia and Biafra, respectively). It is possible to talk about either British nationalism or Scottish nationalism, for instance. The following section focuses on issues arising from the significant national (sometimes understood as ethnic) divisions that exist *within* states. In some cases, **secession** is demanded.

Option 8B: Migration, Identity and Sovereignty

Nationalist movements for secession: Catalonia

Catalonia is an autonomous region of Spain. Many people speak Catalan, rather than Castilian Spanish and the region has its own traditions and food.

- Autonomy existed in the 1930s but Catalonia and Barcelona sided with the losing Republicans during the Spanish Civil War: the victorious Nationalists under General Franco ended dreams of Catalan independence.
- In 1979, the new democratic government of Spain recognised Catalonians as a separate nationality within the Spanish state but only granted autonomy not independence.

Demands for full independence grew, culminating in huge pro-independence demonstrations in 2013–14. Catalonia held a referendum on independence in 2017. However, this was not approved by the government in Madrid and was declared illegal. The turnout was 43 per cent, but with 90 per cent of voters in favour of independence. Although questionable, this was deemed enough by Catalan President Carles Puigdemont to declare independence from Spain.

Figure 17.9 Catalonia map

The government in Madrid stood firm (no EU state recognised the result as valid), and independence was not achieved. Some leaders of the Catalan separatist movement were put on trial, and others fled the county. The trial of some leaders has, if anything, further inflamed tensions with Madrid. Today, support for independence runs at 40–50 per cent so the issue is by no means settled.

Figure 17.10 Catalonian flags on the streets of Barcelona

17 Threats to national sovereignty

Scotland and the UK

Scottish people chose to remain part of the UK in 2014, but only by a small margin. Forty-five per cent of the 3,619,915 people who voted in the referendum on independence indicated they would prefer to be entirely self-governed (Figure 17.11).

- Some Scots resent the way their nation's destiny is controlled by mainly English politicians sitting in Westminster. Scotland's history as part of the UK is complicated. Any mention of ancient conflicts with the English, including the Battle of Culloden in 1746, can stir up strong feelings.
- There is very little support remaining for either the English Conservative Party or Labour Party in Scotland. Many Scots therefore object to being governed by either. In the UK 2019 General Election, 48 of the 59 Scottish seats were won by the Scottish National Party, which supports full independence for Scotland.
- There are economic reasons why some Scots believe they would be better off going it alone. However, much of Scotland's income comes from oil and recent low prices mean independence could come at a greater cost than some nationalists previously supposed.
- In 2020, support for Scottish independence was between 45 and 50 per cent – slightly higher than at any time since 2014.
- Brexit may have helped the cause of Scottish independence as support for the EU is strong in Scotland and leaving the EU is widely viewed as a mistake by 60 per cent or more of Scotland's population.
- Despite the 2014 referendum being billed as 'once in a generation' there are demands for a re-run and now the UK (including Scotland) is outside the EU the result in the future could be very different.

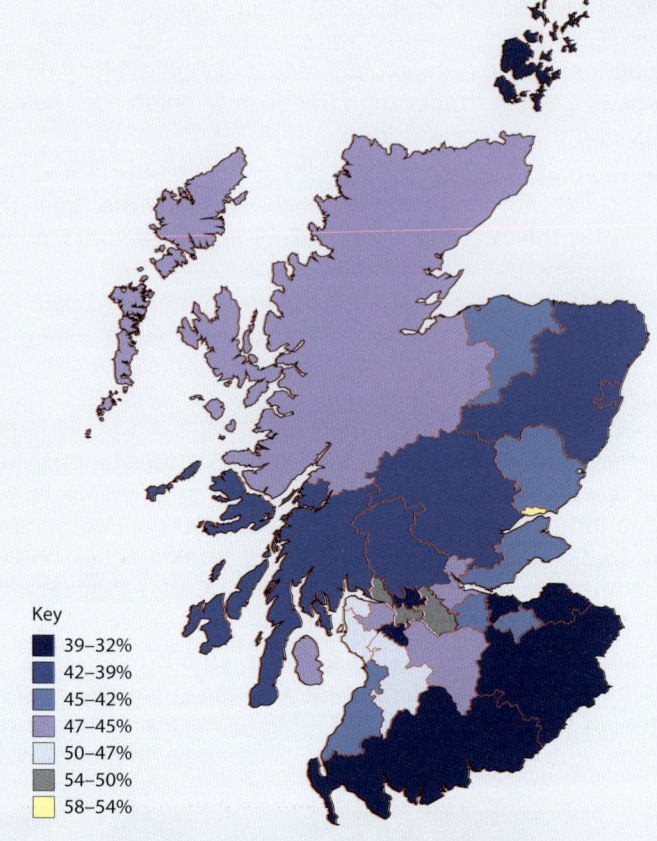

Figure 17.11 The proportion of people in different Scottish regions who voted to leave the UK in 2014

Key
- 39–32%
- 42–39%
- 45–42%
- 47–45%
- 50–47%
- 54–50%
- 58–54%

Option 8B: Migration, Identity and Sovereignty

The internal tensions of the BRIC nations

Many emerging economies, including members of the BRIC group, have experienced significant internal political tensions in recent years. Although they may have prospered overall, the costs and benefits of globalisation have not been distributed evenly among their populations. In some cases, this uneven pattern has clear regional or ethnic dimensions that threaten disunity (Table 17.7).

Figure 17.12 A masked protestor in Hong Kong, 2019

Table 17.7 Evaluating the internal tensions and disunity within selected emerging economies

Emerging economy	Evaluating the internal tensions
Brazil	Brazil's hosting of the Football World Cup in 2014 exposed serious divisions in Brazilian society. Over US$22 billion was spent in preparation for the World Cup, including stadium and infrastructure construction. Large-scale protests took place against this, with campaigners saying the money would have been better spent alleviating the problems that many ordinary Brazilians still face on a day-to-day basis. These include poor public services, high food prices and the need for political reform due to corruption. The displacement of many poorer people from their homes to make room for the new sports stadiums triggered sometimes violent demonstrations in São Paulo and Rio. In March 2016 over 7 million Brazilians protested against the alleged corruption of the Rousseff government. Rousseff's successor, Jair Bolsonaro elected in 2019, has divided Brazilian society on a wide range of issues including support for deforestation and anti-LGBT rights.
China	China has been rocked by two recent nationality issues. In western Xinjiang province some ethnic Uyghurs (who are Muslim) have led small-scale separatist attacks on ethnic Chinese. Since 2014 the Chinese government has increased surveillance, detention and imprisonment of suspected Uyghur 'agitators' including constructing 're-education camps' Some observers have likened the Chinese government action to genocide. In Hong Kong the 'Umbrella Revolution' of pro-democracy protests began in 2019 as a response to moves from Beijing to exert more control of its semi-autonomous region. Sometimes violent protests were met with a hard-line from China culminating in a new National Security Law in 2020, which effectively gives China control over Hong Kong in breach of the Sino-British Joint Declaration of 1985.
India	India has long-running tensions between the Hindi majority and Muslim minority. Riots in Gujarat in 2002 and Muzaffarnagar in 2013 were a foretaste of deadly riots on Delhi is 2020 that killed 53. Hindus and Muslims came to blows over Prime Minster Modi's 2019 Citizenship Amendment Act. This Act has been widely seen as excluding some Indian Muslims from a right to Indian citizenship and its implementation could be viewed as act of Hindu nationalism designed to alienate India's Muslims.

The identity crisis of 'fragile states'

Inevitably, the greatest challenge for unity is found within fragile states: those countries whose government has lost political control and is unable to fulfil its basic responsibilities. As previous chapters have shown, the cultural and religious diversity of some developing countries combined dangerously with a 'power vacuum' in the immediate post–colonial period.

In the 1960s and 1970s, stark differences grew quickly between the prosperity of a politically and economically powerful elite (who allied themselves with foreign investors) and the wider population. The ensuing instability brought so-called fragile state status to Democratic Republic of the Congo, Rwanda, Libya, Somalia and Yemen.

In the twenty-first century, identity crisis problems have persisted for some fragile states:

- Sudan has lacked internal cohesion ever since it became an independent state in 1955, partly due to its large size. An ethnically and culturally diverse nation, Sudan suffered a strong sense of internal separation between its northern and southern regions from the outset, due to its colonial history. Britain divided it along ethnic lines with the Arabs in the north and the mainly black African tribal communities in the south. In 2011, after 50 years of internal strife and 2 million lives lost to conflict, Sudan finally abandoned the struggle for unity. It divided in half when the new state of South Sudan broke away. To the north, the remainder of Sudan retained its name and capital city of Khartoum. South Sudan continues to have some of the worst health and development indicators in the world, including an under-fives infant mortality rate of 381 per 1000 and less than 1 per cent of girls in primary school.
- In 2014, hundreds of Christian families were forced to leave Mosul in northern Iraq after ISIS gave them an ultimatum: convert to Islam or face death. Thousands subsequently fled the country.

For those countries suffering the greatest internal disunity there are no easy solutions. Their borders were drawn under European rule, throwing together disparate tribes to create states where unity remains a distant goal. Yet secession sometimes creates as many problems as it solves, as South Sudan has discovered (see place context, page 287). Somehow, the world's fragile states must find a way to persevere and try to find a common voice as a country.

Review questions

1 Explain what is meant by 'national identity' and what factors reinforce it.
2 To what extent do you think 'British values' are held by most people in the UK?
3 Assess the extent to which sport, culture, politics and the landscape reinforce cultural identity.
4 Explain why cultural cohesion is stronger in some countries than others.
5 Explain why the idea of 'Britishness' might have been undermined by trends in the economy and industry, such as the car industry.
6 Explain what is meant by 'Westernisation'.
7 Using examples, evaluate the extent to which calls for some nations to separate from larger sovereign states have been successful.

Further research

Examine the original source material for David Cameron's 'British values' speech (2014): www.gov.uk/government/news/british-values-article-by-david-cameron

Explore the concept of Britishness: https://constitution-unit.com/2018/12/14/on-the-myth-of-a-growing-sense-of-english-identity/

Research details of the current state of the UK car industry: https://www.smmt.co.uk/industry-topics/uk-automotive/

Explore the issue of devolution within the UK, including Scottish Independence: https://www.instituteforgovernment.org.uk/our-work/devolution

Investigate the issues of China's Uighur people in more depth: https://www.bbc.com/news/world-asia-china-22278037

Exam-style questions

1. Study Table 17.8 below, which shows the distribution of EU migrants across five EU states in 2012.

Table 17.8 Distribution of migrants across EU states, 2012

State	Number of EU migrants living there	Total population
Germany	3,362,600	80,780,000
Spain	2,341,600	46,507,000
Belgium	773,500	11,203,000
Sweden	483,000	9,644,000
Ireland	434,300	4,604,000

   a   Calculate the range of values for migrants in Table 17.8. [1]
   b   Suggest one benefit of using logarithmic graph paper to create a scatter plot of these data. [2]
   c   Name a statistical test that could be used to compare the distribution of migrants living in Germany and Spain. [1]
   d   Suggest reasons for the relationships shown. [6]

2. Explain why sovereign states vary in their level of cultural and political unity. [8]

3. Evaluate the view that global organisations have failed to respond effectively to the most important challenges the world faces. [20]

18 Synoptic themes

By the end of this chapter you should be able to:
- identify links between key concepts and themes across all of the content of your A level course
- recognise that organisations, groups and individuals across the world have different roles to play in causing and managing geographical issues
- understand that attitudes to geographical issues vary enormously and that this influences both the choices people make and their actions
- understand that future geographical issues are affected by choices made today, but that there is uncertainty about the future.

This section of the book explains the three synoptic themes that appear in the specification. The themes are:

1. Players.
2. Actions and attitudes.
3. Futures and uncertainties.

These themes are used to link together different topics and help to show that seemingly very different topics can be linked by common ideas. It also considers some of the specialist geographical concepts in the specification. These are big ideas that cut across many different topics and are another way of being synoptic.

What are synoptic themes?

Being synoptic means seeing 'the big picture' and 'thinking like a geographer' as well as drawing together different elements of geography. Different topics are usually studied in a linear order that allows the development of a detailed understanding of each topic in turn. Being synoptic means:

- seeing links between topics
- making connections between different places and peoples
- recognising that some topics or concepts seem to link to almost all other topics.

Figure 18.1 shows the difference between linear understanding and 'joined-up' understanding. Making links between seemingly separate topics is an important part of being synoptic.

In addition to making links, there are three synoptic themes that appear throughout the specification:

1. **Players:** the people, groups and organisations that cause geographical issues, suffer the consequences of them, make decisions and attempt to manage them.
2. **Attitudes and actions:** the views of different players, why they are held and how they affect decision-making and choice of actions.
3. **Futures and uncertainties:** how different actions taken today will affect people and the planet in the future, and the degree of uncertainty over this.

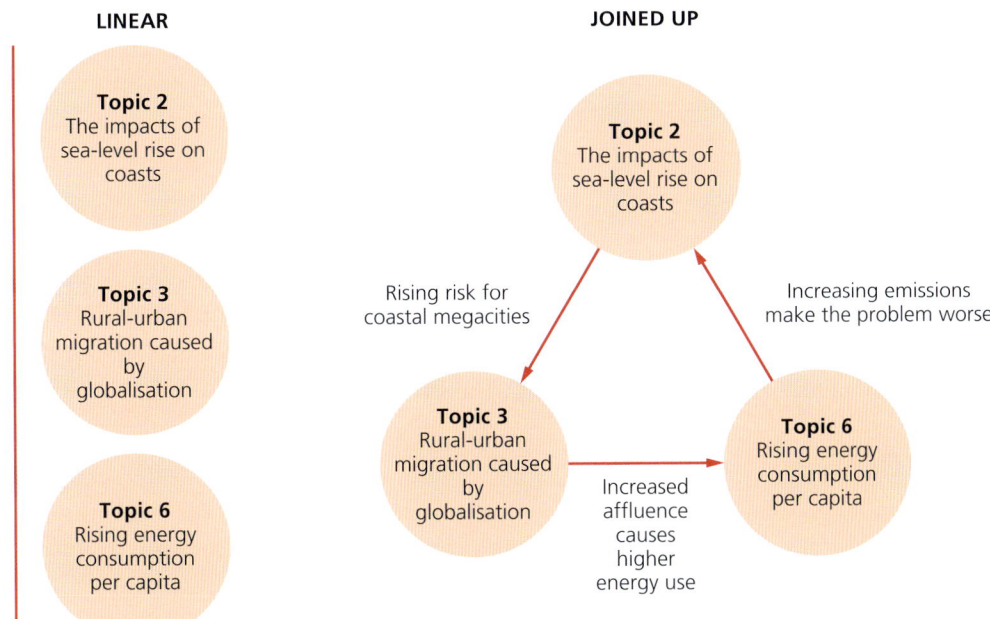

Figure 18.1 Linear versus synoptic understanding

There is a relationship between the three themes, so they should not be seen in isolation (Figure 18.2).

There are also some specialist geographical concepts that apply to different areas of geography. These are covered in the Key concept boxes in this section.

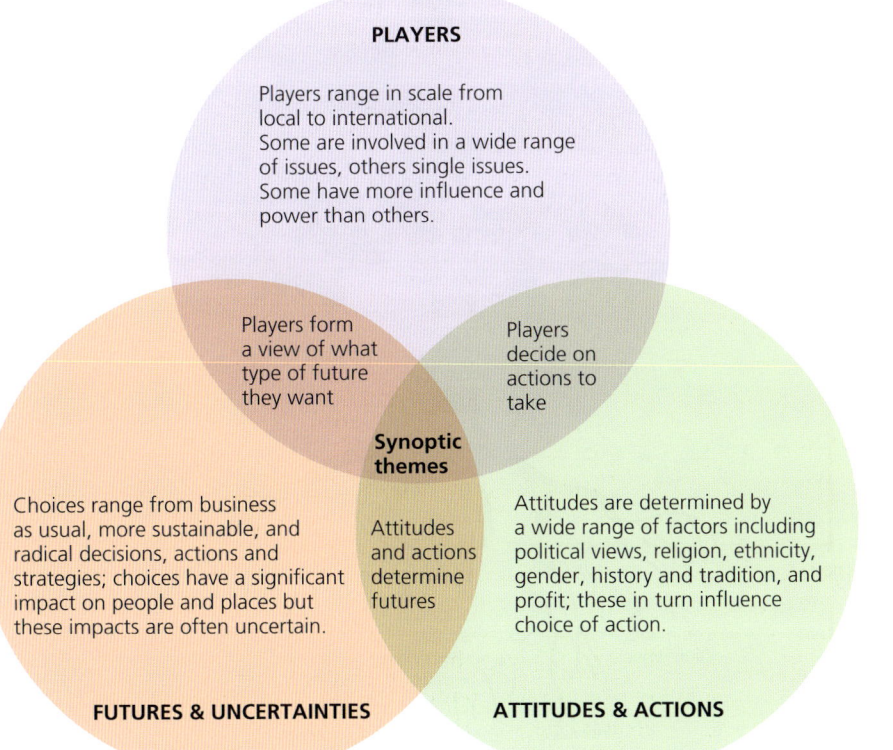

Figure 18.2 Links between the three synoptic themes

18 Synoptic themes

Players

The term 'player' is used to refer to any individual, organisation or group involved in a geographical issue. They could:

- cause the issue, either directly or indirectly
- suffer the consequences of the issue, that is, be affected by it either in a negative or a positive way
- be involved in managing the issue as a decision maker.

Players range from individuals to global organisations. They operate at different scales and, because of this, tend to be involved in slightly different issues. Figure 18.3 shows three broad geographical scales operating within the decision-making tree.

The roots

- Local-scale geographical issues often involve individuals focused on single issues. Individuals may group together, forming community action or pressure groups.
- Sometimes these grow into more formal organisations and join forces with other groups, for instance unions or larger environmental pressure groups.
- More formal groups with a local/national focus are umbrella organisations such as local chambers of commerce and professional bodies.
- Many organisations are national but also campaign on local issues, such as the Campaign to Protect Rural England.

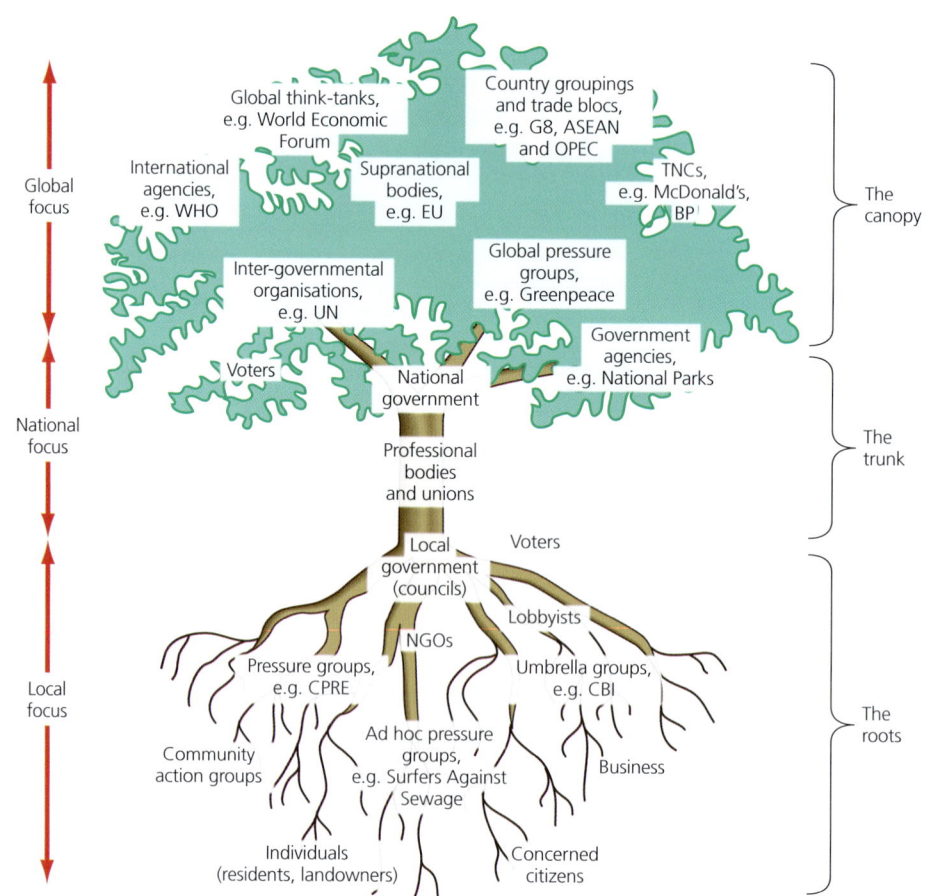

Figure 18.3 Players in the decision-making tree

Option 8B: Migration, Identity and Sovereignty

The trunk
- Many decisions are made by government, both local and national.
- In democracies these are elected bodies and voters can decide to remove them from office; this is a 'safety valve' to guard against unpopular decisions.

The canopy
- Decision makers in the 'trunk' take account of players that have a more global outlook. Government decisions are influenced by international treaties and agreements, such as the 2015 Paris Climate Change Agreement.
- In the UK, the supranational body of the EU has been a powerful player in UK decisions, but post-2020 the UK will need to develop new relationships with European countries and may have opportunities to form different relationships with countries outside the EU.
- Also at a global scale are IGOs, think-tanks, pressure groups, business organisations and TNCs. Many of these have significant power to influence government thinking and decision-making.

Some players have more influence than others, meaning there is **inequality** in terms of power and influence over geographical issues and decision-making.

> **Key concept: Inequality**
>
> At all geographical scales there are differences in opportunity, access to resources and influence between different groups. This is inequality. In terms of players, decision-making power and influence tends to rest with people who:
>
> - have political power, either through being elected or because they have taken political power by force or through corruption
> - have financial resources – these can be used to influence others, fund investments and even 'pay off' opponents
> - have influence through culture and tradition, such as inherited influence.
>
> In some societies, some groups – such as women, minority ethnic and minority religious groups – lack power and influence because of discriminatory laws and cultural prejudice.

Airport expansion

A good example of the involvement of different players in a geographical issue is the long-running debate over airport runway expansion in the southeast of England. This issue has local, national and global dimensions and illustrates how **globalisation** and **interdependence** increasingly affect geographical issues.

There is a strong economic case for airport expansion in the South East because existing runway capacity at Heathrow and Gatwick is fully utilised. In order for London to compete as a 'world city' it needs a global hub airport. The 2015 Davies Report on airport expansion argued that inbound UK tourism was worth £56 billion in 2014, and the aviation industry £12 billion, supporting 160,000 jobs. By 2014, three main options were on the table (Figure 18.4).

Local concerns
Not surprisingly, airport expansion is opposed by many local residents, not least those whose homes would be demolished by Heathrow's expansion. Noise is a major local issue, as is traffic congestion. Many local MPs are against Heathrow expansion too. Local campaign groups, such as Stop Heathrow Expansion (SHE), have campaigned against the proposals; AirportWatch operates as an umbrella campaign organisation uniting many opposed groups, including environmental organisations such as Friends of the Earth.

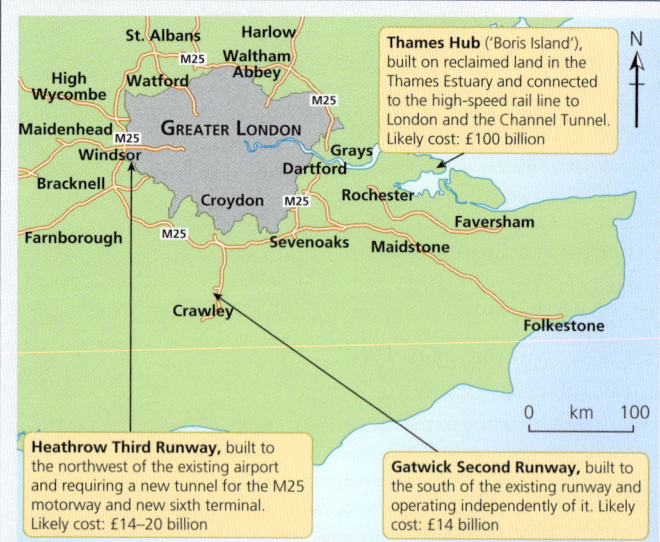

Figure 18.4 Airport expansion options for Southeast England in 2014

especially at Heathrow. These include unions such as the GMB and Unite, whose members would benefit from employment expansion. The Confederation of British Industry (CBI), which represents businesses, is also supportive. In 2018 MPs voted 415 to 119 in favour of a third runway at Heathrow.

Global players

Crucially, many of the major supporters of expansion are global TNCs. These include airlines such as British Airways, easyJet and Virgin Atlantic. Investors in Heathrow Airport Holdings (the owners of Heathrow) include Ferrovial, a Spanish transportation TNC, and sovereign wealth funds from Singapore, China and Qatar. There is also a global consensus from organisations like the OECD, IMF and World Economic Forum that 'globalisation is good', so greater connectivity through airport expansion is seen as desirable.

As of 2020, the future of Heathrow is very uncertain. Concerns about the environmental impact of flying are now widespread. The UK's exit from the UK has altered the economic landscape, and the Covid-19 pandemic has thrown the air travel industry into a period of enormous uncertainty. The case of expansion may be weaker than in the very recent past.

National interests

It is worth noting that the former Mayor of London, Boris Johnson, managed to use his high profile to get his own Thames Hub Airport proposal 'on the table', illustrating the political power of high office. Nationally a range of organisations have supported expansion,

---

Key concept: Globalisation and interdependence

Globalisation is the increasing integration of economies, societies and cultures through global networks of trade, communication, immigration, transportation and political decision-making. It creates a more interconnected world, and also a more interdependent one. This interdependence means that:

- The economic success of one country is tied to and dependent on economic success in other countries.
- Countries have to take into account the fact that many of their own citizens live and work abroad, and that many foreign workers reside in their country.
- Political and economic decisions are not taken in isolation; they have to take account of decisions made in other countries and by other players.

The option to be an 'island' with few ties or links to other countries no longer exists for a country wishing to be economically successful. Many governments pursue policies that enhance global integration, be that through trade deals, global environmental treaties or global financial governance.

Decision-making in most situations tends to flow towards the trunk of the decision-making tree in Figure 18.3 (page 326). Final decisions about geographical issues are usually made either by government at national or local level, or internationally by IGOs, as part of a **system** of decision-making. Players in the canopy are larger and better funded than those at the roots. Consequently, the canopy players are likely to have more power and influence than the root players.

Attitudes and actions

The attitudes of different players and groups vary significantly from place to place. Figure 18.6 (page 330) shows an example of this from the Pew Research Center's 2018 Global Attitudes survey. People in different countries were asked what they considered 'X is a major threat to our country'. The result of this survey revealed wide differences in attitudes and shows the importance of context and **identity** in influencing people's attitudes.

- Climate change is perceived by many as a threat in Greece and France, much less so in the USA where the focus is often more on economic development than environmental protection.
- Japan and the USA, technologically advanced nations, perceive cyberattacks as a threat. Much more so than less developed Kenya.
- Many people in Europe perceive terrorism as a threat: France especially has suffered numerous attacks which have entered the national consciousness.
- Greeks, badly affected by the 2007–8 Global Financial Crisis, perceive the global economy as a threat much more than other countries.

Recent history, cultural attitudes to the environment, physical geography, age, immigration and demographics can all have a major impact on attitudes such that there are significant differences between countries even at the same level of development.

> ### Key concept: Systems
>
> Decision-making concerning geographical issues operates as a system, as shown in Figure 18.5. However, the system varies from place to place:
>
> - In democracies it is often a transparent system, allowing different players to contribute their views and, in many cases, appeal decisions.
> - In authoritarian countries there may be no transparency at all, with decisions made without recourse to those affected or interested.
> - Even an apparently transparent system might be open to abuse and corruption.
> - Even in democratic countries, some decisions concerning security or sensitive business interests may be made 'behind closed doors' and not be subject to public scrutiny.
>
>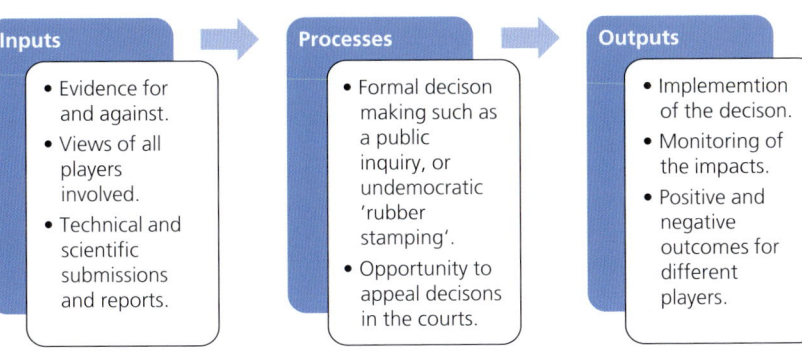
>
> **Figure 18.5** The decision-making system

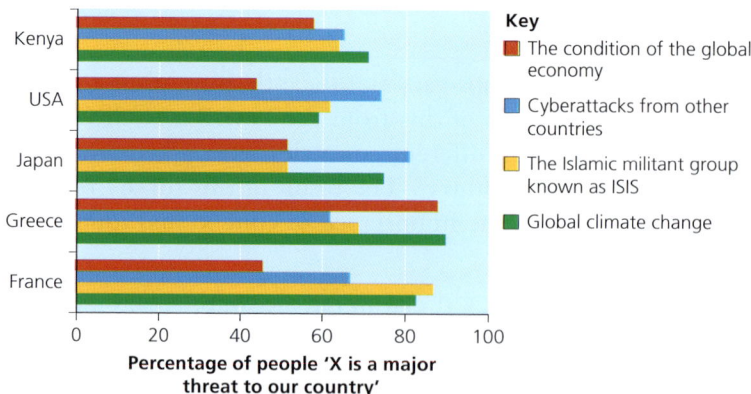

Figure 18.6 Pew Research Center's 2018 Global Attitudes survey: X is a major threat to our country

> ### Key concept: Identity
>
> Identity refers to the beliefs, perceptions and characteristics that make one group of people different from another. This is strongly related to place. Differences in identity mean that we should not expect contrasting groups of people to share the same fears, desires, ambitions or concerns. What is important to one group of people may be largely irrelevant to another.
>
> Identity, and strong attachment to specific places, shapes the attitude of traditional and tribal societies. Although small in number, Native American, Australian indigenous, Amazonian tribes and other indigenous groups share some common characteristics:
>
> - direct connection to the land and landscape as a source of food and shelter
> - religious significance placed on landscape features, plants and animals
> - strong sense of territory.
>
> These groups are likely to resist change and attempts to alter their landscape and territory because of their direct connection to it.
>
> Equally, strongly held religious views can influence attitudes, such as the Amish shunning modern technology (and therefore limiting contact with non-Amish people) or Buddhists' concern for environmental issues.

Political and economic systems

Even within developed countries there is no consensus on how a country should be run. This can be illustrated by examining a small number of very similar countries:

- Nordic-model countries, including Denmark, Sweden, Finland and Norway
- Anglo-Saxon-model countries, including the UK, Ireland, Australia and New Zealand
- the USA, which is considered separately.

All of these countries are advanced capitalist economies with high incomes, as well as being mature democracies. Figure 18.7 shows how the amount of tax you can be expected to pay in relation to your income varies between the three groups.

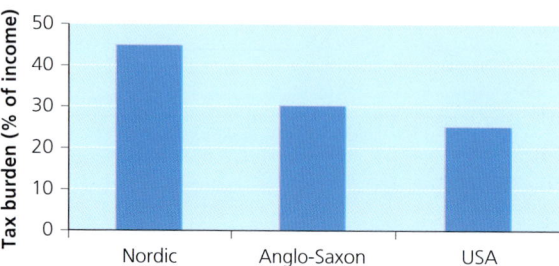

Figure 18.7 Tax burden as a percentage of income

Option 8B: Migration, Identity and Sovereignty

The differences in Figure 18.7 result from the attitudes held and policy choices taken in the three groups of countries. They reflect subtle cultural and identity differences rather than fundamental differences in their way of life:

- Nordic countries use high taxes to fund comprehensive welfare state systems providing top-quality universal health, education and pension systems designed to reduce inequality and promote social mobility. This system has reduced poverty and inequality. In Sweden, people hand over about 50 per cent of all earned income in tax.
- Anglo-Saxon countries have a lower tax burden and, as a result, a welfare system that is more of a safety net to support people temporarily as opposed to the more extensive Nordic system.
- In the USA, a tax burden of about 25 per cent precludes the provision of free, universal healthcare and generous state pensions; the attitude is more 'stand on your own two feet' and provide for yourself rather than expecting the government to provide for you.

All of the countries mentioned above are close allies in the 'West' with similar views on global issues, but they are not the same in terms of attitudes to social justice, equality or even the environment.

Political viewpoint has a strong influence on attitudes and actions. Those who prioritise people, society and responsibility tend to be on the left wing of the political spectrum. Those who rank wealth creation and individual rights highly tend to be on the right wing. Figure 18.8 shows the range of political viewpoints using a diagram called a Nolan chart.

Figure 18.8 A Nolan chart of the spectrum of political viewpoints

It might appear that political viewpoints have little relevance in geography. However, they strongly influence decision-making. Consider an issue such as the need to increase water supplies. The method chosen will be influenced by political viewpoint:

- Authoritarian political systems tend to build large government-planned, top-down schemes such as dams and reservoirs. These will be seen as increasing the prestige of the government and benefiting the whole nation. Environmental and individual social costs will tend not to be considered. The Three Gorges Dam in China is an example of this approach.

- Left-wing political systems might focus on smaller-scale, local schemes using intermediate technology and local decision-making (bottom-up). Individual needs are more likely to be considered. The government would own and build the schemes as they would be seen as necessary for the social good.
- By contrast, right-wing political systems are likely to look for a market-led solution, inviting private businesses and TNCs to increase water supply. These businesses would pick the most economical solution and make a profit. Individual rights would be respected, for instance by paying compensation to landowners. Environmental issues could be largely overlooked.
- In a libertarian system, anyone would be able to get water from anywhere without any interference from government. Many people fear this would lead to chaos and anarchy.

In many European countries, including the UK, politics has drifted toward the centre. Water supply in the UK is privatised but the government sets minimum supply standards and uses Ofwat to regulate prices and protect consumers from excessive price rises. This political centre 'mix' of private economic freedom and government protection is now very common.

Political viewpoint also has an influence on the question of **mitigation** and **adaptation** policies.

> ### Key concept: Mitigation and adaptation
>
> These related terms are most often used in the context of managing global warming.
>
> - **Mitigation:** preventing something from happening, for example, reducing greenhouse gas emissions now to stop future global warming, or attempting to prevent a natural hazard from occurring.
> - **Adaptation:** dealing with the impacts of something, for example, adapting our lifestyles to cope with a warming world, or migration as a solution to poverty.
>
> These terms might be summarised as dealing with the causes (mitigation) or dealing with the consequences (adaptation). Table 18.1 summarises the possible relationship between political standpoint and these two approaches.
>
> **Table 18.1** Mitigation, adaptation and political viewpoint
>
> | How should the problem of global warming be managed? ||
> |---|---|
> | Left-wing or centrist approach ↓ | Right-wing or libertarian approach ↓ |
> | MITIGATION
'Collective' policies, where a whole society chooses to act | ADAPTATION
Individuals adapt as best they can and private enterprise is relied on to provide 'solutions' |
> | Attitudinal 'fix': changing behaviour to reduce the harm done to others, e.g. leading a 'greener' lifestyle to reduce carbon emissions | Technological 'fix': coping with global warming by building flood defences and genetically engineering new crops |

Conservation versus exploitation

A major synoptic theme is the debate over conservation of the environment, landscape and natural resources versus exploitation. Figure 18.9 shows that there are three broad approaches to this, which reflect very different attitudes.

Protection	MOTIVE	Profit

Conservation	Sustainable management	Exploitation
Limiting resource exploitation to minimum requirements of humans, and protecting as much of the natural environment as possible	Exploiting resources in some places, conserving in others; where resources are exploited, they are managed to minimise losses to ecosystems and environment	Using natural resources to maximise profits and economic growth; resources are seen in economic value terms only

Figure 18.9 Contrasting attitudes to natural resources

It's important to recognise that attitudes to the environment are not static. They change as a country develops. This is shown by the environmental Kuznets curve model (Figure 18.10).

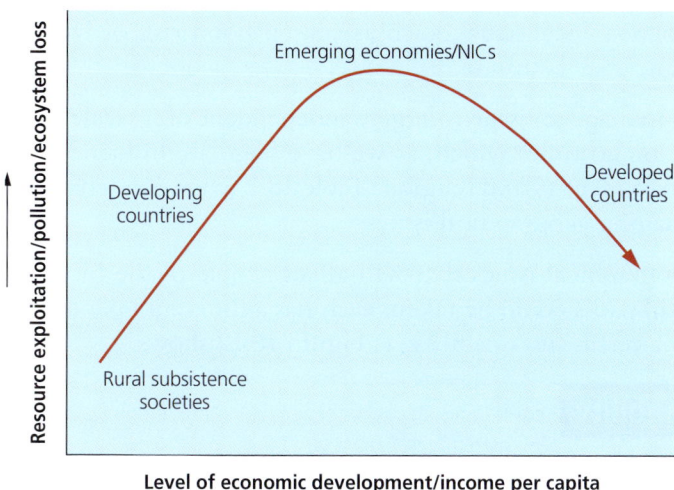

Figure 18.10 The environmental Kuznets curve

Changes in Figure 18.10 can be explained by:

- the limited environmental impact of small-scale, subsistence farming in LDCs/rural subsistence societies
- the processes of industrialisation and urbanisation, which gradually lead to environmental degradation
- the lack of management systems and monitoring in developing and emerging countries, where the priority is economic development.

At some stage a 'turning point' is reached – as development level increases, the environmental situation begins to improve. This could be because:

- people in high-income societies have money, and leisure time, and demand improvements to environmental management so they can enjoy the landscape
- once most people's basic human needs are met (water, shelter, food, healthcare) their attention turns to environmental quality and quality of life
- richer countries have money and governance systems – often democracy – that lead to a demand for change and the means to action change.

However, there are reasons to suspect that Figure 18.10 may not be the full story:

- It could be argued that developed countries have simply 'exported' their pollution to emerging countries; 'made in China' usually means 'pollution in China'.
- Population pressure and poverty in many LDCs is leading to large-scale deforestation, desertification, overhunting for bush meat, unsustainable water use and overfishing – so the argument that poor places have a limited impact on the environment may not be the case.

Figure 18.10 is strongly linked to the concept of **sustainability**, which is a highly contested concept.

Futures and uncertainties

There are many geographical issues that players need to make decisions about in terms of management approach. These range from very obvious global issues, such as our changing climate, to very local issues, such as coastal flood risk or how to regenerate rundown areas.

The decision about how to manage an issue is largely independent of its scale. The choice depends much more on attitudes which, as we have seen, are influenced by factors ranging from politics to identity and religion. Table 18.2 (page 335) contrasts different approaches to managing geographical issues.

Making decisions about geographical issues that will affect people in the future is not an easy task. Three broad paths could be taken: each has a different take on the key question of development versus sustainability, as Figure 18.12 shows.

Key concept: Sustainability

Sustainability has become a geographical buzzword. It was first used in the context of sustainable development in the 1987 UN World Development Report to mean 'development that meets the needs of the present without compromising the ability of future generations to meet their own needs'. Figure 18.11 illustrates the three pillars of sustainability.

Using this model, sustainability is met if economic development is equitable (fair) and does not damage the environment (viable). In addition, the environment itself meets people's needs (is bearable) in terms of the resources it provides.

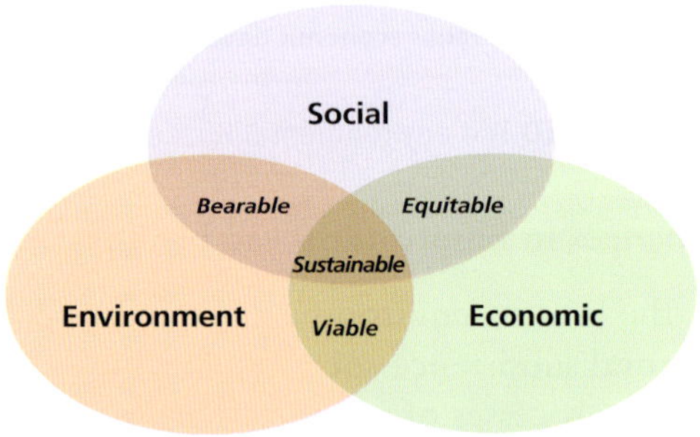

Figure 18.11 The three pillars of sustainability

More recently, sustainability has taken on the wider meaning of 'environmental sustainability'. This implies that humans need to reduce their impact on the planet to a sustainable level – one that prevents irreversible environmental damage.

Sustainability can be quite a difficult concept. A common misconception is to assume sustainability means that a solution will 'last a long time'. This is not necessarily the case. Large dams and reservoirs are designed to last a long time but, because they use huge quantities of concrete, frequently displace people and flood ecosystems, they may not be sustainable. Conversely, a large number of small earth dams designed to last ten years may be much more sustainable if they are built with local earth using human labour. Judging whether something meets the criteria for sustainability (Figure 18.11) is, therefore, a complex task.

Table 18.2 Contrasting management approaches

Finance source	Solutions and management can be led by the free market, i.e. financed by private businesses or TNCs; alternatively, money can come from government and so be state-led; NGOs finance management strategies as do individuals themselves (self-help)
Management control	There is a significant contrast between top-down decision-making and implementation led by large organisations (local and national government, TNCs, IGOs) and bottom-up approaches led by and involving local groups, communities and NGOs
Technology	Some solutions rely on sophisticated, hi-tech equipment and technology, whereas others prefer a low-tech, intermediate technology approach that uses fewer resources
Resources	There is a contrast between management strategies that are capital intensive (high financial costs) and those that are labour intensive

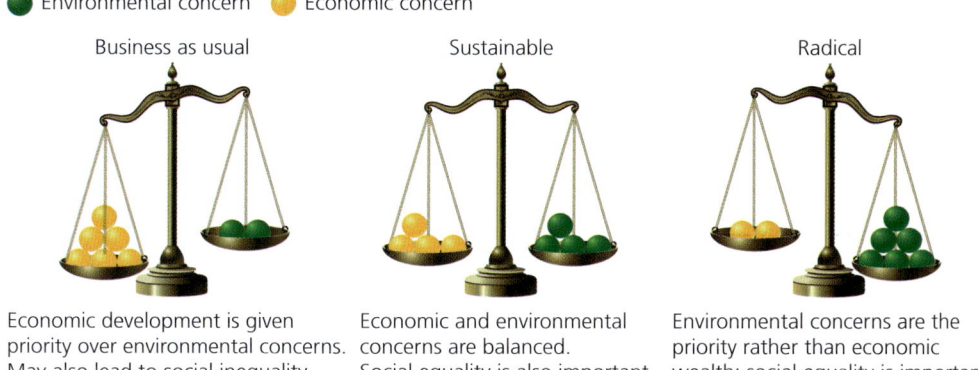

Figure 18.12 Business as usual, sustainable and radical futures

'Business as usual' implies choices being made today that mean the future would be similar to the present. Very broadly, this would mean a world that:

- continues to rely on fossil fuels
- continues to use non-renewable resources and to generate waste and pollution
- moves further down the globalisation path, with increasing wealth for some and relative poverty for others
- continues to cause environmental degradation to water supplies, ecosystems, soils and the atmosphere.

To many people – from individual consumers to some governments and international organisations (for example, the IPCC and WWF) – this type of future is increasingly risky. Many would focus on global climate change as the key threat that needs to be addressed. A more sustainable future could mean:

- a move towards renewable energy sources
- switching to renewable resources, recycling and efficient resource use
- pollution would increasingly have to be paid for
- a more guarded approach to globalisation, using systems like fair trade to ensure greater equality and social good
- a global effort to reduce key environmental concerns, such as carbon emissions and deforestation.

> **Key term**
>
> **Polluter Pays Principle:** The idea that whoever generates pollution should pay the costs of cleaning it up, either through taxes or fines or by being forced to use technology to prevent its emission in the first place.

Some players would argue that even this does not go far enough. Radical 'green growth' policies, advocated by green political parties, environmentalists and some scientists, would go further:

- wholesale switching to renewable energy and low-energy production systems, such as organic farming
- ethical consumerism, advocating locally sourced, renewable, fair-traded products
- widespread use of green taxation to reduce pollution and protect the biosphere
- decentralised government and local decision-making.

This radical approach emphasises the link between ecosystems and human well-being. Activities that disrupt ecosystems will, ultimately, disrupt humans. The **Polluter Pays Principle** and Precautionary Principle are at the heart of green political decision-making. The latter principle states that humans should exercise caution when introducing new technology (such as GM crops) until the technology can be demonstrated to have no negative impact on human or ecosystem well-being.

Many players argue that action on issues such as global warming, water shortages and pollution levels is required today because failure to act will create unacceptable **risks** in the future as key **thresholds** are passed.

Uncertainties

Decisions made today about geographical issues will affect people in the future. However, there are a number of key uncertainties about the future that make decision-making a major challenge. The most fundamental of these is demographic uncertainty. Figure 18.14 shows projected global population growth to 2100.

> ### Key concept: Risk and thresholds
>
> In geography, risk is best thought of as degree of exposure to harm. The risk could involve harm from natural hazards, running out of food or water, falling into poverty or contracting a disease, among others. Human actions can increase or decrease these risks. However, as Figure 18.13 shows, in some circumstances systems can be forced across a threshold or tipping point from one stable state to a new stable state. This new stable state could involve much higher risks than before.
>
>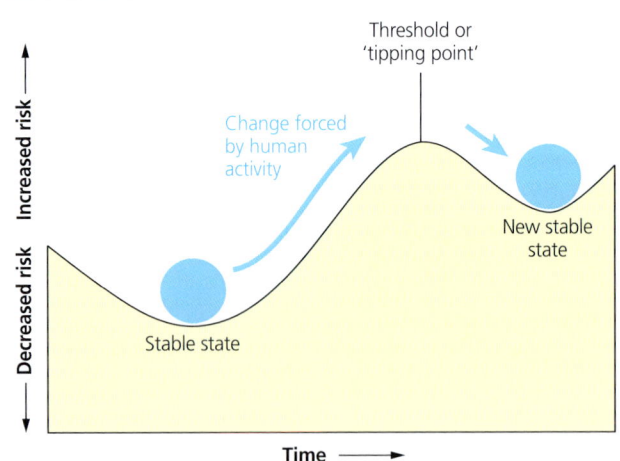
>
> Figure 18.13 The threshold concept
>
> Crossing thresholds could dramatically increase risk. Some possible future thresholds include:
>
> - passing 2°C of global warming, which many scientists argue will result in widespread ecological and climate system changes
> - overexploitation of renewable water resources beyond their annual recharge rate, leading to rapid depletion of water supplies
> - expansion of farmland, combined with deforestation, which passes a critical threshold for the survival of wild species, leading to mass extinction.

Table 18.3 Global warming future choices

	Business as usual choice	Sustainable choice	Radical choice
Decisions taken today about managing global warming	Carbon dioxide levels continue to rise on current trends	Switch to renewable energy, use of recycling, energy intensity falls in developed countries	Fossil-fuel use abandoned, widespread use of carbon taxes, local production replaces globalisation
Possible future	5°C warming by 2100	2–3°C warming by 2100	1–2°C warming by 2100

Estimates range from 9 to 13 billion. Projections made during the period 1990–2000 often concluded that world population would peak at around 9 billion in 2050. Projections made by the UN in 2019 suggest global population will still be slowly rising in 2100, and is most likely to be 11 billion.

Demographers cannot predict future population with accuracy and, in just the same way, scientists cannot predict future global warming and economists struggle to predict future GDP. This uncertainty is important to understand; for example, it is unhelpful for players wishing to prepare communities for the future by building **resilience**, for instance.

Consider an issue such as global warming caused by human-induced climate change. How does uncertainty affect the choices that might be made by players about how to manage this issue? Decision makers today could set one of the objectives shown in Table 18.3.

The exact outcome of each choice, on people in the future, is uncertain for a number of reasons.

1 Political

Some global leadership was shown at the COP21 climate conference in Paris in December 2015, as EU countries especially pushed for a global agreement to reduce emissions. However, will the EU still be a powerful political force in 2030 or 2050? If the world then is dominated politically by China and Russia, agreements made in 2015 might be torn up or simply ignored by future superpowers.

Key concept: Resilience

Resilience is the ability to cope with change or stress. Resilient communities can cope with changes such as disease outbreaks, natural hazards, weather variability or resource shortages and recover quickly. Resilience also implies that communities can cope and recover without depending completely on outside help; that is, they are at least in part self-sufficient.

Building resilience is a key goal of many government, NGO and IGO decision makers and players, because resilient communities will require less short-term help when faced with future climate, ecological, water or hazard stress.

2 Demographic

Slower or faster than expected population growth (see Figure 18.14) in the future would decrease or increase emissions compared to forecasts made today. Future emissions depend on future population size, which cannot be known today, meaning that a decision to be 'radical' today could lead to 'sustainable' levels of warming if population growth is higher than anticipated.

3 Economic

The future size of the world economy is a significant unknown. If global GDP grows from US$80 trillion in 2015 by 1 per cent per year until 2100, it would be US$186 trillion; 3 per cent annual growth raises this to US$987 trillion by 2100. The difference in consumption levels between these two scenarios is huge and would have a major impact on emissions.

4 Scientific

Even the best scientists and computer models cannot predict our future climate with much accuracy. This is because the number of variables is so large, and because some physical processes (such as climate feedback mechanisms and ocean circulation) are not yet fully understood.

This difficulty in predicting what the impacts of choices made today will be on future generations should make players and decision makers cautious. The outcome of most decision-making can only be partially known at best.

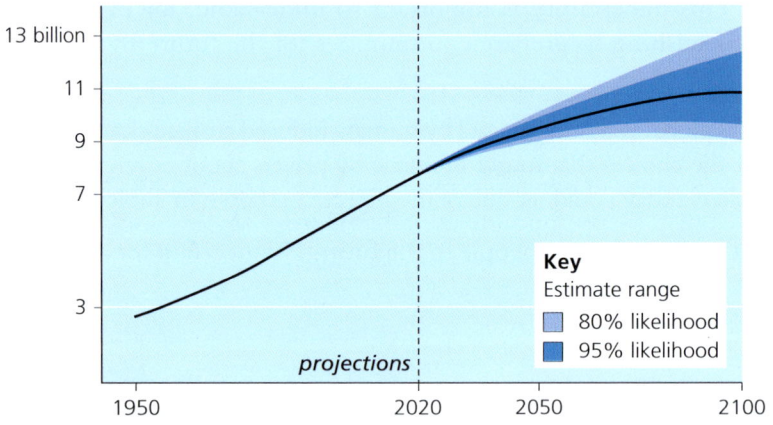

Figure 18.14 UN world population projections

Review questions

1. What are the three key synoptic themes?
2. Explain how players differ in terms of the scale they operate at, and their influence.
3. Explain why some players will be unhappy, regardless of whether Heathrow airport is expanded or not.
4. Study Figure 18.6 (page 330). Suggest why perceptions of climate change as a threat vary between countries.
5. Briefly outline what a 'Nolan chart' shows.
6. Using Figure 18.10 (page 333) explain why attitudes to the environment and pollution might change over time.
7. What is meant by 'sustainability' and why is it a controversial theme?
8. Compare the actions that would need to be taken to manage global warming under 'sustainable' versus 'radical' future objectives.
9. Explain the impact of uncertainty on decision-making.

Further research

The Pew Research Center website allows an exploration of the attitudes of people in different countries on a huge range of issues: www.pewglobal.org

Explore the idea of thresholds and tipping points using the Yale Environment 360 website: https://e360.yale.edu/features/as-climate-changes-worsens-a-cascade-of-tipping-points-looms

Explore the US National Intelligence Council's 'Global Trends' report: https://www.dni.gov/index.php/global-trends-home

19 Independent investigation

> The independent investigation consists of a written report of approximately 3000 to 4000 words. There are a series of sequential stages that require students to:
> - research and define questions or hypotheses that are linked in some way to an area of the specification
> - design and implement an individual fieldwork and research methodology, and then to present the data and information
> - carry out analysis, reach conclusions and critically reflect on the methods and results
> - explain how the results and findings relate to the wider geographical context and help extend geographical understanding.

Introducing the independent investigation

Planning and then undertaking your own individual investigation is a rare opportunity to confirm and test geographical assumptions or established theories. But a personal independent investigation can also present challenges. It requires skills of organisation and motivation, as well as independence. These require a degree of sophistication, although many of the problems that arise can be predicted and managed beforehand. It's all down to good planning.

Successful investigations inevitability end up considering geographical ideas that are linked to changes that occur over *space* and/or *time*. Effective individual investigations will also address the following:

1. Is the work geographical and linked in a meaningful way to some area of content within the specification?
2. Is the investigation and planned work manageable in terms of scale, time, equipment, location and transport? Most importantly, is it local in scale and achievable?
3. Does the initial topic research indicate there will be sufficient high-quality supporting data and information?
4. Does the work provide links to other geographical topics and issues so that it can be framed within a 'bigger picture'?

Figure 19.1 Delivering a successful independent investigation requires careful planning and management of your time and resources

Key terms

Fieldwork: A contested term often meaning different things in different times and places, as well as to different people. Lonergan and Andersen (1988) suggest that it is supervised learning that involves first-hand experience outside the classroom. But there are different types of fieldwork, ranging from observational and data collection, to 'tour' fieldwork. However, the best fieldwork experiences are when students connect with or immerse themselves in an environment in order to make geographical sense of what they have found. Students should be prepared to take responsibility for their own learning, ensuring that they play an active part in the field.

Literature research: This important aspect of the investigation involves finding the current academic or published information around a topic. Be sure to evaluate any research information based on age, author, source, etc., as well as checking whether the research is agreed or supported by other authors.

Understanding fieldwork and the geographical enquiry

The importance of **fieldwork** in geography reflects the influence of early geographers and geographical institutions, who defined modern geography as a field-based discipline.

Fieldwork was seen as an opportunity to discover, explore and find things, as well as to test new geographical ideas.

At the heart of fieldwork often lies a geographical enquiry (or route to enquiry), a series of stages that are undertaken in a similar way to a traditional scientific investigation. In Table 19.1, a model of enquiry is presented that has four central ideas.

Table 19.1 Independent geographical enquiry

1	Creating a need to know	Students decide enquiry questions, framed by teacher input
2	Using data	Students are involved in key decisions about fieldwork procedure and data sources
3	Making sense of geographical information	Students independently analyse evidence and make decisions/reach conclusions
4	Reflecting on learning	Students consider the validity of evidence/reliability of data and methods

Adapted from Margaret Roberts (2003) *Learning Through Enquiry: Making sense of geography in the Key Stage 3 classroom*

Your A level geography has its own similar (in fact more detailed) enquiry model of investigation (Table 19.2). This, like Table 19.1, has distinct stages that are followed sequentially.

Purpose, context and setting the scene

Coming up with your *own* geographical idea or focus can be one of the most challenging aspects of the independent investigation. This might be something that is a topical issue locally, for example, or something that you have read about or studied previously.

Importantly it must have *geographical meaning*, a *purpose* and a *direction* to follow. If you are visiting a field centre, for example, you might be given opportunities to work in a particular environment away from home and school but you will still need to devise your own individual title. Figure 19.2 (page 341) shows examples of geographical themes that could be investigated in an upland environment.

You will be responsible for recognising and selecting a *broad* geographical theme, refining it and making it work as an independent investigation.

Literature research and background information

Having developed an idea, the next stage would most likely be to conduct a **literature research**. The purpose of this is to get additional background information that may be used in a variety of ways:

- to help set the context, e.g. location details (geology map, large-scale OS maps, etc.)
- to explore parallel examples and places
- to get the most up-to-date thinking about a topic or subject
- to show local opinions
- to explore geographical models and theories that may be relevant to the idea of focus.

Table 19.2 The Pearson/Edexcel model of enquiry (adapted from the specification)

	Stage	Description
1	Purpose; identification of a suitable question/aim/hypothesis and developing a focus	Identify appropriate field research questions/aims/hypotheses based on knowledge and understanding of relevant aspects of physical and/or human geography. Research relevant literature sources linked to possible fieldwork opportunities presented by the environment, considering their practicality and relationship to compulsory and optional content. Understand the nature of the current research literature relevant to the focus. This should be clearly and appropriately referenced within the written report.
2	Designing the fieldwork methodologies; research and selection of appropriate equipment	Thinking about how to observe and record geographical ideas in the field, and how to design appropriate data-collection strategies, taking account of sampling and the frequency and timing of observations. Good selection of practical field methodologies (primary) appropriate to the investigation (may include a combination of qualitative and quantitative techniques).
3	Data collation and presentation	Know how to use appropriate diagrams, graphs and maps, and how to use geospatial technologies to select and present relevant aspects of the investigation outcomes.
4	Analysis, interpretation and explanation of results and information	Use appropriate techniques for analysing field data and research information. Write a coherent analysis of fieldwork findings and results linked to a specific geographical focus.
5	Conclusions and critical reflection on methods and results	Use knowledge and understanding to question and interpret meaning from the investigation (theory, concepts, comparisons) – all linked to conclusions. Demonstrate the ability to critically examine field data (including any measurement errors) in order to comment on its accuracy and/or the extent to which it is both representative and reliable.
6	Conclusions and critical reflection on methods and results	Explain how the results relate to the wider geographical context and use the experience to extend geographical understanding. Show an understanding of the ethical dimensions of field research.

Figure 19.2 The range of geographical themes that could be investigated in an upland environment such as the Lake District

19 Independent investigation

> **Key terms**
>
> **Harvard:** A style of referencing primarily used in academic writing to cite (reference) information sources. Each reference should contain in a sequenced order:
> 1. Name of the author(s)
> 2. Year published
> 3. Title
> 4. Publisher
> 5. Pages used
>
> This approach can be modified for newspaper and internet sources:
> - Last name, First initial. (Year published). Article title. *Newspaper,* [online] pages. Available at: URL [Accessed Day Month Year].
> - Last name, First initial. (Year published). Page title. [online] Website name. Available at: URL [Accessed Day Month Year].
>
> **Aim:** A statement of what the project/investigation is setting out to achieve. It must be geographically sound and achievable.
>
> **Question:** A question that is asked (in a question format), that often links with the overall title and can be used as a way of sub-dividing the title.
>
> **Hypothesis:** A statement whose accuracy can be tested objectively using scientific methodology. Null and alternative hypotheses are normally used in connection with statistical tests (Chi-squared and Spearman's rank).

In many respects the background information kick-starts the process of 'searching for answers' before you have even stepped out and started to collect any primary or first-hand fieldwork data.

It is very important to keep an accurate record of the sources you use, and to use a consistent referencing style such as **Harvard** in the individual investigation.

In Stage 1 (Table 19.2, page 341), scene setting should include a general discussion of and introduction to the geographical issue or topic that you are studying. In other words, this is the opportunity to say how the most up-to-date literature (from your research) gives a context and background to the focus. Remember that the literature research helps to develop both a purpose and context. Figure 19.3 shows an example of a technical document from the Department for Environment, Food and Rural Affairs (Defra) and the Environment Agency outlining the costs of coastal protection. Documents like this can be used to signal the local, regional or national significance of the work being investigated.

Table 19.3 Examples of aims, questions and hypotheses

Aim	Question	Hypothesis
An investigation into the reliability and variability of regional weather forecasts through comparisons with local primary data in area T. An investigation into the effectiveness of traffic-calming measures in town X.	How and why do cliff and beach profiles vary along a stretch of coastline between points X and Y? To what extent are golf courses generally an environmental, economic and social asset in rural area C?	Most shoppers purchase goods and services at least once a week. Shingle beaches have a steeper gradient than sand beaches.

The number, volume and range of geographical information available to you means that you must be selective and be prepared to reject information that is not directly linked to the purpose or focus of what you are trying to achieve.

Appropriate geographical scale

A successful independent investigation outcome depends partly on the choice of an appropriate geographical scale for the investigation. For instance, an investigation into a town's flood risk is unlikely to produce significant results if conducted in just once small part of the catchment. Many natural systems show considerable variability, so a larger sample would be required.

Defining the hypothesis or question

Questions and **hypotheses** generally fall into two main types: those that focus on spatial, areal or temporal *differences* and those that focus on *relationships* between variables. These are the main areas of focus for the independent investigation. Whatever is chosen, there must be a clear geographical element and an obvious linkage to the specification.

It is up to the individual student to decide how many questions or hypotheses might be appropriate; it is possible to use a 'mix and match approach' – a single overarching **aim** and then a series of two to three sub-questions, for example.

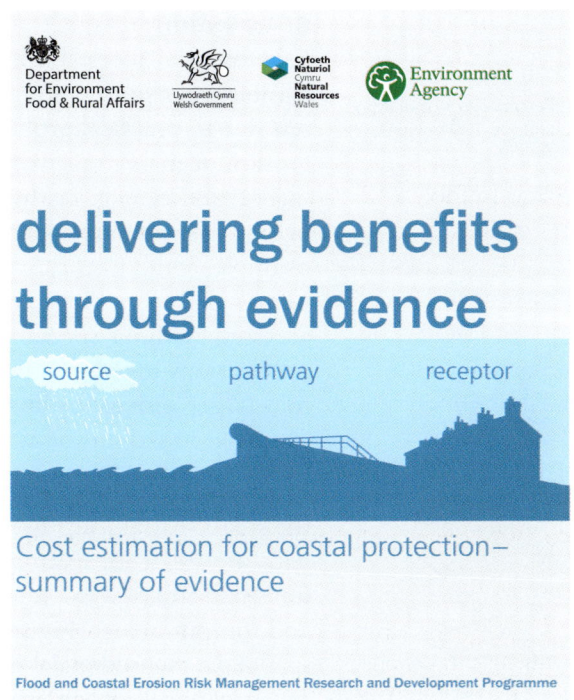

Indicative costs associated with the cost of coastal protection

Option	Significance			Indicative cost (£/m)	
	Enabling costs	Capital costs	Maintenance costs	Scottish Natural Heritage	Environment Agency UCD
Beach recharge and breakwater	Medium	High	Medium	–	2,700–7,300
Beach recharge and groynes	Medium	High	Medium	–	1,600–4,700
Rock armour	Medium	High	Low	–	1,350–6,000
Impermeable revetments and seawalls	Medium	High	Low	2,000–5,000	700–5,400
Timber revetments	Medium	Medium	Medium	20–500	–
Rock revetments	Medium	High	Low	1,000–3,000	650–2,850
Groynes	Medium	Medium	Medium	10,000 to 100,000 per structure	–
Nearshore breakwaters	Medium	Medium	Low	400–1,000	1,750–4,300
Artificial rock dune protection	Low	Medium	Low	200–600	–
Gabion revetments	Medium	Medium	Medium	50–500	–
Beach nourishment	Medium	Medium	Medium	50–2,000	350–6,450
Shingle recycling/re-profiling	Low	Low	Low	10–200	15–120
Dune fencing	Negligible	Low	Low	4–20	–
Dune thatching	Negligible	Low	Low	2–20	–

Notes: The Scottish Natural Heritage (SNH) costs relate to a 2000 cost base and the Environment Agency costs relate to a 2007 cost base. An allowance for inflation using a suitable index is required to update these values to present day costs.

Figure 19.3 Technical documents like these can give a study context and provide the latest up-to-date information

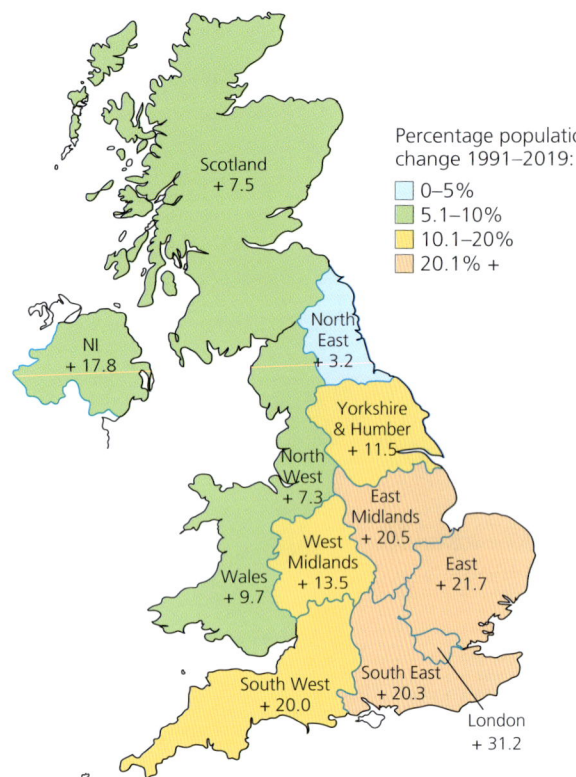

Figure 19.4 The 2019 Index of Multiple Deprivation (IMD) boundaries may not fit exactly with historic or political census areas, e.g. wards or enumeration districts

Feasibility of the research, the hypothesis or question

It is essential to establish that a piece of research work is feasible at the outset. Some hypotheses may be impossible to test due to practical problems of measurement (e.g. rates of erosion or mass movement on river banks or coastal cliffs), inaccessibility (e.g. research locations on private property) or lack of secondary sources that can be freely accessed. For example, some historic river and catchment data from the National Rivers Flow Archive (NRFA) cannot be easily accessed without special permission. Also, some research publications may be difficult to get hold of without going to a larger university library (although many publications can be accessed online with a one-off fee).

Problems can also arise as a result of boundary changes to the small spatial units used to aggregate census population data. In 2001, for example, the old enumeration districts for small areas were replaced by a series of new 'super' and 'output areas' (Figure 19.4). This lack of comparable spatial data would make it extremely difficult to attempt an investigation of population change in a suburb or market town for, say, the period 1971 to 2011.

Linking the focus to a 'core concept'

The Edexcel Specification has 14 core concepts that are included across a range of topics. These core concepts are overlapping geographical ideas that, on some occasions, can provide an interesting and more sophisticated dimension to the individual investigation. Table 19.4 gives examples of three specialised concepts and how they might be linked to fieldwork.

Fieldwork design, methodology and data-collection strategies

For this part of the independent investigation you need to think carefully about how, when and what **primary** and **secondary** data can be collected.

Quantitative and qualitative data

Quantitative data and information is that which includes numbers and numerical data, whereas qualitative is descriptive and can include things such as photographs

> **Key terms**
>
> **Primary data:** Primary data is generally considered to be first-hand data collected by a student themselves (or as part of a group).
>
> **Secondary data:** Secondary data (which may be part of the research) means information that has already been collected by someone else.
>
> In reality there is some 'grey' between these two ideas and approaches, suffice it to say that you should expect to have a reasonable balance between the two types.

Option 8B: Migration, Identity and Sovereignty

Table 19.4 Examples of specialised concepts from the Pearson/Edexcel Specification and how they might be linked to fieldwork

Concept	Definition/interpretation	Examples of individual investigation opportunities
Resilience	The capacity of a system to experience shocks while retaining essentially the same function, structure, feedbacks and identity	Flood risk and resilience Economy of a town and resilience Ecosystem resilience and threats
System	Systems thinking is the process of understanding how those things (parts), which may be regarded as systems, influence one another within a complete entity or larger system (boundary)	Inputs, outputs and stores within a local sub-catchment Economy of an urban centre as a system Understanding carbon flows in a woodland ecosystem
Identity	Identity is about ways in which people connect to various places and the effects of such bonds in identity development, place making, perception and practice; it's to do with belonging, meaning and attachment at a very personalised level	Place identity as seen by tourists The identity of rural versus urban areas Connections to place for different ages/cultures

(Figure 19.5) or written texts. Also see page 351 for more about photography and images.

Both qualitative and quantitative approaches are valuable to an enquiry, but the balance between the two will be very dependent on the choice of topic and its focus.

A zone of urban land use with a canal waterway dissecting the image. The scene appears to show historic buildings, likely Victorian with an industrial purpose. In much of the picture there is evidence of change: improvement and modernisation with a resulting pattern of gentrification most likely. This will have increased the desirability of the area – becoming an attractive waterfront area with flats and/or offices in the foreground near the canal barge. The process of re-urbanisation may also be happening. Such improvements tend to reduce the socio-economic mix and create places and spaces where some residents will be excluded because of housing affordability.

The place is continuing to change and undergo improvements but these are driven by private-sector development, which is far from resilient. A downturn in the economy, house prices and place vitality will create an uncertain future in terms of further enhancements to housing stock.

Figure 19.5 High-quality analysis of photographs is a useful strategy to show 'deeper thinking'. Here is an example of a geographical narrative for an urban place

Physical topics may feel like they lend themselves to more 'counting', but always try to blend in some good qualitative information. More human topics are likely to swing towards qualitative, but the quantitative can often be found in the form of published data and research.

Piloting the work

Pilot surveys are trials, surveys or tests usually carried out in advance of larger-scale surveys. Pilot surveys should give a clear insight into the feasibility and timescale of your investigation, especially where equipment and recording sheets are involved. They need not take long to do and will certainly help with action planning and choosing appropriate sampling strategies.

Such pilot surveys can help in a number of ways; Table 19.5 gives some examples.

Evidence of modification should be documented and presented as part of the final write-up. It demonstrates that you are reflective and that your design is flexible and adaptable, and that you have given good regard to planning.

Other considerations

Careful thought needs to be given to the timing and location of data collection. Questionnaire surveys in towns or rural areas, for example, might want to be timed to maximise the number of potential respondents (visitors or local residents).

Similarly, in a physical geography context, systems and processes will vary from time to time and especially season to season.

So, all investigations need to give careful consideration to this temporal aspect. For example, shopping centre investigations should consider the time of day and whether data collection should take place on a weekday or at the weekend. Permission is also likely to be needed. A similar questionnaire survey in a commuter village might be timed for a weekend when most people are at home. Beach studies need to take account of tide times and are best conducted at low tide or on an ebb tide. Thought should also be given to the location of

Table 19.5 Use of pilot surveys in different contexts

Questionnaires	Trial your proposed set of questions and directed responses on five to ten people. Time how long each completed session takes. Identify poor or weak questions, consider if the question sequence could be improved and if the answers can be easily processed, e.g. within a spreadsheet if they are coded. Modify and test again before using for the full investigation.
Photographs	It is worth taking some sample images relating to your enquiry, especially if you are testing a new digital feature, e.g. 360 panorama. This will familiarise you with the workings and limitations of the device, and alert you to potential lighting or viewpoint problems. Also check for low light problems, especially if just using a phone camera.
Sediment surveys	A pilot survey is important for giving you the overall feel of the study area and helping identify variation in the sediment characteristics. Sites where there is the largest variability are where in real testing a larger number of samples should be taken. This is called the cumulative mean technique – see Figure 19.6.

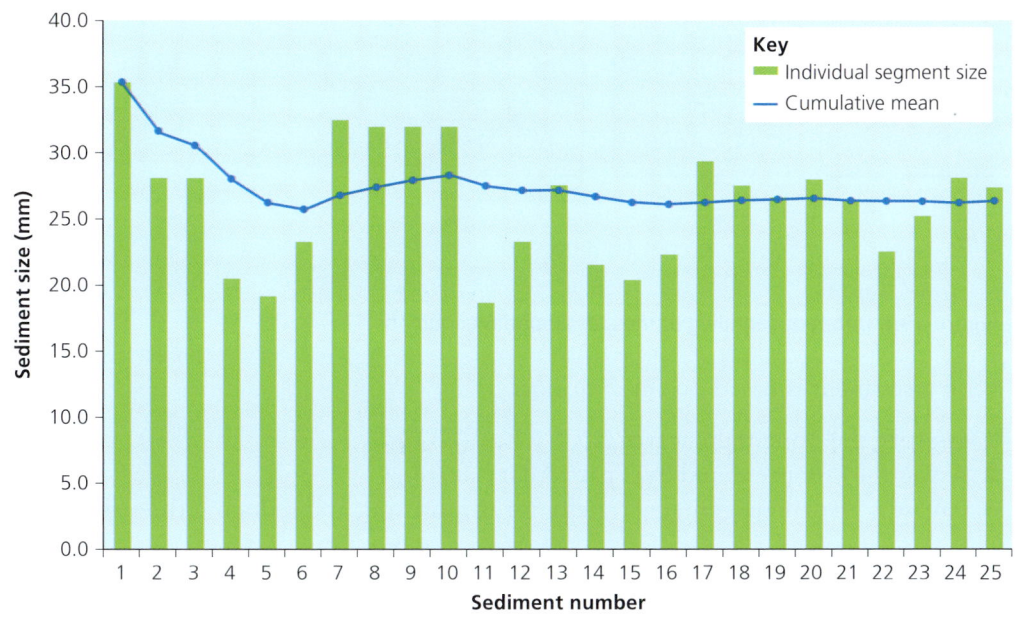

Figure 19.6 Use of a cumulative mean can help with the decision about how many samples to take

on-street interviews. Surveys involving shoppers (Figure 19.7) conducted exclusively at bus stops or in car parks are unlikely to produce representative data.

Access and the problem of private land ownership may create obstacles to successful investigation. A river that is otherwise suitable for investigation of flood risk may be inaccessible due to a lack of public footpaths or unknown private land owners.

Figure 19.7 Where you do your interview survey can introduce an element of bias, so it needs careful consideration

Managing fieldwork risks

While geography fieldwork mostly is very safe, there is still an element of risk no matter which environment you are working in. Your job is to anticipate, minimise and manage any possible risks, starting at the planning stage and extending throughout the fieldwork.

> **Key term**
>
> **Sample:** This is the limited number of measurements that you make. This is different to a population, which is the total number of measurements that you could potentially take if you measured everything and/or everyone in the environment. Larger samples generally mean greater reliability, until a point is reached where increasing sample size has very little effect on quality of outcome (see Figure 19.6).

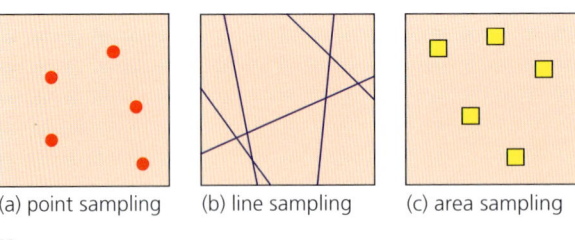

(a) point sampling   (b) line sampling   (c) area sampling

Key

☐ = total population study area

Figure 19.8 Different sampling frames for a spatial enquiry

Risk assessment is based on three things:

1. Identify the potential hazards.
2. Assess the risk presented by a particular hazard.
3. Create a plan/management strategy for minimising and reducing the risk.

Your teacher(s) will be able to offer more advice on the risk assessment procedure, but best practice would suggest that the individual investigation will include a formal risk assessment of some sort. Table 19.6 gives examples of typical safety guidelines.

A meaningful approach to sampling

Collecting a 'good' sample is integral to the research design. The more rigorous and reliable the methods of data collection, the greater the validity of the conclusions.

There are three main aspects of **sampling**:

1. Deciding on an appropriate sample frame.
2. Choosing an appropriate sample size.
3. Selecting the best sampling method.

Sampling frame

The sampling frame (or design frame) is linked to what you know and understand about the population under investigation, for example, a grid placed over a particular residential area, or transect lines across an industrial estate (Figure 19.8). The precise characteristics of the sampling frame should be established, along with any unusual patterns, features or characteristics which might be avoided. This will reduce sampling bias. The area should also be defined as a discrete geographical area, e.g. a section of coast, part of a catchment or two output areas within two different parts of an urban area.

Table 19.6 Examples of actions that should be considered to keep you safe during fieldwork

Let someone know where you are	It makes a lot of sense to always leave a record of where you are going and when you are expected back, especially when working alone.
Check mobile reception	Don't assume your phone can always be used to get help. Make sure you know whether there is a phone signal where you are working.
Carry the right equipment	If working in a remote upland landscape then it would be sensible to have a first-aid kit and survival bag, as well as a torch, for instance. In some instances, e.g. at the bottom of a cliffed coastline, you may need to wear a hard hat.
Carry a map or electronic map	Stick to paths and way-marked signs. If there is any doubt in an urban area, then never work alone.
Traffic	Traffic can be a particular risk. Be aware of your surroundings and the best places to cross.
Clothing	Wear clothing and footwear appropriate to the environment that you are working in; pay particular attention to possible cold or heat exposure.

Sample size and number of sample points

Because most natural populations in geography are very big (and spread over a large area or change over time), investigations are inevitably based on sample data. Decisions have to be made on the quantity of data to collect (i.e. sample size) and objective sampling methods to ensure **representativeness**.

In the context of questionnaire surveys, the aims of the investigation will affect decisions on whether the appropriate strategy is street interviews, door-to-door interviews or postal questionnaires.

The choice of scale also becomes a significant decision. An enquiry into the spatial pattern of consumer spending in a central shopping area requires a town or city large enough to have at least five or six clearly defined shopping streets/areas. A large town or city of 300,000 people is likely to be much more suitable than one of 30,000.

In summary, think about:

1. What is the *minimum* number of replicates/repeats that you need to collect so that you can carry out analysis (including statistics) reliably?
2. What is the *maximum* number of replicates above which the results do not change? In other words, at what point will collecting more data not significantly improve the accuracy of results?

Always remember that you need to get enough samples to draw a graph, do some analysis and then generate a conclusion based on that information.

Sampling method

We have already established the need for appropriate sampling in order to measure the population in a fair and representative way. Table 19.7 outlines the three main spatial sampling methods. In practice it is possible to use all three methods simultaneously to give increased accuracy and improve the data collection process.

Deciding on the correct sampling method requires some knowledge of the population.

> **Key terms**
>
> **Representativeness:** The extent to which a subset of a statistical population accurately reflects the characteristics (or members) of the entire population. When a sample is not representative, the result is known as a sampling error; in other words, the statistical difference between the observed (sample) data and the results that would be obtained from using data involving the entire population from which the sample was derived. In practice the degree of sample error can be complex to estimate.
>
> **Transect:** A transect is a line along which sampling is undertaken. This can be a tape measure, road, river or any other accessible linear feature.

Table 19.7 Sampling methods and particular considerations

Method	Operational description	Particular considerations
Systematic sampling	Samples are chosen in a systematic or regular way, e.g. every 5 m, every hour, or every fifth person. This is used when the environment or population has an expected environmental gradient or change (spatially or temporally) but the degree of change may be uncertain.	Can give good coverage (spatially of an area) and it's straightforward to design and undertake. It has the potential to miss areas when surveying along particular points or lines, however, which can lead to an under- or over-representation of certain groups or features in an area. It often uses a **transect**.
Stratified sampling	Samples are taken at predetermined points or times based on an understanding of the study area in terms of the groups, individuals and sub-groups. This is used when the environment or population has an observed environmental gradient or change (spatially or temporally) and the expected change can be used to inform the sampling procedure.	This approach reduces the potential for bias in areas of variation, but the sampling design frame needs to take account of the underlying characteristics of the area or population in order to make the correct selections. In some instances it can be difficult to get data on groups in order to stratify the sample (e.g. a 'profile' of tourists to a town).
Random sampling	This is sampling using random numbers to generate times and/or co-ordinates for when a sample should be taken. This is used when the environment or population has no known environmental gradient or is thought to occur at random.	This sampling approach should minimise any elements of human bias and, therefore, sample error. However random sampling can leave gaps in a sampling design frame or lead to an undesirable clustering of points. Random sampling can also be time consuming to undertake compared to stratified or systematic.

In reality, constraints of time and resources impose limits on geographical investigations (although these should not be used as 'excuses' as part of a limitation). Most investigations must strike a balance between the amount of data collected and the time and resources expended on data collection.

Considerations connected to time and resources are essentially practical. Two key questions must be examined alongside the fieldwork design and methodology:

1 How much time will be devoted to data collection and how much potential fieldwork time will be spent travelling (and walking) to and from the fieldwork site?
2 Will data be collected by a group or by an individual? If in a group, how many students will be needed for the data collection process? This last consideration is important in questionnaire surveys, where rejection rates are often high and where each interview may take several minutes.

Choice of equipment and production of data-recording sheets

Primary, first-hand data collection relies on three main skills: observation, measurement and interview. Often this data collection, to a greater or lesser extent, requires the use of specialist equipment and recording sheets (or an electronic version).

- Observation and systematic recording is perhaps the simplest data-collection technique. Annotated field sketches and photographs rely on direct observation. Quantitative data on traffic flows, land uses, shop types, environmental assessment and plant species frequency are also likely to be through observation.

- Fieldwork in physical geography often involves data collection through measurement. Slope surveys, for instance, using instruments such as clinometers (Figure 19.9), ranging poles and tapes, can generate numerical data on valley and beach profiles. Equipment including anemometers, thermometers and humidity meters are used to measure microclimate. Soil pH is measured using a BDH soil-testing kit; and callipers measure the long, short and median axes of sediment. In all cases, a well laid out and logical recording sheet makes data processing much more straightforward at a later date. Figure 19.10 shows an example of a recording sheet for measuring soil infiltration, which might form part of a piece of work looking at the way in which water moves through a catchment as a part of a larger system.

Figure 19.9 Measurement (here, using a clinometer) is often important in physical geography, so a well-designed recording sheet can be essential

Date 18 May	Site number 2	Previous weather Dry	
Time from start (min)	Amount lost (cm)	Time taken for loss (sec)	Infiltration rate (cm/min)
4	6	50	1.8
8	5	175	1.6
20	6	100	1.2

Figure 19.10 Getting a logical booking sheet to record data on to makes life easier when it comes to data representation

- Data that relates to people's attitudes, behaviour and perceptions can found by using questionnaires or interviews, or a combination of the two. There are two different ways of collecting this qualitative and semi-quantitative data: on-the-street surveys, which are most often conducted in public spaces (e.g. a shopping street), though occasionally doorstep interviews are undertaken; and remote questioning (e.g. a postal or online survey). In each case the interview or questionnaire will need to be customised for both the target audience and the mode of delivery.

Remember that research suggests people are sometimes economical with the truth in surveys and opinion polls. They often respond with an answer that 'sounds' good, or that they think the interviewer wants to hear. Questionnaires can also fail due to poor design, questioning sequence or if the questions are poorly constructed.

Accurate and **reliable** data are essential to a successful investigation: it is impossible to have confidence in the outcome of an investigation if doubts exist about the quality of the data.

Taking geographical pictures

In the twenty-first century, images and photographs surround us more than ever. They are an essential part of the fabric of our personal life and our work life at school or college. Most of us are also very used to taking photos with a range of devices, and often with a personal mobile phone. There are several key advantages to this: it's quick, easy and convenient, plus it's very easily shared with other people. Photographs can also capture a particular geographical moment or event that can be difficult to describe in words.

In the context of the independent investigation, photographs and imagery are likely to form an important part of the data-collection process. They might, for instance, document how a particular piece of equipment was used, or give an alternative image of a place (away from the tourist gaze). Photographs can be used as a source of evidence for change in both settlements and landscapes. Images are an obvious record of what places were like in the past and comparing old photos with their current counterparts, especially if taken from the same view and perspective, can be very effective. You often find these sorts of comparisons presented in a side-by-side way. A range of special effects can be applied to make images more useful in the context of the individual investigation (see Table 19.8).

> **Key terms**
>
> **Accuracy:** In geography the accuracy of a measurement system/ data is the degree of closeness of measurements of a quantity to that quantity's actual (true/real) value. The further a measurement is from its expected value, the less accurate it is.
>
> **Reliability:** This refers to the consistency or reproducibility of a measurement. A measurement is said to have a high reliability if it produces consistent results under consistent conditions. True reliability cannot be calculated – it can only be estimated based on knowledge and understanding of the topic.

Table 19.8 Some examples of special effects that can be applied in the context of geography and the individual investigation

Effect	What it does	Example application
Black and white	Changes a coloured image to a black and white or greyscale image	To give more emphasis to the contours and topography of a landscape; can also be used for a particular effect, for example, to add contrast
Photo montage	A series of thumbnail-sized images	Very good for showing a group of different people, for example, or a selection of contrasting views of landscapes or places (see Figure 19.11)
360 panorama	A sequence of photos stitched together into an immersive viewer	These are commonly used to show cityscapes and landscapes; they look good when finished, but be careful to hold the camera steady when taking the pictures
Photo-sequence	A series of shots taken in succession, often used to demonstrate change	To show a dynamic event, e.g. a powerful wave crashing on to a cliff or a large number of pedestrians in a high street; longer gaps between photos can show diurnal variations in activities
Annotation	Can show a better understanding of the processes occurring at a particular location	A range of tools will allow you to annotate your photos on a tablet or phone; one of the most popular is the Skitch app from Evernote
Embedding location data on to a picture	Creates pictures that are geo-located using lat-long attributes coded into the 'exif' data attached to an image	There are a number of apps that can support geo-located information attached to pictures

Judgement surveys

Judgement or 'quality' surveys are included here since they often feature within geographical fieldwork. They are useful since they rely on observation and are low tech. They can be seen as poor quality, however, if they are designed and undertaken without due care and attention. There are two main issues:

1 A concern about comparability when lumping together group data and observations made by different people who have not been correctly 'calibrated'. That is, they have not come to a common agreement as to what a '3', for example, represents under a particular category.

Figure 19.11 A photo montage can be useful for showing contrasts between places, in this instance residential housing types

Option 8B: Migration, Identity and Sovereignty

2 The quality survey can suffer from a lack of customisation. In other words, surveys lifted directly from books and articles without appropriate modification for the context of the environment or location being studied. This can also include revising the scales and changing the weightings of particular descriptors.

The majority of judgement surveys are probably best described as semi-quantitative, in that they combine a mixture of numbers linked to a qualitative observation score. However, they are not all semi-qualitative. Figure 19.12, for example, shows an example of two more unusual and customised quality indices. The land capability index is numerically defined only.

One of the more powerful and often overlooked aspects of quality or judgement surveys is that they can often be used to generate data about other people's perceptions of places, landscapes and environments. This could be useful, for example, to see if a particular project or initiative has had a positive effect as seen through the eyes of different stakeholders.

Qualitative techniques

A range of qualitative techniques should be given consideration as a useful way of collecting high-quality primary data. Remember that qualitative data and information

A land capacity index

Class	Altitude (m)	Wetness	Soil quality	Soil fertility (pH)	Slope (°)
1 High quality	Below 100	No limitations Free drainage Rainfall <750 mm	Deep soil 75 cm+ Stone free Loam texture	7+ (neutral)	Level (not above 3)
2	100–150	Imperfectly drained Drainage easily modified by liming	Depth 50–75 cm Slightly stony	6.0–6.5	Slight (not above 7)
3	150–200	Some problems but possible to install drainage system	Depth 25–50 cm Stony – may be sandy or clayey texture	5.5–6.0	Moderate (not above 11)
4	200–350	Poorly drained but can be improved to maintain pasture	Shallow – under 25 cm Very stony	5.0	Significant (11–20)
5 Low quality	Above 350	Poorly drained Drainage almost impossible to install Rainfall >1,250 mm	No humus Very stony – skeletal soil only	Under 4.5	Steep (over 20)

A cinema index

CINEMA QUALITY INDEX						Site 1 The Odeon
Bad features	1	2	3	4	5	Good features
Old films			✓			Latest films
Uncomfortable seats		✓				Comfortable seats
Dirty	✓					Clean
Difficult to get to	✓					Easy to get to
Expensive				✓		Cheap
Sub-totals	2	2	3	4	0	Total = 11

Figure 19.12 Land capability index and cinema quality index recording sheets

Table 19.9 Examples of qualitative data-collection techniques

Technique	Overview
Interview	Usually face-to-face over a period of several minutes with open-ended questions. A recording or transcript is essential to allow future data processing.
Focus group	A small group of people are asked about their perceptions, opinions, beliefs and attitudes towards a place, service, concept, strategy, etc.
Oral history	The collection of historical information using audio/video recordings of interviews with people who have personal knowledge of past events. This will often be conducted by interview but it may be possible to obtain oral histories from sources such as YouTube.
Historic texts and images	These can take a variety of forms, such as brochures, newspapers and even old postcards (especially good for coastal change).
Participant observation	A low-key observation technique – making notes and documenting the type, movements and activities of people.
Perception studies	A mixed bag of qualitative ideas including the possibility of mental maps and interviews to extract attitudes from visitors and local residents, for example.

is non-numerical, often unstructured and typically descriptive. Qualitative data is sometimes overlooked by students as it can be harder to process and analyse. Don't be fooled by thinking that text and images are in any way less valuable as part of the research than numerical data. Table 19.9 gives examples and brief descriptions of qualitative data-collection techniques.

Crowd-sourced and 'big data'

Sometimes it is difficult to find information on people's behaviour and attitudes using traditional large-scale data sources, such as the census and social surveys, because they don't release data at an individual or 'granular scale'. Instead data is only available at a local or regional level for selected attributes. They also only offer a snapshot rather than a continuously changing view.

However, data sources such as that available from social networks contain a wealth of information about people's geographical behaviour at an individual level. These **big data** sources are commonly referred to as geospatial crowd-sourced data (GSD) or volunteered geographical information. GSD is increasingly being used to study individual people's daily behaviours and it has an increasing role when it comes to examining local resilience in the context of responses to natural disasters.

In relation to the individual investigation, platforms such as Twitter/X, Facebook, Flickr and YouTube (Figure 19.13, page 355) have large volumes of information with 'geographic footprints' that can be searched and sometimes utilised to see how people perceive places, events and issues. Such information can provide a useful perception context to the topic under investigation.

Ethical implications

You should have an awareness of the ethical issues that are embedded in any study that involves the collection, analysis and representation of geographical information about human communities. The most common ethical dilemmas in human geography focus around participation, consent, and the safe-guarding/confidentiality of personal information and data. In physical geography the main ethical considerations are about consent/access to study sites and potential damage (i.e. overuse, trampling) or possible pollution (including litter) of study sites.

> **Key term**
>
> **Big data:** Extremely large, dynamic data sets that may be analysed to reveal patterns, trends and associations, especially relating to human behaviour and interactions. It may provide a source of information that could be used to support the individual investigation (Figure 19.14, page 355). Frequently it has GIS information linked to it. Some researchers work with big and complex data sets and produce very interesting map 'mash-ups' – see Figure 19.15 (page 356).

Option 8B: Migration, Identity and Sovereignty

Figure 19.13 Social media can be used as an effective search tool – it's a hybrid type of primary data

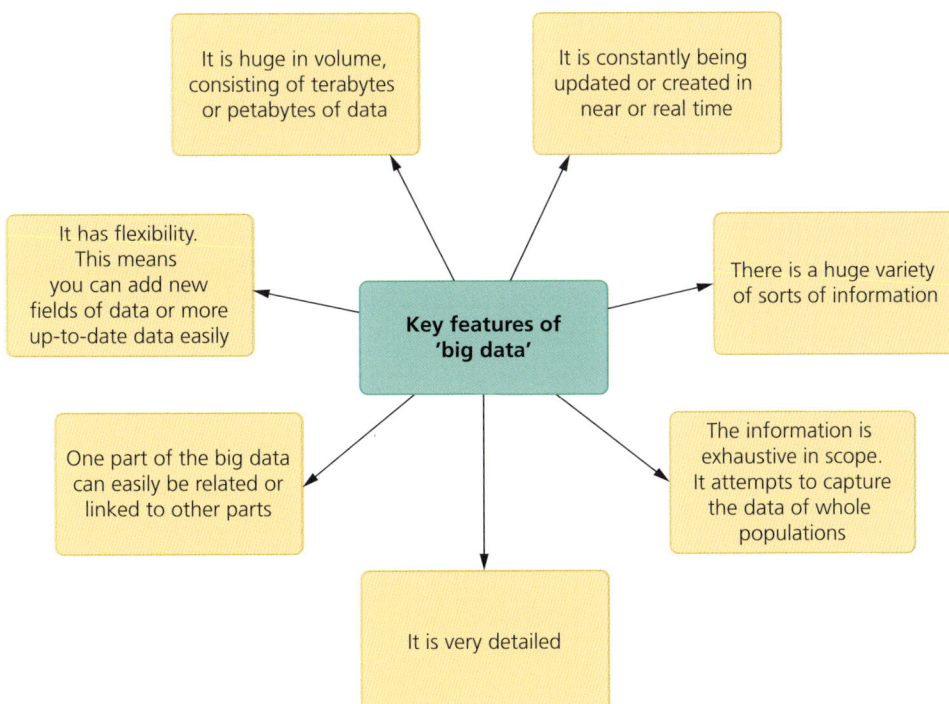

Figure 19.14 Key features of big data

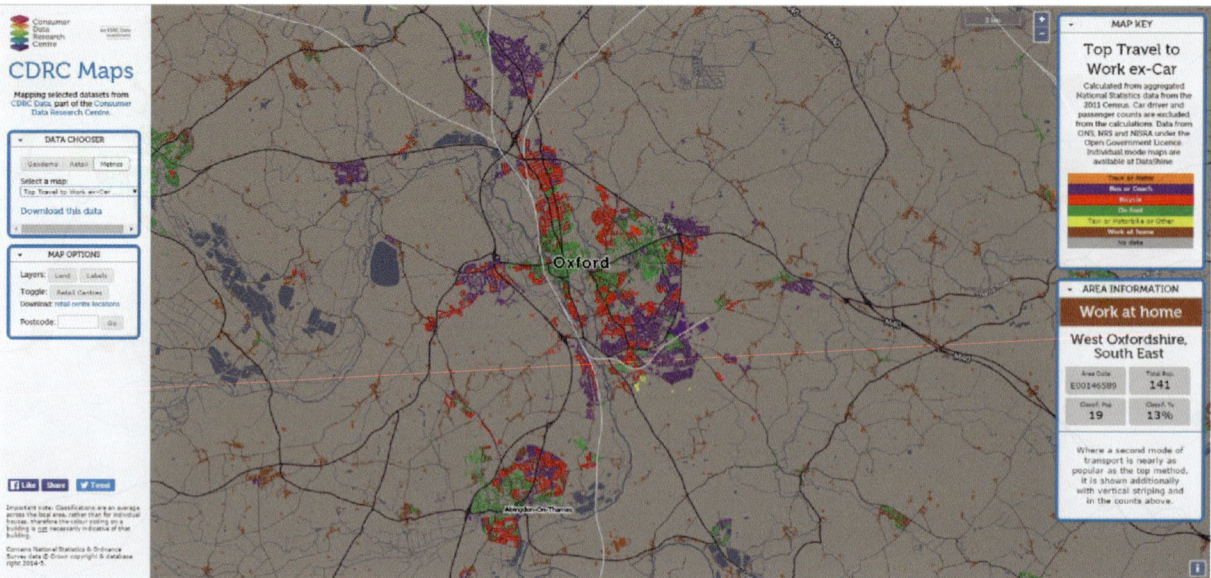

Figure 19.15 People's travel to work patterns (excluding cars) around Oxford

Remember, you need to consider the ethical impact of fieldwork in order to fulfil the requirements of the Pearson/Edexcel mark scheme.

Data and information representation

Before data and information can be represented in a meaningful way, it is likely that it will need to be collated, sorted and selected so that only the data and information that is relevant to the overall focus of the work is used. See Figure 19.16.

You might want to include an annotated example of a completed recording sheet or part of a questionnaire, for example, in part of the report to show how you collated the data, but don't include the actual sheets themselves.

Tables can be used to summarise complex information. They can be any size or shape, as long as they are laid out in a sensible way. It is important to be clear so that the information is easy for the person reading the work to understand. Use brief headings within tables and put the units in the column header. Giving row and column totals may make it easier for anyone reading the work to interpret.

Representing data ready for analysis

Data representation is about getting what you have found ready for analysis. At the initial stage of analysis data are processed and presented as tables, charts and maps. Remember that the purpose of data representation is to clarify meaning, i.e. to make initial sense of what you have found. The questions, aims or hypotheses under scrutiny will determine the collation, grouping and representation of data. There may also be a natural sequence that is followed in the 'making sense of data' part.

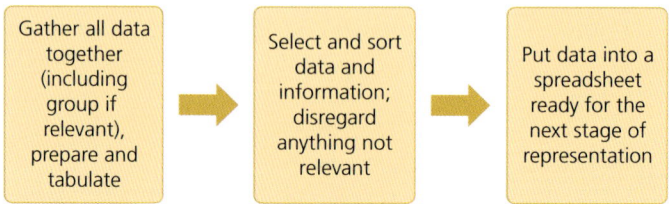

Figure 19.16 The stages that should be undertaken in order to get data and information ready for representation

Option 8B: Migration, Identity and Sovereignty

For example, an enquiry looking at patterns of movement of people might begin by:

1 mapping the location of new housing in an area
2 presenting the data first as a frequency table showing distances moved
3 plotting this same data as a histogram.

Cartographical methods

Cartographic methods of representation relate to maps. There are several types of map, including: dot, choropleth, isopleth, proportional symbol and flow maps. Choosing an appropriate type and scale of map is important to deliver the required outcomes, and there may be a trade-off in terms of level of detail and time taken to construct. GIS can help construct maps quickly and the user can easily change category intervals and colours to create the most appropriate visual effect (see Figure 19.17). Page 362 discusses the use of GIS in analysis.

Graphical methods

Graphical data is information presented as charts and graphs, rather than as tables of information or maps. Graphical data have two important advantages:

1 Firstly, they provide a suitable summary of geographical data.
2 Secondly, and crucially, their visual nature helps in the understanding of patterns and trends in data and information.

The choice of graphical techniques is very important (Table 19.10, page 359). Firstly, it must be appropriate to the data, but also it should aim to achieve a balance between preserving as much detail as possible and allowing generalisation to reveal patterns and trends and to expose anomalies.

It is tempting to use as many different graphical and cartographical techniques as possible. In many instances this leads to similar data and information being presented in two or three different ways, which serves no purpose other than to show that you can use a range of techniques. Sometimes this can be counterproductive, as it can blur meaning in the results. Always use the most appropriate presentation technique, whether hand-drawn or using a computer.

Analysis and explanation of data and information

There can be no doubt that the analysis of numerical data and information will form an important part of the enquiry process. Approached in the correct way it can be both interesting and rewarding, as well as sometimes surprising and revealing. To many students, however, data analysis is the part of the independent investigation that is approached with the most trepidation and concern. Combined with the data representation, data analysis carries 24 out of the 70 marks available for the entire independent investigation.

Analysis usually consists of three linked activities:

1 Description (supported by information, data and other evidence). Part of the description will likely be to undertake statistics, either descriptive or inferential. See below.
2 Explanations (providing likely geographical reasons linked to patterns and trends in the data).
3 Finding, synthesising and suggesting geographical links in the data (and linkage to the theory, literature or comparative studies).

Table 19.11 (page 359) summarises some statistical analysis techniques.

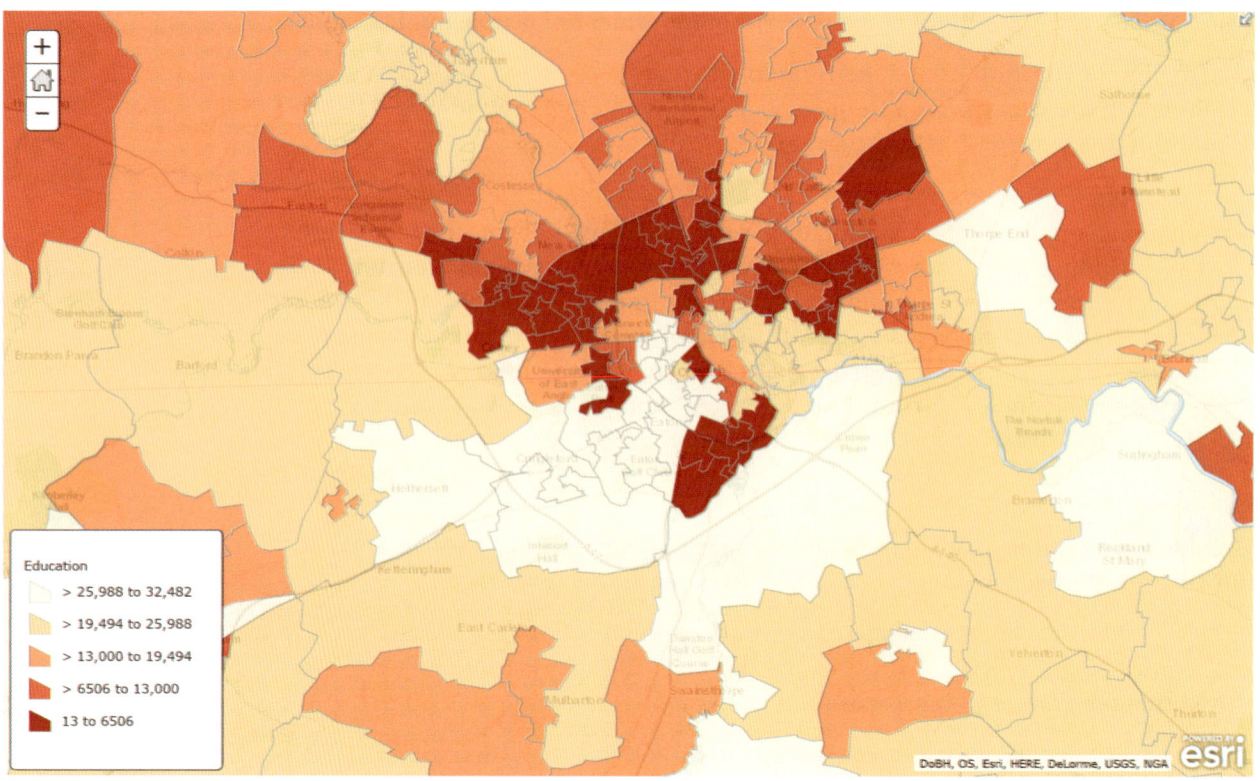

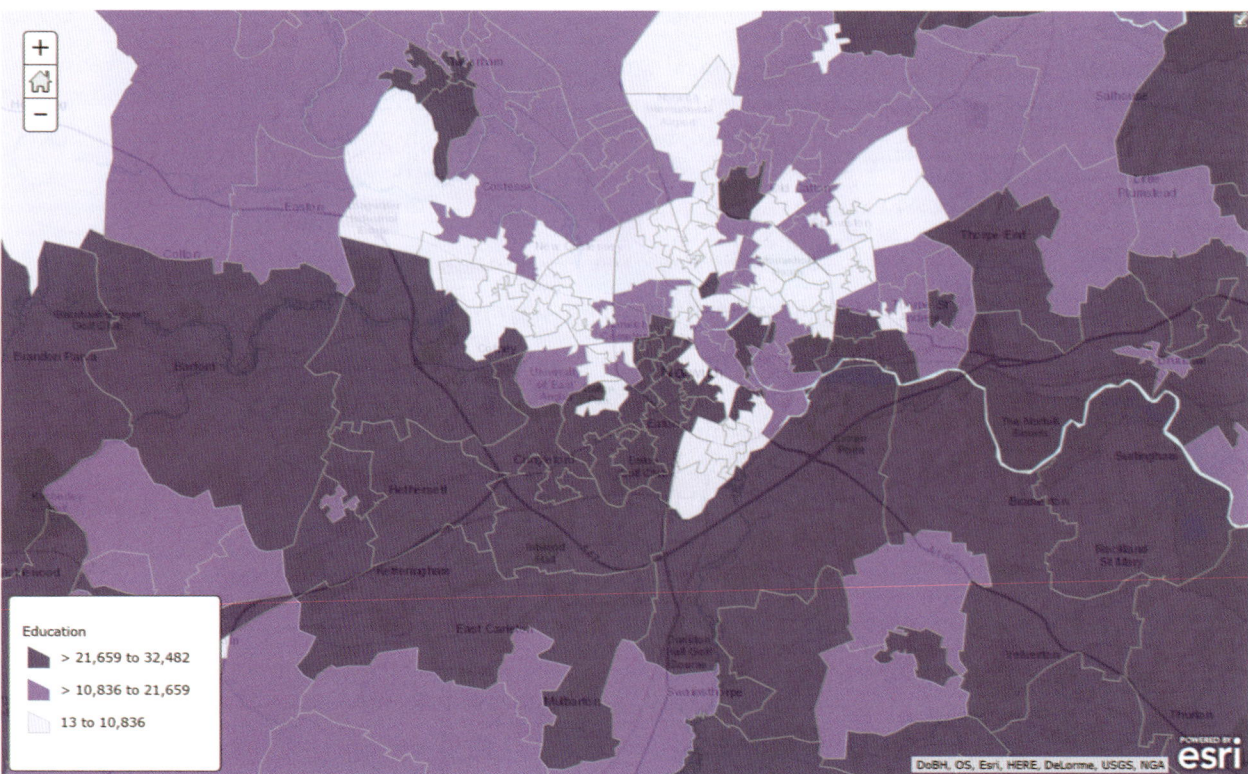

Figure 19.17 Two contrasting GIS maps showing the same data about educational outcomes. Changing the colours and number of categories (or classes) on a choropleth map can create very different visual effects. Which is the better map, do you think? Is there a good use of colour?

Option 8B: Migration, Identity and Sovereignty

Table 19.10 Examples of cartographical and graphical techniques linked to types of information. Note many of the cartographical techniques can be combined with a spatial element, i.e. overlaying located graphs on to maps

Type of information	Graphical (G) and cartographical (C) techniques
Representation of sequential data that changes over time	Line graphs (G) Pictograms (G) Circular graphs/rose diagrams (G)
Recorded data at different sites that has different component categories	Bar charts and histograms (G) Pyramid graphs (G) Pie charts (G) Mirror graphs (G) Multiple/compound bar charts (G)
Where measurements of side views have been taken	Long and cross profiles (G) Cross sections (G)
Data that has been collected to demonstrate spatial variation	Choropleth maps (C) Isopleth maps (C) Dot distribution maps (C) Proportional symbol maps (C)
Representation of data with orientation or bearing	Rose diagrams (G) Polar co-ordinates (G)
Continuous data (along a transect)	Kite diagram (G) Scatter graph (G) Multiple/compound bar charts (G)
Representing linkage or connections between two sets of data	Scatter graphs (G) Mirror graphs (G)

Adapted from John and Richardson (1997) *Methods of Presenting Fieldwork Data.* Geographical Association.

Table 19.11 Techniques for data analysis using quantitative techniques, including the corresponding Excel function command; note there are a large number of written resources available to help you use these procedures

Statistical purpose	Name of method	Excel reference or description, where relevant
To compare and summarise data (central tendency)	Mean Median Mode	=AVERAGE(number1,number2) =MEDIAN(number1,number2) =MODE(number1,number2)
To measure the dispersion, spread and variability of data (range)	Quartiles (upper and lower) Inter-quartile range Standard deviation	=QUARTILE(array,quart) Calculate (Q1, Q3) and get difference =STDEV(number1,number2)
To calculate the degree of statistical correlation (linkage) between two variables	Spearman's rank Line of best fit	Excel doesn't do this directly, but the following is similar: =CORREL(number1,number2)
To test for differences between sets of data (can be more than two)	Chi-squared test Student's t-test Mann-Whitney U test*	Needs a specialist spreadsheet, or Excel plug-in
To mathematically describe the distribution of points in a pattern	Nearest neighbour statistic*	Needs a specialist spreadsheet, or Excel plug-in

*Not specified by Pearson/Edexcel but might be a useful analysis tool for some individual investigations

Key terms

Mode: Strictly speaking, this is the number in the data set that appears most frequently. If a data set is organised into groups or classes it can be displayed as bars, and the highest bar indicates the modal class or category. It can be a useful indicator to see where most numbers are concentrated, but remember that some data sets can be 'bimodal', i.e. have two modes.

Mean: This is the most commonly used measure of central tendency and is the arithmetic or mathematical average of the values in the sample. The advantage is that all values are taken into account, but the mean can be influenced by outliers or extreme values. To calculate the mean you need to add all the values together and then divide by the number of values.

Median: This divides the ordered data set into two halves. If you had 41 observations, for instance, the median would rank 21 in the list; if you had an even number of observations, say 40, then normally you would use the mid-point between 19 and 20).

Describing data presented as tables, charts and maps

Descriptions should not be over-generalised but, at the same time, should give a coherent and clear summary of patterns and trends. Information and data description usually involves a simple sequence:

1 An initial description of the main patterns and trends.
2 Exemplification of the patterns and trends using data/information from the relevant tables, charts or maps.
3 Identification of anomalies or exceptions that deviate from the main patterns and trends.

It is then possible to ask a series of questions that effectively kick-start the analysis process:

- What is the range (or spread) of values within the data set?
- Where are most of the values concentrated (i.e. is there any clustering)?
- Are there any clear gaps between the concentrations?
- What is the shape of the distribution of the data values?
- Are there any extreme values (which may include anomalies and/or outliers)? How far separated are these from the normal range of data?

Descriptive statistical techniques

Descriptive statistical techniques include measures of central tendency and dispersion. They are used to describe or summarise data sets numerically. Central tendency (**mode**, **mean** and **median**) all give a number that can be considered the 'centre' of a set of values.

Under some circumstances it may be appropriate to use other descriptive statistics, including the Gini coefficient and Lorenz curves. The Gini coefficient is usually used to demonstrate income inequality (Figure 19.18) but can be used to measure any form of uneven distribution. There are some online spreadsheets available that allow

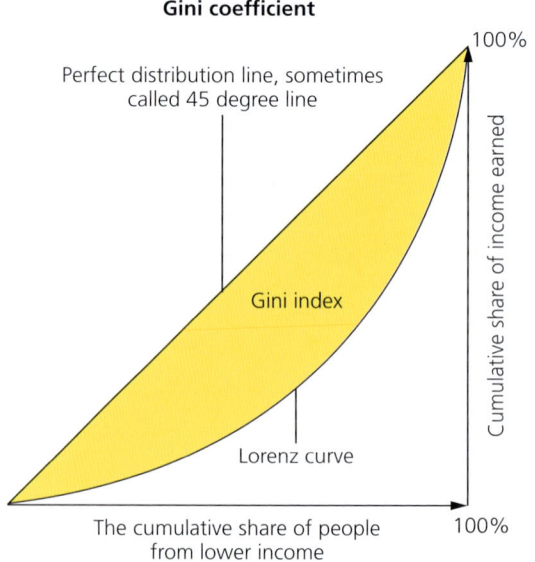

Figure 19.18 Graphical representation of the Gini coefficient and Lorenz curves – this can be used to show inequality at a local level with census output area statistics

Option 8B: Migration, Identity and Sovereignty

users to insert their own local census data (e.g. mortality or health information) and calculate Gini's and Lorenz curves. A good explanation can be found here: www.youtube.com/watch?v=yN1alTAMo3w.

Unfortunately, these measures of central tendency tell us nothing about the dispersion of values in a data set. It is important to have a summary measure of dispersion because very different data sets can yield similar values for the mean, median and mode. The main measures of dispersion are the **range**, **inter-quartile range** and **standard deviation**.

More advanced statistics: inferential techniques

Whatever the results of an enquiry, there is always the possibility that outcomes could be due to chance. This is because most data is from a subset or sample rather than from the entire statistical **population**.

Using inferential statistical tests (see Table 19.11, page 359) we can establish the probability that the results from the data are due to chance. If this probability is small we are able to accept our results with a high level of confidence (and accept the idea statistically). The Pearson/Edexcel Specification names two inferential statistics for testing hypotheses of difference (the chi-squared test and the t-test) and two tests for relationships (Spearman's rank correlation and best-fit trend lines).

You are not required to manually perform statistical tests, instead there should be a brief statement indicating why a particular test might be suitable. If using Excel for example, an annotated screenshot including the cell formula could be included.

Understanding inferential statistical outcomes

We make sense of the results from statistics by assessing their **statistical significance** or statistical 'satisfaction'.

Inferential statistics always need to make use of a **null hypothesis**, as this is what the stats test is actually looking at.

Once we have derived a numerical answer from our test, it can then be compared to a set of significance tables and the result interpreted:

- If the null hypothesis is rejected, the (alternative) hypothesis is accepted.
- If the null hypothesis is accepted, the (alternative) hypothesis is rejected.

The hypothesis is always described in relation to a confidence limit, for example, at 95 per cent we can accept the hypothesis. This is when the calculated value exceeds the theoretical value from the statistical table. Alternatively, a hypothesis that failed to reach the 95 per cent level would be rejected.

Often a statistical answer is combined with a graph to make more sense of the answer and to confirm the statistical result. Figure 19.19, for example, shows different types of correlation from two variables that would be linked to a Spearman's rank test.

Analysis of qualitative data and information

The mass of words and text generated by interviews or data in a textual/photographic form needs to be described, analysed and summarised. The focus and topic may require the researchers to seek relationships between various themes or to relate behaviour or ideas to geo-demographic characteristics of respondents, such as age, gender, ethnicity, etc.

> **Key terms**
>
> **Range:** The difference between the highest and lowest value. For example, if beach sediment samples (stone sizes) were taken along a stretch and the highest value was 141 mm and the lowest 6 mm, the range would be 135 mm. This measure is potentially useful when comparing different areas/data sets as it tells you something about the spread of the data.
>
> **Inter-quartile range:** Perhaps a better measure of data spread. The inter-quartile range is the difference between the 25th and 75th percentile. This means that half of the observations lie in the inter-quartile range – a much more useful measure of the spread of the data around the central value (i.e. the median). To calculate the inter-quartile range, simply put the data into rank order (from highest to lowest) and then divide into four equal parts (quartiles). Subtract quartile 1 from quartile 3. One of the advantages of this method over others (e.g. standard deviation) is that extreme values are ignored since they lie outside of the inter-quartile range.

Key terms

Standard deviation: This measures the spread of the data about the mean value. It can be done 'long hand', but using a spreadsheet is much more realistic for most purposes.

Population: Any complete group with at least one characteristic in common. Populations can consist of any objects or events that can be counted or measured in an environment.

Statistical significance: Means establishing the probability that the results or outcomes are not due to chance. Statistical significance is normally set at the 95 per cent (or 0.05) threshold. At this level, there is only a 5 per cent probability (i.e. one in twenty) that the result has occurred purely by chance.

Null hypothesis: A null hypothesis is the opposite statement to a hypothesis (an idea or tentative theory usually stating that some sort of relationship exists between two or more variables, see page 361). Inferential statistics look to find out whether the null hypothesis is actually correct. On the basis of the results from the inferential test, the hypothesis:
1. can be accepted
2. can be rejected
3. may be inconclusive, so cannot be interpreted reliably.

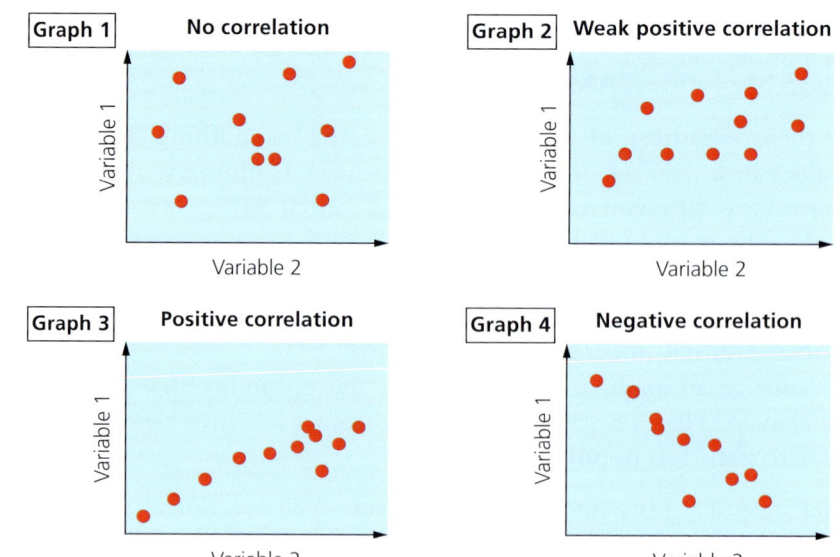

Figure 19.19 Different types of correlation on four scatter graphs

Generally at A level, appropriate analysis would involve the categorisation of qualitative information: in other words, undertaking a count, for example, of negative adjectives that appear in a blog, forum or editorial piece. This then allows the 'count' data to be analysed in a more traditional manner, i.e. as categorical data. You can use the 'find' function in many software programs to semi-automate the counting process.

An alternative approach might be to use a thematic charting approach, sometimes called *framework analysis*. In this process the researcher initially identifies particular themes (these could be initiatives, issues or processes, for example) and then in the rows identifies particular places, texts or observations, for instance. Table 19.12 illustrates the use of different documents and sources to exemplify themes. In such a table it is important to reference sources – see items in green.

Other forms of qualitative analysis can be used, including coding (using highlighting techniques), annotation of photographs (with explanatory text) and narratives. Figure 19.5 (page 345), for example, could be repurposed to include more analytical comments.

GIS as a tool for data analysis

GIS should not only be seen as a tool for data presentation but also as a mechanism to extract, manipulate and analyse data and information that has geospatial attributes. Figure 19.20 (page 363) shows a simple digital catchment map for an area around Skegness on the east coast, for example. GIS can also show interactions between variables. Layers can be interrogated to find out spatially where there might be overlap between features.

Making geographical explanations and connections

If the results of an investigation are consistent with the idea or hypothesis being tested, then the hypothesis and the theory or logic that underpins it would seem to be verified. Sometimes the data provide a general pattern or trend that is clear, but at other times values deviate from those expected. Such anomalies could be explained by the complexity of geographical systems and features. Often their explanation involves recognition and understanding of more than one causal factor, as well as possible problems with data-collection methodology and sampling accuracy (discussed on page 364).

Table 19.12 Qualitative sources, coding and framework analysis

	Theme 1: Changing nature of shops and services	Theme 2: Gentrification leading to higher house prices	Theme 3: Seasonal and low-paid employment
Respondent – interview (1)	Has seen a decrease in the number of affordable cafes and other places to eat *2:33*	Knows many people who cannot afford to move to a bigger house so they are worried about having a family *4:27*	
Local plan – document (2)		Wants to encourage more low-cost housing and identifies possible sites *page 23, para 3*	The authority is working with local businesses to pilot a living wage scheme and better pension provision *page 75, para 2*
Forum – internet research (3)	People are worried about more outdoor shops and lack of high street balance *Dave9066, 3rd comment*		Concerned about lack of flexibility and low wages for mothers who work part time *JanetK, 6th comment*

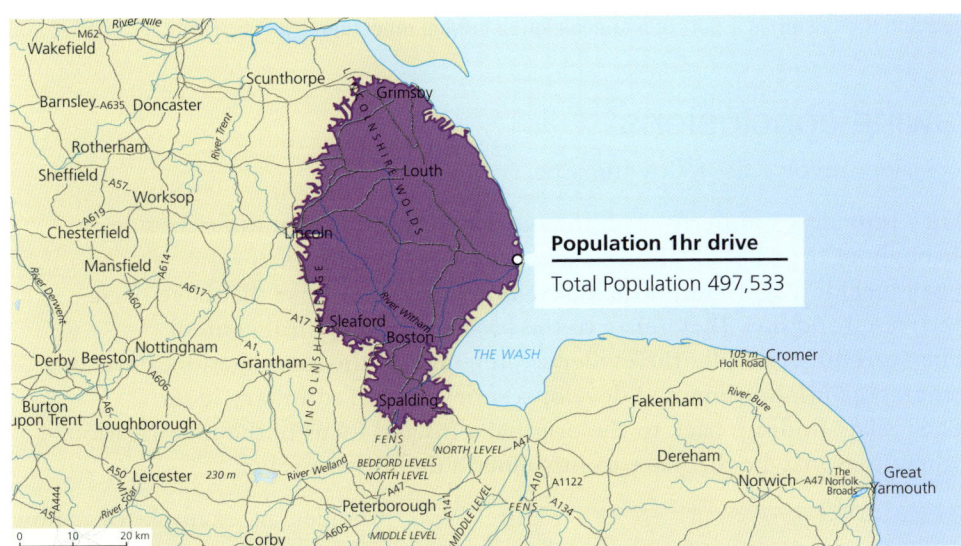

Figure 19.20 GIS can calculate the number of people, for example, who live within a one-hour drive of a particular point on a map; this map also shows the 'drive-time catchment' as a purple area or buffer

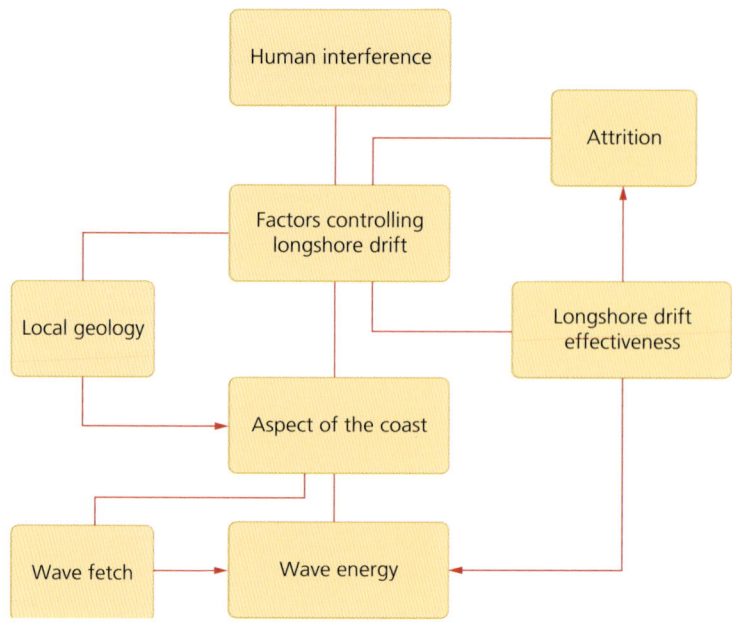

Figure 19.21 A simple concept map can help with the process of explanation; it might be further refined and then used as a part of a conclusion or critical reflection

Drawing conclusions

Conclusions involve the following aspects:

- They provide a summary of all the major findings made at different stages throughout the individual investigation.
- They include a synthesis of the relevant geographical links that have been uncovered and how they relate to the broader geographical content.
- They carefully consider the evidence presented in the investigation, draw out geographic implications and develop a series of brief summary ideas.
- They relate findings back to the original aims and focus of the investigation; conclusions may be the place to introduce a definitive answer.

Figure 19.22 provides an example of a flow diagram which illustrates themes that could be covered as part of a conclusion.

Conclusions are often ambiguous and even unclear (and this should always be acknowledged as part of the write-up using words such as 'partial', 'tentative' or 'inconclusive'). This is not an admission of failure. Most often it arises because geographical environments and processes are complex, and data collection cannot be managed to create a reliable outcome due to the range of other external factors influencing the results.

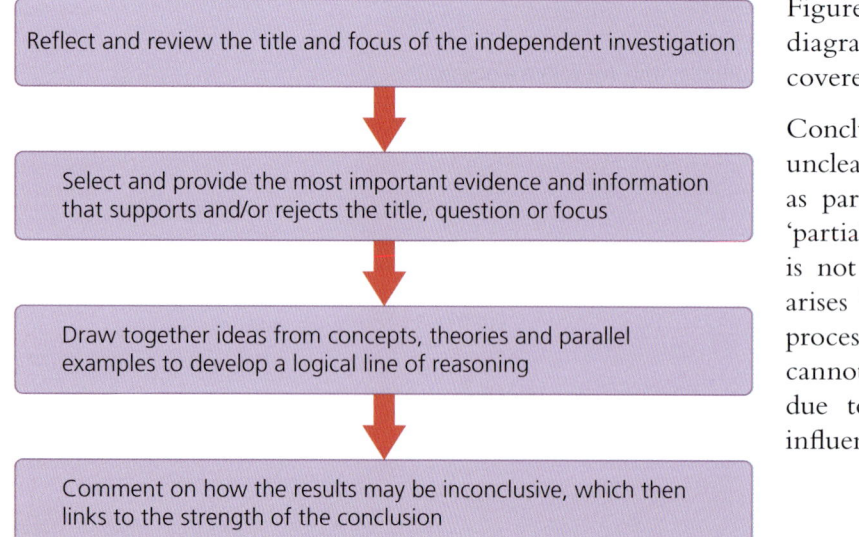

Figure 19.22 An example flow diagram for producing conclusions

Option 8B: Migration, Identity and Sovereignty

Critical reflection

Critical reflection is an opportunity to look back over *all* the fieldwork and research and to identify any shortcomings. It is also the chance to think more widely about the meaning of your results, comparing them to similar studies, noting and explaining similarities and differences.

It is almost certain that all A level investigations will encounter some limitations of methodology. The most obvious concerns the choice and delivery of the sampling strategy, and, as a result, the amount and quality of data collected. In questionnaire surveys, for instance, obtaining high-quality and representative samples can be difficult. This is because this type of survey work should be based on a stratified approach based on the small output area statistics for a local area. Another problem is that rejection rates for street interviews are high and students may resort to interviewing anyone willing to respond. There is also the problem of visitors (as opposed to residents) – a profile of them is likely unknown so it becomes difficult to create a reliable sampling design.

Any deviation from the 'best' sampling strategy will create potential for lower reliability and, ultimately, affect the validity of the investigation. Sometimes the investigation inadvertently introduces **bias**.

Insufficient sample data is another problem for many investigations. Insufficient data and information (primary and/or secondary) makes statistical analysis difficult which, in turn, provides little confidence in the reliability of results. This problem can be tackled at the planning stage by identifying the statistical test to be used for analysis and the minimum sample size needed to obtain statistically significant results. Where there is doubt it is always best to collect a larger rather than a smaller sample.

In summary, it's often useful to critically reflect in terms of:

1. errors linked to sampling and design
2. problems that are operator related (e.g. mishandling equipment)
3. those which are directly linked to equipment (e.g. level of accuracy and resolution – see Figure 19.23).

Your evaluation section may also provide an opportunity to suggest improvements and make recommendations that would improve the overall reliability of the investigation.

> **Key term**
>
> **Bias:** A consistent error in data. Doing more measurements as replicates is one way of reducing or eliminating bias; for example, repeating a measurement of temperature at the same location and at the same time would be more reliable (and therefore reduce bias) compared to a single measurement.

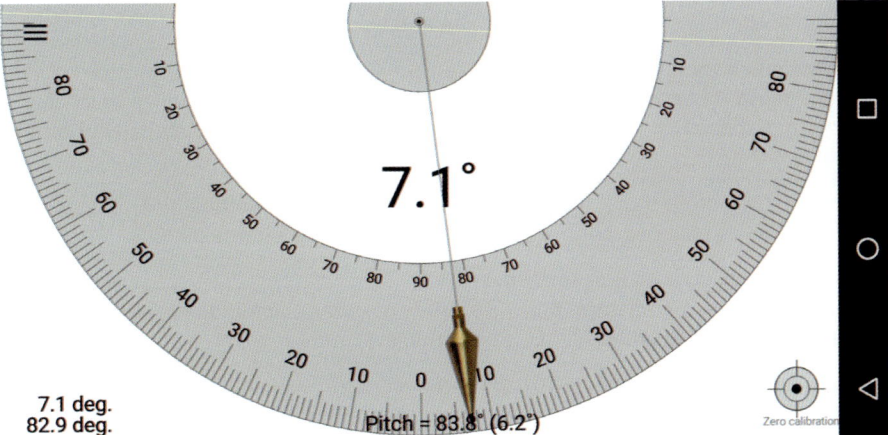

Figure 19.23 Equipment may offer a precise reading, but how can you be sure that it is accurate?

The bigger picture: linking to the wider geographical context and extending understanding

The Pearson/Edexcel mark scheme makes reference to the need to place the study in a wider geographical context as well as to comment on how it helped extend understanding. In order to achieve these objectives, it will be necessary to consider the literature/research elements of the investigation again and then to make other links, some of which might be far reaching.

Table 19.13 shows examples of topics and themes, and then possible wider linkages that could be briefly explored as part of a conclusion and critical evaluation.

Final considerations and project management

The following ideas might be considered a useful set of hints and tips for any independent investigation. Think of them as an integral part of the first proofread (or pre-submission editing) and check that:

- it has a clear and focused geographical argument that is easy and logical to follow
- it follows a logical series of stages, mirroring those identified in the specification, which link to a high-quality route to enquiry
- it has a clear geographical purpose and genuinely tries to uncover some 'new' geography
- it makes use of high-quality and specialised language and is well structured
- it uses a high-quality range of research and literature to contextualise the focus
- it includes an **executive summary** (Figure 19.24), which is found at the start of the work, before the introduction and context
- it is around 3000 to 4000 words long, but not substantially over that (this excludes words in diagrams, tables, etc.)
- it is correctly referenced, including in-text referencing and/or a bibliography
- it introduces research material in a selective way, making sure that large chunks of text are not copied.

> **Key term**
>
> **Executive summary:** An opportunity to provide a short but factual summary of the whole investigation. Such summaries are often called abstracts in academic writing – an example can be seen in Figure 19.24.

Table 19.13 Topics and themes and their wider geographical linkages

Focus/title	Wider geographical linkages
A study on the carbon stores of two contrasting woodland ecosystems	Deforestation; biomass production; land-use change; climate change
Evaluating the evidence for lowland glacial features conforming to an expected model in terms of size and shape	Predicting the characteristics of other glacial features; land-use change and futures; mathematical predictability of other physical geography features
An examination of the factors that make cliffs vulnerable to collapse and subsequent recession	Rising sea levels; coastalisation; changes in the numbers of stormy days
What is the identity of place T as seen by different social and economic groups?	Identity of other places; different social and economic groups
An investigation into the patterns of energy consumption in two contrasting communities	Socio-economic controls; future energy costs (e.g. oil); future demographic changes; government policies

ABSTRACT

Following Gregory and Walling (1971), who described how relatively simple equipment can be used to monitor processes in a small drainage basin, experiments were started in the catchment close to the Leonard Wills Field Centre (grid reference ST 0537) in 1973. The purpose of these experiments was to provide a set of long-term data which could be easily incorporated into sixth form courses and which would illustrate the hydrological principles included in A-level syllabuses. This paper outlines the results collected over the last twelve years and provides an interpretation of both the water balance within the catchment and also the long and short term responses of the stream to rainfall.

Figure 19.24 An example of an abstract (or executive summary) from an academic journal

Review questions

1. Explain the reasons why it is necessary to develop a high-quality route to enquiry.
2. Is crowd-sourced data primary or secondary, and why is it difficult to categorise?
3. What sort of sample methods might be appropriate for the following investigations?
   a. A survey of differing infiltration rates within a drainage basin.
   b. A land-use survey along a coast as a part of an investigation into contrasting schemes.
   c. Sediment size and roundness across and down a pebble beach.
   d. The perceived cultural characteristics of a place.
   e. The variations in the health of the population of two small census areas.
4. Evaluate the advantages and disadvantages of the different types of secondary data.
5. Examine the different limitations under the headings of 'temporal' and 'spatial' that your investigation is likely to encounter.
6. Why should you not always trust the results from statistical testing? What are the potential pitfalls?
7. Examine the potential ethical issues of the fieldwork and research that you are proposing to undertake.

Further research

Find out more about oral histories from this website: **www.ohs.org.uk**

The Royal Geographical Society (RGS) produces a comprehensive guide to fieldwork with information and instructions for a range of techniques that can be used to carry out geographical investigations in different locations and settings: **https://www.rgs.org**

The Field Studies Council (FSC) has a dedicated website for fieldwork; it is constantly updated to match the changing requirements of AS and A level: **www.geography-fieldwork.org**

For many landscape investigations geology is going to play a key role in the research; access the BGS online geology map viewer here: **https://www.bgs.ac.uk/map-viewers/bgs-geology-viewer/**

GLOSSARY

Acculturation: A process of cultural change that takes place when two different cultures meet and interact; it includes the transfer of a dominant culture's ideas on to a subordinate culture.

Accuracy: In geography the accuracy of a measurement system/data is the degree of closeness of measurements of a quantity to that quantity's actual (true/real) value. The further a measurement is from its expected value, the less accurate it is.

Afforestation: The planting of trees in an area that has not been forested in recent times.

Agricultural drought: The rainfall deficiency from meteorological drought leads to deficiency of soil moisture and soil water availability, which has a knock-on effect on plant growth and reduces biomass.

Aim: A statement of what the project/investigation is setting out to achieve. It must be geographically sound and achievable.

Albedo: A measure of the proportion of the incoming solar radiation that is reflected by the surface back into the atmosphere and space.

Albedo flip: When the sunlight reflected by white ice is suddenly absorbed as ice melts, creating a dark surface of open water.

Anthropogenic: Processes and actions associated with human activity.

Aquaculture: The farming of aquatic organisms such as fish, crustaceans, molluscs and aquatic plants.

Arab Spring: A series of pro-democracy, pro-human rights civil uprisings in 2011 that affected Syria, Tunisia, Libya, Egypt, Bahrain and Iran. Some governments were overthrown but, in most cases, protracted instability followed the uprisings.

Arctic barometer: The idea that the sensitive Arctic region provides an early warning of the pressures and environmental changes caused by global warming.

Backwash: Flows of people, investment and resources directed from peripheral to core regions. This process is responsible for the polarisation of regional prosperity between regions within the same country.

Base flow: The normal, day-to-day discharge of the river.

Bias: A consistent error in data. Doing more measurements as replicates is one way of reducing or eliminating bias; for example, repeating a measurement of temperature at the same location and at the same time would be more reliable (and therefore reduce bias) compared to a single measurement.

Big data: Extremely large, dynamic data sets that may be analysed to reveal patterns, trends and associations, especially relating to human behaviour and interactions. It may provide a source of information that could be used to support the individual investigation. Frequently it has GIS information linked to it. Some researchers work with big and complex data sets and produce very interesting map 'mash-ups'.

Bilateral aid: Aid that is delivered on a one-to-one basis between a donor and a recipient country.

Biofuel: A fuel derived immediately from living matter, such as agricultural crops, forestry or fishery products, and various forms of waste (municipal, food shops, catering, etc.). A distinction is made between primary and secondary biofuels:

- **Primary biofuels** include fuelwood, wood chips and pellets, and other organic materials that are used in an unprocessed form, primarily for heating, cooking or electricity generation.
- **Secondary biofuels** are derived from the processing of biomass and include liquid biofuels such as ethanol and biodiesel, which can be used by vehicles and in industrial processes.

Biomass: Organic matter used as a fuel, especially in power stations for the generation of electricity.

Blue water navy: One which can deploy into the open ocean, i.e. with large, ocean-going ships. Many smaller nations only have a green water navy designed to patrol littoral waters, i.e. those close to the nation's coastline.

Blue water: Water is stored in rivers, streams, lakes and groundwater in liquid form (the visible part of the hydrological cycle).

Bottom up: Small-scale development schemes.

Brand value, or brand equity: The value of a brand measured using metrics such as market share, customer opinion of the brand and brand loyalty.

Bretton Woods institutions: The IMF and the World Bank were founded at the Bretton Woods conference in the USA at the end of the Second World War to help rebuild and guide the world economy. The General Agreement on Tariffs and Trade (GATT) was set up soon afterwards and later became the WTO.

Carbon cycle: The biogeochemical cycle by which carbon moves from one sphere to another. It acts as a closed system made up of linked subsystems that have inputs, throughputs and outputs. Carbon stores function as sources (adding carbon to the atmosphere) and sinks (removing carbon from the atmosphere).

Carbon cycle pumps: The processes operating in oceans to circulate and store carbon. There are three sorts: biological, carbonate and physical.

Carbonate rocks: These contain a high proportion of carbonate ions (CO_3^{2-}) within minerals such as calcite, aragonite and siderite. All limestones (including chalk) and dolomites are carbonate rocks.

Catchment: The area of land drained by a river and its tributaries.

Channel flow: The flow of water in streams or rivers.

Channel storage: The storage of water in streams or rivers.

Climate change adaptation: This includes any passive, reactive or anticipatory action taken to adjust to changing climatic conditions. There are two types:

- **hard strategies** require technology, for example, wind farms
- **soft strategies** involve legislation, such as land-use zoning.

Cold War: A period of tension between ideologically rival superpowers the capitalist USA and communist USSR that lasted from 1945 to 1990. It was also the period when nuclear weapons, and systems to deliver them, were perfected, adding to the tension.

Colonial control: The direct militarised control exerted over territories conquered by mainly European powers in the period 1600 to 1900. They were ruled by force, with almost no power or influence being given to the original population.

Condensation: The change from a gas to a liquid, such as when water vapour changes into water droplets.

Convectional rainfall: Often associated with intense thunderstorms, which occur widely in areas with ground heating such as the Tropics and continental interiors.

Core-periphery system: The uneven spatial distribution of national population and wealth between two or more regions of a country, resulting from flows of migrants, trade and investment from periphery to core.

Critical threshold: An abrupt change in an ecological state. Small environmental changes can trigger significant responses. Negative and positive feedback loops reinforce or undermine changes once an alternative stable state has become established.

Cryosphere: Areas of the Earth where water is frozen into snow or ice.

Cultural cohesion: The capacity of different national and ethnic groups to make a mutual commitment to live together as citizens of the same state.

Cultural heterogeneity: A society where there is a high level of cultural and or ethnic diversity among its citizens, often resulting in a multi-lingual and multi-faith community.

Cultural landscape: The distinctive character of a geographical place or region that has been shaped over time by a combination of physical and human processes.

Deforestation: The cutting down and removal of all or most of the trees in a forested area.

Democracy: Countries with a system of government in which power is either held by regularly elected representatives or directly by the people.

Dependency: In the context of economic development it means that the progress of a developing country is influenced by economic, cultural and political forces that are controlled by developed countries.

Deprivation: When an individual's well-being and quality of life fall below a level regarded as a reasonable minimum. Measuring deprivation usually relies on indicators relating to employment, housing, health and education.

Desertification: Land degradation in arid, semi-arid and dry sub-humid regions resulting from various factors, including climatic variations and human activities.

Development gap: The widening income and prosperity gap between the global 'haves' of the developed world and the 'have-nots' of the developing world, especially the least developed countries.

Dew point: The temperature at which dew forms; it is a measure of atmospheric moisture.

Diaspora: A dispersed group of people with a shared cultural background who have spread internationally from their original homeland.

Diplomacy: The negotiation and decision-making that takes place between nations as part of international relations, leading to international agreements and treaties.

Ebola: A highly contagious and fatal disease spread through contact with body fluids infected by a filovirus. Its symptoms are fever and severe internal bleeding. The host species of the virus has not been confirmed but fruit bats and primates have been implicated.

Economic migrant: Most migrants move for economic reasons such as a better job or hope of higher pay. Many economic migrants move voluntarily (they choose to move) but not all. In some cases migrants are forced to move by people traffickers or even family members, and put to work in a new location.

Economic miracle: An informal term commonly used to refer to a period of dramatic and fast economic development that is unexpectedly strong.

Economic restructuring: The shift from primary and secondary industry towards tertiary and quaternary industry as a result of deindustrialisation. It has large social and economic costs.

Economy of scale: If the production of a commodity is expanded then the unit cost price may fall. This is because certain fixed costs (such as the cost of lighting and heating a factory) are spread over more units of output (the product). As a result, the products can be sold more cheaply, which increases revenues further.

Ecosystem resilience: The level of disturbance that ecosystems can cope with while keeping their original state.

Emergency aid: Rapid assistance given by organisations or governments to people in immediate distress following natural or man-made disasters. The aim is to relieve suffering and the aid includes such things as food and water, temporary housing and medical help.

Energy density: The amount of energy stored in a given system, substance, or region of space per unit volume.

Energy mix: The combination of different available energy sources used to meet a country's total energy demand. The exact proportions or mix vary from country to country. It is an important component of energy security.

Energy pathway: The route taken by any form of energy from its source to its point of consumption. The routes involve different forms of transport, such as tanker ships, pipelines and electricity transmission grids.

Enhanced greenhouse effect: The intensification of the natural greenhouse effect by human activities, primarily through fossil fuel combustion and deforestation, causing global warming.

Ethnicity: The shared identity of an ethnic group which may be based on common ancestral roots or cultural characteristics such as language, religion, diet or clothing.

Ethnoscape: A cultural landscape constructed by a minority ethnic group, such as a migrant population. Their culture is clearly reflected in the way they have remade the place where they live.

Eutrophication: Excessive richness of nutrients in a lake or other body of water, frequently due to run-off from farming land, which causes a dense growth of plant life and death of animal life from lack of oxygen.

Evaporation: The change in state of water from a liquid to a gas.

Evapotranspiration (ET or EVT): The combined effect of evaporation and transpiration.

Exclusive economic zone (EEZ): The area of ocean extending 200 nautical miles beyond the coastline (or to the edge of the continental shelf), over which a nation controls the sea and sub-sea resources. EEZ borders are decided by the UN in the event of a dispute.

Executive summary: An opportunity to provide a short but factual summary of the whole investigation. Such summaries are often called abstracts in academic writing – an example can be seen in Figure 19.24.

Expatriate: Someone who has migrated to live in another state but remains a citizen of the state where they were born.

Falling or recessional limb: The part of a storm hydrograph in which the discharge starts to decrease.

Famine drought: A humanitarian crisis in which the widespread failure of agricultural systems leads to food shortages and famines with severe social, economic and environmental impacts.

Fieldwork: A contested term often meaning different things in different times and places, as well as to different people. Lonergan and Andersen (1988) suggest that it is supervised learning that involves first-hand experience outside the classroom. But there are different types of fieldwork, ranging from observational and data collection, to 'tour' fieldwork. However, the best fieldwork experiences are when students connect with or immerse themselves in an environment in order to make geographical sense of what they have found. Students should be prepared to take responsibility for their own learning, ensuring that they play an active part in the field.

Flash flooding: A flood with an exceptionally short lag time – often minutes or hours.

Fluxes: The rate of flow between the stores.

Fossil water: Ancient, deep groundwater from former pluvial (wetter) periods.

Fracking: Hydraulic fracking or oil/gas well stimulation is a technique in which rock is fractured by a pressurised liquid.

Fragile state: A country with weak governance, limited rule of law, humanitarian crises and social tensions (often the legacy of armed conflict or civil war). The basic social and economic needs of most people are not being met.

Free trade: The exchange of goods and services free of import/export taxes and tariffs or quotas on trade volume. Taxes, tariffs and quotas are forms of protectionism designed to make imports more expensive than locally produced goods (thus protecting local producers).

Geographical Information System (GIS): Maps with 'layers' of information are an important tool in analysing place characteristics.

Geo-strategic policies: Policies that attempt to meet the global and regional policy aims of a country by combining diplomacy with the movement and positioning of military assets.

Global citizenship: A way of living wherein a person identifies strongly with global-scale issues, values and culture rather than (or in addition to) narrower place-based identity.

Global commons: Global resources so large in scale that they lie outside of the political reach of any one state. International law identifies four global commons: the oceans, the atmosphere, Antarctica and outer space.

Global governance: The term 'governance' suggests broader notions of steering or piloting rather than the direct form of control associated with 'government'. 'Global governance' therefore describes the steering rules, norms, codes and regulations used to regulate human activity at an international level. At this scale, regulation and laws can be tough to enforce, however.

Green Revolution: The use of high yield varieties (HYVs) of crops along with the use of agrochemicals and irrigation to increase yields and improve food supplies; begun in the 1960s.

Green water: Water stored in the soil and vegetation (the invisible part of the hydrological cycle).

Grey water: Refers to waste bath, sink or washing water. It can be recycled, resulting in savings in water usage.

Groundwater flooding: Flooding that occurs after the ground has become saturated from prolonged heavy rainfall.

Groundwater flow: The slow transfer of percolated water underground through pervious or porous rocks.

Harvard: A style of referencing primarily used in academic writing to cite (reference) information sources. Each reference should contain in a sequenced order:

1 Name of the author(s)
2 Year published
3 Title
4 Publisher
5 Pages used

This approach can be modified for newspaper and internet sources:

- Last name, First initial. (Year published). Article title. Newspaper, [online] pages. Available at: URL [Accessed Day Month Year].
- Last name, First initial. (Year published). Page title. [online] Website name. Available at: URL [Accessed Day Month Year].

Hegemonic power: The ability of a powerful state or player to influence outcomes without reverting to 'hard power' tactics such as military force. Instead, control is exercised using a range of 'soft' strategies including diplomacy, aid, and the work of the media and educational institutions.

HIPC policies: The Heavily Indebted Poor Countries (HIPC) Initiative was launched in 1996 by the IMF and World Bank, with the aim of ensuring that no poor country faces a debt burden it cannot manage. Countries must meet certain criteria, commit to poverty reduction through policy changes, and demonstrate a good track record over time.

Homogeneous culture: A society where there is very little cultural or ethnic diversity and most people share cultural traits with one another, including language, religion, dress and diet.

Hydrological drought: Associated with reduced stream flow and groundwater levels, which decrease because of reduced inputs of precipitation and continued high rates of evaporation. It results in reduced storage in lakes and reservoirs, often with marked salinisation and poorer water quality.

Hydroponics: A method of growing plants using mineral nutrient solutions without soil.

Hyperpower: An unchallenged superpower that is dominant in all aspects of power (political, economic, cultural, military); examples include the USA from 1990 to 2010 and Britain from 1850 to 1910.

Hypothesis: A statement whose accuracy can be tested objectively using scientific methodology. Null and alternative hypotheses are normally used in connection with statistical tests (Chi-squared and Spearman's rank).

Ideology: A set of beliefs, values and opinions held by many people in a society. These determine what is considered normal or acceptable behaviour. Superpowers project their ideology on others. In the case of the USA this includes 'Western values' of free speech, individual liberty, free-market economics and consumerism.

Infiltration: The movement of water from the ground surface into the soil.

Infiltration capacity: The maximum rate at which rain can be absorbed by a soil.

Integrated water resource management (IWRM): A process which promotes the co-ordinated development and management of water, land and related resources in order to maximise economic and social welfare in an equitable manner without compromising the sustainability of vital ecosystems.

Integration: The process by which migrants (individuals and groups) become accepted into society. It is a two-way process of adaptation by migrants and host societies involving understanding and respecting cultural values. It involves migrants entering the labour market, finding places to live and accessing services.

Interception loss: This is water that is retained by plant surfaces and later evaporated or absorbed by the vegetation and transpired. When the rain is light, for example, drizzle, or of short duration, much of the water will never reach the ground and will be recycled by this process (it's the reason you can stand under trees when it is raining and not get wet).

Inter-governmental organisations (IGOs): Regional or global organisations whose members are nation states. They uphold treaties and international law, as well as allowing co-operation on issues such as trade, economic policy, human rights, conservation and military operations.

Intergovernmental Panel on Climate Change (IPCC): The leading international organisation for the scientific assessment of climate change.

Inter-quartile range: Perhaps a better measure of data spread. The inter-quartile range is the difference between the 25th and 75th percentile. This means that half of the observations lie in the inter-quartile range – a much more useful measure of the spread of the data around the central value (i.e. the median). To calculate the inter-quartile range, simply put the data into rank order (from highest to lowest) and then divide into four equal parts (quartiles). Subtract quartile 1 from quartile 3. One of the advantages of this method over others (e.g. standard deviation) is that extreme values are ignored since they lie outside of the inter-quartile range.

Inter-tropical convergence zone (ITCZ): A concentration of warm air that produces rainfall as part of a global circulation system (the Hadley cell). It moves north and south across the equator seasonally. Small shifts in its location can cause drought.

Ions: Ions are atoms or molecules with a positive or negative charge. Chemical weathering of carbonate rocks (limestone) by solution produces carbonate ions (HCO_3^-) whereas hydrolysis of silicate minerals such as feldspar produces calcium ions (Ca^{2+}) and carbonate ions (CO_3^{2-}). These ions are transported in solution through the hydrological cycle to the oceans.

IS (Islamic State, also known as ISIS, ISIL and Daesh): A terrorist organisation that rose to prominence in 2013 during the Syrian civil war, occupying parts of the Middle East and carrying out terrorist attacks worldwide.

Israel–Palestine conflict: The Israel–Palestine conflict between the Jewish state of Israel and Arabs in Palestine (who claim the same territory) is one of the world's longest running and most intractable conflicts. The involvement of Colonial, Cold War and contemporary superpowers has often complicated a religious and territorial dispute into a wider geostrategic conflict over political and economic influence. Numerous peace initiatives brokered since 1948 have largely failed to find a sustainable solution to the ongoing conflict.

Jökulhlaup: A type of glacial outburst flood that occurs when the dam containing a glacial lake fails.

Lag time: The time interval between peak rainfall and peak discharge.

Land conversion: Any change from natural ecosystems to an alternative use; it usually reduces carbon and water stores and soil health.

Land grabbing: A contentious issue involving the acquisition of large areas of land in developing countries by domestic and transnational companies, governments and individuals. In some instances, land is simply taken over and not paid for.

Land reform: Most often this involves the redistribution of property and agricultural land as result of government-initiated or government-backed actions.

Literature research: This important aspect of the investigation involves finding the current academic or published information around a topic. Be sure to evaluate any research information based on age, author, source, etc., as well as checking whether the research is agreed or supported by other authors.

Maternal mortality: The death of a woman while pregnant or within 42 days from the end of the pregnancy.

Mean: This is the most commonly used measure of central tendency and is the arithmetic or mathematical average of the values in the sample. The advantage is that all values are taken into account, but the mean can be influenced by outliers or extreme values. To calculate the mean you need to add all the values together and then divide by the number of values.

Median: This divides the ordered data set into two halves. If you had 41 observations, for instance, the median would rank 21 in the list; if you had an even number of observations, say 40, then normally you would use the mid-point between 19 and 20.

Meteorological drought: Defined by shortfalls in precipitation as a result of short-term variability within the longer-term average overall, as shown in many semi-arid and arid regions such as the Sahel. Drought has become almost a perennial problem in recent years as longer-term trends have shown a downward movement in both rainfall totals and the duration and predictability of the rainy season, or the occurrence of megadroughts in California in 2017–19.

Middle class: Globally, the middle class are defined as people with discretionary income. They can spend this on consumer goods and perhaps holidays. The global middle class can be defined as people with an annual income of over US$10,000.

Millennium Ecosystem Assessment (MEA): The UN Millennium Ecosystem Assessment was the first major global audit of the health of ecosystems in 2005, highlighting their degradation (the loss of natural productivity through overuse and destruction).

Mitigation: Involves the reduction or prevention of GHG emissions by new technologies and low-carbon energies (renewables, nuclear), becoming more energy efficient, or changing attitudes and behaviour.

Mode: Strictly speaking, this is the number in the data set that appears most frequently. If a data set is organised into groups or classes it can be displayed as bars, and the highest bar indicates the modal class or category. It can be a useful indicator to see where most numbers are concentrated, but remember that some data sets can be 'bimodal', i.e. have two modes.

Morbidity: A state of ill health.

Multilateral aid: Aid (usually financial, sometimes technical) given by donor countries to international aid organisations such as the World Bank or Oxfam. These organisations distribute the aid to what they deem to be deserving causes.

Nationalism: The belief held by people belonging to a particular nation that their own interests are much more important than those of people belonging to other nations.

Nationalist: A political movement focused on national independence or the abandonment of policies that are viewed by some people as a threat to national sovereignty or national culture.

Negative externalities: Costs suffered by people and places as a result of changing economic activity. These may be unintended social or environmental impacts, such as unemployment or pollution.

Neo-colonial: The indirect actions by which developed countries exercise a degree of control over the development of their former colonies. This can be achieved through varied means including conditions attached to aid and loans, cultural influence and military or economic support (either overt or covert) for particular political groups or movements within a developing country.

Net migration: Net migration is the difference between the number of immigrants (arrivals) and emigrants (departures) usually over a period of a year. If immigration is higher than emigration then there is positive net migration. The term can be applied to a country, or a region within a country.

Nimbyism: 'Not in my backyard' – people protesting about developments which they see as detrimental to their own neighbourhood.

Null hypothesis: A null hypothesis is the opposite statement to a hypothesis (an idea or tentative theory usually stating that some sort of relationship exists between two or more variables). Inferential statistics look to find out whether the null hypothesis is actually correct. On the basis of the results from the inferential test, the hypothesis:

1 can be accepted
2 can be rejected
3 may be inconclusive, so cannot be interpreted reliably.

Nutrition transition: A change in diet from staple carbohydrates towards protein (meat, fish), dairy products and fat. It often includes eating more processed food. It occurs as people transition from rural poverty to being urban, middle-class workers.

Ocean acidification: The decrease in the pH of the Earth's oceans caused by the uptake of carbon dioxide from the atmosphere.

Official Development Assistance (ODA): A term used by the OECD to measure aid. It is widely used as an indicator of flows of international aid. Flows are transfers of resources, either in cash or in the form of commodities or services.

Parent company: The original business that a global TNC has developed around and whose directors still make decisions that affect the organisation as a whole. Both Starbucks and Google are parent companies to global networks of subsidiary businesses, including Ritea Ltd (Starbucks Coffee Company, Ireland) and Google Ireland Ltd.

Peak discharge: The time when the river reaches its highest flow.

Percolation: The transfer of water from the surface or from the soil into the bedrock beneath.

Percolines: Lines of concentrated water flow between soil horizons to the river channel.

Petagrams (Pg) or Gigatonnes (Gt): The units used to measure carbon; one petagram (Pg), also known as a gigatonne (Gt), is equal to a trillion kilograms, or 1 billion tonnes.

pH: A logarithmic measure of acidity or alkalinity. A value of 7 means neutral; above this the pH is alkaline, below this it is more acidic.

Players: Individuals, groups or organisations with an involvement or interest in a particular issue.

Polluter Pays Principle: The idea that whoever generates pollution should pay the costs of cleaning it up, either through taxes or fines or by being forced to use technology to prevent its emission in the first place.

Population: Any complete group with at least one characteristic in common. Populations can consist of any objects or events that can be counted or measured in an environment.

Positive feedback: This occurs when a small change to a system imbalances that system and this leads to further changes which are amplified over time. In terms of climate change, small increases in temperature lead to further changes to the climate system, and even higher temperatures.

Post-colonial migrants: People who moved to the UK from former colonies of the British Empire during the 1950s, 1960s and 1970s.

Potential evapotranspiration (PET or PEVT): The water loss that would occur if there was an unlimited supply of water in the soil for use by vegetation.

Precipitation: The movement of water in any form from the atmosphere to the ground.

Primary data: Primary data is generally considered to be first-hand data collected by a student themselves (or as part of a group).

In reality there is some 'grey' between these two ideas and approaches, suffice it to say that you should expect to have a reasonable balance between the two types.

Processes: The physical mechanisms that drive the fluxes of water between the stores.

Question: A question that is asked (in a question format), that often links with the overall title and can be used as a way of sub-dividing the title.

Range: The difference between the highest and lowest value. For example, if beach sediment samples (stone sizes) were taken along a stretch and the highest value was 141 mm and the lowest 6 mm, the range would be 135 mm. This measure is potentially useful when comparing different areas/data sets as it tells you something about the spread of the data.

Rare earth minerals (or rare earth elements, REE): A group of metal elements crucial to modern communication, medical and laser technology. Found dispersed in rocks, they are hard to mine, costly and supplies are limited.

Reforestation: Planting trees in places with recent tree cover, replacing lost primary forests.

Reliability: This refers to the consistency or reproducibility of a measurement. A measurement is said to have a high reliability if it produces consistent results under consistent conditions. True reliability cannot be calculated – it can only be estimated based on knowledge and understanding of the topic.

Remote sensing: Surveillance by satellites such as Landsat generates data that can authenticate, or refute, official government data.

Rendition: The practice of sending a foreign criminal or terrorist suspect covertly to be interrogated in a country where there is less concern about the humane treatment of prisoners.

Representative Concentration Pathway (RCP): The IPCC has a range of very different views or scenarios, called RCPs, of how the world may look in 2100, based on the level of CO_2 in the atmosphere. Their numbers show different radiative forcing, measured in watts per square metre, by 2100. This means the difference in atmospheric energy inputs and outputs since the Industrial Revolution.

Representativeness: The extent to which a subset of a statistical population accurately reflects the characteristics (or members) of the entire population. When a sample is not representative, the result is known as a sampling error; in other words, the statistical difference between the observed (sample) data and the results that would be obtained from using data involving the entire population from which the sample was derived. In practice the degree of sample error can be complex to estimate.

Reservoir turnover: The rate at which carbon enters and leaves a store is measured by the mass of carbon in any store divided by the exchange flux.

Residence time: The average times a water molecule will spend in a reservoir or store.

Rising limb: The part of a storm hydrograph in which the discharge starts to rise.

River regime: The annual variation in discharge or flow of a river at a particular point or gauging station, usually measured in cumecs.

Sample: This is the limited number of measurements that you make. This is different to a population, which is the total number of measurements that you could potentially take if you measured everything and/or everyone in the environment. Larger samples generally mean greater reliability, until a point is reached where increasing sample size has very little effect on quality of outcome.

Sanctions: These can be diplomatic, such as ordering staff at a foreign embassy home, or economic, such as banning trade between countries. Military sanctions ban trade in weapons and military co-operation, while sporting sanctions can be used to prevent a country taking part in global sporting events. The aim is to force a country back to the negotiating table without using military force.

Saturated overland flow: The upward movement of the water table into the evaporation zone.

Schengen Agreement: An international agreement that aims to make it easier for people to move freely within the EU. Passports do not usually have to be shown by citizens at the borders of the 26 EU and non-EU countries that have agreed to this.

Secession: The act of separation for part of a state to create a new and fully independent country.

Secondary data: Secondary data (which may be part of the research) means information that has already been collected by someone else.

Sequestering: The natural storage of carbon by physical or biological processes such as photosynthesis.

Silicate rocks: Igneous, metamorphic and sedimentary rocks that contain large quantities of silicate minerals such as feldspar, mica, clays and quartz can be called silicate rocks. Examples include granite (igneous), schist (metamorphic) and sandstone (sedimentary). Chemical weathering of silicate minerals frequently produced calcium ions as well as other weathering products.

Solar radiation management (SRM): Often called geo-engineering, this would involve planet-scale engineering to reflect more solar radiation (sunlight) back into outer space so less reaches the Earth's surface. This could offset warming caused by greenhouse gas emissions. It is controversial and would require global co-operation. SRM does not reduce emissions, so works in a different way to other mitigation methods, but the outcome would be similar by slowing temperature rise or even reversing it. In contrast, adaptation methods try to cope with rising temperature and other changes to climate.

Sovereign wealth fund (SWF): Government-owned investment vehicles that invest the wealth of nations in companies and projects around the world. They are typically associated with China and countries that have large revenues from oil, such as Qatar. Many SWFs are worth hundreds of billions of dollars. They are controversial because they allow foreign governments to effectively own assets in another country.

Sovereignty: The ability of a place and its people to self-govern without any outside interference.

Sphere of influence: A geographical area over which a country believes it has economic, military, cultural or political rights. Spheres of influence extend beyond the borders of the country and represent a region where the country believes it has a right to influence the policies of other countries.

Standard deviation: This measures the spread of the data about the mean value. It can be done 'long hand', but using a spreadsheet is much more realistic for most purposes.

Staple foods: Carbohydrates relied on in large quantity and eaten regularly, such as potatoes and wheat for bread in Europe and North America, maize in Latin America, and rice in Asia. Stable supply at affordable prices is important to regional food security.

Statistical significance: Means establishing the probability that the results or outcomes are not due to chance. Statistical significance is normally set at the 95 per cent (or 0.05) threshold. At this level, there is only a 5 per cent probability (i.e. one in twenty) that the result has occurred purely by chance.

Stem flow: This is when water trickles along twigs and branches and then down the trunk.

Stores: Reservoirs where water is held, such as the oceans.

Structural adjustment programmes (SAPs): Policies promoted by the World Bank and IMF to help developing countries overcome their debt problems. These are now superseded by poverty reduction strategy papers (PRSPs) as for many countries SAPs resulted in unacceptable hardship and little progress with solutions to debts.

Superpower: A nation with the ability to project its influence anywhere in the world and be a dominant global force.

Surface run-off: The movement of water that is unconfined by a channel across the surface of the ground. Also known as overland flow.

Surface water flooding: Flooding that occurs when intense rainfall has insufficient time to infiltrate the soil, so flows overland.

Sustainable management: The environmentally appropriate, socially beneficial and economically viable use of ecosystems for present and future generations.

Systems approach: Systems approaches study hydrological phenomena by looking at the balance of inputs and outputs, and how water is moved between stores by flows.

Tax haven: A country or territory with a nil or low rate of corporation tax, such as Bermuda.

Teleconnection: In atmospheric science, refers to climate anomalies which relate to each other at large distances.

Thermohaline circulation: The global system of surface and deep water ocean currents is driven by temperature (thermo) and salinity (haline) differences between areas of oceans.

Throughfall: This is when the rainfall persists or is relatively intense, and the water drops from the leaves, twigs, needles, etc.

Throughflow: The lateral transfer of water down slope through the soil via natural pipes and percolines.

Top down: Large-scale capital intensive development schemes, usually developed by government.

Totalitarian regime: A system of government that is centralised and dictatorial; it requires complete subservience to the state with control being in the hands of elites. These may be the military or powerful families or tribes. For some, 'totalitarian' and 'authoritarian' are taken to mean more or less the same thing.

Transboundary water: A water resource, including rivers, lakes and aquifers, that occupies a territory shared by more than one state.

Transect: A transect is a line along which sampling is undertaken. This can be a tape measure, road, river or any other accessible linear feature.

Transfer pricing: A financial flow occurring when one division of a TNC based in one country charges a division of the same firm based in another country for the supply of a product or a service. It can lead to less corporation tax being paid.

Transpiration: The diffusion of water from vegetation into the atmosphere, involving a change from a liquid to a gas.

Trickle-down: The positive impacts on the peripheral region of wealth creation in core regions. These may include investment (in the form of back offices and branch plants), regional aid and grants, and the diffusion of innovations, technology and infrastructure from the core to the periphery.

Unilateral intervention: Military intervention undertaken by a state (or a group of states) outside the umbrella of the UN.

Urbanisation: The increase in the number of people living in towns and cities compared to the number of people living in the countryside.

Virtual water: The hidden flow of water when food or other commodities are traded.

War on terror: The ongoing campaign by the USA and its allies to counter international terrorism, initiated by al-Qaeda's attacks on the World Trade Center in New York and the Pentagon on 11 September 2001.

Watershed: The high land which divides and separates waters flowing to different rivers.

Wetland: An area of marsh, fen, peatland or water, whether natural or artificial, permanent or temporary, with water that is static or flowing, fresh, brackish or salt.

INDEX

acculturation 152
acidification 128–9
adaptation 332
afforestation 12–13, 127, 142
Afghanistan 224–5, 253, 298
Africa
   China–Africa Research Initiative 183–4
   drought 29–30
   Ebola outbreak 249
   economic development 251–2
   human rights 216–17, 238, 251–2
   land grabbing 239
   Niger Delta 238–9
   political boundaries 278–81
   Sahel region 27–8
   water balance 15
   water sharing 81–2
   water shortages 60
agricultural drought 24–5, 55
agriculture 13, 27, 31, 55, 61–2, 67, 77–8, 126
   see also farming
aid *see* development aid
airport expansion 327–8
albedo 11, 134–5
Amazonia
   changing climate 130
   deforestation 13
Americanisation 316
Amnesty International 233
Antarctica
   ice sheets 2, 4, 133
   managing 306–8
Antarctic Treaty (1959) 306
anthropogenic influences 87
Apple 166, 168, 179, 289, 316
aquaculture 136
aquifers 53, 55, 61
Arab Spring (2011) 185–6, 255
Aral Sea 62–3
Arctic
   albedo 134–5
   barometer 133
   carbon stores 133–4
   resources 179
   water cycle 133–4
arsenic 55
Asia
   economic development 184–5, 189–90
   energy production 113–14
   flooding 33
   human rights 216–18, 219–20
   nationalism 321
   water balance 15
   water transfer schemes 72–5
Association of Southeast Asian Nations (ASEAN) 172, 231
asylum seekers 267–8
   see also migration
atmosphere governance 304
atmospheric water 3–4
Australia
   drought 28–9
   human rights 226–7
   life expectancy 205–6
   migration rules 226–7, 264
   water balance 15
backwash effects 261, 269–70
base flow 16–17
Berlin Wall 247
bias 365
big data 354–5
bilateral aid 229
biodiversity management 304–5
biofuels 64, 120–2
biological carbon cycle 88, 96–7
biomass energy 120, 122
biosphere governance 304
Black Swan events 191
blue water 3
blue water navy 148
Bolivia 198, 226
borders 278–82
   see also migration; nation states
borrowing, global 301
Botswana 251–2
brands 165–7
Brazil 158, 205, 321
Bretton Woods institutions 299–300
Brexit 157
BRICS countries 114, 156, 175, 321
British Empire 151, 283–4
British values 311–12
   see also identity
calcium 91
Canada 224–5
capitalism 163–4, 292
carbon 86
   balance 97
   crossroads 140
   cycle 86–105, 126, 133–4, 142
   density 101

emissions   122–5, 135, 175
fluxes   86, 89
fossil fuels   90, 103
mitigation methods   142
origins   89
sequestering   88, 93–7
soil balance   101–2
stores   88–90, 99–101, 133, 138
taxation   142
carbon capture and storage (CCS)   122–4, 142
carbon cycle pumps   93–4
carbon dioxide (CO2)   86, 88–91, 97, 100, 103, 123, 133, 135, 137–9, 174–5
car industry   315
Catalonia   319
channel flow   10
chemical weathering   91
China
BRI   150, 182
counterfeit products   179
economic development   248
foreign direct investment (FDI)   183–4
human rights   219
nationalism   321
rural-urban migration   262
sphere's of influence   181–2
as superpower   149, 156–8, 159, 219
and Taiwan   282
China–Africa Research Initiative   183–4
civil liberties   216
see also human rights
clear cutting   126
climate
and biological carbon   97
change   14, 40–5, 129–44
and drainage basin system   11
energy consumption   110
forcing   103, 140
global warming   40, 103–5, 132–4, 137
greenhouse effect   98–100, 129
model maps   135
predictions   103
zones   130
clouds   3–4, 7
coal   113, 115–16
coasts, storm surges   34, 55
Coca-Cola   64, 78, 166
Cold War   151, 154–5, 163, 180, 247, 255
colonial control   152–3, 222–3, 283–4
neo-colonialism   154–5, 183–4, 283
post-colonialism   153–4, 285–9
condensation   6–7
Conference on Environment and Development (1992)   296
conservation   332–4
water management   77–8

consumption
energy   107–12, 115
food   176
middle classes   175–6
resources   176
by superpowers   175–6
water   176
convectional rainfall   7
Convention on International Trade in Endangered Species of Wild Fauna and Flora (CITES)   304
core–periphery systems   261, 269–70
corruption *see* political corruption
counterfeit products   179
countryside, representations of   313
Covid-19 pandemic   187, 263
crisis response   170
critical reflection   365
critical threshold   129
Crop Moisture Index (CMI)   26
cryosphere   2–4, 133
cultural cohesion   310–14
cultural heterogeneity   288
cultural landscape   317
cultural power   149, 166–7
cultural unity   278–9
culture   277–8
cumecs   16
cumulus cloud   7
dams   73–5
data
accuracy   351
analysis   357–9
big data   354–5
primary   344, 350
qualitative   344–6, 353–4
quantitative   344–6
recording   350–1
reliability   7, 351
representation   356–60
secondary   344
sources   344
deepwater oil   118–19
deforestation   12–13, 34, 126–7, 175
degassing   91
democracy   216, 217–20, 247, 285
Democratic Republic of Congo (DRC)   285–6, 298
dependency theory   160
deprivation   202
desalination   75–6
descriptive statistics   360–4
desertification   27–8, 30
development aid   229–30, 233–7, 248–51
development gap   209
dew point   6
diaspora   267

diplomacy 148
disaster relief 170
discrimination 222–7, 252
disease 235–6, 249–51
Disney 156, 166, 167, 316–17
diversity 271–3
   see also culture; ethnicity
Doctors Without Borders 233
Doha Development Agenda 231
drainage basin system 5–15
   water transfer schemes 72–5
drought 24–31, 41–5
Ebola 249–50
ecological footprint 195
economic alliances 172
economic development 161–2, 163–4, 184–5, 215–16, 248–52
   see also superpowers
economic inequality 250–2
economic power 148, 150
economic restructuring 187
economic support 230–1, 237
   see also development aid
economic systems 330–2
economy, global 163–8, 289–91, 314–15
economy of scale 302
ecosystems 61, 88, 97, 100–1, 103–4, 128, 131, 304–7
education 197–9, 206–9
El Niño–Southern Oscillation (ENSO) 25–6, 30, 41, 43–4, 105
embargoes 231
emergency aid 250
   see also development aid
Emergency Events Database (EM-DAT) 36
empire building 151–2, 283–4
energy
   consumption 107–12, 115, 176
   efficiency 142
   intensity 107
   mix 108, 121
   pathways 110–11, 115
   production 63–4, 113–15
   recyclable 120–2
   renewable 120, 142
   security 106–7
   see also fossil fuels
enhanced greenhouse effect 129
environmental governance **304**
equality *see* discrimination; human rights
erosion, soil 10
ethical issues, research 354
Ethiopia 239
ethnicity 204, 209, 222–3, 271, 281–2, 317
ethnoscape 317
European Convention on Human Rights (ECHR) 215

European Union (EU)
   free movement 263, 269–71
   Schengen Agreement (1995) 263, 269
   as superpower 149–50, 187–8
   weaknesses 150, 157
eutrophication 37
evaporation 3–5, 10–11, 12, 41
evapotranspiration (EVT) 11, 12, 15, 17, 41
exclusive economic zones (EEZs) 178–9
expatriation 290
exploitation 332–4
export processing zones (EPZs) 261
Fairtrade Foundation 231
famine drought 24–5
farming 13–14, 25, 50, 55, 77–8
   see also agriculture
fieldwork 340, 347
financial crisis (2007-9) 187, 291–2
fishing 136
flash flooding 32
flooding 18–19, 31–9, 41–5
food
   consumption 176
   nutrition 64
   nutrition transition 176
foreign direct investment (FDI) 60, 183–4
forests 12–13, 96, 126–7, 130–2
   see also deforestation
fossil fuels 90, 103, 113
   alternatives 119–25
   coal 113, 115–16
   gas 114, 116–17, 179
   oil 113–14, 115–16, 179
   unconventional 118–19
fossil water 4
fracking 50
fragile state 298–9, 321–2
Freedom House 216
freedom ratings 216
   see also human rights
free movement 263, 269–71
   see also migration
free trade 163, 172
freshwater 5, 48, 50, 75–6
fuel cells 123–4
gas 114, 116–17, 179
gender
   discrimination 222–6
   and life expectancy 200–1, 204
Gender Inequality Index (GII) 222
General Agreement on Tariffs and Trade (GATT) (1948) 208, 299
Gene Revolution 62
genetically-modified (GM) crops 62, 78
Geneva Conventions (1949) 215

geographical images   351–2, 358
Geographical Information System (GIS)   128
geological carbon cycle   88
geopolitics   151–2, 228–32, 246–8
geo-strategic policies   150–1
gigatonnes (Gt)   87
Gini index   250, 252
glacial outburst floods (GOFs)   33–4
glaciers   42
   melt water   17
global citizenship   310
global commons   305–6
global governance   304–8
   see also United Nations
globalisation   261, 328
global warming   40, 103–5, 132–4, 137
   see also climate change
global water balance   14
government see politics and government
grasslands   128
Great Ruaha River   67
greenhouse effect   98–100, 129
greenhouse gases (GHGs)   123, 138–43
Greenland   2, 4, 133
Green Revolution   53, 61–2
green water   3
gross domestic product (GDP)   157, 159, 189, 194, 207, 209, 229–30
gross national income (GNI)   194
groundwater   3–5, 13, 42
   contamination   55
   flooding   31
   flow   3–4
Haiti   234–5
Happy Planet Index (HPI)   195–6
Harvard referencing   342
health   197, 201, 249–51
healthcare   203
Heartland Theory   150–1
Heavily Indebted Poor Countries (HIPC)   301
hegemony   156, 300–1
high yield variety (HYV) crops   53
homogeneous culture   278
human development
   development gap   209
   health   197, 201
   life expectancy   197, 200–6
   measuring   194–200
   targets   206–11, 236
Human Development Index (HDI)   199–200
human rights   297
   discrimination   222–7
   versus economic development   215–16
   geopolitical intervention   228–32, 246–8
   global   213–21, 238–44, 251–2

legislation   213–15
military intervention   239–44
national   222–7
see also United Nations (UN)
Human Rights Act (1998)   215
Human Rights Watch   233
hydroelectric power (HEP)   64, 120
hydrogen fuel cells   123–4
hydrographs   16–17
hydrological cycle
   accessible water   4–5, 48–50
   climate change impacts   40–5
   deforestation   126
   drainage basin   5–15
   drought   24–31, 41–5
   fluxes   3–4
   global system   2–5
   global warming   104–5
   invisible   3
   rainforest   12
   semi-arid areas   12
   surplus   31–9, 41–5
   urbanisation impacts   18–20
   water balance   14–15, 48–52
   water budgets   14
   water stores   2–4, 10, 42, 79–80
hydrological drought   24–5
hydrophonics   78
hyperpowers   148
   see also superpowers
Ice Age   2
icecaps   3, 42
Iceland   279
ice sheets   2, 4, 133
identity   310–22, 330
ideology   149, 187
India   158, 219, 321
indigenous people
   Americas   224–5
   Australia   205–6
Indonesia   158
industrialisation   63–4
industrial waste   55
inequality of outcomes   195
infiltration   8–10, 13
innovation   167–8
integrated water resource management (IWRM)   80–1
intellectual property (IP)   178–80
interception   8, 12
interdependence   328
inter-governmental organisations (IGOs)   163, 208–9, 232, 295, 299–302
Intergovernmental Panel on Climate Change (IPCC)   87, 140

internally displaced people (IDPs) 297
internal migration 262, 270
international migration 261, 263–6
International Monetary Fund (IMF) 60, 160, 164, 208, 295, 299–301
inter-tropical convergence zone (ITCZ) 130
investment 231
   see also foreign direct investment (FDI)
Iraq 282
irrigation 14, 50, 61–2, 77–8
Islamic law 197
Islamic State (IS) 185–6, 242–3
Japan 159, 264
jökulhlaup 34
Jonglei Canal Project 31
judgement surveys 352–3
Kuznets curve 132
Kyoto Protocol (1997) 143–4, 175, 296
labour flows 261
lag time 16–17
lakes 3, 5, 42, 54
land conversion 126
land grabbing 239
Latin America, populism 293
latitudes 130
League of Nations 295
Libya 240, 253
life expectancy 195, 197, 200–6
   Australia 205–6
   Brazil 205
   UK 204
limestone 89–90
literature research 340
living standards 50, 202
   see also health
low-tax states 289–91
Madrid Protocol (1991) 306
malaria 235–6
marine ecosystems 104, 136, 306–7
McDonald's 167, 279, 316
mean 360
median 360
Médicins sans frontières (Doctors Without Borders) (MSF) 233
mega dams 73–5
meteorological drought 24–5
Mexico 158, 274
middle classes 175–6
Middle East 185–7, 242–3
migration
   causes of 267–8, 275–6
   consequences of 271–3
   free movement 263, 269–71
   global patterns 265–6, 275–6
   internal 262, 270

international 261, 263–6
   policies 264
   population sizes 263
   post-colonial 267, 288–9
   rural-urban 262
   tensions 273–4
military intervention 231–2, 239–44, 253–5, 298–9
military power 148, 151–3, 170–1, 190
Millennium Development Goals (MDGs) 48, 64, 209–10, 236, 296
Millennium Ecosystem Assessment (MEA) 101
minerals 91
   see also carbon
mitigation 332
mode 360
modernisation theory 159
monsoon 17, 61
Montreal Protocol (1987) 304
morbidity 36
multilateral aid 229
Myanmar 221
nationalism 283–8, 310–22
nation states 277–8
   borders 278–82
   disunity 318–19
   economic models 291–3
   fragile states 298–9, 321–2
   low-tax states 289–91
   nationalism 283–8
   strategies within global systems 289–93
NATO (North Atlantic Treaty Organization) 117, 151, 170–2
neo-colonialism 154–5, 183–4, 283
net primary production (NPP) 100
newly industrialised countries (NICs) 161
New Zealand 313
nimbyism 66
non-governmental organizations (NGOs) 82, 232–5, 249, 337
North Atlantic Drift (NAD) 103
North Atlantic Oscillation (NAO) 44
North Korea 217–18
nuclear power 120
nuclear weapons 149, 151, 170, 231
nutrition 64
nutrition transition 176
obesity 202–3
oceans 3–4, 17, 42, 93–5, 99–100, 128–30, 132–3, 136–7, 305–6
Official Development Assistance (ODA) 229–30
oil 113–14, 115–16, 118–19, 179
oil shale 118
Organisation for Economic Co-operation and Development (OECD) 208

Organization of the Petroleum Exporting
    Countries (OPEC)   111
outgassing   91
Oxfam   233, 291
ozone layer   304
Palmer Drought Severity Index (PDSI)   26
parent company   290
Paris climate change agreement (2015)   304–5
patents   167–8
peacekeeping   168–9, 254–7, 296–7
peak discharge   16
peatlands   96
percolation   10
permafrost   5, 42, 103
petagrams (Pg)   87
pH   128–9
photosynthesis   99–100
pie charts   208
pilot surveys   346
politics and government   330–2
    corruption   218–21, 237–9
    democracy   216, 217–20, 247, 285
    developmental policies   206–7
    geopolitics   151–2, 228–32, 246–8
    global governance   295–9
    inter-governmental organisations (IGOs)   163, 208–9, 232, 295, 299–302
    political corruption   218–21, 237–9
    political power   148
    systems of   207, 217–18
    totalitarianism   207
    totalitariansim   217–18
Polluter Pays Principle   336
pollution   53–5, 64–5, 238
population   50, 157
populism   293
positive feedback (climate)   103
post-colonialism   153–4, 285–9
post-colonial migration   267
potential evapotranspiration (PEVT)   11
poverty   202, 236
power   150, 156, 159
    see also superpowers
precipitation   3–10, 15, 41, 43
    convectional   7
    cyclonic   7
    data reliability   7
    and flooding   33, 43
    human impact on   12
    intensity   7
    monsoon   17
    Sahel region   27–8
    seasonality   7
primary data   344, 350
propaganda   313–14

property markets   317–18
proportional circles   208
qualitative data   344–6, 353–4
quantitative data   344–6
rain see precipitation
rainwater harvesting   79
Ramsar Convention on Wetlands (1991)   31
rare earth elements (REE)   176
recyclable energy   120–2
recycling, water   78–9
reforestation   127
refugees   297
rendition   242–4
renewable energy   120, 142
Representative Concentration Pathway (RCP)   135–6
reservoirs   12, 25, 28, 42, 66, 88
residence times, of water   4
resources
    consumption   176
    contested   178–81
    and power   149
rising limb   16
risk   336–8
risk assessment   347–8
River Nile   66–71
rivers   3, 5–6, 14, 16–18, 54, 67–71, 79–80, 306
runaway global warming   134
rural-urban migration   262
Russia   158
    contested borders   281
    contested resources   179–81
    gas production   115, 117
    human rights   241
    see also USSR
Rwanda   223
salinisation   13
sampling   348–50
sanctions   169, 231, 297
sanitation   48, 56–7, 64–5
saturated overland flow   10
Saudi Arabia   241–2
scatter graphs   203
Schengen Agreement (1995)   263, 269
Scotland, independence   320
sea ice   133
sea level   2
secession   318–19
secondary data   344
sequestering carbon   88, 93–7
shale   89–90
shale gas   118
Sharia   197
Singapore   264–5, 279
slash-and-burn agriculture   126
snow   42

melt water   5, 10, 17, 33–4
social progress   207
soil
   carbon content   97
   erosion   10
   health   101–2
   management   97
   moisture   3–4, 8–10, 11, 13, 15, 34, 41
solar radiation management (SRM)   141
South Africa, drought   29–30
South America
   biofuels   122
   water balance   15
South Korea   217–18
South Sudan   287–8
sovereignty   277–82
sovereign wealth funds (SWFs)   314
Soviet Union see USSR
Spain
   rural-urban migration   262
   secession   319
spheres of influence   180–2
staple foods   176
stem flow   8
storm hydrographs   16–18
storm surges   34, 55
streams   3, 6, 54
structural adjustment programmes (SAPs)   60, 301
Sudan   287–8
superpowers
   contested resources   178–81
   costs of   189
   decision-making   168–9
   definition   148
   and democracy   218–20
   emerging   149, 156–7, 175–6, 183–5, 219
   hegemony   156
   ideology   149, 187
   index   149
   objectives   252
   polarity   151–2, 190
   resource consumption   174
   sphere's of influence   180–2
   status   148
   theories of power   159–62
   uncertainty   187–8
   see also war and conflict
surface run-off   8–9
surveys   352–3
sustainability   334–6
   water management   76–80
Sustainable Development Goals (SDGs)   48, 64, 210–11, 236, 296
sustainable management   131
synoptic themes   324–5

Syria   253–4, 282
system feedback   86
systems approach   2
Taiwan   282
Tanzania   302
tar sands   118
tax avoidance   290–1
tax havens   290–1
teleconnections   30
terrorism   173, 187, 242–4
thresholds   336–8
throughfall   8
throughflow   10
thunderstorms   7–8
torture   242–4
totalitarianism   207, 217–18
trade   163–5, 172, 230–1, 299–302
   see also economic alliances
trade blocs   302–3
trade winds   26
trafficking   267
Transatlantic Trade and Investment Partnership (TTIP)   302
transboundary water   306
transfer pricing   290
transnational corporations (TNCs)   60, 111, 165–7, 172, 289–91, 316
Trans-Pacific Partnership (TPP)   302
transpiration   3–4, 11
tropical rainforests   96
Türkiye   158
Ukraine   241, 253, 281
uncertainty   336–8
unconventional fossil fuels   118–19
unilateral intervention   298–9
United Nations (UN)
   Convention for the Law of the Sea (UNCLOS)   306–7
   Convention on the Rights of the Child (CRC)   296
   Development Programme (UNDP)   210–11
   Educational, Scientific and Cultural Organization (UNESCO)   199, 208
   global governance   295–9
   history   296
   Mission for Ebola Emergency Response (UNMEER)   249
   peacekeeping   254–6, 296–7
   Security Council   168–70
United States
   Americanisation   316
   anti-communist policies   151
   cold war   151, 154–5
   culture and history   277–8, 313
   energy consumption   112
   as 'global police'   169
   hegemonic power   300–1
   Mexico-US border   274

migration   274
as superpower   148, 153, 187–8, 252
unilateral intervention   298–9
Universal Declaration of Human Rights (UDHR) (1948)   213–14, 296–7
urbanisation, and flooding   18–20, 35
USSR
cold war   151, 154–5
collapse of   163, 247
as superpower   153, 247
see also Russia
vapour transport   3–4
vegetation
and flooding   34
infiltration   8–10
transpiration   11
as water store   3, 5, 11
Viet Nam   286–7
virtual water   51
volcanoes   91–2
war and conflict   150, 151, 170, 185–7, 267
alliances   170–1
Cold War   151, 154–5, 163, 180, 247, 255
contested resources   178–81
crisis response   170
decision-making   168–9
geo-strategic policies   150–1
'global police'   168–9
global security   172–3
Middle East   185–7
military intervention   231–2, 239–44, 253–5, 298–9
military power   148, 151–3, 170–1, 191
nuclear weapons   149, 151, 170, 231
peacekeeping   168–9, 254–7, 296–7
rendition   242–4
Rwanda   223
sanctions   168, 297
Syria   253–4
terrorism   173, 187, 242–4
torture   242–4
unilateral intervention   298–9
war crimes trials   297
war on terror   242–4, 298–9
see also colonial control; superpowers
water
access to   4–5, 48–50, **56**
atmospheric   3–4

availability gap   51–2
balance   14–15, 48–52
blue   3
budgets   14
conflicts   66–71
conservation   77–8
consumption   176
cryosphere   2–4
fossil   4
freshwater   5, 48, 50, 55
green   3
management   44–5
over-abstraction   55
pollution   53–5, 64–5
price   58–60
quality   53–5
recycling   78–9
rising demand   50
security   **52**, 65–71
sharing   81–2
shortages   56–60
stores   2–4, 10, 42, 79–80
supply   52–6, 61–82, 302
sustainability   76–80
virtual   51
see also drainage basin; hydrological cycle
WaterAid   65, 72
waterlogging   13–14
water poverty index (WPI)   56
watershed   5–6
water transfer schemes   72–5
weathering, chemical   91
weather stations   7
well-being   195
and water supply   64
Westernisation   316–17
wetlands   30–1, 42, 96
whaling   306–7
wind   26
woodlands   34
see also forests
World Bank   60, 160, 164, 208, 295, 299–302
World Economic Forum (WEF)   164
World Intellectual Property Organization (WIPO)   178
world systems theory   160–2
World Trade Organization (WTO)   164, 208, 299, 302
World Wide Fund for Nature (WWF)   136
Zimbabwe   221